5e

BIOLOGY

TODAY & TOMORROW
WITHOUT PHYSIOLOGY

BIOLOGY 5e

TODAY & TOMORROW
WITHOUT PHYSIOLOGY

Cecie Starr | Christine A. Evers | Lisa Starr

CENGAGE
Learning®

Australia • Brazil • Mexico • Singapore • United Kingdom • United States

**Biology Today and Tomorrow Without Physiology,
Fifth Edition**
Cecie Starr, Christine A. Evers, Lisa Starr

Product Director: Mary Finch

Senior Product Team Manager: Yolanda Cossio

Managing Content Developer: Trudy Brown

Product Assistant: Victor Luu

Content Developer: Lauren Oliveira

Associate Content Developers: Kellie N. Petruzzelli,
Casey J. Lozier

Senior Market Development Manager: Tom Ziolkowski

IP Analyst: Christine M. Myaskovsky

IP Project Manager: John N. Sarantakis

Content Project Manager: Hal Humphrey

Senior Art Director: Bethany Casey

Manufacturing Planner: Karen Hunt

Production Service: Grace Davidson & Associates

Photo Researchers: Cheryl DuBois and Megan Cooper,
Lumina Datamatics

Text Researcher: Kavitha Balasundaram, Lumina Datamatics

Copy Editor: Anita Wagner Hueftle

Illustrators: Lisa Starr, ScEYEnce Studios, Precision
Graphics, Gary Head

Interior Designer: Michael Stratton, Stratton Design

Cover Designer: Michael Stratton, Stratton Design

Cover Image: © Shutterstock, Inc./Felix Rohan

Compositor: Lachina Publishing Services

For product information and technology assistance, contact us at
Cengage Learning Customer & Sales Support, 1-800-354-9706.

For permission to use material from this text or product,
submit all requests online at **www.cengage.com/permissions**.
Further permissions questions can be e-mailed to
permissionrequest@cengage.com.

Library of Congress Control Number: 2014958337

ISBN-13: 978-1-305-11739-6

Cengage Learning
20 Channel Center Street
Boston, MA 02210
USA

Cengage Learning is a leading provider of customized learning solutions with office locations around the globe, including Singapore, the United Kingdom, Australia, Mexico, Brazil, and Japan. Locate your local office at
www.cengage.com/global.

Cengage Learning products are represented in Canada by Nelson Education, Ltd.

To learn more about Cengage Learning Solutions, visit **www.cengage.com**.

Purchase any of our products at your local college store or at our preferred online store **www.cengagebrain.com**.

Printed in the United States of America
Print Number: 01 Print Year: 2015

BC

BRIEF CONTENTS

CONTENTS

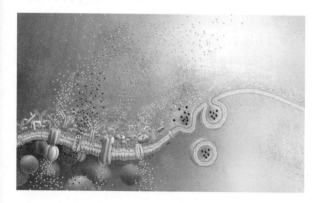

UNIT 1 HOW CELLS WORK

UNIT 2 GENETICS

UNIT 3 EVOLUTION AND DIVERSITY

15 Animal Evolution

UNIT 4 ECOLOGY

16 Population Ecology

Biology is a huge field, with a wealth of new discoveries being made every day, and biology-related issues such as climate change, stem cell research, and personal genetics often making headlines. This avalanche of information can be intimidating to non-scientists. This book was designed and written specifically for students who most likely will not become biologists and may never again take another science course. It is an accessible and engaging introduction to biology that provides future decision-makers with an understanding of basic biology and the process of science.

A Wealth of Applications This book is packed with everyday applications of biological processes. At every opportunity, we enliven discussions of biological processes with references to their effects on human health and the environment. This edition also continues to focus on real world applications pertaining to the field of biology, including social issues arising from new research and developments. Descriptions of current research, along with photos of scientists who carry it out, underscore the concept that biology is an ongoing endeavor carried out by a diverse community of people. Discussions include not only what was discovered, but also how the discoveries were made, how our understanding has changed over time, and what remains to be discovered. These discussions are provided in the context of an accessible introduction to well-established concepts that underpin modern biology. Every topic is examined from an evolutionary perspective, emphasizing the connections between all forms of life.

Accessible Text Understanding stems from making connections between concepts and details, so a text with too little detail reads as a series of facts that beg to be memorized. However, excessive detail can overwhelm the introductory student. Thus, we constantly strive to strike the perfect balance between level of detail and accessibility. We once again revised the text to eliminate details that do not contribute to a basic understanding of essential concepts. We also know that English is a second language for many introductory students, so we avoid idioms and aim for a clear, straightforward style.

Analogies to familiar objects and phenomena will help students understand abstract concepts. For example, in the discussion of transpiration in Chapter 27 (Plant Form and Function), we explain that a column of water is drawn upward through xylem as a drinker draws fluid up through a straw.

In-Text Learning Tools To emphasize connections between biological topics, each chapter begins with an APPLICATION section that explores a current event or controversy directly related to the chapter's content. For example, a discussion of binge drinking on college campuses introduces the concept of metabolism in Chapter 4. This section presents an overview of the metabolic pathway that breaks down alcohol, linking the function of enzymes in the pathway to hangovers, alcoholism, and cirrhosis. The section is illustrated with a photo of a tailgate party that preceded a recent Notre Dame–Alabama football game, and also a photo of Gary Reinbach just before he died at age 22 of alcoholic liver disease. (In the index, you'll find health-related applications denoted by red squares and environmental applications by green squares.)

To strengthen a student's analytical skills and offer insight into contemporary research, each chapter includes an exercise called DIGGING INTO DATA that is placed in a section with relevant content. The exercise consists of a short text passage—usually about a published scientific experiment—and a table, chart, or other graphic that presents experimental data. A student can use information in the text and graphic to answer a series of questions. For example, the exercise in Chapter 2 asks students to interpret results of a study that examined the effect of dietary fat intake on "good" and "bad" cholesterol levels.

The chapter itself consists of several numbered sections that contain a manageable chunk of information. Every section ends with a boxed TAKE-HOME MESSAGE in which we pose a question that reflects the critical content of the section, and then answer the question in bulleted list format. Every chapter has at least one FIGURE IT OUT QUESTION with an answer immediately following. These questions allow students to quickly check their understanding as they read. Mastering scientific vocabulary challenges many students, so we have included an ON-PAGE GLOSSARY of key terms introduced in each two-page spread, in addition to a complete glossary at the book's end. The end-of-chapter material features a VISUAL SUMMARY that reinforces each chapter's key concepts. A SELF-QUIZ poses multiple choice and other short answer questions for self-assessment (answers are in Appendix I). A set of more challenging CRITICAL THINKING QUESTIONS provides thought-provoking exercises for the motivated student. The end matter of several chapters now includes a VISUAL QUESTION that reinforces learning in a nonverbal style.

Design and Content Revisions Throughout the book, text and art have been revised to help students grasp difficult concepts. The following list highlights some of the revisions to each chapter.

Introduction

1 **Invitation to Biology** Renewed and updated emphasis on the relevance of new species discovery and the process of science.

Unit 1 How Cells Work

2 **Molecules of Life** New graphic illustrates radioactive decay.

3 **Cell Structure** Application section updated with current statistics and 'pink slime' story. Micrograph comparisons now feature *Paramecia* and include a confocal image. Essay about the nature of life expanded to add Gerald Joyce's "life is squishy" concept.

4 **Energy and Metabolism** Application section now illustrated with a real-life example. Diffusion illustrated with a tea bag in hot water.

5 **Capturing and Releasing Energy** Application section updated with current statistics and illustrated with a current photo of air pollution in China. Yogurt production added to fermentation section.

Unit 2 Genetics

6 **DNA Structure and Function** Content reorganized: material on cloning folded into Application section for concept connection, and chromosome structure now appears after DNA structure. New art demonstrates how replication errors become mutations.

7 **Gene Expression and Control** Ricin discussion revised to include medical applications. New material includes hairlessness mutation (in cats), evolution of lactose tolerance, heritability of DNA methylations, telomeres.

8 **How Cells Reproduce** New material on telomeres, asexual vs. sexual mud snails. New micrograph shows multiple crossovers.

9 **Patterns of Inheritance** Epistasis is now illustrated with human skin color. New material about environmentally-triggered hemoglobin production in *Daphnia*; continuous variation in dog face length arising from short tandem repeats foreshadows DNA fingerprinting in chapter 10.

10 **Biotechnology** Updated coverage of personal genetic testing includes social impact of Angelina Jolie's response to her test. New photos illustrate genetically modified animals. New "who's the daddy" critical thinking question offers students an opportunity to analyze a paternity test based on SNPs.

Unit 3 Evolution and Diversity

11 **Evidence of Evolution** Photos of 19th century naturalists added to emphasize the process of science that led to natural selection theory. How banded iron formations provide evidence of the evolution of photosynthesis added to fossil section. Plate tectonics art updated to reflect new evidence of lava lamp mantle movements.

12 **Processes of Evolution** New opening essay on resistance to antibiotics as an outcome of agricultural overuse (warfarin material now exemplifies directional selection). New art illustrates founder effect, and hypothetical example in text replaced with reduced diversity of ABO alleles in Native Americans. New art illustrates stasis in coelacanths.

13 **Early Life Forms and the Viruses** New introductory essay about study of the human microbiome, new coverage of Ebola, and new figure depicting mechanisms of gene exchange in prokaryotes.

14 **Plants and Fungi** Additional coverage of fungal ecology, including information about white-nose syndrome in bats.

15 **Animal Evolution** New introductory essay about invertebrates as a source of medicines. Updated information about Neanderthals and added coverage of the newly discovered Dennisovans.

Unit 4 Ecology

16 **Population Ecology** Updated coverage of human demographics.

17 **Communities and Ecosystems** New photos illustrate species interactions; updated coverage of the increases in greenhouse gases.

18 **The Biosphere and Human Effects** New essay about dispersion of the radioactive material released at Fukushima and new Digging Into Data about bioaccumulation of this material in tuna.

We owe a special debt to the members of our advisory board, listed below. They helped us shape the book's design and to choose appropriate content. We appreciate their guidance.

Andrew Baldwin, *Mesa Community College*
Charlotte Borgeson, *University of Nevada, Reno*
Gregory A. Dahlem, *Northern Kentucky University*
Gregory Forbes, *Grand Rapids Community College*
Hinrich Kaiser, *Victor Valley Community College*
Lyn Koller, *Pierce College*
Terry Richardson, *University of North Alabama*

We also wish to thank the reviewers listed below.

Idris Abdi, *Lane College*
Meghan Andrikanich, *Lorain County Community College*
Lena Ballard, *Rock Valley College*
Barbara D. Boss, *Keiser University, Sarasota*
Susan L. Bower, *Pasadena City College*
James R. Bray Jr., *Blackburn College*
Mimi Bres, *Prince George's Community College*
Randy Brewton, *University of Tennessee*
Evelyn K. Bruce, *University of North Alabama*
Steven G. Brumbaugh, *Green River Community College*
Chantae M. Calhoun, *Lawson State Community College*
Thomas F. Chubb, *Villanova University*
Julie A. Clements, *Keiser University, Melbourne*
Francisco Delgado, *Pima Community College*
Elizabeth A. Desy, *Southwest Minnesota State University*
Brian Dingmann, *University of Minnesota, Crookston*
Josh Dobkins, *Keiser University, online*
Hartmut Doebel, *The George Washington University*
Pamela K. Elf, *University of Minnesota, Crookston*
Johnny El-Rady, *University of South Florida*
Patrick James Enderle, *East Carolina University*
Jean Engohang-Ndong, *BYU Hawaii*
Ted W. Fleming, *Bradley University*
Edison R. Fowlks, *Hampton University*
Martin Jose Garcia Ramos, *Los Angeles City College*
J. Phil Gibson, *University of Oklahoma*
Judith A. Guinan, *Radford University*
Carla Guthridge, *Cameron University*
Laura A. Houston, *Northeast Lakeview–Alamo College*
Robert H. Inan, *Inver Hills Community College*
Dianne Jennings, *Virginia Commonwealth University*
Ross S. Johnson, *Chicago State University*
Susannah B. Johnson Fulton, *Shasta College*
Paul Kaseloo, *Virginia State University*
Ronald R. Keiper, *Valencia Community College West*
Dawn G. Keller, *Hawkeye Community College*
Ruhul H. Kuddus, *Utah Valley State College*
Dr. Kim Lackey, *University of Alabama*
Vic Landrum, *Washburn University*
Lisa Maranto, *Prince George's Community College*
Catarina Mata, *Borough of Manhattan Community College*
Kevin C. McGarry, *Keiser University, Melbourne*
Timothy Metz, *Campbell University*
Ann J. Murkowski, *North Seattle Community College*
Alexander E. Olvido, *John Tyler Community College*
Joshua M. Parke, *Community College of Southern Nevada*
Elena Pravosudova, *Sierra College*
Nathan S. Reyna, *Howard Payne University*
Carol Rhodes, *Cañada College*
Todd A. Rimkus, *Marymount University*
Laura H. Ritt, *Burlington County College*
Lynette Rushton, *South Puget Sound Community College*
Erik P. Scully, *Towson University*

Marilyn Shopper, *Johnson County Community College*
Jennifer J. Skillen, *Community College of Southern Nevada*
Jim Stegge, *Rochester Community and Technical College*
Lisa M. Strain, *Northeast Lakeview College*
Jo Ann Wilson, *Florida Gulf Coast University*

We were also fortunate to have conversations with the following workshop attendees. The insights they shared proved invaluable.

Robert Bailey, *Central Michigan University*
Brian J. Baumgartner, *Trinity Valley Community College*
Michael Bell, *Richland College*
Lois Borek, *Georgia State University*
Heidi Borgeas, *University of Tampa*
Charlotte Borgenson, *University of Nevada*
Denise Chung, *Long Island University*
Sehoya Cotner, *University of Minnesota*
Heather Collins, *Greenville Technical College*
Joe Conner, *Pasadena Community College*
Gregory A. Dahlem, *Northern Kentucky University*
Juville Dario-Becker, *Central Virginia Community College*
Jean DeSaix, *University of North Carolina*
Carolyn Dodson, *Chattanooga State Technical Community College*
Kathleen Duncan, *Foothill College, California*
Dave Eakin, *Eastern Kentucky University*
Lee Edwards, *Greenville Technical College*
Linda Fergusson-Kolmes, *Portland Community College*
Kathy Ferrell, *Greenville Technical College*
April Ann Fong, *Portland Community College*
Kendra Hill, *South Dakota State University*
Adam W. Hrincevich, *Louisiana State University*
David Huffman, *Texas State University, San Marcos*
Peter Ingmire, *San Francisco State*
Ross S. Johnson, *Chicago State University*
Rose Jones, *NW-Shoals Community College*
Thomas Justice, *McLennan Community College*
Jerome Krueger, *South Dakota State University*
Dean Kruse, *Portland Community College*
Dale Lambert, *Tarrant County College*
Debabrata Majumdar, *Norfolk State University*
Vicki Martin, *Appalachian State University*
Mary Mayhew, *Gainesville State College*
Roy Mason, *Mt. San Jacinto College*
Alexie McNerthney, *Portland Community College*
Brenda Moore, *Truman State University*
Alex Olvido, *John Tyler Community College*
Molly Perry, *Keiser University*
Michael Plotkin, *Mt. San Jacinto College*
Amanda Poffinbarger, *Eastern Illinois University*
Johanna Porter-Kelley, *Winston-Salem State University*
Sarah Pugh, *Shelton State Community College*
Larry A. Reichard, *Metropolitan Community College*
Darryl Ritter, *Okaloosa-Walton College*
Sharon Rogers, *University of Las Vegas*
Lori Rose, *Sam Houston State University*
Matthew Rowe, *Sam Houston State University*
Cara Shillington, *Eastern Michigan University*
Denise Signorelli, *Community College of Southern Nevada*
Jennifer Skillen, *Community College of Southern Nevada*
Jim Stegge, *Rochester Community and Technical College*
Andrew Swanson, *Manatee Community College*
Megan Thomas, *University of Las Vegas*
Kip Thompson, *Ozarks Technical Community College*
Steve White, *Ozarks Technical Community College*
Virginia White, *Riverside Community College*
Lawrence Williams, *University of Houston*
Michael L. Womack, *Macon State College*

Cengage Learning Testing Powered by Cognero
is a flexible, online system that allows you to:
- author, edit, and manage test bank content from multiple Cengage Learning solutions
- create multiple test versions in an instant
- deliver tests from your LMS, your classroom, or wherever you want

Instructor Companion Site Everything you need for your course in one place! This collection of book-specific lecture and class tools is available online via www.cengage.com/login. Access and download Power-Point presentations, images, instructor's manual, videos, and more.

Cooperative Learning Cooperative Learning: Making Connections in General Biology, 2nd Edition, authored by Mimi Bres and Arnold Weisshaar, is a collection of separate, ready-to-use, short cooperative activities that have broad application for first year biology courses. They fit perfectly with any style of instruction, whether in large lecture halls or flipped classrooms. The activities are designed to address a range of learning objectives such as reinforcing basic concepts, making connections between various chapters and topics, data analysis and graphing, developing problem solving skills, and mastering terminology. Since each activity is designed to stand alone, this collection can be used in a variety of courses and with any text.

MindTap A personalized, fully online digital learning platform of authoritative content, assignments, and services that engages students with interactivity while also offering instructors their choice in the configuration of coursework and enhancement of the curriculum via web-apps known as MindApps. MindApps range from ReadSpeaker (which reads the text out loud to students) to Kaltura (which allows you to insert inline video and audio into your curriculum). MindTap is well beyond an eBook, a homework solution or digital supplement, a resource center website, a course delivery platform, or a Learning Management System. It is the first in a new category—the Personal Learning Experience.

New for this edition! MindTap has an integrated Study Guide, expanded quizzing and application activities, and an integrated Test Bank.

Aplia for Biology The Aplia system helps students learn key concepts via Aplia's focused assignments and active learning opportunities that include randomized, automatically graded questions, exceptional text/art integration, and immediate feedback. Aplia has a full course management system that can be used independently or in conjunction with other course management systems such as MindTap, D2L, or Blackboard.

Acknowledgments

Writing, revising, and illustrating a biology textbook is a major undertaking for two full-time authors, but our efforts constitute only a small part of what is required to produce and distribute this one. We are truly fortunate to be part of a huge team of very talented people who are as committed as we are to creating and disseminating an exceptional science education product.

Biology is not dogma; paradigm shifts are a common outcome of the fantastic amount of research in the field. Ideas about what material should be taught and how best to present that material to students changes from one year to the next. It is only with the ongoing input of our many academic reviewers and advisors (previous page) that we can continue to tailor this book to the needs of instructors and students while integrating new information and models. We continue to learn from and be inspired by these dedicated educators.

On the production side of our team, the indispensable Grace Davidson orchestrated a continuous flow of files, photos, and illustrations while managing schedules, budgets, and whatever else happened to be on fire at the time. Grace, thank you as always for your patience and dedication. Thank you also to Cheryl DuBois, John Sarantakis, and Christine Myaskovsky for your help with photoresearch. Copyeditor Anita Hueftle and proofreader Diane Miller, your valuable suggestions kept our text clear and concise.

Yolanda Cossio, thank you for continuing to support us and for encouraging our efforts to innovate and improve. Thanks also to Cengage Production Manager Hal Humphrey, Marketing Manager Tom Ziolkowski, and to Lauren Oliveira, who creates our exciting technology package, Associate Content Developers Casey Lozier and Kellie Petruzzelli, and Product Assistant Victor Luu.

Lisa Starr and Christine Evers, November 2014

BIOLOGY 5e

TODAY & TOMORROW

1

INVITATION TO BIOLOGY

Application ⊗ 1.1 The Secret Life of Earth

A. Paul Oliver discovered this tree frog perched on a sack of rice during a rainy campsite lunch in New Guinea's Foja Mountains. The explorers dubbed the new species "Pinocchio frog" after the Disney character because the male frog's long nose inflates and points upward during times of excitement.

B. Dr. Jason Bond holds a new species of trapdoor spider he discovered in sand dunes of California beaches in 2008. Bond named the spider *Aptostichus stephencolberti*, after TV personality Stephen Colbert.

Figure 1.1 Newly discovered species.
Each of the thousands of species discovered every year is a reminder that we do not yet know all of the organisms living on our own planet. We don't even know how many to look for. Information about the 1.8 million species we do know about is being collected in The Encyclopedia of Life, an online database maintained by collaborative effort (www.eol.org).

(A) Tim Laman/National Geographic Stock; (B) Courtesy East Carolina University.

In this era of detailed satellite imagery and cell phone global positioning systems, could there possibly be any places left on Earth that humans have not yet explored? Actually, there are plenty of them. In 2005, for example, helicopters dropped a team of scientists into the middle of a vast and otherwise inaccessible cloud forest atop New Guinea's Foja Mountains. Within a few minutes, the explorers realized that their landing site, a dripping, moss-covered swamp, had been untouched by humans. Team member Bruce Beehler remarked, "Everywhere we looked, we saw amazing things we had never seen before. I was shouting. This trip was a once-in-a-lifetime series of shouting experiences."

How did the explorers know they had landed in uncharted territory? For one thing, the forest was filled with plants and animals previously unknown even to native peoples that have long inhabited other parts of the region. During the next month, the team members discovered many new species, including a rhododendron plant with flowers the size of a plate and a frog the size of a pea. They also came across hundreds of species that are on the brink of extinction in other parts of the world, and some that supposedly had been extinct for decades. The animals had never learned to be afraid of humans, so they could easily be approached. A few were discovered as they casually wandered through campsites (Figure 1.1A).

New species are discovered all the time, often in places much more mundane than Indonesian cloud forests (Figure 1.1B). How do we know what species a particular organism belongs to? What is a species, anyway, and why should discovering a new one matter to anyone other than a scientist? You will find the answers to such questions in this book. They are part of the scientific study of life, **biology**, which is one of many ways we humans try to make sense of the world around us.

Trying to understand the immense scope of life on Earth gives us some perspective on where we fit into it. For example, hundreds of new species are discovered every year, but about 20 species become extinct every minute in rain forests alone—and those are only the ones we know about. The current rate of extinctions is about 1,000 times faster than normal, and human activities are responsible for the acceleration. At this rate, we will never know about most of the species that are alive on Earth today. Does that matter? Biologists think so. Whether or not we are aware of it, humans are intimately connected with the world around us. Our activities are profoundly changing the entire fabric of life on Earth. These changes are, in turn, affecting us in ways we are only beginning to understand.

Ironically, the more we learn about the natural world, the more we realize we have yet to learn. But don't take our word for it. Find out what biologists know, and what they do not, and you will have a solid foundation upon which to base your own opinions about how humans fit into this world. By reading this book, you are choosing to learn about the human connection—your connection—with all life on Earth.

1.2 Life Is More Than the Sum of Its Parts

What, exactly, is the property we call "life"? We may never actually come up with a good definition, because living things are too diverse, and they consist of the same basic components as nonliving things. When we try to define life, we end up with a long list of properties that differentiate living from nonliving things. These properties often emerge from the interactions of basic components. To understand how that works, take a look at these groups of squares:

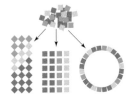

A property called "roundness" emerges when the squares are organized one way, but not other ways. The idea that different structures can be assembled from the same basic building blocks is a recurring theme in our world, and also in biology.

Life has successive levels of organization, with new properties emerging at each level (Figure 1.2). This organization begins with interactions between **atoms**, which are fundamental units of matter—the building blocks of all substances ①. Atoms bond together to form **molecules** ②. There are no atoms unique to living things, but there are unique molecules. In today's natural world, only living things make the "molecules of life," which are lipids, proteins, DNA, RNA, and complex carbohydrates. The emergent property of "life" appears at the next level, when many molecules of life become organized as a cell ③. A **cell** is the smallest unit of life. Cells survive and reproduce themselves using energy, raw materials, and information in their DNA.

Some cells live and reproduce independently; others do so as part of a multicelled organism ④. An **organism** is an individual that consists of one or more cells. In most multicelled organisms, cells are organized as tissues, organs, and organ systems that interact to keep the body working properly.

A **population** is a group of interbreeding individuals of the same type, or species, living in a given area ⑤. At the next level, a **community** consists of all populations living in a given area ⑥. Communities may be large or small, depending on the area defined.

The next level of organization is the **ecosystem**, which is a community interacting with its physical and chemical environment ⑦. The most inclusive level, the **biosphere**, encompasses all regions of Earth's crust, waters, and atmosphere in which organisms live ⑧.

Figure 1.2 Levels of organization in nature.

① Atoms are fundamental units of matter.

② Molecules consist of atoms.

③ Cells consist of molecules.

④ Organisms consist of cells.

⑤ Populations consist of organisms.

⑥ Communities consist of populations.

⑦ Ecosystems consist of communities interacting with their environment.

⑧ The biosphere consists of all ecosystems on Earth.

Take-Home Message 1.2

How do living things differ from nonliving things?

- All things, living or not, consist of the same building blocks: atoms. Atoms bond together to form molecules.
- In today's natural world, only living things make lipids, proteins, DNA, RNA, and complex carbohydrates. The unique properties of life emerge as these molecules become organized into cells.
- Higher levels of life's organization include multicelled organisms, populations, communities, ecosystems, and the biosphere.

atom Fundamental building block of all matter.

biology The scientific study of life.

biosphere All regions of Earth where organisms live.

cell Smallest unit of life.

community All populations of all species in a given area.

ecosystem A community interacting with its environment.

molecule Two or more atoms bonded together.

organism Individual that consists of one or more cells.

population Group of interbreeding individuals of the same species that live in a given area.

1.3 How Living Things Are Alike

Even though we cannot precisely define "life," we can intuitively understand what it means because all living things share a particular set of key features. All require ongoing inputs of energy and raw materials; all sense and respond to change; and all pass DNA to offspring.

Organisms Require Energy and Nutrients Not all living things eat, but all require energy and nutrients on an ongoing basis. Inputs of both are essential to maintain the functioning of individual organisms and the organization of life in general. A **nutrient** is a substance that an organism needs for growth and survival but cannot make for itself.

Organisms spend a lot of time acquiring energy and nutrients (Figure 1.3). However, the source of energy and the type of nutrients acquired differ among organisms. These differences allow us to classify living things into two categories: producers and consumers. A **producer** makes its own food using energy and simple raw materials it obtains from nonbiological sources. Plants are producers; by a process called **photosynthesis**, they use the energy of sunlight to make sugars from water and carbon dioxide (a gas in air). Consumers, by contrast, cannot make their own food. A **consumer** obtains energy and nutrients by feeding on other organisms. Animals are consumers. So are decomposers, which feed on the wastes or remains of other organisms. The leftovers from consumers' meals end up in the environment, where they serve as nutrients for producers. Said another way, nutrients cycle between producers and consumers.

Unlike nutrients, energy is not cycled. It flows through the world of life in one direction: from the environment, through organisms, and back to the environment. This flow maintains the organization of every living cell and body, and it also influences how individuals interact with one another and their environment. The energy flow is one-way, because with each transfer, some energy escapes as heat, and cells cannot use heat as an energy source. Thus, energy that enters the world of life eventually leaves it (we return to this topic in Chapter 5).

Organisms Sense and Respond to Change An organism cannot survive for very long in a changing environment unless it adapts to the changes. Thus, every living thing has the ability to sense and respond to change both inside and outside of itself (Figure 1.4). Consider how, after you eat, the sugars from your meal enter your bloodstream. The added sugars set in motion a series of events that causes cells throughout the body to take up sugar faster, so the sugar level in your blood quickly falls. This response keeps your blood sugar level within a certain range, which in turn helps keep your cells alive and your body functioning properly.

All of the fluids outside of cells make up a body's internal environment. That environment must be kept within certain ranges of temperature and other conditions, or the cells that make up the body will die. By sensing and adjusting to change, organisms keep conditions in the internal environment within a range that favors survival. **Homeostasis** is the name for this process, and it is one of the defining features of life.

Organisms Grow and Reproduce With little variation, the same types of molecules perform the same basic functions in every organism. For example, information in an organism's **DNA** (deoxyribonucleic acid) guides ongoing functions that sustain the individual through its lifetime. Such functions include **development**:

producer acquiring energy and nutrients from its environment

consumer acquiring energy and nutrients by eating a producer

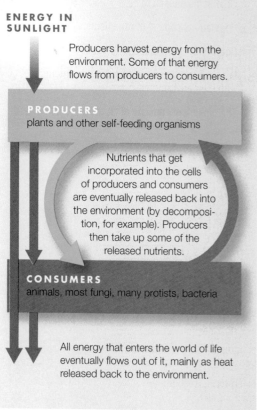

ENERGY IN SUNLIGHT

Producers harvest energy from the environment. Some of that energy flows from producers to consumers.

PRODUCERS
plants and other self-feeding organisms

Nutrients that get incorporated into the cells of producers and consumers are eventually released back into the environment (by decomposition, for example). Producers then take up some of the released nutrients.

CONSUMERS
animals, most fungi, many protists, bacteria

All energy that enters the world of life eventually flows out of it, mainly as heat released back to the environment.

Figure 1.3 The one-way flow of energy and the cycling of materials in the world of life.

Top, © Victoria Pinder, www.flickr.com/photos/vixstarplus.

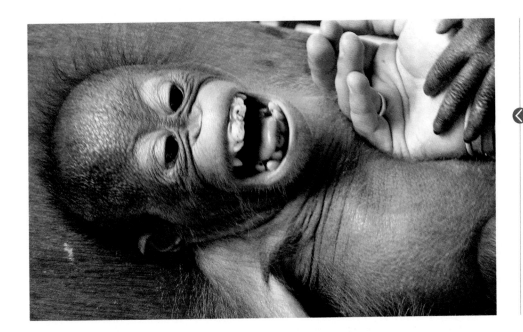

Figure 1.4 Organisms sense and respond to stimulation.
This baby orangutan is laughing in response to being tickled. Apes and humans make different sounds when being tickled, but the airflow patterns are so similar that we can say apes really do laugh.

© Dr. Marina Davila Ross, University of Portsmouth.

the process by which the first cell of a new individual becomes a multicelled adult; **growth**: increases in cell number, size, and volume; and **reproduction**: processes by which individuals produce offspring.

Individuals of every natural population are alike in certain aspects of their body form and behavior because their DNA is very similar: Orangutans look like orangutans and not like caterpillars because they inherited orangutan DNA, which differs from caterpillar DNA in the information it carries. **Inheritance** refers to the transmission of DNA to offspring. All organisms receive their DNA from one or more parents.

DNA is the basis of similarities in form and function among organisms. However, the details of DNA molecules differ, and herein lies the source of life's diversity. Small variations in the details of DNA's structure give rise to differences among individuals, and also among types of organisms. As you will see in later chapters, these differences are the raw material of evolutionary processes.

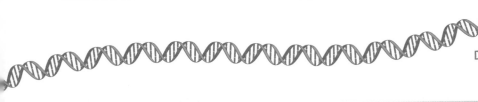

DNA

Take-Home Message 1.3

How are all living things alike?

- A one-way flow of energy and a cycling of nutrients sustain life's organization.
- Organisms sense and respond to conditions inside and outside themselves. They make adjustments that keep conditions in their internal environment within a range that favors cell survival, a process called homeostasis.
- All organisms use information in the DNA they inherited from their parent or parents to develop, grow, and reproduce. DNA is the basis of similarities and differences in form and function among organisms.

consumer Organism that gets energy and nutrients by feeding on the tissues, wastes, or remains of other organisms.

development Multistep process by which the first cell of a new multicelled organism gives rise to an adult.

DNA Deoxyribonucleic acid; carries hereditary information that guides development and other activities.

growth In multicelled species, an increase in the number, size, and volume of cells.

homeostasis Process in which an organism keeps its internal conditions within tolerable ranges by sensing and responding to change.

inheritance Transmission of DNA to offspring.

nutrient Substance that an organism needs for growth and survival but cannot make for itself.

photosynthesis Process by which a producer uses light energy to make sugars from carbon dioxide and water.

producer Organism that makes its own food using energy and nonbiological raw materials from the environment.

reproduction Process by which parents produce offspring.

animal Multicelled consumer that develops through a series of stages and moves about during part or all of its life.

archaea Group of single-celled organisms that lack a nucleus but are more closely related to eukaryotes than to bacteria.

bacteria The most diverse and well-known group of single-celled organisms that lack a nucleus.

biodiversity Scope of variation among living organisms.

eukaryote Organism whose cells characteristically have a nucleus.

fungus Single-celled or multicelled eukaryotic consumer that breaks down material outside itself, then absorbs nutrients released from the breakdown.

plant A multicelled, typically photosynthetic producer.

prokaryote Single-celled organism with no nucleus.

protists A group of diverse, simple eukaryotes.

species Unique type of organism.

taxonomy Practice of naming and classifying species.

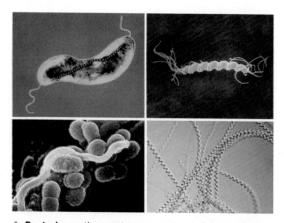

A. Bacteria are the most numerous organisms on Earth. Clockwise from upper left, a bacterium with a row of iron crystals that acts like a tiny compass; a common resident of cat and dog stomachs; spiral cyanobacteria; types found in dental plaque.

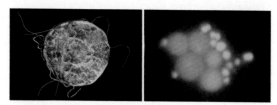

B. Archaea may resemble bacteria, but they are more closely related to eukaryotes. These are two types of archaea from a hydrothermal vent on the seafloor.

Figure 1.5 A few representative prokaryotes.

(A) top left, Dr. Richard Frankel; top right, Science Source; bottom left, www.zahnarzt-stuttgart .com; bottom right, © Susan Barnes; (B) left, Dr. Terry Beveridge, Visuals Unlimited/Corbis; right, © Dr. Harald Huber, Dr. Michael Hohn, Prof. Dr. K.O. Stetter, University of Regensburg, Germany.

1.4 How Living Things Differ

Living things differ tremendously in their observable characteristics. Various classification schemes help us organize what we understand about the scope of this variation, which we call Earth's **biodiversity**.

For example, organisms can be grouped on the basis of whether they have a nucleus, which is a saclike structure containing a cell's DNA. **Bacteria** (singular, bacterium) and **archaea** (singular, archaeon) are organisms whose DNA *is not* contained within a nucleus. All bacteria and archaea are single-celled, which means each organism consists of one cell (Figure 1.5). Collectively, these organisms are the most diverse representatives of life. Different kinds are producers or consumers in nearly all regions of Earth. Some inhabit such extreme environments as frozen desert rocks, boiling sulfurous lakes, and nuclear reactor waste. The first cells on Earth may have faced similarly hostile conditions.

Traditionally, organisms without a nucleus have been called **prokaryotes**, but the designation is now used only informally. This is because, despite the similar appearance of bacteria and archaea, the two types of cells are less related to one another than we once thought. Archaea turned out to be more closely related to **eukaryotes**, which are organisms whose DNA *is* contained within a nucleus. Some eukaryotes live as individual cells; others are multicelled (Figure 1.6). Eukaryotic cells are typically larger and more complex than bacteria or archaea.

Protists are the simplest eukaryotes, but as a group they vary dramatically, from single-celled consumers to giant, multicelled producers.

Fungi (singular, fungus) are eukaryotic consumers that secrete substances to break down food externally, then absorb nutrients released by this process. Many fungi are decomposers. Most fungi, including those that form mushrooms, are multicellular. Fungi that live as single cells are called yeasts.

Plants are multicelled eukaryotes, and the vast majority of them are photosynthetic producers that live on land. Besides feeding themselves, plants also serve as food for most other land-based organisms.

Animals are multicelled eukaryotic consumers that ingest tissues or juices of other organisms. Unlike fungi, animals break down food inside their body. They also develop through a series of stages that lead to the adult form. All animals actively move about during at least part of their lives.

What Is a Species? Each time we discover a new **species**, or unique kind of organism, we name it. **Taxonomy**, the practice of naming and classifying species, began thousands of years ago, but naming species in a consistent way did not become a priority until the eighteenth century. At the time, European explorers who were just discovering the scope of life's diversity started having more and more trouble communicating with one another because species often had multiple names. For example, the dog rose (a plant native to Europe, Africa, and Asia) was alternately known as briar rose, witch's briar, herb patience, sweet briar, wild briar, dog briar, dog berry, briar hip, eglantine gall, hep tree, hip fruit, hip rose, hip tree, hop fruit, and hogseed—and those are only the English names! Species often had multiple scientific names too, in Latin that was descriptive but often cumbersome. The scientific name of the dog rose was *Rosa sylvestris inodora seu canina* (odorless woodland dog rose), and also *Rosa sylvestris alba cum rubore, folio glabro* (pinkish white woodland rose with smooth leaves).

An eighteenth-century naturalist, Carolus Linnaeus, standardized a two-part naming system that we still use. By the Linnaean system, every species is given a

Protists are a group of extremely diverse eukaryotes that range from giant multicelled seaweeds to microscopic single cells.

Plants are multicelled eukaryotes. Almost all plants are photosynthetic producers, and most of them have roots, stems, and leaves.

Fungi are eukaryotic consumers that secrete substances to break down food outside their body. Most are multicelled (left), but some are single-celled (above).

Animals are multicelled consumers that ingest tissues or juices of other organisms. All actively move about during at least part of their life.

Figure 1.6 A few representative eukaryotes.

genus A group of species that share a unique set of traits.

taxon Group of organisms that share a unique set of traits.

unique two-part scientific name. The first part of a scientific name is the **genus** (plural, genera), a group of species that share a unique set of features. The second part is the specific epithet. Together, the genus name and the specific epithet designate one species. Thus, the dog rose now has one official name, *Rosa canina*, that is recognized worldwide.

Genus and species names are always italicized. For example, *Panthera* is a genus of big cats. Lions belong to the species *Panthera leo*. Tigers belong to a different species in the same genus (*Panthera tigris*), and so do leopards (*P. pardus*). Note how the genus name may be abbreviated after it has been spelled out once.

A Rose by Any Other Name The individuals of a species share a unique set of inherited traits. For example, giraffes normally have very long necks, brown spots on white coats, and so on. These are morphological (structural) traits. Individuals of a species also share biochemical traits (they make and use the same molecules) and behavioral traits (they respond the same way to certain stimuli, as when hungry giraffes feed on tree leaves). We can rank species into ever more inclusive categories based on some subset of traits it shares with other species. Each rank, or **taxon** (plural, taxa), is a group of organisms that share a unique set of traits. Each category above species—genus, family, order, class, phylum (plural, phyla), kingdom, and domain—consists of a group of the next lower taxon (Figure 1.7). Using this system, we can sort all life into a few categories (Figure 1.8).

It is easy to tell that orangutans and caterpillars are different species because they appear very different. Distinguishing between species that are more closely related may be much more challenging (Figure 1.9). In addition, traits shared by members of a species often vary a bit among individuals, as eye color does among

> A "species" is a convenient but artificial construct of the human mind.

	wild carrot	marijuana	apple	prickly rose	dog rose
domain	Eukarya	Eukarya	Eukarya	Eukarya	Eukarya
kingdom	Plantae	Plantae	Plantae	Plantae	Plantae
phylum	Magnoliophyta	Magnoliophyta	Magnoliophyta	Magnoliophyta	Magnoliophyta
class	Magnoliopsida	Magnoliopsida	Magnoliopsida	Magnoliopsida	Magnoliopsida
order	Apiales	Rosales	Rosales	Rosales	Rosales
family	Apiaceae	Cannabaceae	Rosaceae	Rosaceae	Rosaceae
genus	*Daucus*	*Cannabis*	*Malus*	*Rosa*	*Rosa*
species	*carota*	*sativa*	*domestica*	*acicularis*	*canina*

Figure 1.7 Taxonomic classification of five species that are related at different levels.
Each species has been assigned to ever more inclusive groups, or taxa: in this case, from genus to domain.

From the left, Joaquim Gaspar; © kymkemp.com; Sylvie Bouchard/Shutterstock.com; Courtesy of Melissa S. Green, www.flickr.com/photos/henkimaa; © Grodana Sarkotic.

Figure It Out: Which of the plants shown here are in the same order?

Answer: Marijuana, apple, prickly rose, and dog rose

**Figure 1.8
Two little ways
to see the big
picture of life.**
Lines in such
diagrams indicate
evolutionary
connections.

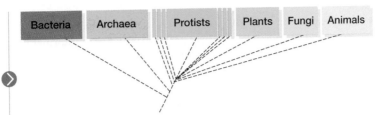

A. Six-kingdom classification system. The protist kingdom includes the most ancient multicelled and all single-celled eukaryotes.

B. Three-domain classification system. The Eukarya domain includes protists, plants, fungi, and animals.

people. How do we decide whether similar-looking organisms belong to the same species? The short answer to that question is that we rely on whatever information we have. Early naturalists studied anatomy and distribution—essentially the only methods available at the time—so species were named and classified according to what they looked like and where they lived. Today's biologists are able to compare traits that the early naturalists did not even know about, including biochemical ones.

The discovery of new information sometimes changes the way we distinguish a particular species or how we group it with others. For example, Linnaeus grouped plants by the number and arrangement of reproductive parts, a scheme that resulted in odd pairings such as castor-oil plants with pine trees. Having more information today, we place these plants in separate phyla.

Evolutionary biologist Ernst Mayr defined a species as one or more groups of individuals that potentially can interbreed, produce fertile offspring, and do not interbreed with other groups. This "biological species concept" is useful in many cases, but it is not universally applicable. For example, we may never know whether two widely separated populations could interbreed if they got together. As another example, populations often continue to interbreed even as they diverge, so the exact moment at which two populations become two species is often impossible to pinpoint. We return to speciation and how it occurs in Chapter 12, but for now it is important to remember that a "species" is a convenient but artificial construct of the human mind.

Figure 1.9 Four butterflies, two species: Which are which?
The top row shows two forms of the species *Heliconius melpomene*; the bottom row, two forms of *H. erato*.

H. melpomene and *H. erato* never cross-breed. Their alternate but similar patterns of coloration evolved as a shared warning signal to predatory birds that these butterflies taste terrible.

© 2006 Axel Meyer, "Repeating Patterns of Mimicry." *PLoS Biology* Vol. 4, No. 10, e341 doi:10.1371/journal.pbio.0040341. Used with Permission.

Take-Home Message 1.4

How do organisms differ from one another?

- Organisms differ in their details; they show tremendous variation in observable characteristics.
- We divide Earth's biodiversity into broad groups based on traits such as having a nucleus or being multicellular.
- Each species is given a unique, two-part scientific name.
- Classification systems group species on the basis of shared traits.

1.5 The Science of Nature

Most of us assume that we do our own thinking, but do we, really? You might be surprised to find out how often we let others think for us. Consider how a school's job (which is to impart as much information to students as quickly as possible)

control group Group of individuals identical to an experimental group except for the independent variable under investigation.

critical thinking Evaluating information before accepting it.

data Experimental results.

experiment A test designed to support or falsify a prediction.

experimental group In an experiment, a group of individuals who have a certain characteristic or receive a certain treatment.

hypothesis Testable explanation of a natural phenomenon.

model Analogous system used for testing hypotheses.

prediction Statement, based on a hypothesis, about a condition that should exist if the hypothesis is correct.

science Systematic study of the observable world.

scientific method Making, testing, and evaluating hypotheses.

variable In an experiment, a characteristic or event that differs among individuals or over time.

How do my own biases affect what I'm learning?

meshes perfectly with a student's job (which is to acquire as much knowledge as quickly as possible). In this rapid-fire exchange of information, it can be very easy to forget about the quality of what is being exchanged. Any time you accept information without questioning it, you let someone else think for you.

Thinking About Thinking **Critical thinking** is the deliberate process of judging the quality of information before accepting it. "Critical" comes from the Greek *kriticos* (discerning judgment). When you use critical thinking, you move beyond the content of new information to consider supporting evidence, bias, and alternative interpretations. How does the busy student manage this? Critical thinking does not necessarily require extra time, just a bit of extra awareness. There are many ways to do it. For example, you might ask yourself some of the following questions while you are learning something new:

> What message am I being asked to accept?
> Is the message based on facts or opinion?
> Is there a different way to interpret the facts?
> What biases might the presenter have?
> How do my own biases affect what I'm learning?

Such questions are a way of being conscious about learning. They can help you decide whether to allow new information to guide your beliefs and actions.

How Science Works Critical thinking is a big part of **science**, the systematic study of the observable world and how it works. A scientific line of inquiry usually begins with curiosity about something observable, such as (for example) a decrease in the number of birds in a particular area. Typically, a scientist will read about what others have discovered before making a **hypothesis**, a testable explanation for a natural phenomenon. An example of a hypothesis would be, "The number of birds is decreasing because the number of cats is increasing."

A **prediction**, or statement of some condition that should exist if the hypothesis is correct, comes next. Making predictions is often called the if–then process, in which the "if" part is the hypothesis, and the "then" part is the prediction: *If* the number of birds is decreasing because the number of cats is increasing, *then* reducing the number of cats should stop the decline.

Next, a researcher will test the prediction. Tests may be performed on a **model**, or analogous system, if working with an object or event directly is not possible. For

A. Studying the ecological benefits of weedy buffer zones on farms.

B. Measuring how much wood is produced by extremely old trees.

example, animal diseases are often used as models of similar human diseases. Careful observations are one way to test predictions that flow from a hypothesis. So are **experiments**: tests designed to support or falsify a prediction. A typical experiment explores a cause-and-effect relationship using **variables**, which are characteristics or events that can differ among individuals or over time.

Biological systems are typically complex, with many interdependent variables. It can be difficult to study one variable separately from the rest. Thus, biology researchers often test two groups of individuals simultaneously. An **experimental group** is a set of individuals that have a certain characteristic or receive a certain treatment. An experimental group is tested side by side with a **control group**, which is identical to the experimental group except for one independent variable: the characteristic or the treatment being tested. Any differences in experimental results between the two groups is likely to be an effect of changing the variable. Test results—**data**—that are consistent with the prediction are evidence in support of the hypothesis. Data inconsistent with the prediction are evidence that the hypothesis is flawed and should be revised.

A necessary part of science is reporting one's results and conclusions in a standard way, such as in a peer-reviewed journal article. The communication gives other scientists an opportunity to evaluate the information for themselves, both by checking the conclusions drawn and by repeating the experiments. Forming a hypothesis based on observation, and then systematically testing and evaluating the hypothesis, are collectively called the **scientific method** (Table 1.1).

Examples of Experiments in Biology There are many different ways to do research, particularly in biology (Figure 1.10). Some biologists survey, simply observing without making or testing hypotheses. Others make hypotheses based on observations, and leave the testing to others. However, despite a broad range of approaches, scientific experiments are typically designed in a consistent way, so the effects of changing one variable at a time can be measured. To give you a sense of how biology experiments work, we summarize two published studies here.

In 1996 the U.S. Food and Drug Administration (FDA) approved Olestra®, a fat replacement manufactured from sugar and vegetable oil, as a food additive. Potato chips were the first Olestra-containing food product to be sold in the United States. Controversy about the chip additive soon raged. Many people complained of intestinal problems after eating the chips, and thought that the Olestra was at fault. Two

Table 1.1 The Scientific Method

Observe some aspect of nature.

Think of an explanation for your observation (in other words, form a hypothesis).

Test the hypothesis.
  a. Make a prediction based on the hypothesis.
  b. Test the prediction using experiments or surveys.
  c. Analyze the results of the tests (data).

Decide whether the results of the tests support your hypothesis or not (form a conclusion).

Report your results to the scientific community.

Figure 1.10 A few examples of scientific research in the field of biology.

(A) Photo by Scott Bauer, USDA/ARS; (B) MICHAEL NICHOLS/National Geographic Creative; (C) © Roger W. Winstead, NC State University; (D) National Cancer Institute; (E) Courtesy of Susanna López-Legentil.

C. Improving efficiency of biofuel production from agricultural waste.

D. Devising a vaccine that helps prevent cancer.

E. Discovering medically active natural products made by marine animals.

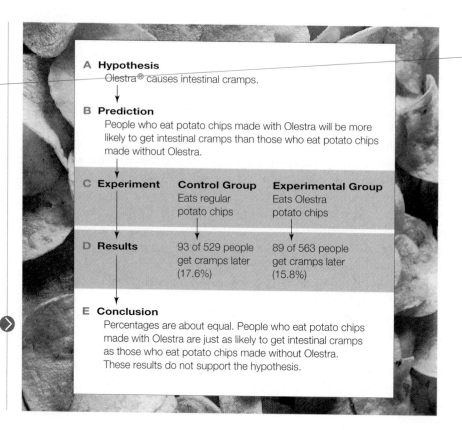

A **Hypothesis**
Olestra® causes intestinal cramps.

B **Prediction**
People who eat potato chips made with Olestra will be more likely to get intestinal cramps than those who eat potato chips made without Olestra.

C **Experiment**	**Control Group**	**Experimental Group**
	Eats regular potato chips	Eats Olestra potato chips
D **Results**	93 of 529 people get cramps later (17.6%)	89 of 563 people get cramps later (15.8%)

E **Conclusion**
Percentages are about equal. People who eat potato chips made with Olestra are just as likely to get intestinal cramps as those who eat potato chips made without Olestra. These results do not support the hypothesis.

Figure 1.11 The steps in a scientific experiment to determine whether Olestra causes intestinal cramps. A report of this study was published in the *Journal of the American Medical Association* in January 1998.

Left, © Bob Jacobson/Corbis; background right, © SuperStock.

years later, researchers at the Johns Hopkins University School of Medicine designed an experiment to test whether Olestra causes cramps. The researchers made the following prediction: *if* Olestra causes cramps, *then* people who eat Olestra should be more likely to get cramps than people who do not eat it. To test the prediction, they used a Chicago theater as a "laboratory." They asked 1,100 people between the ages of thirteen and thirty-eight to watch a movie and eat their fill of potato chips. Each person received an unmarked bag containing 13 ounces of chips. In this experiment, the individuals who received Olestra-laden potato chips were the experimental group, and the individuals who received regular chips were the control group.

A few days after the movie, the researchers contacted all of the people who participated in the experiment and collected any reports of post-movie gastrointestinal problems. Of 563 people making up the experimental group, 89 (15.8 percent) reported having cramps. However, so did 93 of the 529 people (17.6 percent) making up the control group—who had eaten the regular chips. People were about as likely to get cramps whether or not they ate chips made with Olestra. These results did not support the prediction, so the researchers concluded that eating Olestra does not cause cramps (Figure 1.11).

A different experiment that took place in 2005 investigated whether certain behaviors of peacock butterflies help the insects avoid predation by birds. The researchers performing this experiment began with two observations. First, when a peacock butterfly rests, it folds its wings, so only the dark underside shows (Figure 1.12A). Second, when a butterfly sees a predator approaching, it repeatedly flicks its wings open, while also moving them in a way that produces a hissing sound and a series of clicks (Figure 1.12B).

The researchers were curious about why the peacock butterfly flicks its wings. After they reviewed earlier studies, they came up with two hypotheses that might explain the wing-flicking behavior.

1. Wing-flicking probably attracts predatory birds, but it also exposes brilliant spots that resemble owl eyes. Anything that looks like owl eyes is known to startle small, butterfly-eating birds, so exposing the wing spots might scare off predators.

2. The hissing and clicking sounds produced when the peacock butterfly moves its wings may be an additional defense that deters predatory birds.

The researchers then used their hypotheses to make the following predictions:

1. *If* exposing brilliant wing spots startles butterfly-eating birds, *then* peacock butterflies missing their spots will be more likely to get eaten.

2. *If* hissing and clicking sounds deter birds butterfly-eating birds, *then* peacock butterflies unable to make these sounds will be more likely to get eaten.

The next step was the experiment. The researchers used a black marker to cover up the wing spots of some butterflies, and scissors to cut off the sound-making part of the wings of others. A third group had both treatments, their wings painted and also cut. The researchers then put each butterfly into a large cage with a hungry blue tit (Figure 1.12C) and watched the pair for thirty minutes.

Figure 1.12D lists the results of the experiment. All butterflies with unmodified wing spots survived, regardless of whether they made sounds. By contrast, only half of the butterflies that had spots painted out but could make sounds survived. Most

Figure 1.12 **Testing peacock butterfly defenses.**

(A) © Matt Rowlings, www.eurobutterflies.com; (B) © Adrian Vallin; (C) © Antje Schulte; (D) *Proceedings of the Royal Society of London,* Series B (2005) 272: 1203–1207.

Figure It Out: What percentage of butterflies with spots painted and wings cut survived the test?

Answer: 20 percent

A. With wings folded, a resting peacock butterfly resembles a dead leaf, so it is appropriately camouflaged from predatory birds.

B. When a predatory bird approaches, a butterfly flicks its wings open and closed, revealing brilliant spots and producing hissing and clicking sounds.

C. Researchers tested whether the wing-flicking behavior of peacock butterflies affected predation by blue tits.

Experimental Treatment	Number of Butterflies Eaten (of Total)
Spots painted out	5 of 10
Wings cut	0 of 8
Spots painted, wings cut	8 of 10
None	0 of 9

D. The researchers painted out the spots of some butterflies, cut the sound-making part of the wings on others, and did both to a third group; then exposed each butterfly to a hungry blue tit for 30 minutes. Results are listed on the right.

Digging Into Data

Peacock Butterfly Predator Defenses The photographs below represent the experimental and control groups used in the peacock butterfly experiment. Identify the experimental groups, and match them up with the relevant control group(s). *Hint:* Identify which variable is being tested in each group (each variable has a control).

Adrian Vallin, Sven Jakobsson, Johan Lind and Christer Wiklund, *Proc. R. Soc. B* (2005: 272, 1203, 1207). Used with permission of The Royal Society and the author.

A. Wing spots painted out

B. Wing spots visible; wings silenced

C. Wing spots painted out; wings silenced

D. Wings painted but spots visible

E. Wings cut but not silenced

F. Wings painted, spots visible; wings cut, not silenced

of the silenced butterflies with painted-out spots were eaten quickly. The test results confirmed both predictions, so they support the hypotheses. Predatory birds are indeed deterred by peacock butterfly wing-flicking behavior.

The scientific community consists of critically thinking people trying to poke holes in one another's ideas.

1.6 The Nature of Science

Bias in Interpreting Experimental Results Experimenting with a single variable apart from all others is not often possible, particularly when studying humans. For example, remember that the people who participated in the Olestra experiment were chosen randomly, which means the study was not controlled for gender, age, weight, medications taken, and so on. These variables may well have influenced the experiment's results.

Humans are by nature subjective, and scientists are no exception. Researchers risk interpreting their results in terms of what they want to find out. That is

Figure 1.13 **Example of how generalizing from a subset can lead to a conclusion that is incorrect.**

(A) Tim Laman/ National Geographic Stock; (B) © Bruce Beehler/ Conservation International.

A. The cloud forest that covers about 2 million acres of New Guinea's Foja Mountains is extremely remote and difficult to access, even for natives of the region. The first major survey of this forest occurred in 2005.

B. In science, discovery of an error is not always bad news. Kris Helgen holds a golden-mantled tree kangaroo he found during the 2005 Foja Mountains survey. This kangaroo species is extremely rare in other areas, so it was thought to be critically endangered prior to the expedition.

why they typically design experiments that will yield quantitative results, which are counts or some other data that can be measured or gathered objectively. Quantitative results minimize the potential for bias, and also give other scientists an opportunity to repeat the experiments and check the conclusions drawn from them. This last point gets us back to the role of critical thinking in science. Scientists expect one another to recognize and put aside bias in order to test hypotheses in ways that may prove them wrong. If a scientist does not, then others will, because exposing errors is just as useful as applauding insights. The scientific community consists of critically thinking people trying to poke holes in one another's ideas. Ideally, their collective efforts make science a self-correcting endeavor.

Sampling Error Researchers cannot always observe all individuals of a group. For example, the explorers you read about in Section 1.1 did not—and could not—survey every uninhabited part of the Foja Mountains. The cloud forest alone cloaks more than 2 million acres (Figure 1.13A), so surveying all of it would take unrealistic amounts of time and effort.

When researchers cannot directly observe all individuals of a population, all instances of an event, or some other aspect of nature, they may test or survey a subset. Results from the subset are then used to make generalizations about the whole. However, generalizing from a subset is risky because subsets are not necessarily representative of the whole. Consider the golden-mantled tree kangaroo, an animal first discovered in 1993 on a single forested mountaintop in New Guinea. For more than a decade, the species was never seen outside of that habitat, which is getting smaller every year because of human activities. Thus, the golden-mantled tree kangaroo was considered to be one of the most endangered animals on the planet. Then, in 2005, the New Guinea explorers discovered that this kangaroo species is fairly common in the Foja Mountain cloud forest (Figure 1.13B). As a result, biologists now believe its future is secure, at least for the moment.

Sampling error is a difference between results obtained from a subset, and results from the whole (Figure 1.14A). Sampling error may be unavoidable, but knowing how it can occur helps researchers design their experiments to minimize it. For example, sampling error can be a substantial problem with a small subset, so experimenters try to start with a relatively large sample, and they repeat their experiments (Figure 1.14B). To understand why these practices reduce the risk of sampling error, think about flipping a coin. There are two possible outcomes of each flip: The coin lands heads up, or it lands tails up. Thus, the chance that the coin will land heads up is one in two (1/2), or 50 percent. However, when you flip a coin repeatedly, it often lands heads up, or tails up, several times in a row. With just 3 flips, the proportion of times that the coin actually lands heads up may not even be close to 50 percent. With 1,000 flips, however, the overall proportion of times the coin lands heads up is much more likely to approach 50 percent.

Probability is the measure, expressed as a percentage, of the chance that a particular outcome will occur. That chance depends on the total number of possible outcomes. For instance, if 10 million people enter a drawing, each has the same probability of winning: 1 in 10 million, or (an extremely improbable) 0.00001 percent. Analysis of experimental data often includes probability calculations. If there is a very low probability that a result has occurred by chance alone, the result is said to be **statistically significant**. In this context, the word "significant" does not refer to the result's importance. Rather, it means that a rigorous statistical analysis has shown a very low probability (usually 5 percent or less) of the result being incorrect because of sampling error.

A. Natalie chooses a random jelly bean from a jar. She is blindfolded, so she does not know that the jar contains 120 green and 280 black jelly beans.

The jar is hidden from Natalie's view before she removes her blindfold. She sees one green jelly bean in her hand and assumes that the jar must hold only green jelly beans. This assumption is incorrect: 30 percent of the jelly beans in the jar are green, and 70 percent are black. The small sample size has resulted in sampling error.

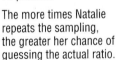

B. Still blindfolded, Natalie randomly chooses 50 jelly beans from the jar. She ends up choosing 10 green and 40 black beans.

The larger sample leads Natalie to assume that one-fifth of the jar's jelly beans are green (20 percent) and four-fifths are black (80 percent). The larger sample more closely approximates the jar's actual green-to-black ratio of 30 percent to 70 percent.

The more times Natalie repeats the sampling, the greater her chance of guessing the actual ratio.

Figure 1.14 How sample size affects sampling error.
© Gary Head.

probability The chance that a particular outcome of an event will occur; depends on the total number of outcomes possible.

sampling error Difference between results derived from testing an entire group of events or individuals, and results derived from testing a subset of the group.

statistically significant Refers to a result that is statistically unlikely to have occurred by chance alone.

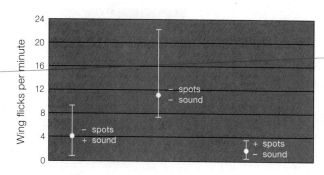

Figure 1.15 Example of error bars in a graph.
This graph was adapted from the peacock butterfly research described in Section 1.5.

 The researchers recorded the number of times each butterfly flicked its wings in response to an attack by a bird.

The squares represent average frequency of wing flicking for each sample set of butterflies. The error bars that extend above and below the dots indicate the range of values—the sampling error.

Figure It Out: What was the fastest rate at which a butterfly with no spots or sound flicked its wings? Answer: 22 times per minute

Science helps us communicate our experiences without bias.

Variation in data is often shown as error bars on a graph (Figure 1.15). Depending on the graph, error bars may indicate variation around an average for one sample set, or the difference between two sample sets.

Scientific Theories Suppose a hypothesis stands even after years of tests. It is consistent with all data ever gathered, and it has helped us make successful predictions about other phenomena. When a hypothesis meets these criteria, it is considered to be a **scientific theory** (Table 1.2). To give an example, all observations to date have been consistent with the hypothesis that matter consists of atoms. Scientists no longer spend time testing this hypothesis for the compelling reason that, since we started looking 200 years ago, no one has discovered matter that consists of anything else. Thus, scientists use the hypothesis, now called atomic theory, to make other hypotheses about matter.

Scientific theories are our best objective descriptions of the natural world. However, they can never be proven absolutely, because to do so would necessitate testing under every possible circumstance. For example, in order to prove atomic theory, the atomic composition of all matter in the universe would have to be checked—an impossible task even if someone wanted to try.

Like all hypotheses, a scientific theory can be disproven by a single observation or result that is inconsistent with it. For example, if someone discovers a form of matter that does not consist of atoms, atomic theory would have to be revised. The potentially falsifiable nature of scientific theories means that science has a built-in system of checks and balances. A theory is revised until no one can prove it to be incorrect. The theory of evolution, which states that change occurs in a line of descent over time, still holds after a century of observations and testing. As with all other scientific theories, no one can be absolutely sure that it will hold under all possible conditions, but it has a very high probability of not being wrong. Few other theories have withstood as much scrutiny.

You may hear people apply the word "theory" to a speculative idea, as in the phrase "It's just a theory." This everyday usage of the word differs from the way it is used in science. Speculation is an opinion, belief, or personal conviction that is not necessarily supported by evidence. A scientific theory is different. By definition, a scientific theory is supported by a large body of evidence, and it is consistent with all known data.

A scientific theory also differs from a **law of nature**, which describes a phenomenon that has been observed to occur in every circumstance without fail, but for which we do not have a complete scientific explanation. The laws of thermodynamics, which describe energy, are examples. We understand *how* energy behaves, but not exactly *why* it behaves the way it does.

The Scope of Science Science helps us be objective about our observations in part because of its limitations. For example, science does not address many questions, such as "Why do I exist?" Answers to such questions can only come from within as an integration of the personal experiences and mental connections that shape our consciousness. This is not to say subjective answers have no value, because no human society can function for long unless its individuals share standards for making judgments, even if they are subjective. Moral, aesthetic, and philosophical standards vary from one society to the next, but all help people decide what is important and good. All give meaning to our lives.

Neither does science address the supernatural, or anything that is "beyond nature." Science neither assumes nor denies that supernatural phenomena occur, but scientists often cause controversy when they discover a natural explanation for something that was thought to have none. Such controversy arises when a society's moral standards are interwoven with its understanding of nature. Nicolaus Copernicus proposed in 1540 that Earth orbits the sun. Today that idea is generally accepted, but the prevailing belief system had Earth as the immovable center of the universe. In 1610, Galileo Galilei published evidence for the Copernican model of the solar system, an act that resulted in his imprisonment. He was publicly forced to recant his work, spent the rest of his life under house arrest, and was never allowed to publish again.

As Galileo's story illustrates, exploring a traditional view of the natural world from a scientific perspective is often misinterpreted as a violation of morality. As a group, scientists are no less moral than anyone else, but they follow a particular set of rules that do not necessarily apply to others: Their work concerns only the natural world, and their ideas must be testable by other scientists.

Science helps us communicate our experiences of the natural world without bias. As such, it may be as close as we can get to a universal language. We are fairly sure, for example, that the laws of gravity apply everywhere in the universe. Intelligent beings on a distant planet would likely understand the concept of gravity. Thus, we might well use gravity or another scientific concept to communicate with them, or anyone, anywhere. The point of science, however, is not to communicate with aliens. It is to find common ground here on Earth.

© Raymond Gehman/Corbis.

Theory	Main Premises
Atomic theory	All matter consists of atoms.
Big bang	The universe originated with an explosion and continues to expand.
Cell theory	All organisms consist of one or more cells, the cell is the basic unit of life, and all cells arise from existing cells.
Evolution	Change occurs in the inherited traits of a population over generations.
Global warming	Human activities are causing Earth's average temperature to increase.
Plate tectonics	Earth's crust is cracked into pieces that move in relation to one another.

Table 1.2 Examples of Scientific Theories

Take-Home Message 1.6

Why does science work?

- Researchers minimize sampling error by using large sample sizes and by repeating their experiments. Probability calculations can show whether a result is unlikely to have occurred by chance alone.
- Science is concerned only with testable ideas about observable aspects of nature.
- Ideally, science is a self-correcting process because it is carried out by a community of people who systematically check one another's work and conclusions.
- Because a scientific theory is thoroughly tested and revised until no one can prove it wrong, it is our best way of objectively describing the natural world.

law of nature Generalization that describes a consistent natural phenomenon for which there is incomplete scientific explanation.

scientific theory Hypothesis that has not been disproven after many years of rigorous testing.

Summary

Section 1.1 **Biology** is the scientific study of life. We know about only a fraction of the organisms that live on Earth, in part because we have explored only a fraction of its inhabited regions.

Section 1.2 Biologists think about life at different levels of organization, with new properties emerging at successively higher levels. All matter consists of **atoms**, which bond together to form **molecules**. **Organisms** are individuals that consist of one or more **cells**, the organizational level at which life emerges. A **population** is a group of interbreeding individuals of a species in a given area; a **community** is all populations of all species in a given area. An **ecosystem** is a community interacting with its environment. The **biosphere** includes all regions of Earth that hold life.

Section 1.3 Life has underlying unity in that all living things have similar characteristics: (1) All organisms require energy and **nutrients** to sustain themselves. **Producers** harvest energy from the environment to make their own food by processes such as **photosynthesis**; **consumers** ingest other organisms, or their wastes or remains. (2) Organisms keep the conditions in their internal environment within ranges that their cells tolerate—a process called **homeostasis**. (3) **DNA** contains information that guides an organism's **growth**, **development**, and **reproduction**. The passage of DNA from parents to offspring is **inheritance**.

Section 1.4 The many types of organisms that currently exist on Earth differ greatly in details of body form and function. **Biodiversity** is the sum of differences among living things. **Bacteria** and **archaea** are **prokaryotes**, single-celled organisms whose DNA is not contained within a nucleus. The DNA of single-celled or multicelled **eukaryotes** (**protists**, **plants**, **fungi**, and **animals**) is contained within a nucleus.

Each **species** has a two-part name. The first part is the **genus** name. When combined with the specific epithet, it designates the particular species. With **taxonomy**, species are ranked into ever more inclusive **taxa** on the basis of shared traits.

Section 1.5 **Critical thinking**, the self-directed act of judging the quality of information as one learns, is an important part of **science**. Generally, a researcher observes something in nature, forms a **hypothesis** (testable explanation) for it, then makes a **prediction** about what might occur if the hypothesis is correct. Predictions are tested with observations, **experiments**, or both.

Experiments typically are performed on an **experimental group** as compared with a **control group**, and sometimes on **model** systems. Conclusions are drawn from **data**. A hypothesis that is not consistent with data is modified or discarded. The **scientific method** consists of making, testing, and evaluating hypotheses, and sharing results with the scientific community.

Biological systems are usually influenced by many interacting **variables**. Research approaches differ, but experiments are designed in a consistent way, in order to study a single cause-and-effect relationship in a complex natural system.

Section 1.6 Small sample size increases the potential for **sampling error** in experimental results. In such cases, a subset may be tested that is not representative of the whole. Researchers design experiments carefully to minimize sampling error and bias, and they use **probability** calculations to check the **statistical significance** of their results.

Science helps us be objective about our observations because it is concerned only with testable ideas about observable aspects of nature. Opinion and belief have value in human culture, but they are not addressed by science. A **scientific theory** is a long-standing hypothesis that is useful for making predictions about other phenomena. It is our best way of objectively describing nature. A **law of nature** is a phenomenon that occurs without fail, but has an incomplete scientific explanation.

Self-Quiz

Answers in Appendix I

1. _____ are fundamental building blocks of all matter.
   a. Cells
   b. Atoms
   c. Organisms
   d. Molecules

2. The smallest unit of life is the _____ .
   a. atom
   b. molecule
   c. cell
   d. organism

3. _____ is the transmission of DNA to offspring.
   a. Reproduction
   b. Development
   c. Homeostasis
   d. Inheritance

4. A process by which an organism produces offspring is called _____ .
   a. reproduction
   b. development
   c. homeostasis
   d. inheritance

5. Organisms require _____ and _____ to maintain themselves, grow, and reproduce.
   a. DNA; energy
   b. food; sunlight
   c. nutrients; energy
   d. DNA; cells

6. _____ move around for at least part of their life.

7. By sensing and responding to change, an organism keeps conditions in its internal environment within ranges that its cells can tolerate. This process is called _____ .
   a. sampling error
   b. development
   c. homeostasis
   d. critical thinking

8. DNA _____ .
   a. guides form and function
   b. is the basis of traits
   c. is transmitted from parents to offspring
   d. all of the above

9. A butterfly is a(n) _____ (choose all that apply).
   a. organism
   b. domain
   c. species
   d. eukaryote
   e. consumer
   f. producer
   g. prokaryote
   h. trait

10. A bacterium is _____ (choose all that apply).
    a. an organism
    b. single-celled
    c. an animal
    d. a eukaryote

11. Bacteria, Archaea, and Eukarya are three _____ .

12. A control group is _____ .
    a. a set of individuals that have a characteristic under study or receive an experimental treatment
    b. the standard against which an experimental group is compared
    c. the experiment that gives conclusive results

13. Science addresses only that which is _____ .
    a. alive
    b. observable
    c. variable
    d. indisputable

14. Match the terms with the most suitable description.
    _____ life
    _____ probability
    _____ species
    _____ scientific theory
    _____ hypothesis
    _____ prediction
    _____ producer

    a. if–then statement
    b. unique type of organism
    c. emerges with cells
    d. testable explanation
    e. measure of chance
    f. makes its own food
    g. time-tested hypothesis

15. In one survey, fifteen randomly selected students were found to be taller than 6 feet. This data led to the conclusion that the average height of a student is greater than 6 feet. This is an example of _____ .
    a. experimental error
    b. sampling error
    c. a subjective opinion
    d. experimental bias

Critical Thinking

1. A person is declared to be dead upon the irreversible ceasing of spontaneous body functions: brain activity, or blood circulation and respiration. However, only about 1% of a person's cells have to die in order for all of these things to happen. How can someone be dead when 99% of his or her cells are still alive?

2. Explain the difference between a one-celled organism and a single cell of a multicelled organism.

3. Why would you think twice about ordering from a restaurant menu that lists only the second part of the species name (not the genus) of its offerings? *Hint:* Look up *Ursus americanus, Ceanothus americanus, Bufo americanus, Homarus americanus, Lepus americanus,* and *Nicrophorus americanus.*

4. Once there was a highly intelligent turkey that had nothing to do but reflect on the world's regularities. Morning always started out with the sky turning light, followed by the master's footsteps, which were always followed by the appearance of food. Other things varied, but food always followed footsteps. The sequence of events was so predictable that it eventually became the basis of the turkey's theory about the goodness of the world. One morning, after more than 100 confirmations of this theory, the turkey listened for the master's footsteps, heard them, and had its head chopped off.

   Any scientific theory is modified or discarded upon discovery of contradictory evidence. The absence of absolute certainty has led some people to conclude that "theories are irrelevant because they can change." If that is so, should we stop doing scientific research? Why or why not?

5. In 2005, researcher Woo-suk Hwang reported that he had made immortal stem cells from human patients. His research was hailed as a breakthrough for people affected by degenerative diseases, because stem cells may be used to repair a person's own damaged tissues. Hwang published his results in a peer-reviewed journal. In 2006, the journal retracted his paper after other scientists discovered that Hwang's group had faked their data. Does the incident show that results of scientific studies cannot be trusted? Or does it confirm the usefulness of a scientific approach, because other scientists discovered and exposed the fraud?

2

MOLECULES OF LIFE

Application ❯ 2.1 **Fear of Frying**

The human body requires only about a tablespoon of fat each day to stay healthy, but most people in developed countries eat far more than that. The average American eats about 70 pounds of fat per year, which may be part of the reason why the average American is overweight. Being overweight increases one's risk for many chronic illnesses. However, the total quantity of fat in the diet may have less impact on health than the types of fats. Fats are more than inert molecules that accumulate in strategic areas of our bodies. They are the main constituents of cell membranes, and as such they have powerful effects on cell function.

The typical fat molecule has three fatty acid tails, each a long chain of carbon atoms that can vary a bit in structure. Fats with a certain arrangement of hydrogen atoms around those carbon chains are called *trans* fats. Small amounts of *trans* fats occur naturally in red meat and dairy products, but the main source of these fats in the American diet is an artificial food product called partially hydrogenated vegetable oil. Hydrogenation is a manufacturing process that adds hydrogen atoms to oils in order to change them into solid fats. In 1908, Procter & Gamble Co. developed partially hydrogenated soybean oil as a substitute for the more expensive solid animal fats they had been using to make candles. By 1911, more households in the United States became wired for electricity, so the demand for candles was waning. P & G needed another way to sell its proprietary fat. Partially hydrogenated vegetable oil looks a lot like lard, so the company began aggressively marketing it as a revolutionary new food: a solid cooking fat with a long shelf life, mild flavor, and lower cost than lard or butter.

Figure 2.1 Unhealthy *trans* fats are abundant in partially hydrogenated oils commonly used to make manufactured and fast foods.

© Kentoh/Shutterstock.com.

By the mid-1950s, hydrogenated vegetable oil had become a major part of the American diet. For decades, it was considered to be healthier than animal fats because it was made from plants, but we now know otherwise. *Trans* fats, which are abundant in hydrogenated vegetable oils, raise the level of cholesterol in our blood more than any other fat, and they directly alter the function of our arteries and veins. The effects of such changes are quite serious. Eating as little as 2 grams per day (about 0.4 teaspoon) of hydrogenated vegetable oil measurably increases one's risk of atherosclerosis (hardening of the arteries), heart attack, and diabetes. A small serving of french fries made with hydrogenated vegetable oil contains about 5 grams of *trans* fat (Figure 2.1). At this writing, hydrogenated oil is still a component of many manufactured and fast foods: french fries, stick margarines, ready-to-use frostings, cookies, crackers, cakes and pancakes, peanut butter, pies, doughnuts, muffins, chips, microwave popcorn, pizzas, burritos, chicken nuggets, fish sticks, and so on.

All organisms consist of the same kinds of molecules, but small differences in the way those molecules are put together can have big effects. With this concept, we introduce you to the chemistry of life. This is your chemistry. It makes you far more than the sum of your body's molecules.

2.2 Start With Atoms

REMEMBER: Atoms are fundamental units of matter—the building blocks of all substances (Section 1.2).

Even though atoms are about 20 million times smaller than a grain of sand, they consist of even smaller subatomic particles. Positively charged **protons** (p⁺) and uncharged **neutrons** occur in an atom's core, or **nucleus**. Negatively charged **electrons** (e⁻) move around the nucleus (Figure 2.2). **Charge** is an electrical property: Opposite charges attract, and like charges repel. A typical atom has about the same number of electrons and protons. The negative charge of an electron is the same magnitude as the positive charge of a proton, so the two charges cancel one another. Thus, an atom with the same number of electrons and protons carries no charge.

All atoms have protons. The number of protons in the nucleus is called the **atomic number**, and it determines the type of atom, or element. **Elements** are pure substances, each consisting only of atoms with the same number of protons in their nucleus. For example, the element carbon has an atomic number of 6 (Figure 2.3). All atoms with six protons in their nucleus are carbon atoms, no matter how many electrons or neutrons they have. Elemental carbon (the substance) consists only of carbon atoms, and all of those atoms have six protons. Each of the 118 known elements has a symbol that is typically an abbreviation of its Latin or Greek name (see Appendix II). Carbon's symbol, C, is from *carbo*, the Latin word for coal. Coal is mostly carbon.

All atoms of an element have the same number of protons, but they can differ in the number of other subatomic particles. Those that differ in the number of neutrons are called **isotopes**. The total number of neutrons and protons in the nucleus of an isotope is its **mass number**. Mass number is written as a superscript to the left of the element's symbol. For example, the most common isotope of hydrogen has one proton and no neutrons, so it is designated ^1H. Other hydrogen isotopes include deuterium (^2H, one proton and one neutron) and tritium (^3H, one proton and two neutrons).

The most common carbon isotope has six protons and six neutrons (^{12}C). Another naturally occurring carbon isotope, ^{14}C, has six protons and eight neutrons (6 + 8 = 14). Carbon 14 is an example of a **radioisotope**, or radioactive isotope. Atoms of a radioisotope have an unstable nucleus that breaks up spontaneously. As a nucleus breaks up, it emits radiation (subatomic particles, energy, or both), a process called **radioactive decay**. The atomic nucleus cannot be altered by ordinary means, so radioactive decay is unaffected by external factors such as temperature, pressure, or whether the atoms are part of molecules.

Each radioisotope decays at a predictable rate into predictable products. For example, when carbon 14 decays, one of its neutrons splits into a proton and an electron. The nucleus emits the electron as radiation. Thus, a carbon atom with eight neutrons and six protons (^{14}C) becomes a nitrogen atom, with seven neutrons and seven protons (^{14}N):

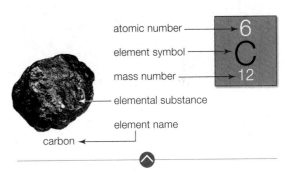

nucleus of ^{14}C, with 6 protons, 8 neutrons

nucleus of ^{14}N, with 7 protons, 7 neutrons

This process is so predictable that we can say with certainty that about half of the atoms in any sample of ^{14}C will be ^{14}N atoms after 5,730 years. The predictability of

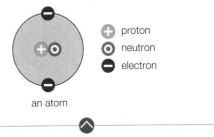

proton
neutron
electron

an atom

Figure 2.2 Atoms consist of subatomic particles. Models such as this do not show what atoms really look like. Electrons move in defined, three-dimensional spaces about 10,000 times bigger than the nucleus. Protons and neutrons occur in the nucleus.

atomic number
element symbol
mass number
elemental substance
element name
carbon

Figure 2.3 Example of an element: carbon.

Left, Theodore Gray/Visuals Unlimited, Inc.

atomic number Number of protons in the atomic nucleus; determines the element.

charge Electrical property; opposite charges attract, and like charges repel.

electron Negatively charged subatomic particle.

element A pure substance that consists only of atoms with the same number of protons.

isotopes Forms of an element that differ in the number of neutrons their atoms carry.

mass number Of an isotope, the total number of protons and neutrons in the atomic nucleus.

neutron Uncharged subatomic particle in the atomic nucleus.

nucleus Core of an atom; occupied by protons and neutrons.

proton Positively charged subatomic particle that occurs in the nucleus of all atoms.

radioactive decay Process by which atoms of a radioisotope emit energy and subatomic particles when their nucleus spontaneously breaks up.

radioisotope Isotope with an unstable nucleus.

Figure 2.4 PET scans.
PET scans use radioactive tracers to form a digital image of a process in the body's interior. These two PET scans reveal the activity of a molecule called MAO-B in the body of a nonsmoker (left) and a smoker (right). The activity is color-coded from red (highest activity) to purple (lowest). Low MAO-B activity is associated with violence, impulsiveness, and other behavioral problems.

Brookhaven National Laboratory.

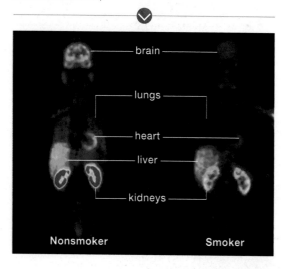

radioactive decay makes it possible for scientists to estimate the age of a rock or fossil by measuring its isotope content (we return to this topic in Section 11.4).

Radioisotopes are often used in **tracers**, which are substances with a detectable component. For example, a molecule in which an atom (such as ^{12}C) has been replaced with a radioisotope (such as ^{14}C) can be used as a radioactive tracer. When delivered into a biological system, a radioactive tracer may be followed as it moves through the system with instruments that detect radiation (Figure 2.4).

Why Electrons Matter The more we learn about electrons, the weirder they seem. Consider that an electron has mass but no size, and its position in space is described as more of a smudge than a point. It carries energy, but only in incremental amounts (this concept will be important to remember when you learn how cells harvest and release energy). An electron gains energy only by absorbing the precise amount needed to boost it to the next energy level. Likewise, it loses energy only by emitting the exact difference between two energy levels.

Imagine that an atom is a multilevel apartment building with a nucleus in the basement. Each "floor" of the building corresponds to a certain energy level, and each has a certain number of "rooms" available for rent. Two electrons can occupy each room. Pairs of electrons populate rooms from the ground floor up. The farther an electron is from the nucleus in the basement, the greater its energy. An electron can move to a room on a higher floor if an energy input gives it a boost, but it immediately emits the extra energy and moves back down.

A. The first shell corresponds to the first energy level, and it can hold up to 2 electrons. Hydrogen has one proton, so it has 1 electron and one vacancy. A helium atom has 2 protons, 2 electrons, and no vacancies.

first shell

one proton
one electron
hydrogen (H)

helium (He)

B. The second shell corresponds to the second energy level, and it can hold up to 8 electrons. Carbon has 6 electrons, so its first shell is full. Its second shell has 4 electrons and four vacancies. Oxygen has 8 electrons and two vacancies. Neon has 10 electrons and no vacancies.

second shell

carbon (C)

oxygen (O)

neon (Ne)

C. The third shell corresponds to the third energy level, and it can hold up to 8 electrons. A sodium atom has 11 electrons, so its first two shells are full; the third shell has one electron. Thus, sodium has seven vacancies. Chlorine has 17 electrons and one vacancy. Argon has 18 electrons and no vacancies.

third shell

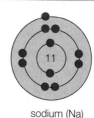
sodium (Na)

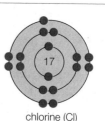
chlorine (Cl)

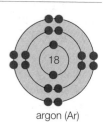
argon (Ar)

Figure 2.5 Shell models.
Each circle (shell) represents one energy level. To make these models, we fill the shells with electrons from the innermost shell out, until there are as many electrons as the atom has protons. The number of protons in each model is indicated.

Figure It Out: Which of these models have unpaired electrons in their outer shell?

Answer: Hydrogen, carbon, oxygen, sodium, and chlorine

A **shell model** helps us visualize how electrons populate atoms (Figure 2.5). In this model, nested "shells" correspond to successively higher energy levels. Thus, each shell includes all of the rooms on one floor (energy level) of our atomic apartment building. We draw a shell model of an atom by filling it with electrons from the innermost shell out, until there are as many electrons as the atom has protons. There is only one room on the first floor, and it fills up first. In hydrogen, the simplest atom, a single electron occupies that room (Figure 2.5A). Helium, with two protons, has two electrons that fill the room—and the first shell. In larger atoms, more electrons rent the second-floor rooms (Figure 2.5B). When the second floor fills, more electrons rent third-floor rooms (Figure 2.5C), and so on.

When an atom's outermost shell is filled with electrons, we say that it has no vacancies, and it is in its most stable state. Helium, neon, and argon are examples of elements with no vacancies. Atoms of these elements are chemically stable, which means they have very little tendency to interact with other atoms. Thus, these elements occur most frequently in nature as solitary atoms. By contrast, when an atom's outermost shell has room for another electron, it has a vacancy. Atoms with vacancies tend to get rid of them by interacting with other atoms; in other words, they are chemically active. For example, the sodium atom (Na) in Figure 2.5C has one electron in its outer (third) shell, which can hold eight. With seven vacancies, we can predict that this atom is chemically active. In fact, this particular sodium atom is not just active, it is extremely so. Why? The shell model shows that a sodium atom has an unpaired electron, but in the real world, electrons really like to be in pairs when they populate atoms. Atoms that have unpaired electrons are called **free radicals**. With a few exceptions, free radicals are very unstable, easily forcing electrons upon other atoms or ripping electrons away from them. This property makes free radicals dangerous to life. A sodium atom with 11 electrons (a sodium radical) quickly evicts the one unpaired electron, so that its second shell—which is full of electrons—becomes its outermost, and no vacancies remain. This is the atom's most stable state. The vast majority of sodium atoms on Earth are like this one, with 11 protons and 10 electrons.

Atoms with an unequal number of protons and electrons are ions. An **ion** carries a net (or overall) charge. Sodium ions (Na⁺) offer an example of how atoms gain a positive charge by losing an electron (Figure 2.6A). Other atoms gain a negative charge by accepting an electron (Figure 2.6B).

© Michael S. Yamashita / Corbis.

Figure 2.6 Ion formation.

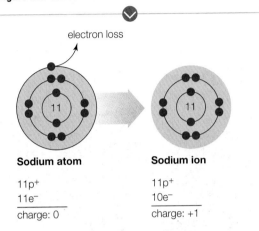

Sodium atom

11p⁺
11e⁻
charge: 0

Sodium ion

11p⁺
10e⁻
charge: +1

A. A sodium atom (Na) becomes a positively charged sodium ion (Na⁺) when it loses the single electron in its third shell. The atom's full second shell is now its outermost, so it has no vacancies.

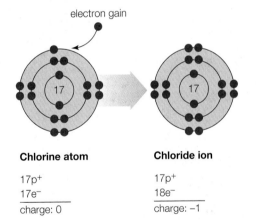

Chlorine atom

17p⁺
17e⁻
charge: 0

Chloride ion

17p⁺
18e⁻
charge: –1

B. A chlorine atom (Cl) becomes a negatively charged chloride ion (Cl⁻) when it gains an electron and fills the vacancy in its third, outermost shell.

Figure It Out: Does a chloride ion have an unpaired electron?   Answer: No

---

Take-Home Message 2.2

What are atoms?

- Atoms consist of electrons moving around a nucleus of protons and neutrons. The number of protons determines the element. Isotopes are forms of an element that have different numbers of neutrons.
- Unstable nuclei of radioisotopes emit radiation as they spontaneously break down (decay). Radioisotopes decay at a predictable rate to form predictable products.
- An atom's electrons are the basis of its chemical behavior. When an atom's outermost shell is not full of electrons, it has a vacancy and it is chemically active. Atoms that get rid of vacancies by gaining or losing electrons become ions (charged).

---

free radical Atom with an unpaired electron.

ion Atom or molecule that carries a net charge.

shell model Model of electron distribution in an atom.

tracer A substance that can be traced via its detectable component.

Figure 2.7 The water molecule.
Each water molecule has two hydrogen atoms bonded to the same oxygen atom.

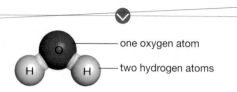

— one oxygen atom

— two hydrogen atoms

Figure 2.8 Ionic bonds in table salt, or NaCl.

(A) left, Francois Gohier/Science Source; top right, Melica/Shutterstock.

Na⁺ Cl⁻

A. Above, tiny crystals of sodium chloride compose table salt. Right, each crystal consists of many sodium and chloride ions locked together in a cubic lattice by ionic bonds.

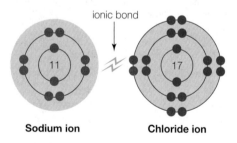

ionic bond

11

17

Sodium ion   **Chloride ion**

B. The strong mutual attraction of opposite charges holds a sodium ion and a chloride ion together in an ionic bond.

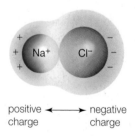

positive ⟷ negative
charge   charge

C. Ions taking part in an ionic bond retain their charge, so the molecule itself is polar. One side is positively charged (represented by a blue overlay); the other side is negatively charged (red overlay).

2.3 From Atoms to Molecules

REMEMBER: The same building blocks, arranged different ways, form different products; atoms bond together to form molecules (Section 1.2).

A **chemical bond** is an attractive force that arises between two atoms, and it is one way that atoms rid themselves of vacancies. Chemical bonds make molecules out of atoms. A molecule consists of atoms held together in a particular number and arrangement by chemical bonds. For example, a water molecule consists of three atoms: two hydrogen atoms bonded to the same oxygen atom (Figure 2.7). Because a water molecule has atoms of two or more elements, it is called a **compound**. Other molecules, including molecular oxygen (a gas in air), have atoms of one element only.

The term "bond" applies to a continuous range of atomic interactions. However, we can categorize most bonds into distinct types based on their different properties. Which type forms depends on the atoms taking part in the molecule.

Ionic Bonds Two ions may stay together by the mutual attraction of their opposite charges, an association called an **ionic bond**. Ionic bonds can be quite strong. Ionically bonded sodium and chloride ions make sodium chloride (NaCl), which we know as table salt; a crystal of this substance consists of a lattice of sodium and chloride ions interacting in ionic bonds (Figure 2.8A).

Ions retain their respective charges when participating in an ionic bond (Figure 2.8B). Thus, one "end" of an ionically bonded molecule has a positive charge, and the other "end" has a negative charge. Any such separation of charge into distinct positive and negative regions is called **polarity** (Figure 2.8C).

Covalent Bonds In a **covalent bond**, two atoms share a pair of electrons, so each atom's vacancy becomes partially filled (Figure 2.9). Sharing electrons links the two atoms, just as sharing a pair of earphones links two friends (left). Covalent bonds can be stronger than ionic bonds, but they are not always so.

Table 2.1 shows some of the different ways we represent molecules that are held together with covalent bonds. In structural formulas, a line between two atoms represents a single covalent bond, in which two atoms share one pair of electrons. For example, molecular hydrogen (H_2) has one covalent bond between hydrogen atoms (H—H).

Two, three, or even four covalent bonds may form between atoms when they share multiple pairs of electrons. For example, two atoms sharing two pairs of electrons are connected by two covalent bonds. Such double bonds are represented by a double line between the atoms. A double bond links the two oxygen atoms in molecular oxygen (O=O). Three lines indicate a triple bond, in which two atoms share three pairs of electrons. A triple covalent bond links the two nitrogen atoms in molecular nitrogen (N≡N).

Double and triple bonds are not distinguished from single bonds in structural models, which show positions and relative sizes of the atoms in three dimensions. The bonds are shown as one stick connecting two balls, which represent atoms. Elements are usually coded by color:

carbon   hydrogen   oxygen   nitrogen   phosphorus

Table 2.1 Representing Covalent Bonds in Molecules

Common name	Water	Familiar term
Chemical name	Dihydrogen monoxide	Describes elemental composition.
Chemical formula	H_2O	Indicates unvarying proportions of elements. Subscripts show number of atoms of an element per molecule. The absence of a subscript means one atom.
Structural formula	H—O—H	Represents each covalent bond as a single line between atoms.
Structural model		Shows relative sizes and positions of atoms in three dimensions.
Shell model		Shows how pairs of electrons are shared in covalent bonds.

Atoms share electrons unequally in a polar covalent bond. A bond between an oxygen atom and a hydrogen atom in a water molecule is an example: One atom (the oxygen, in this case) pulls the electrons a little more toward its side of the bond, so that atom bears a slight negative charge. The atom at the other end of the bond (the hydrogen, in this case) bears a slight positive charge. Covalent bonds in most compounds are polar. By contrast, atoms participating in a nonpolar covalent bond share electrons equally. There is no difference in charge between the two ends of such bonds. Molecular hydrogen (H_2), oxygen (O_2), and nitrogen (N_2) are examples.

Molecular hydrogen (H—H)
Two hydrogen atoms, each with one proton, share two electrons in a nonpolar covalent bond.

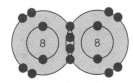

Molecular oxygen (O═O)
Two oxygen atoms, each with eight protons, share four electrons in a double covalent bond.

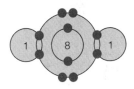

Water (H—O—H)
Two hydrogen atoms share electrons with an oxygen atom in two covalent bonds. The bonds are polar because the oxygen exerts a greater pull on the shared electrons than the hydrogens do.

Figure 2.9 Covalent bonds, in which atoms fill vacancies by sharing electrons.
Two electrons are shared in each covalent bond. When sharing is equal, the bond is nonpolar. When one atom exerts a greater pull on the electrons, the bond is polar.

Take-Home Message 2.3

How do atoms interact in chemical bonds?

- A chemical bond forms between atoms when their electrons interact. Depending on the atoms taking part in it, the bond may be ionic or covalent.
- An ionic bond is a strong mutual attraction between ions of opposite charge.
- Atoms share a pair of electrons in a covalent bond. When the atoms share electrons unequally, the bond is polar.

2.4 Hydrogen Bonds and Water

Life evolved in water. All living organisms are mostly water, many of them still live in it, and all of the chemical reactions of life are carried out in water-based fluids. What makes water so fundamentally important for life?

Water has unique properties that arise from the two polar covalent bonds in each water molecule. Overall, the molecule has no charge, but the oxygen atom

chemical bond An attractive force that arises between two atoms when their electrons interact. Links atoms into molecules.

compound Molecule that has atoms of more than one element.

covalent bond Type of chemical bond in which two atoms share a pair of electrons.

ionic bond Type of chemical bond in which a strong mutual attraction links ions of opposite charge.

polarity Any separation of charge into distinct positive and negative regions.

Figure 2.10 Hydrogen bonds and water.

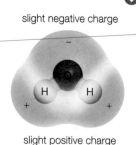

slight negative charge

slight positive charge

A. Polarity of the water molecule. Each of the hydrogen atoms in a water molecule bears a slight positive charge (represented by a blue overlay). The oxygen atom carries a slight negative charge (red overlay).

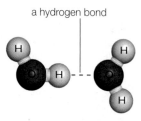

a hydrogen bond

B. A hydrogen bond is an attraction between a hydrogen atom and another atom taking part in a separate polar covalent bond.

C. The many hydrogen bonds that form among water molecules impart special properties to liquid water.

carries a slight negative charge, and the two hydrogen atoms carry a slight positive charge. Thus, the molecule itself is polar (Figure 2.10A).

The polarity of individual water molecules attracts them to one another. The slight positive charge of a hydrogen atom in one water molecule is drawn to the slight negative charge of an oxygen atom in another. This type of interaction is called a hydrogen bond. A **hydrogen bond** is an attraction between a covalently bonded hydrogen atom and another atom taking part in a separate polar covalent bond (Figure 2.10B). Like ionic bonds, hydrogen bonds form by the mutual attraction of opposite charges. However, unlike ionic bonds, hydrogen bonds do not make molecules out of atoms, so they are not chemical bonds.

Hydrogen bonds lie on the weaker end of the spectrum of atomic interactions; they form and break much more easily than covalent or ionic bonds. Even so, many of them form, and collectively they can be quite strong. As you will see, hydrogen bonds stabilize the characteristic structures of biological molecules such as DNA and proteins. They also form in tremendous numbers among water molecules (Figure 2.10C). Extensive hydrogen bonding among water molecules gives liquid water the special properties that make life possible.

Water Is an Excellent Solvent The polarity of the water molecule and its ability to form hydrogen bonds make water an excellent **solvent**, which means that many other substances can dissolve in it. Substances that dissolve easily in water are **hydrophilic** (water-loving). Ionic solids such as sodium chloride (NaCl) dissolve in water because the slight positive charge on each hydrogen atom in a water molecule attracts negatively charged ions (Cl⁻), and the slight negative charge on the oxygen atom attracts positively charged ions (Na⁺). Hydrogen bonds among many water molecules are collectively stronger than an ionic bond between two ions, so the solid dissolves as water molecules tug the ions apart and surround each one (right).

When a substance such as NaCl dissolves, its component ions disperse uniformly among the molecules of liquid, and it becomes a **solute**. Sodium chloride is called a **salt** because it releases ions other than H⁺ and OH⁻ when it dissolves in water (more about this in the next section). A uniform mixture such as salt dissolved in water is called a **solution**. Chemical bonds do not form between molecules of solute and solvent, so the proportions of the two substances in a solution can vary. The amount of a solute that is dissolved in a given volume of fluid is its **concentration**.

Many nonionic solids also dissolve easily in water. Sugars are examples. Molecules of these substances have one or more polar covalent bonds, and atoms participating in a polar covalent bond can form hydrogen bonds with water molecules. Hydrogen bonding with water pulls individual molecules of the solid away from one another and keeps them apart. Unlike ionic solids, these substances retain their molecular integrity when they dissolve, which means they do not dissociate into atoms.

Water does not interact with **hydrophobic** (water-dreading) substances such as oils. Oils consist of nonpolar molecules, and hydrogen bonds do not form between nonpolar molecules and water. When you mix oil and water, the water breaks into small droplets, but quickly begins to cluster into larger drops as new hydrogen bonds form among its molecules. The bonding excludes molecules of oil and pushes them together into drops that rise to the surface of the water. The very same interactions occur at the thin, oily membrane that separates the watery fluid inside cells from the watery fluid outside of them. Such interactions give rise to the structure of cell membranes.

Water Has Cohesion Molecules of some substances resist separating from one another, and the resistance gives rise to a property called **cohesion**. Water has cohesion because hydrogen bonds collectively exert a continuous pull on its individual molecules. You can see cohesion in water as surface tension, which means that the surface of liquid water behaves a bit like a sheet of elastic (left).

Cohesion is a part of many processes that sustain multicelled bodies. Consider how sweating helps keep your body cool during hot, dry weather. Sweat, which is about 99 percent water, cools the skin as it evaporates. Why? **Evaporation** is the process in which molecules escape from the surface of a liquid and become vapor. Evaporation of water is resisted by hydrogen bonding among individual water molecules. In other words, overcoming water's cohesion takes energy. Thus, evaporation sucks energy (in the form of heat) from liquid water, and this lowers the water's surface temperature.

Another example of cohesion's importance to life involves plants. Water molecules evaporate from leaves, and replacements are pulled upward from roots. Cohesion makes it possible for columns of liquid water to rise from roots to leaves inside narrow pipelines of vascular tissue. In some trees, these pipelines extend hundreds of feet above the soil (Section 27.4 returns to this topic).

Water Stabilizes Temperature All atoms jiggle nonstop, so the molecules they make up jiggle too. We measure the energy of this motion as degrees of **temperature**. Adding energy (in the form of heat, for example) makes the jiggling faster, so the temperature rises. Hydrogen bonding keeps water molecules from moving as much as they would otherwise, so it takes more heat to raise the temperature of water compared with other liquids. Temperature stability is an important part of homeostasis because most of the molecules of life function properly only within a certain range of temperature.

Below 0°C (32°F), water molecules do not jiggle enough to break hydrogen bonds between them, and they become locked in the rigid, lattice-like bonding pattern of ice (Figure 2.11). Individual water molecules pack less densely in ice than they do in water, which is why ice floats on water. Sheets of ice that form on the surface of ponds, lakes, and streams can insulate the water under them from subfreezing air temperatures. Such "ice blankets" protect aquatic organisms during long, cold winters.

Figure 2.11 Ice.
Top, hydrogen bonds lock water molecules in a rigid lattice in ice. The molecules in this lattice pack less densely than in liquid water (compare Figure 2.10C), so ice floats on water. Bottom, a covering of ice can insulate water underneath it, thus keeping aquatic organisms from freezing during harsh winters.

Bottom, www.flickr.com/photos/roseofredrock.

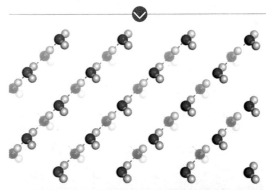

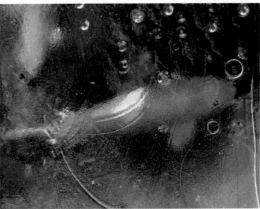

Take-Home Message 2.4

What gives water the special properties that make life possible?

- Extensive hydrogen bonding among water molecules arises from the polarity of the individual molecules.
- Hydrogen bonding among water molecules imparts cohesion to liquid water, and gives it the ability to stabilize temperature and dissolve many substances.

cohesion Property of a substance that arises from the tendency of its molecules to resist separating from one another.

concentration Amount of solute per unit volume of solution.

evaporation Transition of a liquid to a vapor.

hydrogen bond Attraction between a covalently bonded hydrogen atom and another atom taking part in a separate covalent bond.

hydrophilic Describes a substance that dissolves easily in water.

hydrophobic Describes a substance that resists dissolving in water.

salt Ionic compound that releases ions other than H^+ and OH^- when it dissolves in water.

solute A dissolved substance.

solution Uniform mixture of solute completely dissolved in a solvent.

solvent Liquid in which other substances dissolve.

temperature Measure of molecular motion.

Figure 2.12 A pH scale.
Here, red dots signify hydrogen ions (H⁺) and blue dots signify hydroxyl ions (OH⁻). Also shown are the approximate pH values for some common solutions.

This pH scale ranges from 0 (most acidic) to 14 (most basic). A change of one unit on the scale corresponds to a tenfold change in the amount of H⁺ ions.

Photos, © JupiterImages Corporation.

Figure It Out: What is the approximate pH of cola?

Answer: 2.5

2.5 Acids and Bases

A hydrogen atom, remember, is just a proton and an electron. When a hydrogen atom participates in a polar covalent bond, the electron is pulled away from the proton, just a bit. Hydrogen bonding in water tugs on that proton even more, so much that the proton can be pulled right off of the molecule. The electron stays with the rest of the molecule, which becomes negatively charged (ionic), and the proton becomes a hydrogen ion (H⁺). For example, a water molecule that loses a proton becomes a hydroxyl ion (OH⁻). The loss is more or less temporary, because these two ions easily get back together to form a water molecule (H_2O). With other molecules, the loss of a hydrogen ion in water is essentially permanent.

We use a value called **pH** to measure of the number of hydrogen ions in a water-based fluid. In pure water, the number of H⁺ ions is the same as the number of OH⁻ ions, and the pH is 7, or neutral. The higher the number of hydrogen ions, the lower the pH. A one-unit decrease in pH corresponds to a tenfold increase in the number of H⁺ ions (Figure 2.12). One way to get a sense of the pH scale is to taste dissolved baking soda (pH 9), distilled water (pH 7), and lemon juice (pH 2).

An **acid** is a substance that gives up hydrogen ions in water. Acids can lower the pH of a solution and make it acidic (below pH 7). **Bases** accept hydrogen ions from water, so they can raise the pH of a solution and make it basic (above pH 7). Nearly all of life's chemistry occurs near pH 7. Under normal circumstances, fluids inside cells and bodies stay within a certain range of pH because they are buffered. A **buffer** is a set of chemicals that can keep pH stable by alternately donating and accepting ions that affect pH. For example, two chemicals, carbonic acid and bicarbonate, are part of a homeostatic mechanism that normally keeps your blood pH between 7.3 and 7.5. Carbonic acid forms when carbon dioxide gas dissolves in the fluid portion of blood. It can dissociate into a hydrogen ion and a bicarbonate ion, which in turn recombine to form carbonic acid:

$$H_2CO_3 \longrightarrow H^+ + HCO_3^- \longrightarrow H_2CO_3$$
$$\text{carbonic acid} \qquad \text{bicarbonate} \qquad \text{carbonic acid}$$

An excess of OH⁻ ions in the blood causes the carbonic acid in it to release H⁺ ions. These combine with the excess OH⁻ ions to form water, which does not affect pH. Excess H⁺ in blood combines with the bicarbonate, so it does not affect pH either.

Any buffer can neutralize only so many ions. Even slightly more than that limit and the pH of the fluid will change dramatically. Buffer failure can be catastrophic in a biological system because most biological molecules can function properly only within a narrow range of pH. Consider what happens when breathing is impaired suddenly. Carbon dioxide gas accumulates in tissues, and too much carbonic acid forms in blood. If the excess acid reduces blood pH below 7.3, a dangerous level of unconsciousness called coma can be the outcome.

Take-Home Message 2.5

Why are hydrogen ions important in biological systems?

• The number of hydrogen ions in a fluid determines its pH. Most biological systems function properly only within a narrow range of pH. Buffers help keep pH stable.

• Acids release hydrogen ions in water; bases accept them.

2.6 Organic Molecules

The same elements that make up a living body also occur in nonliving things, but their proportions differ. For example, compared to sand or seawater, a human body has a much larger proportion of carbon atoms. Why? Unlike sand or seawater, a body contains a lot of the molecules of life—complex carbohydrates and lipids, proteins, and nucleic acids—and these molecules consist of a high proportion of carbon atoms. Compounds that consist primarily of carbon and hydrogen atoms are said to be **organic**. The term is a holdover from a time when such molecules were thought to be made only by living things, as opposed to the "inorganic" molecules that formed by nonliving processes. We now know that organic compounds were present on Earth long before organisms were.

Carbon's importance to life arises from its versatile bonding behavior. Carbon has four vacancies in its outer shell, so it can form four covalent bonds with other atoms, including other carbon atoms. Many organic molecules have a chain of carbon atoms, and this backbone often forms rings (Figure 2.13). Small molecular groups that attach to the backbone impart chemical properties to the molecule. For example, carboxyl groups (—COOH) make amino acids and fatty acids acidic; hydroxyl groups (—OH) make sugars polar. Carbon's ability to form chains and rings, and also to bond with many other elements, means that atoms of this element can be assembled into a wide variety of organic compounds.

As you will see shortly, the function of an organic molecule depends on its structure. The structure of even a small organic molecule can be quite complicated (Figure 2.14A), so representations are typically simplified. Hydrogen atoms and some of the bonds may not be shown, but are understood to exist where they should. Carbon rings such as the ones that occur in glucose and other sugars are often depicted as polygons (Figure 2.14B). If no atom is shown at a corner or at the end of a bond, a carbon is implied there. Ball-and-stick models are used to depict an organic molecule's three-dimensional arrangement of atoms (Figure 2.14C); space-filling models are used to show its overall shape (Figure 2.14D). Proteins and nucleic acids are often modeled as ribbon structures, which, as you will see in Section 2.9, show how the molecule folds and twists.

A. Carbon's versatile bonding behavior allows it to form a variety of structures, including rings.

B. Carbon rings form the backbone of many sugars, starches, and fats, such as those found in foods.

Figure 2.13 Carbon rings.

(B) Getty Images.

What Cells Do to Organic Compounds All biological systems are based on the same organic molecules, but the details of those molecules differ among organisms. Just as atoms bonded in different numbers and arrangements form different molecules, simple organic building blocks bonded in different numbers and arrangements form different versions of the molecules of life.

Cells assemble complex carbohydrates, lipids, proteins, and nucleic acids from small organic molecules. These small organic molecules—sugars, fatty acids, amino acids, and nucleotides—are called **monomers** when they are used as subunits of larger molecules. A molecule that consists of multiple monomers is a **polymer**. Cells build polymers from monomers, and break down polymers to release

Figure 2.14 Structural models of an organic molecule. All of these models represent the same molecule: glucose.

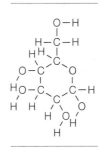

A. A structural formula for an organic molecule—even a simple one—can be very complicated. The overall structure is obscured by detail.

B. Structural formulas of organic molecules are often simplified by using polygons as symbols for rings, and omitting some bonds and element labels.

C. A ball-and-stick model shows the arrangement of atoms and bonds in three dimensions.

D. A space-filling model can be used to show a molecule's overall shape. Individual atoms are visible in this model. Space-filling models of larger molecules often show only the surface contours.

acid Substance that releases hydrogen ions in water.

base Substance that accepts hydrogen ions in water.

buffer Set of chemicals that can keep the pH of a solution stable by alternately donating and accepting ions that contribute to pH.

monomer Molecule that is a subunit of polymers.

organic Describes a compound that consists mainly of carbon and hydrogen atoms.

pH Measure of the amount of hydrogen ions in a fluid.

polymer Molecule that consists of multiple monomers.

A. Metabolism refers to processes by which cells acquire and use energy as they make and break down molecules. Humans and other consumers break down the molecules in food. Their cells use energy and raw materials from the breakdown to maintain themselves and to build new components.

Exactostock/SuperStock.

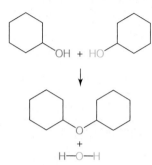

B. Cells often build a large molecule from small ones by condensation. In this reaction, an enzyme removes a hydroxyl group from one molecule and a hydrogen atom from another. A covalent bond forms between the two molecules; water forms too.

C. Cells use a water-requiring reaction called hydrolysis to split a large molecule into smaller ones. An enzyme attaches a hydroxyl group and a hydrogen atom (both from water) at the cleavage site.

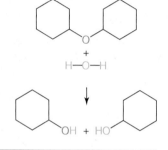

Figure 2.15 Metabolism.
Two common reactions by which cells build and break down organic molecules are shown.

monomers. These and other processes of molecular change are called **reactions**. Cells constantly run reactions as they acquire and use energy to stay alive, grow, and reproduce—activities collectively called **metabolism** (Figure 2.15A). Metabolism also requires **enzymes**, which are organic molecules (usually proteins) that speed up reactions without being changed by them. For example, in a common reaction called condensation, an enzyme covalently bonds two monomers together (Figure 2.15B). In hydrolysis, the reverse of condensation, an enzyme splits an organic polymer into its component monomers (Figure 2.15C).

Take-Home Message 2.6

How are all of the molecules of life alike?

• The molecules of life (carbohydrates, lipids, proteins, and nucleic acids) are organic, which means they consist mainly of carbon and hydrogen atoms.

• The structure of an organic molecule starts with a chain of carbon atoms (the backbone) that may form one or more rings.

• By processes of metabolism, cells assemble the molecules of life from monomers. They also break apart polymers into component monomers.

2.7 Carbohydrates

A **carbohydrate** is an organic compound that consists of carbon, hydrogen, and oxygen in a 1:2:1 ratio. The term can apply to a sugar molecule or a polymer of them, so these compounds are also called saccharides (saccharide means sugar). Cells use different kinds for fuel, as structural materials, and for storing energy.

Monosaccharides (one sugar) are the simplest carbohydrates, and common types have a backbone of five or six carbon atoms. Glucose, shown in Figure 2.14, is a monosaccharide. Sucrose, which is our table sugar, is a disaccharide (two sugars) that consists of glucose and fructose monomers. Monosaccharides and disaccharides are very soluble in water, so they can move easily through the water-based internal environments of all organisms.

Breaking the bonds of a monosaccharide releases energy that can be harnessed to power other reactions (Chapter 5 returns to this topic). Monosaccharides are also remodeled into other important compounds. For example, cells of plants and many animals make vitamin C from glucose. Human cells are unable to make vitamin C, so we need to get it from our food.

Foods that we call "complex" carbohydrates consist mainly of polysaccharides, which are chains of hundreds or thousands of monosaccharide monomers. The chains may be straight or branched, and can have one or many types of monosaccharides. The most common polysaccharides are cellulose, starch, and glycogen. All consist only of glucose monomers, but as substances their properties are very different. Why? The answer begins with differences in patterns of covalent bonding that link their monomers.

Cellulose, the major structural material of plants, is the most abundant organic molecule on Earth. Hydrogen bonding locks its long, straight chains of covalently bonded glucose monomers into tight, sturdy bundles (Figure 2.16A). The bundles form tough fibers that act like reinforcing rods inside stems and other plant parts, helping these structures resist wind and other forms of mechanical stress. Cellulose

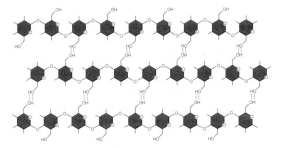

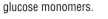

A. Cellulose is the main structural component of plants. Above, in cellulose, hydrogen bonds stabilize chains of glucose monomers in tight bundles that form long fibers. Few organisms can digest this tough, insoluble material.

B. Starch is the main energy reserve in plants, which store it in their roots, stems, leaves, seeds, and fruits. Below, starch consists of long, coiled chains of glucose monomers.

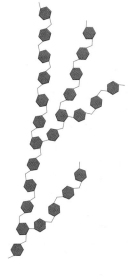

C. Glycogen functions as an energy reservoir in animals, including people. It is especially abundant in the liver and muscles. Above, glycogen consists of highly branched chains of glucose monomers.

Figure 2.16 Three of the most common complex carbohydrates and their locations in a few organisms. Each polysaccharide consists only of glucose units, but different bonding patterns that link the subunits result in substances with very different properties.

Middle photo, © JupiterImages Corporation.

is insoluble (it does not dissolve) in water, and it is not easily broken down. Some bacteria and fungi make enzymes that can break it apart into its component sugars, but humans and other mammals do not. Dietary fiber, or "roughage," usually refers to the indigestible cellulose in our vegetable foods. Bacteria that live in the guts of termites and grazers such as cattle and sheep help these animals digest the cellulose in plants. (Chapter 23 returns to the topic of animal digestion.)

In starch, a different covalent bonding pattern between glucose monomers makes a chain that coils up into a spiral (Figure 2.16B). Like cellulose, starch does not dissolve readily in water, but it is easier to break down. These properties make the molecule ideal for storing sugars in the watery, enzyme-filled interior of plant cells. Most plant leaves make glucose during the day, and their cells store it by building starch. At night, hydrolysis enzymes break the bonds between starch's glucose monomers. The released glucose can be broken down immediately for energy, or converted to sucrose that is transported to other parts of the plant. Humans also have hydrolysis enzymes that break down starch, so this carbohydrate is an important component of our food.

Animals store sugars in the form of glycogen, a polysaccharide that consists of highly branched chains of glucose monomers (Figure 2.16C). Muscle and liver cells contain most of the body's glycogen. When the blood sugar level falls, liver cells break down the glycogen, and the released glucose subunits enter the blood.

Take-Home Message 2.7

What is a carbohydrate?

- Cells use simple carbohydrates (sugars) for energy and to build other molecules.
- Sugar monomers, bonded different ways, form complex carbohydrates such as cellulose, starch, and glycogen.

carbohydrate Molecule that consists primarily of carbon, hydrogen, and oxygen atoms in a 1:2:1 ratio.

cellulose Tough, insoluble carbohydrate that is the major structural material in plants.

enzyme Organic molecule that speeds up a reaction without being changed by it.

metabolism All the enzyme-mediated chemical reactions by which cells acquire and use energy as they build and break down organic molecules.

reaction Process of molecular change.

2.8 Lipids

Lipids are fatty, oily, or waxy organic compounds. They vary in structure, but all are hydrophobic. Many lipids incorporate **fatty acids**, which are small organic molecules that consist of a carbon chain "tail" of variable length, and a carboxyl group "head" (Figure 2.17). The tail is hydrophobic (hence the name "fatty"); the carboxyl group makes the head hydrophilic (and acidic). You are already familiar with the properties of fatty acids because these molecules are the main component of soap. The hydrophobic tails of fatty acids in soap attract oily dirt, and the hydrophilic heads dissolve the dirt in water.

Saturated fatty acids have only single bonds linking the carbons in their tails. In other words, their carbon chains are fully saturated with hydrogen atoms (Figure 2.17A). The tail of a saturated fatty acid is flexible and it wiggles freely. Double bonds between carbons limit the flexibility of the tails of an **unsaturated fatty acid** (Figure 2.17B,C). These bonds are *cis* or *trans*, depending on the way the hydrogens are arranged around them (Figure 2.17D,E).

Fats The carboxyl group head of a fatty acid can easily form a covalent bond with another molecule. When it bonds to a glycerol, a type of alcohol, it loses its hydrophilic character. Three fatty acids bonded to the same glycerol form a **triglyceride**, a molecule that is entirely hydrophobic and therefore does not dissolve in water. Triglycerides are the most abundant and richest energy source in vertebrate bodies; gram for gram, they store more energy than carbohydrates.

A **fat** is a substance that consists mainly of triglycerides. Butter and other fats derived from animals have a high proportion of triglycerides in which all three fatty acid tails are saturated. These triglycerides are commonly called saturated fats, and substances that consist of them are solid at room temperature because floppy saturated fatty acid tails can pack tightly together. Vegetable oils, by contrast, have a high proportion of unsaturated fats, the common term for triglycerides in which at least one of the three fatty acid tails is unsaturated. Each double bond makes a rigid kink, and kinky tails cannot pack tightly. This is why most substances that consist of unsaturated fats are liquid at room temperature. The partially hydrogenated vegetable oils that you learned about in Section 2.1 are an exception. They are solid at room temperature because the special *trans* double bond keeps their fatty acid tails straight, allowing them to pack tightly just like saturated fatty acid tails do.

Phospholipids A **phospholipid** has two fatty acid tails and a head that contains a phosphate group (Figure 2.18A). The tails are hydrophobic, but the phosphate group is highly polar and it makes the head very hydrophilic. These opposing properties give rise to the basic structure of cell membranes,

Figure 2.17 Fatty acids.
Double bonds in the tails are highlighted in red.

A. The tail of stearic acid is fully saturated with hydrogen atoms.

B. Linoleic acid, with two double bonds, is unsaturated. The first double bond occurs at the sixth carbon from the end of the tail, so linoleic acid is called an omega-6 fatty acid. Omega-6 and **C** omega-3 fatty acids are "essential fatty acids," which means your body does not make them and they must come from food.

D. The hydrogen atoms around the double bond in oleic acid are on the same side of the tail. Most other naturally occurring unsaturated fatty acids have these *cis* bonds.

E. Hydrogenation creates abundant *trans* bonds, with hydrogen atoms on opposite sides of the tail.

Figure It Out: Are the double bonds in linolenic acid *cis* or *trans*?

Answer: *cis*

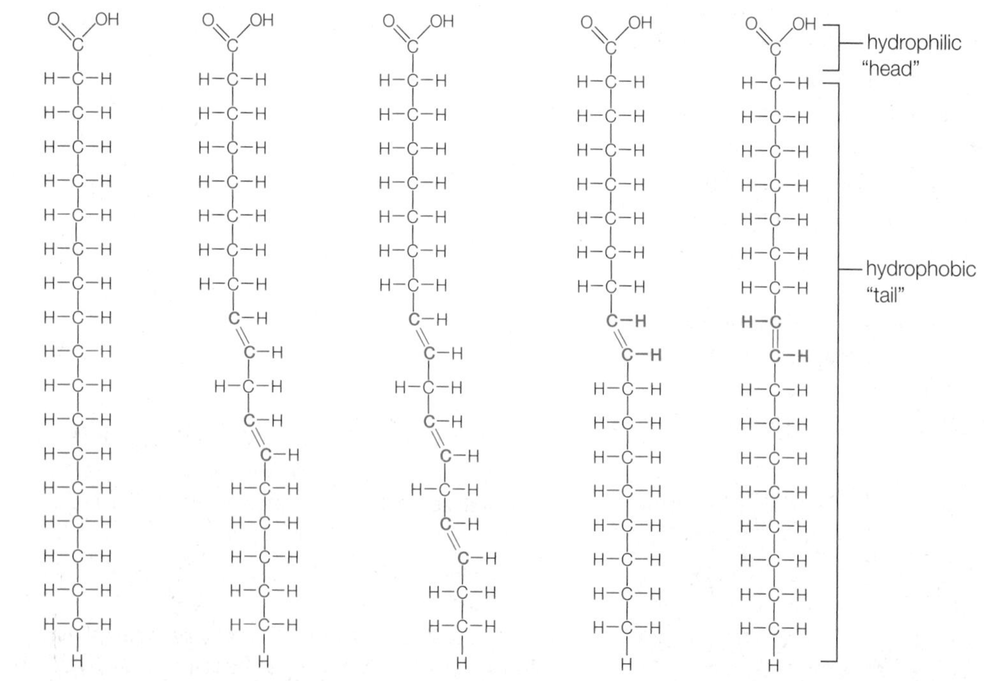

A. stearic acid (saturated) B. linoleic acid (omega-6) C. linolenic acid (omega-3) D. oleic acid (*cis*) E. elaidic acid (*trans*)

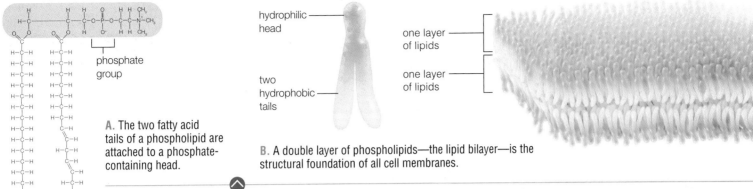

A. The two fatty acid tails of a phospholipid are attached to a phosphate-containing head.

B. A double layer of phospholipids—the lipid bilayer—is the structural foundation of all cell membranes.

Figure 2.18 Phospholipids as components of cell membranes.

which consist mainly of phospholipids. In a cell membrane, phospholipids are arranged in two layers—a **lipid bilayer** (Figure 2.18B). The heads of one layer are dissolved in the cell's watery interior, and the heads of the other layer are dissolved in the cell's fluid surroundings. All of the hydrophobic tails are sandwiched between the hydrophilic heads. Section 3.3 returns to the structure of cell membranes.

Waxes A **wax** is a complex, varying mixture of lipids with long fatty acid tails bonded to carbon rings or other structures. These molecules pack tightly, so waxes are firm and water-repellent. Plants secrete waxes onto their exposed surfaces to  restrict water loss and keep out parasites and other pests. Other types of waxes protect, lubricate, and soften skin and hair. Waxes, together with fats and fatty acids, make feathers waterproof. Bees store honey and raise new generations of bees inside a honeycomb of secreted beeswax.

Steroids **Steroids** are lipids with no fatty acid tails; they have a rigid backbone that consists of twenty carbon atoms arranged in a characteristic pattern of four rings (Figure 2.19). As you will see in later chapters, these molecules serve varied and important physiological functions in plants, fungi, and animals. Cholesterol, the most common steroid in animal tissues, is remodeled into other molecules such as bile salts (which help digest fats), vitamin D (required to keep teeth and bones strong), and steroid hormones.

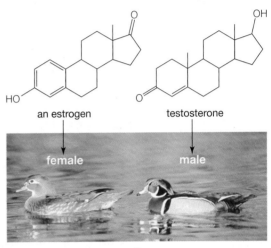

Figure 2.19 Examples of steroids.
Estrogen and testosterone are steroid hormones that govern reproduction and secondary sexual traits in animals. These hormones give rise to the gender-specific traits of many species, including wood ducks.

Bottom, Tim Davis/Science Source.

Take-Home Message 2.8

What are lipids?

- Lipids are fatty, waxy, or oily organic compounds.
- Fats are substances that consist primarily of triglycerides, which have three fatty acid tails. Triglycerides are called unsaturated fats if there are double bonds in one or more of their fatty acid tails, and saturated fats if there are none.
- Phospholipids arranged in a lipid bilayer are the main component of cell membranes.
- Waxes have complex, varying structures. They are components of water-repelling and lubricating secretions.
- Steroids serve varied and important physiological roles in plants, fungi, and animals.

fat Substance that consists mainly of triglycerides.

fatty acid Organic compound with an acidic carboxyl group "head" and a long carbon chain "tail."

lipid Fatty, oily, or waxy organic compound.

lipid bilayer Double layer of lipids arranged tail-to-tail; structural foundation of all cell membranes.

phospholipid A lipid with a phosphate group in its hydrophilic head, and two nonpolar fatty acid tails.

saturated fatty acid Fatty acid with only single bonds linking the carbons in its tail.

steroid A type of lipid with four carbon rings and no fatty acid tails.

triglyceride A molecule with three fatty acid tails that is entirely hydrophobic; main component of fats.

unsaturated fatty acid Fatty acid with one or more carbon–carbon double bonds in its tail.

wax Water-repellent substance that is a complex, varying mixture of lipids.

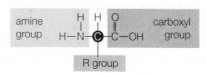

2 A protein's primary structure consists of a linear sequence of amino acids (a polypeptide). Each type of protein has a unique primary structure.

1 A condensation reaction joins the carboxyl group of one amino acid and the amine group of another to form a peptide bond. In this example, a peptide bond forms between the amino acids methionine and valine.

Figure 2.20 How protein structure arises.

(3–5) 1BBB, A third quaternary structure of human hemoglobin A at 1.7-A resolution. Silva, M.M., Rogers, P.H., Arnone, A., Journal: (1992) J.Biol.Chem. 267: 17248-17256.

2.9 Proteins

Cells make the thousands of different **proteins** they need from only twenty kinds of amino acid monomers. An **amino acid** is a small organic compound with an amine group (—NH_2), a carboxyl group (—COOH, the acid), and a side chain called an "R group" that defines the kind of amino acid. In most amino acids, all three groups are attached to the same carbon atom:

The covalent bond that links amino acids in a protein is called a **peptide bond**. During protein synthesis, a peptide bond forms between the carboxyl group of the first amino acid and the amine group of the second (Figure 2.20 **1**). Another peptide bond links a third amino acid to the second, and so on (you will learn more about the details of protein synthesis in Chapter 7). A short chain of amino acids is called a peptide; as the chain lengthens, it becomes a polypeptide. Polypeptides can be hundreds or even thousands of amino acids long.

The idea that structure dictates function is particularly appropriate as applied to proteins, because the diversity in biological activity among these molecules arises from differences in their three-dimensional shape. Protein structure begins with the linear series of amino acids composing a polypeptide **2**. The order of the amino acids, which is called primary structure, defines the type of protein.

The molecule begins to take on three-dimensional shape during protein synthesis, as hydrogen bonds that form between amino acids cause the lengthening polypeptide to twist and fold. Hydrogen bonding holds sections of the polypeptide in loops, helices (coils), or flat sheets, and these patterns constitute the protein's secondary structure **3**. The primary structure of each type of protein is unique, but most proteins have similar patterns of secondary structure.

Much as an overly twisted rubber band coils back upon itself, hydrogen bonding also makes the loops, helices, and sheets of a protein fold up into even more compact domains. These domains are called tertiary structure. Tertiary structure is what makes a protein a working molecule. For example, the helices and loops in a polypeptide called a globin chain fold up together to form a pocket **4**. This pocket

Figure 2.21 An example of a protein domain.
This barrel domain is part of a rotary mechanism in a larger protein. The protein is a molecular motor that pumps hydrogen ions through cell membranes.

pdb ID2W5J, Vollmar, M., Shlieper, D., Winn, M., Buechner, C., Groth, G. "Structure of the C14 rotor ring of the proton translocating chloroplast ATP synthase." (2009) J. Biol. Chem. 284:18228.

3 Secondary structure arises as a polypeptide chain twists into a helix (coil), loop, or sheet held in place by hydrogen bonds.

4 Tertiary structure arises when loops, helices, and sheets fold up into a domain. In this example, the helices of a globin chain form a pocket.

5 Many proteins have two or more polypeptides (quaternary structure). Hemoglobin, shown here, consists of four globins (green and blue). The pocket of each globin now holds a heme group (red).

6 Some types of proteins can aggregate into much larger structures. As an example, organized arrays of keratin, a fibrous protein, compose very long filaments that make up your hair.

holds a heme, which is a small compound essential to the function of the finished protein—hemoglobin. In other proteins, sheets, loops, and helices come together as complex structures that resemble barrels, propellers, sandwiches, and so on. Barrel domains often form tunnels through cell membranes, allowing small molecules to cross. Some proteins have barrel domains that rotate like motors in small molecular machines (Figure 2.21). A protein may have several domains, each contributing a particular structural or functional property to the molecule.

Many proteins also have quaternary structure, which means they consist of two or more polypeptides that are closely associated or covalently bonded together. Hemoglobin is like this **5**. So are most enzymes, which have multiple polypeptides that collectively form a roughly spherical shape.

Fibrous proteins aggregate by many thousands into much larger structures, with their polypeptides organized into strands or sheets. The keratin in your hair is an example **6**. Other fibrous proteins are part of the mechanisms that help cells, cell parts, and multicelled bodies move.

Carbohydrates, lipids, or both may get attached to a protein after synthesis. A protein with carbohydrates attached to it is called a glycoprotein. Molecules that allow a body to recognize its own cells are glycoproteins, as are other molecules that help cells interact in immunity. A protein with one or more lipids attached to it is called a lipoprotein. Some lipoproteins are aggregate structures that consist of variable amounts and types of proteins and lipids (Figure 2.22).

The Importance of Protein Structure Protein shape depends on hydrogen bonds and other interactions that heat, some salts, shifts in pH, or detergents can disrupt. Such disruption can cause proteins to lose their three-dimensional shape, or **denature**. Once a protein's shape unravels, so does its function. You can see denaturation in action when you cook an egg. A protein called albumin is a major component of egg white. Cooking does not disrupt the covalent bonds of albumin's primary structure, but it does destroy the hydrogen bonds that maintain the protein's shape. When a translucent egg white turns opaque, the albumin has been denatured. For a very few proteins, denaturation is reversible if normal conditions return, but albumin is not one of them. There is no way to uncook an egg.

Diseases such as bovine spongiform encephalitis (BSE, or mad cow disease) in cattle, Creutzfeldt–Jakob disease in humans, and scrapie in sheep, are the dire aftermath of a protein that changes shape. These diseases may be inherited, but

Figure 2.22 A lipoprotein particle.
The one depicted here (HDL, which is often called "good" cholesterol) consists of thousands of lipids lassoed into a clump by two proteins.

Castrignanò T, De Meo PD, Cozzetto D, Talamo IG, Tramontano A. (2006). The PMDB Protein Model Database. Nucleic Acids Research, 34: D306–D309.

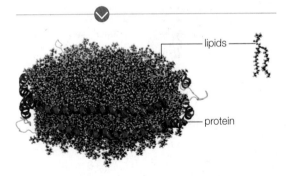

lipids

protein

amino acid Small organic compound that is a subunit of proteins. Consists of a carboxyl group, an amine group, and a characteristic side group (R), all typically bonded to the same carbon atom.

denature To unravel the shape of a protein or other large biological molecule.

peptide bond A bond between the amine group of one amino acid and the carboxyl group of another. Joins amino acids in proteins.

protein Polymer of amino acids; an organic molecule that consists of one or more polypeptides.

Digging Into Data

Effects of Dietary Fats on Lipoprotein Levels

Cholesterol that is made by the liver or that enters the body from food does not dissolve in blood, so it is carried through the bloodstream by lipoproteins. Low-density lipoprotein (LDL) carries cholesterol to body tissues such as artery walls, where it can form deposits associated with cardiovascular disease. Thus, LDL is often called "bad" cholesterol. High-density lipoprotein (HDL) carries cholesterol away from tissues to the liver for disposal, so HDL is often called "good" cholesterol. In 1990, Ronald Mensink and Martijn Katan published a study that tested the effects of different dietary fats on blood lipoprotein levels. Their results are shown in Figure 2.23.

1. In which group was the level of LDL ("bad" cholesterol) highest?
2. In which group was the level of HDL ("good" cholesterol) lowest?
3. An elevated risk of heart disease has been correlated with increasing LDL-to-HDL ratios. Rank the three diets according to their predicted effect on cardiovascular health.

| | Main Dietary Fats | | | |
	cis fatty acids	*trans* fatty acids	saturated fats	optimal level
LDL	103	117	121	<100
HDL	55	48	55	>40
ratio	1.87	2.44	2.2	<2

Figure 2.23 Effect of diet on lipoprotein levels.
Researchers placed 59 men and women on a diet in which 10 percent of their daily energy intake consisted of *cis* fatty acids, *trans* fatty acids, or saturated fats. Blood LDL and HDL levels were measured after three weeks on the diet; averaged results are shown in mg/dL (milligrams per deciliter of blood). All subjects were tested on each of the diets. The ratio of LDL to HDL is also shown.

Source, Mensink RP, Katan MB, "Effect of dietary trans fatty acids on high-density and low-density lipoprotein cholesterol levels in healthy subjects." *NEJM* 323(7):439–45.

Figure 2.24 Variant Creutzfeldt–Jakob disease (vCJD).

A. Charlene Singh was one of the three people who developed symptoms of vCJD disease while living in the United States. Singh, like the others, most likely contracted the disease elsewhere; she spent her childhood in Britain. Diagnosed in 2001, she died in 2004.

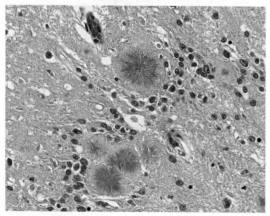

B. Slice of brain tissue from a person with vCJD. Fibers of prion proteins (amyloid fibrils) radiating from several deposits are visible.

more often they arise spontaneously. All are characterized by relentless deterioration of mental and physical abilities that eventually causes death (Figure 2.24A), and all begin with a glycoprotein called PrPC that occurs normally in cell membranes of the mammalian body. This protein is especially abundant in brain cells, but we still know very little about what it does. Sometimes, a PrPC protein misfolds so that part of the molecule forms a sheet instead of a helix. One misfolded molecule should not pose much of a threat, but when this particular protein misfolds it becomes a **prion**, or infectious protein. The shape of a misfolded PrPC protein causes normally folded PrPC proteins to misfold too. Each protein that misfolds becomes infectious, so the number of prions increases exponentially.

The shape of misfolded PrPC proteins allows them to align tightly into long, insoluble fibers that are called amyloid fibrils. Amyloid fibrils grow from their ends as more PrPC proteins misfold (Figure 2.24B). They form patches in the brain that disrupt its function, causing symptoms such as confusion, memory loss, and lack of coordination. Holes form in the brain as its cells die. Eventually, the brain becomes so riddled with holes that it looks like a sponge.

In the mid-1980s, an epidemic of mad cow disease in Britain was followed by an outbreak of a new variant of Creutzfeldt–Jakob disease (vCJD) in humans. Researchers isolated a prion similar to the one in scrapie-infected sheep from cows with BSE, and also from humans affected by the new type of Creutzfeldt–Jakob disease. How did the prion get from sheep to cattle to people? Prions resist denaturation, so treatments such as cooking that inactivate other types of infectious agents have little effect on them. The cattle became infected by the prion after eating feed prepared from the remains of scrapie-infected sheep, and people became infected by eating beef from the infected cattle.

Two hundred people have died from vCJD since 1990. The use of animal parts in livestock feed is now banned in many countries, and the number of cases of BSE and vCJD has since declined. Cattle with BSE still turn up, but so rarely that they pose little threat to human populations.

Take-Home Message 2.9

Why is protein structure important?

- A protein consists of one or more polypeptides, each a chain of amino acids. The order of amino acids in the polypeptide(s) dictates the type of protein.
- During protein synthesis, polypeptides twist and fold into coils, sheets, and loops, which fold and pack further into functional domains.
- A protein's function arises from and depends on its three-dimensional shape.

2.10 Nucleic Acids

Nucleotides are small organic molecules that function as energy carriers, enzyme helpers, chemical messengers, and subunits of DNA and RNA. Each consists of a monosaccharide ring bonded to a nitrogen-containing base and one, two, or three phosphate groups (Figure 2.25A). The monosaccharide is a five-carbon sugar, either ribose or deoxyribose, and the base is one of five compounds with a flat ring structure (we return to the structure of nucleotide bases in Section 6.2). When the third phosphate group of a nucleotide is transferred to another molecule, energy is transferred along with it. You will read about such phosphate-group transfers and their important metabolic role in Section 4.4. The nucleotide **ATP** (adenosine triphosphate) serves an especially important role as an energy carrier in cells.

Nucleic acids are chains of nucleotides in which the sugar of one nucleotide is joined to the phosphate group of the next (Figure 2.25B). An example is **RNA**, or ribonucleic acid, named after the ribose sugar of its component nucleotides. An RNA molecule is a chain of four kinds of nucleotide monomers (one of which is ATP). There are different types of RNA, and they work together to carry out protein synthesis. **DNA**, or deoxyribonucleic acid, is a nucleic acid named after the deoxyribose sugar of its component nucleotides. A DNA molecule consists of two chains of nucleotides twisted into a double helix (Figure 2.25C). Hydrogen bonds hold the chains together (Chapter 6 returns to DNA structure). Each cell starts life with DNA inherited from a parent cell. That DNA contains all of the information necessary to build a new cell and, in the case of multicelled organisms, an entire individual. The cell uses the order of nucleotide bases in DNA—the DNA sequence—to guide production of RNA and proteins (Chapter 7 returns to this topic).

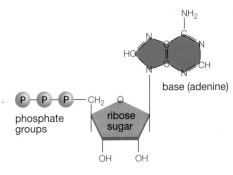

A. Example of a nucleotide: ATP. ATP is a monomer of RNA, and also a participant in many metabolic reactions.

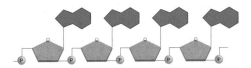

B. A chain of nucleotides is a nucleic acid. The sugar of one nucleotide is covalently bonded to the phosphate group of the next, forming a sugar–phosphate backbone.

C. DNA consists of two chains of nucleotides, twisted into a double helix. Hydrogen bonding maintains the three-dimensional structure of this nucleic acid.

Figure 2.25 Nucleic acid structure.

Take-Home Message 2.10

What are nucleotides and nucleic acids?

- Nucleotides are monomers of the nucleic acids DNA and RNA. Some have additional roles. ATP, for example, is an important energy carrier in cells.
- DNA holds information necessary to build cells and multicelled individuals.
- RNAs carry out protein synthesis.

ATP Nucleotide monomer of RNA; also serves an important role as an energy carrier in cells.

DNA Deoxyribonucleic acid. Nucleic acid that consists of two chains of nucleotides twisted into a double helix; holds hereditary information.

nucleic acid Chain of nucleotides; DNA or RNA.

nucleotide Small molecule with a five-carbon sugar, a nitrogen-containing base, and one, two, or three phosphate groups. Monomer of nucleic acids; some have additional roles.

prion Infectious protein.

RNA Ribonucleic acid. Nucleic acid that consists of a chain of nucleotides. Carries out protein synthesis.

Summary

Section 2.1 All organisms consist of the same kinds of molecules. Seemingly small differences in the way those molecules are put together can have big effects inside a living organism.

Section 2.2 Atoms consist of **electrons**, which carry a negative **charge**, moving about a **nucleus** of positively charged **protons** and uncharged **neutrons**. The number of protons (the **atomic number**) determines the **element**.

Isotopes of an element have the same number of protons but differ in the number of neutrons. The total number of protons and neutrons in the atomic nucleus is called **mass number**.

Tracers can be made with **radioisotopes**, which, by **radioactive decay**, emit particles and energy when their nucleus spontaneously breaks up.

A **shell model** of an atom represents the energy levels of its electrons as concentric circles. Atoms are in their most stable state when all of their shells are full of electrons, so they tend to get rid of vacancies. Many can do so by gaining or losing electrons, thereby becoming **ions**. Atoms with unpaired electrons are **free radicals**. Most free radicals are unstable and dangerous to life.

Section 2.3 A **chemical bond** unites two atoms in a molecule. A **compound** is a molecule that consists of two or more elements. An **ionic bond** is a strong association of two oppositely charged ions. **Polarity** is a separation of charge into positive and negative regions. Atoms share a pair of electrons in a **covalent bond**, which is nonpolar if the sharing is equal, and polar if it is not.

Section 2.4 Two polar covalent bonds give each water molecule an overall polarity. **Hydrogen bonds** that form among water molecules in tremendous numbers are the basis of water's unique life-sustaining properties: **cohesion**, resistance to **temperature** changes, and a capacity to act as a **solvent** for **salts** and other polar **solutes**. The amount of solute in a given volume of a **solution** is the solute's **concentration**. **Hydrophilic** substances dissolve easily in water; **hydrophobic** substances do not. **Evaporation** is the transition of a liquid to a vapor.

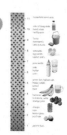

Section 2.5 **pH** is a measure of the number of hydrogen ions (H^+) in a liquid. At neutral pH (7), there are an equal number of H^+ and OH^- ions. **Acids** release hydrogen ions in water; **bases** accept them. A **buffer** can keep the pH of solution consistent. Most cell and body fluids are buffered because most molecules of life work only within a narrow range of pH.

Section 2.6 The molecules of life are **organic**, and chains or rings of carbon atoms form their backbones. All are **polymers** of sugar, fatty acid, or amino acid **monomers**. **Enzymes** carry out **reactions** of **metabolism**.

Section 2.7 Cells use **carbohydrates** for energy, and as structural materials. Enzymes assemble polysaccharides such as **cellulose** from monosaccharide monomers.

Section 2.8 All **lipids** are nonpolar. A **fatty acid** is a lipid with an acidic head and a long carbon chain tail. Only single bonds link the carbons in the tail of a **saturated fatty acid**; the tail of an **unsaturated fatty acid** has one or more double bonds. **Fats** consist mostly of **triglycerides**, which have three fatty acid tails and are nonpolar. When all of a triglyceride's fatty acid tails are unsaturated, it is called an unsaturated fat; if one or more of the tails is saturated, it is a saturated fat. A **lipid bilayer** (of **phospholipids**) is the basic structure of all cell membranes. **Waxes** are part of water-repellent and lubricating secretions. Some **steroids** function as hormones.

Section 2.9 The shape of a **protein** is the source of its function. Protein structure begins as a sequence of **amino acids** linked by **peptide bonds** into a polypeptide. Polypeptides twist into loops, sheets, and coils that can pack further into functional domains. Many proteins, including most enzymes, consist of two or more polypeptides. Fibrous proteins aggregate into much larger structures. A protein that **denatures**, or loses its shape, also loses its function. **Prion** diseases are a fatal consequence of misfolded proteins.

Section 2.10 **Nucleotides** consist of a five-carbon sugar, a nitrogen-containing base, and one, two or three phosphate groups. Nucleotides are monomers of **nucleic acids**, and some, especially **ATP**, have additional functions. **DNA** holds heritable information; **RNA** carries out protein synthesis.

Self-Quiz

1. Which of the following statements is *in*correct?
   a. Isotopes have the same atomic number and different mass numbers.
   b. Atoms have about the same number of electrons as protons.
   c. All molecules consist of atoms.
   d. Free radicals are dangerous because they emit energy.

2. Which element has only one proton?

3. The mutual attraction of opposite charges holds atoms together as molecules in a(n) _____ bond.
   a. ionic
   b. hydrogen
   c. polar covalent
   d. nonpolar covalent

4. A salt does not release _____ in water.
   a. ions
   b. energy
   c. H^+

5. A(n) _____ substance repels water.
   a. acidic
   b. basic
   c. hydrophobic
   d. polar

6. When dissolved in water, a(n) _____ donates H^+ and a(n) _____ accepts H^+.
   a. acid; base
   b. base; acid
   c. buffer; solute
   d. base; buffer

7. _____ is a monosaccharide.
   a. Glucose
   b. Sucrose
   c. Ribose
   d. Starch
   e. a and c
   f. a, b, and c

8. Unlike saturated fatty acids, the tails of unsaturated fatty acids incorporate one or more _____ .
   a. phosphate groups
   b. glycerols
   c. double bonds
   d. single bonds

9. Which of the following is a class of molecules that encompasses all of the other molecules listed?
   a. triglycerides
   b. fatty acids
   c. waxes
   d. steroids
   e. lipids
   f. phospholipids

10. _____ are to proteins as _____ are to nucleic acids.
    a. Amino acids; hydrogen bonds
    b. Amino acids; nucleotides
    c. Sugars; lipids
    d. Sugars; proteins

11. A denatured protein has lost its _____ .
    a. hydrogen bonds
    b. shape
    c. function
    d. all of the above

12. Match the terms with their most suitable description.
    _____ hydrophilic
    _____ atomic number
    _____ hydrogen bonds
    _____ positive charge
    _____ temperature
    _____ negative charge
    _____ solution
    a. protons > electrons
    b. number of protons in nucleus
    c. polar; dissolves easily in water
    d. collectively strong
    e. protons < electrons
    f. measure of molecular motion
    g. solute dissolved in solvent

13. Which of the following are *not* found in DNA?
    a. amino acids
    b. sugars
    c. nucleotides
    d. phosphate groups

14. Match the molecules with the best description.
    _____ wax
    _____ starch
    _____ triglyceride
    _____ DNA
    _____ polypeptide
    _____ ATP
    a. protein primary structure
    b. an energy carrier
    c. water-repellent secretions
    d. carries heritable information
    e. sugar storage in plants
    f. richest energy source in animals

15. Match each molecule with its component(s).
    _____ protein
    _____ phospholipid
    _____ triglyceride
    _____ nucleic acid
    _____ cellulose
    _____ nucleotide
    _____ wax
    _____ glycoprotein
    a. glycerol, fatty acids, phosphate
    b. glycerol, three fatty acids
    c. nucleotides
    d. glucose only
    e. sugar, phosphate, base
    f. amino acid monomers
    g. amino acids, sugars
    h. fatty acids, carbon rings

Critical Thinking

1. Alchemists were the forerunners of modern-day chemists. Many of these medieval scholars and philosophers spent their lives trying to transform lead (atomic number 82) into gold (atomic number 79). Explain why they never succeeded.

2. Draw a shell model of a lithium atom (Li), which has 3 protons, then predict whether the majority of lithium atoms on Earth are uncharged, positively charged, or negatively charged.

3. Polonium is a rare element with 33 radioisotopes. The most common one, ^{210}Po, has 82 protons and 128 neutrons. When ^{210}Po decays, it emits an alpha particle, which is a helium nucleus (2 protons and 2 neutrons). ^{210}Po decay is tricky to detect because alpha particles do not carry very much energy compared to other forms of radiation. For example, they can be stopped by a single sheet of paper or a few inches of air. That is one reason that authorities failed to discover toxic amounts of ^{210}Po in the body of former KGB agent Alexander Litvinenko until after he died suddenly and mysteriously in 2006. What element does an atom of ^{210}Po become after it emits an alpha particle?

4. In the following list, identify the carbohydrate, the fatty acid, the amino acid, and the polypeptide:
   a. $NH_2-CHR-COOH$
   b. $C_6H_{12}O_6$
   c. $(methionine)_{20}$
   d. $CH_3(CH_2)_{16}COOH$

3 CELL STRUCTURE

Application ❯ 3.1 Food for Thought

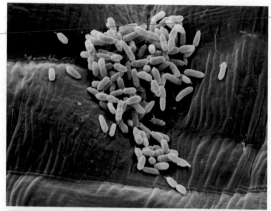

Figure 3.1 Toxin-producing bacteria can contaminate foods.
Top, *Escherichia coli* cells sticking to the surface of a lettuce leaf. Some strains of this bacteria can cause a serious intestinal illness when they contaminate human food (bottom).

Top, © Custom Medical Stock Photo/Getty Images; bottom, Getty Images.

Cell for cell, microorganisms that live in and on a human body outnumber the person's own cells by about ten to one. Most are bacteria that live in the digestive tract, but these cells are not just stowaways. Gut bacteria help with digestion, make vitamins that mammals cannot, prevent the growth of dangerous germs, and shape the immune system. One of the most common intestinal bacteria of warm-blooded animals (including humans) is *Escherichia coli*. Most of the hundreds of types, or strains, of *E. coli*, are helpful, but a few make a toxic protein that can severely damage the lining of the intestine. After ingesting as few as ten cells of a toxic strain, a person may become ill with severe cramps and bloody diarrhea that lasts up to ten days. In some people, complications of infection result in kidney failure, blindness, paralysis, and death. Each year, about 265,000 people in the United States become infected with toxin-producing *E. coli*.

Strains of *E. coli* that are toxic to people live in the intestines of other animals—mainly cattle, deer, goats, and sheep—apparently without sickening them. Humans are exposed to the bacteria when they come into contact with animals that harbor them, for example, by eating fresh fruits and vegetables that have contacted animal feces (Figure 3.1). People also become infected with toxic *E. coli* by eating contaminated ground meat. An animal's feces can contaminate its meat during slaughter. Bacteria in the feces stick to the meat, then get thoroughly mixed into it during the grinding process. Unless the contaminated meat is cooked to at least 71°C (160°F), live bacteria will enter the digestive tract of anyone who eats it.

The United States Department of Agriculture (USDA) recalls food products in which toxic bacteria are discovered. Recalled meat is not necessarily discarded; often, it is cooked and processed into ready-to-eat products such as canned chili. Sterilization by cooking or other means kills bacteria, and it is one way to ensure food safety. Raw beef trimmings, which have a high risk of contact with fecal matter during the butchering process, are effectively sterilized when sprayed with ammonia. Ground to a paste and formed into pellets or blocks, the resulting product is termed "lean finely textured beef" or "boneless lean beef trimmings." This product is routinely used as a filler in prepared food products such as hamburger patties, fresh ground beef, hot dogs, lunch meats, sausages, frozen entrees, canned foods, and other items sold to quick service restaurants, hotel and restaurant chains, institutions, and school lunch programs. In 2012, a series of news reports that nicknamed the product "pink slime" provoked public outrage at its widespread use. Meat industry organizations and the USDA agree that lean finely textured beef, appetizing or not, is perfectly safe to eat because it has been sterilized.

3.2 What, Exactly, Is a Cell?

REMEMBER: The cell is the smallest unit with the properties of life (Section 1.2). A tracer is a substance that can be tracked via a detectable component (2.2). Phospholipids arranged in a lipid bilayer are the main component of cell membranes (2.8).

The Cell Theory Hundreds of years of observations led to the way we now answer the question, What is a cell? Today we know that a cell carries out metabolism and homeostasis, and reproduces either on its own or as part of a larger organism. By this definition, each cell is alive even if it is part of a multicelled body, and all living organisms consist of one or more cells. We also know that cells reproduce by dividing, so it follows that all existing cells must have arisen by division of other cells (later chapters discuss processes by which cells reproduce). As a cell divides, it passes its hereditary material—its DNA—to offspring. Taken together, these generalizations constitute the **cell theory**, which is one of the foundations of modern biology (Table 3.1).

Components of All Cells Cells vary in shape and in what they do, but all share certain organizational and functional features: a plasma membrane, cytoplasm, and DNA (Figure 3.2). A cell's **plasma membrane** separates it from the external environment. Like all other cell membranes, it consists mainly of a phospholipid bilayer, and it is selectively permeable, which means that only certain materials can cross it. Thus, a plasma membrane controls exchanges between the cell and its environment. As you will see in Section 3.3, many different proteins embedded in a lipid bilayer or attached to one of its surfaces carry out membrane functions.

The plasma membrane encloses a jellylike mixture of water, sugars, ions, and proteins called **cytoplasm**. Some or all of a cell's metabolism occurs in cytoplasm, and the cell's internal components, including organelles, are suspended in it. **Organelles** are structures that carry out special metabolic functions inside a cell. Those with membranes compartmentalize substances and activities.

Every cell starts out life with DNA. In nearly all bacteria and archaea, that DNA is suspended directly in cytoplasm. By contrast, the DNA of a eukaryotic cell is contained in a **nucleus** (plural, nuclei), an organelle with a double membrane. All protists, fungi, plants, and animals are eukaryotes. Some of these organisms are independent, free-living cells; others consist of many cells working together as a body.

Constraints on Cell Size A living cell must exchange substances with its environment at a rate that keeps pace with its metabolism. These exchanges occur across the plasma membrane, which can handle only so many exchanges at a time. The rate of exchange across a plasma membrane depends on its surface area: The bigger it is, the more substances can cross it during a given interval. Thus, cell size is limited by a physical relationship called the **surface-to-volume ratio**. By this ratio, an object's volume increases with the cube of its diameter, but its surface area increases only with the square.

Apply the surface-to-volume ratio to a round cell. As Figure 3.3 shows, when a cell expands in diameter, its volume increases faster than its surface area does. Imagine that our round cell expands until it is four times its original diameter. The volume of the cell has increased 64 times (4^3), but its surface area has increased only 16 times (4^2). Each unit of plasma membrane must now handle exchanges for four times as much cytoplasm ($64 \div 16 = 4$). If the cell gets too big, the inward flow of nutrients and the outward flow of wastes across that membrane will not be fast enough to keep the cell alive.

Table 3.1 The Cell Theory

1. Each organism consists of one or more cells.
2. The cell is the structural and functional unit of all organisms. A cell is the smallest unit of life, individually alive even as part of a multicelled organism.
3. All living cells arise by division of preexisting cells.
4. Cells contain hereditary material (DNA), which they pass to their offspring when they divide.

Figure 3.2 General organization of a cell.
All cells start out life with a plasma membrane, cytoplasm, and DNA. This one is a cell from a plant.

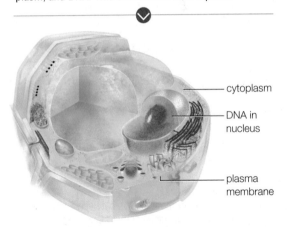

cytoplasm

DNA in nucleus

plasma membrane

cell theory Theory that all organisms consist of one or more cells, which are the basic unit of life; all cells come from division of preexisting cells; and all cells pass DNA to offspring.

cytoplasm Semifluid substance enclosed by a cell's plasma membrane.

nucleus Of a cell, an organelle with two membranes that holds the cell's DNA.

organelle Structure that carries out a special metabolic function inside a cell.

plasma membrane Membrane that encloses a cell and separates it from the external environment.

surface-to-volume ratio A relationship in which the volume of an object increases with the cube of the diameter, and the surface area increases with the square.

Figure 3.3 Examples of surface-to-volume ratio. This physical relationship between increases in volume and surface area limits the size and influences the shape of cells.

Diameter (cm)	2	3	6
Surface area (cm²)	12.6	28.2	113
Volume (cm³)	4.2	14.1	113
Surface-to-volume ratio	3:1	2:1	1:1

Surface-to-volume limits also affect the form of colonial organisms such as strandlike algae, in which small cells attach end to end so each one can interact directly with the environment. It also affects the form of cells in multicelled bodies. For example, some types of muscle cells in your thighs run the length of your upper leg. Each of these cells is thin, so it exchanges substances efficiently with fluids in the surrounding tissue.

How Do We See Cells?   No one even knew cells existed until well after the first microscopes were invented. This is because typical cells are in the micrometer range of size—much smaller than the unaided human eye can perceive (Figure 3.4). One micrometer (μm) is one-thousandth of a millimeter, which is one-thousandth of a meter (Table 3.2). In a light microscope, visible light illuminates a sample. As you will learn in Chapter 5, all light travels in waves. This property of light causes it to bend when passing through a curved glass lens. Inside a light microscope, such lenses focus light that passes through a specimen, or bounces off of one, into a magnified image (Figure 3.5A). Microscopes that use polarized light can yield images in which the edges of some structures appear in three-dimensional relief (Figure 3.5B). Photographs of images enlarged with a microscope are called micrographs; those taken with visible light are called light micrographs (LM).

Most cells are nearly transparent, so their internal details may not be visible unless they are first stained, or exposed to dyes that only some cell parts soak up. Parts that absorb the most dye appear darkest. Staining results in an increase in contrast (the difference between light and dark) that allows us to see a greater range of detail. Researchers often use light-emitting tracers to pinpoint the location of a molecule of interest within a cell. When illuminated with a laser, these tracers fluoresce (emit light), and an image of the emitted light can be captured with a fluorescence microscope (Figure 3.5C). Such images are called fluorescence micrographs.

Structures smaller than about 200 nanometers across appear blurry under light microscopes. To observe objects of this size range clearly, we would have to switch to an electron microscope. There are two types of electron microscope; both use magnetic fields as lenses to focus a beam of electrons onto a sample. A transmission electron microscope directs electrons through a thin specimen, and the specimen's internal details appear as shadows in the resulting image, which is called a transmission electron micrograph, or TEM (Figure 3.5D). A scanning electron microscope directs a beam of electrons back and forth across the surface of a specimen that has been coated with a thin layer of gold or other metal. The irradiated metal emits

Table 3.2 Common Units of Length

Unit	Equivalent	
	Meter	**Inch**
centimeter (cm)	1/100	0.394
millimeter (mm)	1/1000	0.0394
micrometer (μm)	1/1,000,000	0.0000394
nanometer (nm)	1/1,000,000,000	0.0000000394
meter (m)	100 cm 1,000 mm 1,000,000 μm 1,000,000,000 nm	39.4

Figure 3.4 Relative sizes.
Below, the diameter of most cells is between 1 and 100 micrometers. Table 3.2 shows conversions among units of length; also see Units of Measure, Appendix IV.

Louse, Edward S. Ross; Ant, Vladimir Davydov/iStock/360/Getty Images; Frog, © A Cotton Photo/Shutterstock; Rat, © Pakhnyushcha/Shutterstock; Goose, panbazil/Shutterstock.com; Boy, © Piotr Marcinski/Shutterstock; Giraffe, © Valerie Kalyuznnyy/Photos.com; Whale, Dorling Kindersley/Getty Images.

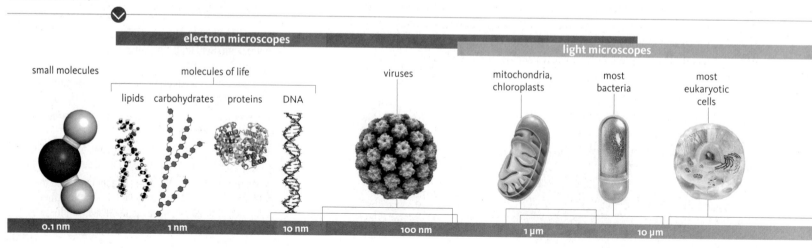

electron microscopes — light microscopes

small molecules — molecules of life — lipids   carbohydrates   proteins   DNA — viruses — mitochondria, chloroplasts — most bacteria — most eukaryotic cells

0.1 nm   1 nm   10 nm   100 nm   1 μm   10 μm

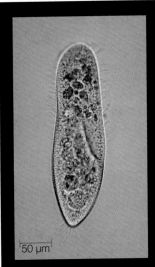 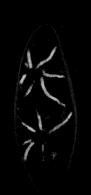 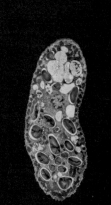 

50 µm

A. Green blobs visible in this light micrograph (LM) of a living cell are ingested algae. Hairlike structures on the cell's surface are waving cilia that propel this motile organism through fluid surroundings.

B. A light micrograph taken with polarized light shows edges in relief. This technique reveals ingested algae, and also some internal structures that are not visible in **A**.

C. In this fluorescence micrograph, yellow pinpoints the location of a particular protein in the membrane of organelles called contractile vacuoles. These organelles are also visible in **B**.

D. A colorized transmission electron micrograph (TEM) reveals several types of internal structures in a plane (slice). Ingested algae are being broken down inside food vacuoles.

E. A scanning electron micrograph (SEM) shows details of the cell's surface, including its thick coat of cilia. The indentation (also visible in **A**) is where the cell takes in food.

electrons and x-rays, which are converted into an image (a scanning electron micrograph, or SEM) of the surface (Figure 3.5E). SEMs and TEMs are always black and white; colored versions have been digitally altered to highlight specific details.

Figure 3.5 Different microscopes reveal different characteristics of the same organism, a single-celled protist called *Paramecium*.

(A) Nancy Nehring/iStockphoto.com; (B) Michael Abbey/Science Source; (C) © Dennis Kunkel Microscopy, Inc./PhototakeUSA.com; (D) © Microworks/PhototakeUSA.com; (E) Steve Gschmeissner/Science Source.

Take-Home Message 3.2

How are all cells alike?

- All cells start life with a plasma membrane, cytoplasm, and a region of DNA. In eukaryotic cells only, the DNA is contained within a nucleus.
- The surface-to-volume ratio limits cell size and influences cell shape.
- Different types of microscopes reveal different aspects of cell structure.

human eye (no microscope)

frog eggs

small animals

largest organisms

100 µm   1 mm   1 cm   10 cm   1 m   10 m   100 m

Figure 3.6 Cell membrane structure. Organization of phospholipids in cell membranes (A) and examples of common membrane proteins (B–E). For clarity, these proteins are often modeled as blobs or geometric shapes; their structure can be extremely complex.

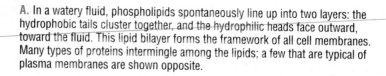

A. In a watery fluid, phospholipids spontaneously line up into two layers: the hydrophobic tails cluster together, and the hydrophilic heads face outward, toward the fluid. This lipid bilayer forms the framework of all cell membranes. Many types of proteins intermingle among the lipids; a few that are typical of plasma membranes are shown opposite.

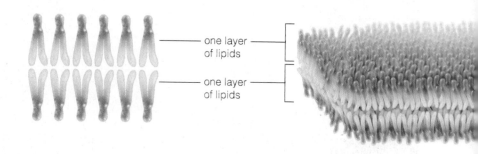

one layer of lipids

one layer of lipids

3.3 Cell Membrane Structure

REMEMBER: Phospholipids have a phosphate-containing head and two fatty acid tails; the tails can vary in length and in saturation (Section 2.8).

A cell's basic structure is essentially a lipid bilayer bubble filled with fluid.

The foundation of all cell membranes is a lipid bilayer that consists mainly of phospholipids (Figure 3.6A). The head of a phospholipid is highly polar and hydrophilic, which means that it interacts with water molecules. Its two long hydrocarbon tails are very nonpolar and hydrophobic, so they do not interact with water molecules. As a result of these opposing properties, phospholipids swirled into water will spontaneously organize themselves into lipid bilayer sheets or bubbles, with hydrophobic tails together, hydrophilic heads facing the watery surroundings. A cell's basic structure is essentially a lipid bilayer bubble filled with fluid (left).

fluid

Other molecules, including cholesterol, proteins, glycoproteins, and glycolipids, are embedded in or attached to the lipid bilayer of a cell membrane. Many of these molecules can move around the membrane more or less freely. We use the term **fluid mosaic** to describe a membrane of a eukaryotic or bacterial cell because it behaves like a two-dimensional liquid of mixed composition. Membrane fluidity occurs because phospholipids in the bilayer are not chemically bonded to one another; they stay organized as a result of collective hydrophobic and hydrophilic attractions. These interactions are, on an individual basis, relatively weak. Thus, individual phospholipids in the bilayer drift sideways and spin around their long axis, and their tails wiggle.

A cell membrane's properties vary depending on the types and proportions of molecules composing it. For example, membrane fluidity decreases with increasing cholesterol content. A membrane's fluidity also depends on the length and saturation of its phospholipids' fatty acid tails. Archaea do not even use fatty acids to build their phospholipids. Instead, they use molecules with reactive side chains, so the tails of archaeal phospholipids form covalent bonds with one another. As a result of this rigid crosslinking, archaeal phospholipids do not drift, spin, or wiggle in a

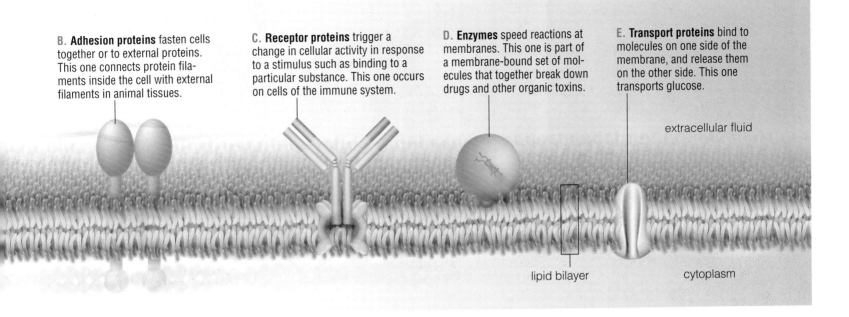

B. Adhesion proteins fasten cells together or to external proteins. This one connects protein filaments inside the cell with external filaments in animal tissues.

C. Receptor proteins trigger a change in cellular activity in response to a stimulus such as binding to a particular substance. This one occurs on cells of the immune system.

D. Enzymes speed reactions at membranes. This one is part of a membrane-bound set of molecules that together break down drugs and other organic toxins.

E. Transport proteins bind to molecules on one side of the membrane, and release them on the other side. This one transports glucose.

extracellular fluid

lipid bilayer

cytoplasm

bilayer. Thus, membranes of archaea are stiffer than those of bacteria or eukaryotes, a characteristic that may help these cells survive in extreme habitats.

Membrane Proteins A cell membrane physically separates an external environment from an internal one, but that is not its only task. Many types of proteins are associated with a cell membrane, and each type adds a specific function to it. Thus, different cell membranes can carry out different tasks depending on which proteins are associated with them. A plasma membrane incorporates certain proteins that no internal cell membrane has, so it has functions that no other membrane does. For example, cells stay organized in animal tissues because **adhesion proteins** in their plasma membranes fasten them together and hold them in place (Figure 3.6B). Plasma membranes and some internal membranes incorporate **receptor proteins**, which trigger a change in the cell's activities in response to a stimulus (Figure 3.6C). Each type of receptor protein receives a particular stimulus, such as binding to a certain hormone. Each receptor protein also triggers a specific response inside the cell, which may involve metabolism, movement, division, or even cell death.

All cell membranes incorporate enzymes (Figure 3.6D). Some membrane enzymes act on other proteins or lipids that are part of the lipid bilayer. All membranes also have **transport proteins**, which move specific substances across the bilayer (Figure 3.6E). These proteins are important because lipid bilayers are impermeable to ions and polar molecules (we return to this topic in Section 4.5).

Take-Home Message 3.3

What is a cell membrane?

• The foundation of all cell membranes is the lipid bilayer: two layers of phospholipids, tails sandwiched between heads.

• Proteins that associate with lipid bilayers add various functions to a membrane.

adhesion protein Plasma membrane protein that fastens cells together in animal tissues.

fluid mosaic Model of a cell membrane as a two-dimensional fluid of mixed composition.

receptor protein Membrane protein that triggers a change in cell activity in response to a stimulus such as binding a certain substance.

transport protein Protein that moves specific ions or molecules across a membrane.

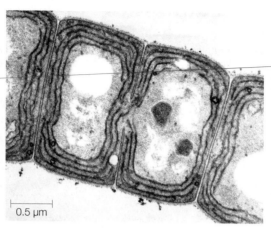

A. *Escherichia coli* is a common bacterial inhabitant of human intestines. Short, hairlike structures are pili; longer ones are flagella. This one is harmless; others can cause disease in humans.

B. *Oscillatoria* are a type of cyanobacteria, an ancient lineage of photosynthetic bacteria. Photosynthesis occurs at internal membranes (green). The multi-sided structures (pink) are protein-enclosed organelles called carboxysomes that assist photosynthesis.

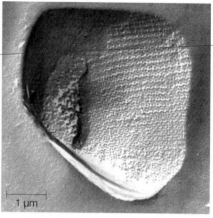

C. *Ferroglobus placidus* is an archaeon that thrives in superheated water spewing from the ocean floor. The durable composition of its lipid bilayers (note the gridlike texture) keeps them intact at extreme heat and pH.

Figure 3.7 Some representative prokaryotes.

(A, B) © Biophoto Associates/Science Source; (C) © K.O. Stetter & R. Rachel, Univ. Regensburg; (D) Cryo-EM image of *Haloquadratum walsbyi*, isolated from Australia. Courtesy of Zhuo Li (City of Hope, Duarte, California, USA), Mike L. Dyall-Smith (Charles Sturt University, Australia), and Grant J. Jensen (California Institute of Technology, Pasadena, California, USA); (E) Biomedical Imaging Unit, Southhampton General Hospital/Science Photo Library; (F) Archivo Angels Tapias y Fabrice Confalonieri.

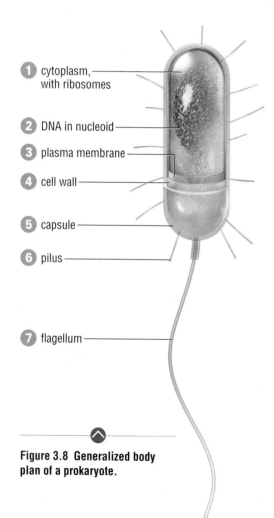

1. cytoplasm, with ribosomes
2. DNA in nucleoid
3. plasma membrane
4. cell wall
5. capsule
6. pilus
7. flagellum

Figure 3.8 Generalized body plan of a prokaryote.

3.4 Introducing Prokaryotic Cells

REMEMBER: A polysaccharide is a long chain of monosaccharides (Section 2.7); a peptide is a short chain of amino acids (2.9).

All bacteria and archaea are single-celled organisms, although individual cells of many species cluster in filaments or colonies (Figure 3.7). Outwardly, cells of the two groups appear so similar that archaea were once presumed to be an unusual group of bacteria. Both were classified as prokaryotes, a word that means "before the nucleus." By 1977, it had become clear that archaea are more closely related to eukaryotes than to bacteria, so they were given their own separate domain. The term "prokaryote" is now an informal designation only.

Bacteria and archaea are the smallest and most metabolically diverse forms of life that we know about. Chapter 13 revisits them in more detail; here we present an overview of structures shared by both groups (Figure 3.8).

Compared with eukaryotic cells, prokaryotes have little in the way of internal framework, but they do have protein filaments under the plasma membrane that reinforce the cell's shape and act as scaffolding for internal structures. The cytoplasm of these cells 1 contains many **ribosomes** (organelles upon which polypeptides are assembled), and in some species, additional organelles. The cytoplasm also contains plasmids, which are small circles of DNA that carry a few genes (units of inheritance). The cell's remaining genes typically occur on one large circular molecule of DNA located in an irregularly shaped region of cytoplasm called the nucleoid 2. In a few species, the nucleoid is enclosed by a membrane. Other internal membranes carry out special metabolic processes such as photosynthesis (Figure 3.7B).

Like all cells, bacteria and archaea have a plasma membrane 3. In nearly all prokaryotes, a rigid **cell wall** 4 surrounding the plasma membrane protects the cell and supports its shape. Archaeal cell walls and bacterial cell walls differ, but both types are permeable to water, so dissolved substances easily cross. Many species of bacteria, including the ones shown in Figure 3.7, have a second membrane

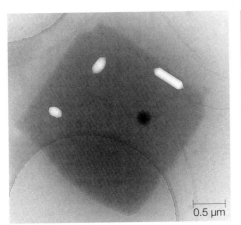

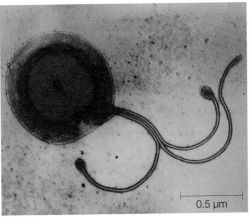

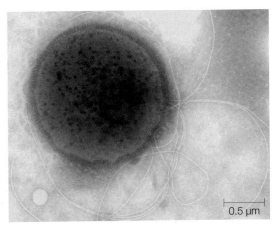

D. The square archaeon *Haloquadratum walsbyi* prefers brine pools saltier than soy sauce. Gas-filled organelles (white structures) buoy these highly motile cells, which can aggregate into flat sheets a bit like floor tiles.

E. *Helicobacter pylori*, a bacterium that can cause stomach ulcers when it infects the lining of the stomach, takes on a ball-shaped form (shown) that offers protection from environmental challenges such as antibiotic treatment.

F. The archaeon *Thermococcus gammatolerans* lives under extreme conditions of salt, temperature, and pressure. It is by far the most radiation-resistant organism ever discovered, capable of withstanding thousands of times more radiation than humans can.

around the cell wall. Outside the cell wall (or the second membrane) is a thick, gelatinous capsule ⑤ and/or a loosely attached layer of slime.

Protein filaments called **pili** (singular, pilus) ⑥ project from the wall of some prokaryotes. Pili help these cells move across or cling to surfaces. One kind, a "sex" pilus, attaches to another bacterium and then shortens. The attached cell is reeled in, and DNA is transferred from one cell to the other. Many types of bacteria and archaea also have one or more flagella projecting from their surface ⑦. **Flagella** (singular, flagellum) are long, slender cellular structures used for motion. A prokaryotic flagellum rotates like a propeller that drives the cell through fluid habitats.

Biofilms Bacterial cells often live so close together that an entire community shares a layer of slime. A communal living arrangement in which single-celled organisms occupy a shared mass of slime is called a **biofilm**. A biofilm is often attached to a solid surface, and may include bacteria, algae, fungi, protists, and/or archaea. Participating in a biofilm allows the cells to linger in a favorable spot rather than be swept away by fluid currents, and to reap the benefits of living communally. For example, rigid or netlike secretions of some species serve as permanent scaffolding for others; species that break down toxic chemicals allow more sensitive ones to thrive in habitats that they could not withstand on their own; and waste products of some serve as raw materials for others. Later chapters discuss some medical implications of biofilms, including dental plaque (Figure 3.9).

Figure 3.9 Oral bacteria in dental plaque, a biofilm. This micrograph shows two species of bacteria (tan, green) and a yeast (red) sticking to one another and to teeth via a gluelike mass of shared, secreted polysaccharides (pink). Other secretions of these organisms cause cavities and periodontal disease.

© Dennis Kinkel Microscopy, Inc./Phototake

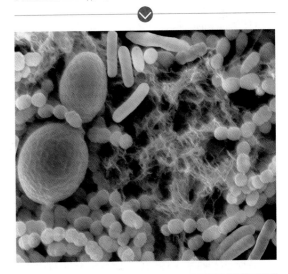

Take-Home Message 3.4

How are bacteria and archaea alike?

- Bacteria and archaea do not have a nucleus. Most kinds have a cell wall around their plasma membrane. The permeable wall reinforces and imparts shape to the cell body.
- The structure of bacteria and archaea is relatively simple, but as a group these organisms are the most diverse forms of life.

biofilm Community of microorganisms living within a shared mass of slime.

cell wall Semirigid but permeable structure that surrounds the plasma membrane of some cells.

flagellum Long, slender cellular structure used for movement.

pilus A protein filament that projects from the surface of some prokaryotic cells.

ribosome Organelle of protein synthesis.

3.5 Introducing Eukaryotic Cells

REMEMBER: Hydrogen bonding makes the loops, helices, and sheets of a polypeptide fold up into functional domains; many proteins consist of two or more polypeptides; fibrous proteins aggregate by many thousands into much larger structures (Section 2.9). ATP has an important metabolic role as an energy carrier (2.10).

In addition to a nucleus, a typical eukaryotic cell has many other membrane-enclosed organelles (Figure 3.10). An enclosing membrane allows an organelle to regulate the types and amounts of substances that enter and exit. Through this control, the organelle maintains a special internal environment that allows it to carry out a particular function—for example, isolating toxic or sensitive substances from the rest of the cell, moving substances through cytoplasm, maintaining fluid balance, or providing a favorable environment for a special process.

The Nucleus A cell nucleus serves two important functions. First, it keeps the cell's genetic material—its one and only copy of DNA—away from metabolic processes that might damage it. Isolated in its own compartment, the DNA stays separated from the bustling activity of the cytoplasm ❶. Second, a nucleus controls the passage of certain molecules across its membrane. The nucleus has a special membrane, the **nuclear envelope**, that carries out this function. A nuclear envelope consists of two lipid bilayers folded together. Proteins embedded in the two lipid bilayers aggregate into thousands of tiny nuclear pores that span the envelope (Figure 3.11). Some bacteria have membranes around their DNA, but we do not consider the bacteria to have nuclei because there are no pores in these membranes.

Large molecules, including RNA and proteins, cannot cross lipid bilayers on their own. Nuclear pores function as gateways for these molecules to enter and exit a nucleus. Protein synthesis offers an example of why this movement is important. Protein synthesis occurs in cytoplasm, and it requires the participation of many molecules of RNA. RNA is produced in the nucleus. Thus, RNA molecules must move from nucleus to cytoplasm, and they do so through nuclear pores. Proteins that carry out RNA synthesis must move in the opposite direction, because this process occurs in the nucleus.

The Endomembrane System The endomembrane system is a series of interacting organelles between the nucleus and the plasma membrane. Its main function is to make lipids, enzymes, and proteins for insertion into the cell's membranes or secretion to the external environment. The endomembrane system also destroys toxins, recycles wastes, and has other special functions. Components of the system vary among different types of cells, but here we present an overview of the most common ones.

Small, membrane-enclosed sacs called **vesicles** ❷ form by budding from other organelles or when a patch of plasma membrane sinks into the cytoplasm. Many types carry substances from one organelle to another, or to and from the

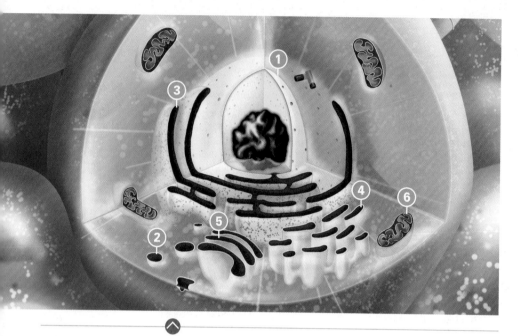

Figure 3.10 Common components of eukaryotic cells. An animal cell is illustrated here.

❶ The nucleus protects and controls access to the cell's DNA.

❷ Vesicles form by budding from other components of the endomembrane system or from the plasma membrane. Some transport substances among organelles of the ER, and to and from the plasma membrane. Others store or break down substances.

❸ Ribosomes attached to rough endoplasmic reticulum (ER) assemble polypeptides that thread into the ER's interior, where they take on tertiary structure and assemble with other polypeptides.

❹ Many proteins made in rough ER migrate through the ER compartment to smooth ER. Some of these proteins stay in smooth ER, as enzymes that assemble lipids and break down carbohydrates, wastes, and toxins. Others are packaged in vesicles for transport to Golgi bodies.

❺ Golgi bodies modify proteins and lipids, then sort and repackage the finished molecules into new vesicles. Some of the new vesicles become lysosomes. Others carry proteins to the plasma membrane for insertion into the lipid bilayer or secretion.

❻ Mitochondria specialize in efficient production of ATP.

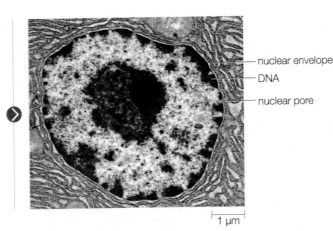

**Figure 3.11
The nucleus.**
This organelle is the defining characteristic of eukaryotic cells. TEM shows a liver cell nucleus.

© Kenneth Bart.

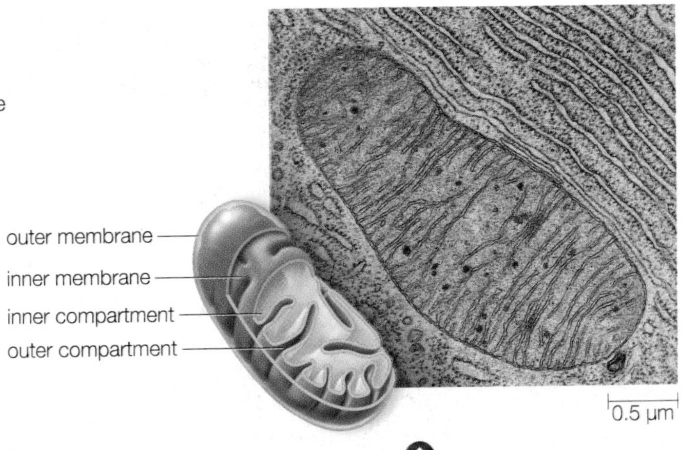

nuclear envelope
DNA
nuclear pore

1 μm

outer membrane
inner membrane
inner compartment
outer compartment

0.5 μm

Figure 3.12 The mitochondrion.
Two membranes, one folded inside the other, form the ATP-making machinery of this eukaryotic organelle. The TEM shows a mitochondrion in a cell from bat pancreas.

Keith R. Porter.

Figure It Out: What organelle is visible to the upper right in the micrograph?   Answer: Rough ER

plasma membrane. Some are a bit like trash cans that collect and dispose of waste, debris, or toxins. Enzymes in vesicles called **peroxisomes** break down fatty acids, amino acids, and poisons such as alcohol. They also break down hydrogen peroxide, a toxic by-product of fatty acid breakdown. Even more powerful enzymes in vesicles called **lysosomes** break down cellular debris and wastes. Ingested bacteria, cell parts, and other particles are delivered to lysosomes for breakdown.

Vacuoles are sacs that form by the fusion of multiple vesicles. Many isolate or break down waste, debris, toxins, or food. Plant cells have a large central vacuole that collects amino acids, sugars, ions, wastes, and toxins. Fluid pressure in a central vacuole keeps plant cells plump, so stems, leaves, and other plant parts stay firm (a central vacuole is illustrated to the left of the nucleus in Figure 3.2).

Endoplasmic reticulum (**ER**) is a system of tubes and flattened sacs. The membrane of the ER extends from the nuclear envelope, and it encloses a single, continuous compartment. Two kinds of ER, rough and smooth, are named for their appearance in electron micrographs. Thousands of ribosomes attached to the outer surface of rough ER give this organelle its "rough" appearance ③. These ribosomes make polypeptides that thread into the interior of the ER as they are assembled. Inside the ER, the polypeptides take on their tertiary structure, and many assemble with other polypeptides. Cells that make, store, and secrete proteins have a lot of rough ER.

Some proteins made in rough ER become part of its membrane. Others migrate through the ER compartment to smooth ER. Smooth ER has no ribosomes, so it does not make its own proteins ④. Some proteins that arrive in smooth ER are immediately packaged into vesicles for delivery elsewhere. Others are enzymes that stay and become part of the smooth ER. Enzymes of the smooth ER make lipids for the cell's membranes, and they break down carbohydrates, fatty acids, and some drugs and poisons.

A **Golgi body** has a folded membrane that often looks like a stack of pancakes ⑤. Enzymes inside of Golgi bodies put finishing touches on proteins and lipids that have been delivered from ER. The finished products are sorted and packaged in new vesicles. Some of the vesicles deliver their cargo to the plasma membrane; others become lysosomes.

Mitochondria The **mitochondrion** (plural, mitochondria) ⑥ is a eukaryotic organelle that specializes in making ATP by aerobic respiration (Chapter 5 details this metabolic pathway). A mitochondrion has two membranes, one highly folded inside the other (Figure 3.12), that form its ATP-making machinery. The organelle

endoplasmic reticulum (**ER**) Membrane-enclosed organelle that is a continuous system of sacs and tubes extending from the nuclear envelope. Smooth ER makes lipids and breaks down carbohydrates and fatty acids; ribosomes on the surface of rough ER make polypeptides.

Golgi body Organelle that modifies polypeptides and lipids, then packages the finished products into vesicles.

lysosome Enzyme-filled vesicle that breaks down cellular wastes and debris.

mitochondrion Eukaryotic organelle that produces ATP by aerobic respiration.

nuclear envelope A double membrane that constitutes the outer boundary of the nucleus. Nuclear pores in the membrane control the entry and exit of large molecules.

peroxisome Enzyme-filled vesicle that breaks down amino acids, fatty acids, and toxic substances.

vesicle Small, membrane-enclosed organelle; different kinds store, transport, or break down their contents.

Digging Into Data

Organelles and Cystic Fibrosis

A plasma membrane transport protein called CFTR moves chloride ions out of cells lining cavities and ducts of the lungs, liver, pancreas, intestines, and reproductive system. Water that follows the ions creates a thin film that allows mucus to slide easily through these structures.

People with cystic fibrosis (CF) have too few copies of the CFTR protein in the plasma membranes of their cells. Not enough chloride ions leave the cells, and so not enough water leaves them either. The result is thick, dry mucus that clogs the airways to the lungs and other passages. Symptoms include difficulty breathing and chronic lung infections. In 2000, researchers tracked the cellular location of the CFTR protein as it was being produced in cells from people with CF (Figure 3.13).

1. Which organelle contains the least amount of CFTR protein in normal cells? In CF cells?
2. In which organelle is the amount of CFTR protein most similar in both types of cells?
3. Where is the CFTR protein getting held up in cells of people who have CF?

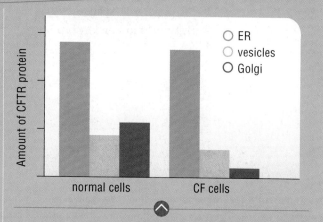

Figure 3.13 Cellular location of the CFTR protein.
Graph compares the amounts of CFTR protein found in endoplasmic reticulum, vesicles traveling from ER to Golgi, and Golgi bodies in CF cells and normal cells.

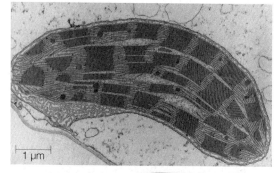

1 μm

two outer membranes
stroma
inner membrane

Figure 3.14 The chloroplast.
Each chloroplast has two outer membranes. Photosynthesis occurs at a third, much-folded inner membrane. TEM shows a chloroplast from a cell in a corn leaf.

Science Source.

resembles a bacterium in size, form, and biochemistry. Mitochondria have their own DNA, which is circular and otherwise similar to bacterial DNA. They divide independently of the cell, and have their own ribosomes. Such clues led to a theory that mitochondria evolved from aerobic bacteria that took up permanent residence inside a host cell (we return to this topic in Section 13.3).

Chloroplasts   Photosynthetic cells of plants and many protists have **chloroplasts**, which are organelles specialized for photosynthesis. Most chloroplasts are oval or disk-shaped (Figure 3.14). Each has two outer membranes enclosing a semifluid interior, the stroma, that contains enzymes and the chloroplast's own DNA. In the stroma, a third, highly folded membrane forms a single, continuous compartment. Photosynthesis occurs at this inner membrane. In many ways, chloroplasts resemble the photosynthetic bacteria from which they evolved.

The Cytoskeleton   Between the nucleus and plasma membrane of all eukaryotic cells is a system of protein filaments collectively called the **cytoskeleton**. Elements of the cytoskeleton reinforce, organize, and move cell structures, and often the whole cell. Some are permanent; others form only at certain times.

Microtubules are long, hollow cylinders that consist of subunits of the protein tubulin (Figure 3.15A). They form a dynamic scaffolding for many cellular processes, rapidly assembling when they are needed, disassembling when they are not. For example, before a eukaryotic cell divides, microtubules assemble, separate the cell's duplicated DNA molecules, then disassemble. As another example, microtubules that form in the growing end of a young nerve cell support its lengthening in a particular direction (Figure 3.15C).

Microfilaments are fine fibers that consist primarily of subunits of a protein called actin (Figure 3.15B). These fibers strengthen or change the shape of eukaryotic cells, and have a critical function in cell migration, movement, and contraction. Crosslinked, bundled, or gel-like arrays of them make up the cell cortex, a reinforcing mesh under the plasma membrane. Microfilaments also connect plasma membrane proteins to other proteins inside the cell.

Intermediate filaments are the most stable elements of the cytoskeleton, forming a framework that lends structure and resilience to cells and tissues in multicelled organisms. Several types of intermediate filaments are assembled from different proteins. For example, intermediate filaments that make up your hair consist of keratin, a fibrous protein. Intermediate filaments of lamins (another fibrous protein) support the nuclear envelope, and also help regulate processes inside the nucleus such as DNA replication.

Motor proteins that associate with cytoskeletal elements move cell parts when energized by a phosphate-group transfer from ATP. A cell is like a bustling train station, with molecules and structures being moved continuously throughout its interior. Motor proteins are a bit like freight trains, dragging cellular cargo along tracks of microtubules and microfilaments (Figure 3.16). One motor protein, myosin, brings about muscle cell contraction by interacting with microfilaments. Dynein, another motor protein, interacts with microtubules to bring about movement of eukaryotic flagella and cilia. Eukaryotic flagella propel sperm ⟶ and other motile cells through fluid by whipping back and forth, a motion that differs from the propeller-like rotation of prokaryotic flagella. **Cilia** (singular, cilium) are short, hairlike structures that project from the surface of some eukaryotic cells. The coordinated waving of many cilia can propel a cell through fluid, and stir fluid around a stationary cell. Cilia on thousands of cells lining your airways sweep inhaled particles away from your lungs.

Some eukaryotic cells, including the amoeba at left, form **pseudopods**, or "false feet." As these temporary, irregular lobes bulge outward, they can move the entire cell or engulf a target such as prey. Elongating microfilaments force the lobe to advance in a steady direction. Motor proteins attached to the microfilaments drag the plasma membrane along with them.

tubulin
— subunit

actin
— subunit

|← 25 nm →|

|← 6–7 nm →|

A. Microtubule.   **B.** Microfilament.

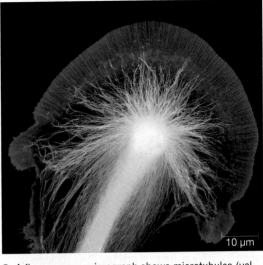

10 µm

C. A fluorescence micrograph shows microtubules (yellow) and microfilaments (blue) in the growing end of a nerve cell. These cytoskeletal elements support and guide the cell's lengthening in a particular direction.

Figure 3.15 Cytoskeletal elements.

(C) © Dylan T. Burnette and Paul Forscher.

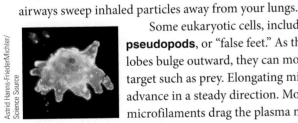

Astrid Hanns-FriederMichler/
Science Source

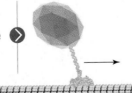

Figure 3.16 Motor proteins.
Here, kinesin (tan) drags a pink vesicle as it inches along a microtubule.

chloroplast Organelle of photosynthesis in the cells of plants and photosynthetic protists.

cilia Short, movable structures that project from the plasma membrane of some eukaryotic cells.

cytoskeleton Network of protein filaments that support, organize, and move eukaryotic cells and their internal structures.

intermediate filament Stable cytoskeletal element that structurally supports cells and tissues.

microfilament Reinforcing cytoskeletal element that functions in cell movement; a fiber of actin subunits.

microtubule Cytoskeletal element involved in movement; hollow filament of tubulin subunits.

motor protein Type of energy-using protein that interacts with cytoskeletal elements to move the cell's parts or the whole cell.

pseudopod A temporary protrusion that helps some eukaryotic cells move and engulf prey.

Figure 3.17 A plant ECM.
This section through a plant leaf shows the cuticle, a protective covering of deposits secreted by living cells.

George S. Ellmore.

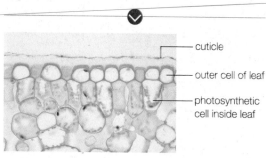

— cuticle

— outer cell of leaf

— photosynthetic cell inside leaf

Figure 3.18 Three types of cell junctions common in animal tissues: tight junctions, gap junctions, and adhering junctions.
The micrograph shows how a profusion of tight junctions (green) seals abutting surfaces of kidney cell membranes to form a waterproof tissue. The DNA in each cell nucleus appears red.

© ADVANCELL (Advanced In Vitro Cell Technologies; S.L.) www.advancell.com.

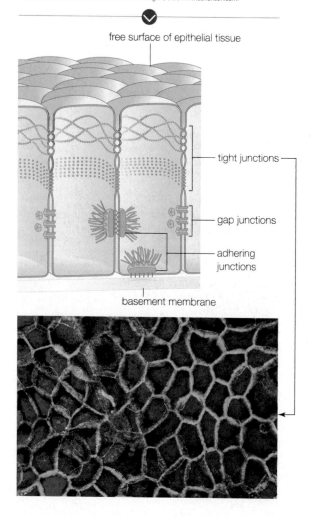

free surface of epithelial tissue

tight junctions

gap junctions

adhering junctions

basement membrane

Extracellular Matrix Many cells secrete an **extracellular matrix** (**ECM**), a complex mixture of molecules that varies by cell type, but often includes polysaccharides and fibrous proteins. A cell wall is an example of ECM. You learned in Section 3.4 that many prokaryotes have walls. Plants have them too, as do fungi and some protists. Wall composition differs among these groups. Animal cells have no walls, but some types secrete an extracellular matrix called basement membrane. Despite the name, basement membrane is not a cell membrane because it does not consist of lipids. Rather, it is a sheet of fibrous material that structurally supports and organizes tissues. A **cuticle** is a type of ECM secreted by cells at a body surface. In plants, a cuticle of waxes and other hydrophobic compounds helps stems and leaves fend off insects and retain water (Figure 3.17). Crabs, spiders, and other arthropods have a cuticle that consists mainly of chitin, a tough polysaccharide.

Cell Junctions In multicelled species, cells can interact with one another and their surroundings by way of cell junctions. **Cell junctions** are structures that connect a cell directly to other cells or to its environment. Cells send and receive substances and signals through some junctions. Other junctions help cells recognize and stick to each other and to ECM.

Three types of cell junctions are common in animal tissues (Figure 3.18). In tissues that line body surfaces and internal cavities, rows of tight junctions fasten the plasma membranes of adjacent cells and prevent body fluids from seeping between them. For example, the lining of the stomach is leakproof because tight junctions seal its cells together. Adhering junctions consist of adhesion proteins, and they make a tissue quite strong by connecting cytoskeletal elements of adjacent cells, and cytoskeletal elements to basement membrane. Contractile tissues (such as heart muscle) have a lot of adhering junctions, as do tissues subject to abrasion or stretching (such as skin). Gap junctions are closable channels that connect the cytoplasm of adjoining animal cells. When open, they permit water, ions, and small molecules to pass directly from the cytoplasm of one cell to another. These channels allow entire regions of cells to respond to a single stimulus. Heart muscle and other tissues in which the cells perform a coordinated action have many gap junctions.

In plants, plasmodesmata (singular, plasmodesma) are open channels that connect the cytoplasm of adjoining cells. These cell junctions extend across the cell walls, and, like gap junctions, they allow substances to flow quickly from cell to cell.

Take-Home Message 3.5

What structures are common in eukaryotic cells?

- A typical eukaryotic cell has many membrane-enclosed organelles.
- A nucleus protects and controls access to the cell's DNA. Organelles of the endomembrane system make, modify, and transport proteins and lipids. Mitochondria produce ATP; chloroplasts carry out photosynthesis.
- An extensive cytoskeleton of protein filaments reinforces a eukaryotic cell's shape, and is the basis of movement of the cell and its parts.
- Many cells secrete an extracellular matrix (ECM) such as a cell wall. Plant cells, fungi, and some protists are walled, but not animal cells.
- Cell junctions structurally and functionally connect cells in tissues. In animal tissues, cell junctions also connect cells with basement membrane.

3.6 The Nature of Life

Carbon, hydrogen, oxygen, and other atoms of organic molecules are the stuff of you, and us, and all of life. Yet it takes more than organic molecules to complete the picture. Life continues only as long as an ongoing flow of energy sustains its organization, because assembling molecules and cells requires energy. Life is no more and no less than a marvelously complex system for prolonging order. With energy and the hereditary codes of DNA, matter becomes organized, generation after generation.

In this chapter, you learned about the structure of cells, which have at minimum a plasma membrane, cytoplasm, and DNA. We often use differences in other cellular components—the presence or absence of a particular organelle, for example—to categorize life's diversity. What about life's commonality? The cell is the smallest unit with the properties of life, but what is it, exactly, that makes a cell, or an organism that consists of them, alive? According to evolutionary biologist Gerald Joyce, the simplest definition of life might well be "that which is squishy." He says, "Life, after all, is protoplasmic and cellular. It is made up of cells and organic stuff and is undeniably squishy."

Defining life more unambiguously than "squishy" is challenging, if not impossible. Even deciding what sets the living apart from the nonliving can be tricky. For example, living things have a high proportion of the organic molecules of life, but so do the remains of dead organisms in seams of coal. Living things use energy to reproduce themselves, but computer viruses, which are arguably not alive, can do that too.

So how do biologists, who study life as a profession, define it? The short answer is that their best definition is a long list of properties that collectively apply to living things, and not to nonliving things. You already know about two of these properties:

1. Living things make and use the organic molecules of life . . .

2. . . . and they consist of one or more cells.

The remainder of this book details the other properties of life:

3. Living things engage in self-sustaining biological processes such as metabolism and homeostasis . . .

4. . . . and they change over their lifetime, for example by maturing and aging . . .

5. . . . and they use DNA as their hereditary material when they reproduce . . .

6. . . . and they have the collective capacity to change over successive generations, for example by adapting to environmental pressures.

Kevin Schafer/Getty Images.

"Life . . . is made up of cells and organic stuff and is undeniably squishy."

Take-Home Message 3.6

What, exactly, is life?

- We describe the characteristic of "life" in terms of properties that are collectively unique to living things.
- Organisms make and use the organic molecules of life. DNA is their hereditary material.
- In living things, the molecules of life are organized as one or more cells that engage in self-sustaining biological processes.
- Living things change over lifetimes, and also over successive generations.

cell junction Structure that connects a cell to another cell or to extracellular matrix.

cuticle Secreted covering at a body surface.

extracellular matrix (ECM) Complex mixture of cell secretions, the composition and function of which vary by cell type.

Summary

Section 3.1 A huge number of bacteria live in and on the human body. Most of them are helpful; only a few types can cause disease. Contamination of food with disease-causing bacteria can result in illness that is sometimes fatal.

Section 3.2 By the **cell theory**, all organisms consist of one or more cells; the cell is the smallest unit of life; each new cell arises from another, preexisting cell; and a cell passes hereditary material to its offspring. All cells start out life with **cytoplasm**, DNA, and a **plasma membrane** that controls the types and kinds of substances that cross it. Most have many additional components. A eukaryotic cell's DNA is contained within a **nucleus**, which is a membrane-enclosed **organelle**. The **surface-to-volume ratio** limits cell size and influences cell shape. Different types of microscopes and techniques reveal different internal and external details of cells.

Section 3.3 A cell membrane is a lipid bilayer (of mainly phospholipids) with many other molecules attached or embedded in it. A bacterial or eukaryotic cell membrane can be described as a **fluid mosaic**; archaeal membranes are not fluid. Proteins carry out most membrane functions. All cell membranes have enzymes, and all have **transport proteins** that help substances move across the membrane. Plasma membranes also incorporate **adhesion proteins** that lock cells together in tissues. Plasma membranes and some internal membranes have **receptor proteins** that trigger a change in cell activities in response to a stimulus.

Section 3.4 Bacteria and archaea (prokaryotes) are single-celled organisms with no nucleus. All have DNA and **ribosomes**. Many also have a protective, rigid **cell wall** and a sticky capsule, and some have motile structures (**flagella**) and other projections (**pili**). They often have plasmids in addition to the single circular molecule of DNA. Bacteria and other microorganisms may live together in a shared mass of slime as a **biofilm**.

Section 3.5 The nucleus in cells of eukaryotes (protists, fungi, plants, and animals) protects and controls access to the cell's DNA. The nucleus has a **nuclear envelope**, a double membrane studded with pores through which molecules pass into and out of the nucleus. A typical eukaryotic cell has many other membrane-enclosed organelles, including the **endoplasmic reticulum** (**ER**), a continuous system of sacs and tubes extending from the nuclear envelope. Polypeptides made in ribosome-studded rough ER pass to smooth ER, which makes lipids and breaks down carbohydrates

and fatty acids. **Golgi bodies** modify proteins and lipids before sorting them into **vesicles**. Enzymes in **peroxisomes** break down substances such as amino acids, fatty acids, and toxins. Enzymes in **lysosomes** break down cellular wastes and debris. **Mitochondria** produce ATP by aerobic respiration; **chloroplasts** carry out photosynthesis.

Elements of a **cytoskeleton** reinforce, organize, and move cell structures and often the entire cell. Cytoskeletal elements include microtubules, microfilaments, and intermediate filaments. Interactions between ATP-driven **motor proteins** and **microtubules** bring about movement of **cilia** and eukaryotic flagella. Elongating **microfilaments** bring about movement of **pseudopods**. **Intermediate filaments** lend structural support to cells and tissues, and they help support the nuclear membrane.

A secreted mixture of materials forms **extracellular matrix** (**ECM**) that has different functions depending on the cell type. In animals, a secreted basement membrane supports and organizes cells in tissues. Among the eukaryotes, plant cells, fungi, and many protists secrete a cell wall around their plasma membrane. Many eukaryotic cell types also secrete a protective **cuticle**.

Cell junctions connect cells to one another and to their environment. Plasmodesmata (in plants) and gap junctions (in animals) connect the cytoplasm of adjacent cells. Also in animal cells, adhering junctions that connect to cytoskeletal elements fasten cells to one another and to basement membrane. Tight junctions form a waterproof seal between cells.

Section 3.6 All living things share a characteristic set of features. They make and use the molecules of life; they consist of one or more cells that engage in self-sustaining biological processes; they change over their lifetime; and they pass their DNA to offspring that can change over generations.

Self-Quiz

Answers in Appendix I

1. All cells have these three things in common:
   a. cytoplasm, DNA, and organelles with membranes
   b. a plasma membrane, DNA, and a nuclear envelope
   c. cytoplasm, DNA, and a plasma membrane
   d. a cell wall, cytoplasm, and DNA

2. Unlike eukaryotic cells, prokaryotic cells _____ .
   a. have no plasma membrane   c. have no nucleus
   b. have RNA but not DNA   d. a and c

3. Every cell is descended from another cell. This idea is part of _____ .
   a. evolution
   b. the theory of heredity
   c. the cell theory
   d. cell biology

4. The surface-to-volume ratio _____ .
   a. does not apply to prokaryotic cells
   b. is part of the cell theory
   c. constrains cell size
   d. b and c

5. True or false? Some protists start out life with no nucleus.

6. Cell membranes consist mainly of _____ and _____ .
   a. lipids; carbohydrates
   b. phospholipids; proteins
   c. lipids; carbohydrates
   d. phospholipids; ECM

7. Which of the following statements is correct?
   a. Ribosomes are only found in bacteria and archaea.
   b. Some animal cells are prokaryotic.
   c. Only eukaryotic cells have mitochondria.
   d. The plasma membrane is the outermost boundary of all cells.
   e. Most membrane functions are carried out by phospholipids.

8. In a lipid bilayer, the _____ of all the lipid molecules are sandwiched between all of the _____ .
   a. hydrophilic tails; hydrophobic heads
   b. hydrophilic heads; hydrophilic tails
   c. hydrophobic tails; hydrophilic heads
   d. hydrophobic heads; hydrophilic tails

9. The main function of the endomembrane system is _____ .
   a. building and modifying proteins and lipids
   b. isolating DNA from toxic substances
   c. secreting extracellular matrix onto the cell surface
   d. producing ATP by aerobic respiration

10. Enzymes contained in _____ break down worn-out organelles, bacteria, and other particles.
    a. lysosomes
    b. mitochondria
    c. endoplasmic reticulum
    d. peroxisomes

11. Put the following structures in order according to the pathway of a secreted protein:
    a. plasma membrane
    b. Golgi bodies
    c. endoplasmic reticulum
    d. post-Golgi vesicles

12. No animal cell has a _____ .
    a. plasma membrane
    b. flagellum
    c. lysosome
    d. cell wall

13. _____ connect the cytoplasm of plant cells.
    a. Plasmodesmata
    b. Adhering junctions
    c. Tight junctions
    d. Adhesion proteins

14. Which of the following organelles contains no DNA? Choose all that are correct.
    a. nucleus
    b. Golgi body
    c. mitochondrion
    d. chloroplast

15. Match each cell component with its main function.
    _____ mitochondrion   a. connection
    _____ chloroplast   b. protective covering
    _____ ribosome   c. ATP production
    _____ nucleus   d. protects DNA
    _____ cell junction   e. protein synthesis
    _____ flagellum   f. photosynthesis
    _____ cuticle   g. movement

Critical Thinking

1. In a classic episode of *Star Trek*, a gigantic amoeba engulfs an entire starship. Spock blows the cell to bits before it has a chance to reproduce. Think of at least one problem a biologist would have with this particular scenario.

2. In plants, the cell wall forms as a young plant cell secretes polysaccharides onto the outer surface of its plasma membrane. Being thin and pliable, this primary wall allows the cell to enlarge and change shape. In mature woody plants, cells in some tissues deposit material onto the primary wall's inner surface. Why doesn't this secondary wall form on the outer surface of the primary wall?

Visual Question

1. What type of micrograph is shown below? Is the organism pictured prokaryotic or eukaryotic? How can you tell?

P. L. Walne and J. H. Arnott, Planta, 77:325–354, 1967.

4

ENERGY AND METABOLISM

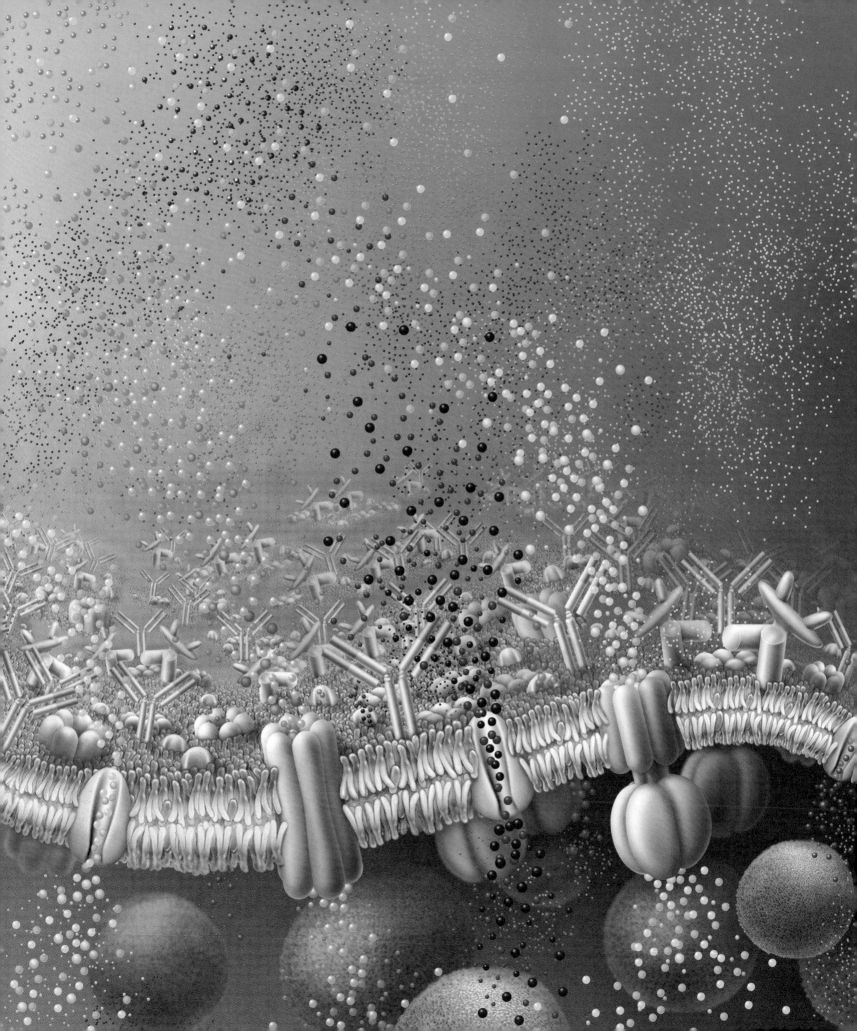

Application ❯ 4.1 A Toast to Alcohol Dehydrogenase

Most college students are under the legal drinking age, but alcohol abuse continues to be the most serious drug problem on college campuses throughout the United States. Recent surveys polled tens of thousands of undergraduates about their drinking habits, and more than half of them reported regularly consuming five or more alcoholic beverages within a two-hour period—a self-destructive behavior called binge drinking. Drinking large amounts of alcohol in a brief period of time is an extremely risky behavior, both for the drinkers and people around them. Every year, around 600,000 students injure themselves while under the influence of alcohol; intoxicated students physically assault 690,000 people and sexually assault 97,000 others. Binge drinking is responsible for killing or causing the death of 1,825 students per year.

Before you drink, consider what you are consuming. All alcoholic drinks—beer, wine, hard liquor—contain the same psychoactive ingredient: ethanol. Almost all ingested ethanol ends up in the liver, a large organ in the abdomen with many important functions. Liver cells make an enzyme, alcohol dehydrogenase (ADH), which is part of a metabolic pathway that detoxifies alcohols (Figure 4.1). This pathway evolved long before humans began to consume alcoholic beverages. Its main function in our bodies is to break down the tiny amount of alcohol that we encounter naturally: Foods such as ripe fruit contain it, and it also forms in our bodies as a metabolic by-product of body cells and gut bacteria.

ADH converts ethanol to acetaldehyde, an organic compound even more toxic than ethanol and the most likely source of various hangover symptoms. A second enzyme converts the toxic acetaldehyde to acetate, which is a nontoxic salt. In the average healthy adult human, these two enzymes can detoxify between 7 and 14 grams of ethanol per hour. The average alcoholic beverage contains between 10 and 20 grams of ethanol, which is why having more than one drink in any two-hour interval may result in a hangover.

Putting more alcohol into your body than your enzymes can detoxify damages it more permanently than a hangover, however. Ethanol breakdown harms liver cells, so the more a person drinks, the fewer liver cells are left to do the breaking down. Ethanol also interferes with normal processes of metabolism. For example, oxygen that would ordinarily take part in breaking down fatty acids is diverted to breaking down the ethanol. As a result, fats tend to accumulate as large globules in the tissues of heavy drinkers. Long-term heavy drinking causes alcoholic hepatitis,

Figure 4.1 Alcohol dehydrogenase.
Left, this enzyme helps the body break down toxic molecules such as ethanol, making it possible for humans to drink beer, wine, and other alcoholic beverages.

Right, a tailgate party at a Notre Dame–Alabama football game. Indiana State police arrested 138 Notre Dame students for underage drinking at tailgate parties during 2012.

© Al Diaz/Miami Herald/MCT via Getty Images.

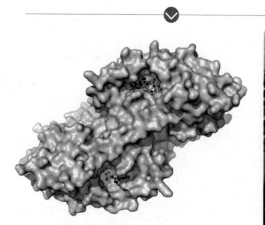

alcohol dehydrogenase

a disease characterized by inflammation and destruction of liver tissue. It also causes cirrhosis, a condition in which the liver becomes so scarred, hardened, and filled with fat that it loses its function. (The term cirrhosis is from the Greek *kirros*, meaning "orange-colored," after the abnormal skin color of people with the disease.) A cirrhotic liver can no longer produce the protein albumin, so the solute balance of body fluids is disrupted, and the legs and abdomen swell with watery fluid. It can no longer remove drugs and other toxins from the blood, so they accumulate in the brain—which impairs mental functioning and alters personality. Restricted blood flow through a cirrhotic liver causes veins to enlarge and rupture, so internal bleeding is a risk. The damage to the body results in a heightened susceptibility to diabetes and liver cancer. Once cirrhosis has been diagnosed, a person has about a 50 percent chance of dying within 10 years (Figure 4.2).

4.2 Life Runs on Energy

REMEMBER: A one-way flow of energy and a cycling of nutrients sustain life's organization (Section 1.3). A law of nature describes something that occurs without fail, but our explanation of why it occurs is incomplete (1.6).

Energy is formally defined as the capacity to do work, but this definition is not very satisfying. Even the brilliant physicists who study it cannot say exactly what it is. However, we do have an intuitive understanding of energy just by thinking about familiar forms of it, such as light, heat, electricity, and motion. We also understand intuitively that one form of energy can be converted to another. Think about how an automobile changes the chemical energy of gasoline into the energy of motion (kinetic energy), or how a lightbulb changes electricity into light.

The formal study of heat and other forms of energy is called thermodynamics (*therm* means heat; *dynam* means power). By making careful measurements, thermodynamics researchers discovered that the total amount of energy before and after every conversion is always the same. In other words, energy cannot be created or destroyed—a phenomenon that is the **first law of thermodynamics**. Energy also tends to spread out, or disperse, until no part of a system holds more than another part. In a kitchen, for example, heat always flows from a hot pan to cool air until the temperature of both is the same. We never see cool air raising the temperature of a hot pan. The tendency of energy to spread out spontaneously is the **second law of thermodynamics**.

Work occurs as a result of energy transfers. Consider how it takes work to push a box across a floor. In this case, a body (you) transfers energy to another body (the box) to make it move. Similarly, a plant cell works to make sugars. Inside the cell, one set of molecules harvests energy from light, then transfers it to another set of molecules. A second set of molecules uses the energy to build sugars from carbon dioxide and water. This particular energy transfer involves the conversion of light energy to chemical energy. Most other types of cellular work occur by the transfer of chemical energy from one molecule to another.

As you learn about such processes, remember that every time energy is transferred, a bit of it disperses—usually in the form of heat. As a simple example, a typical incandescent lightbulb converts only about 5 percent of the energy of electricity into light. The remaining 95 percent of the energy ends up as heat that disperses from the bulb.

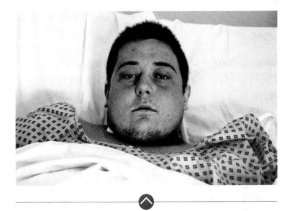

Figure 4.2 Gary Reinbach.
The 22-year-old died from alcoholic liver disease shortly after this photograph was taken, in 2009. The odd color of his skin is a symptom of cirrhosis.

Transplantation is a last-resort treatment for a failed liver, but there are not enough liver donors for everyone who needs a transplant. Reinbach was refused a transplant that may have saved his life because he had not abstained from drinking for the prior 6 months.

Stuart Clark/The Sunday Times/nisyndication.

© Piotr Marcinski/Shutterstock.com.

energy The capacity to do work.

first law of thermodynamics Energy cannot be created or destroyed.

second law of thermodynamics Energy tends to disperse spontaneously.

A. Energy In
Sunlight reaches environments on Earth. Producers in those environments capture some of its energy and convert it to other forms that can drive cellular work.

PRODUCERS

B. Some of the energy captured by producers ends up in the tissues of consumers.

CONSUMERS

C. Energy Out
With each energy transfer, some energy escapes into the environment, mainly as heat. Living things do not use heat to drive cellular work, so energy flows through the world of life in one direction overall.

Figure 4.3 Energy flow through the world of life.

Figure 4.4 Illustration of potential energy.
By opposing gravity's downward pull, the rope keeps the man from falling. Similarly, a chemical bond that attaches two atoms keeps them from flying apart.

© Greg Epperson/Shutterstock.com.

Dispersed heat is not useful for doing work, and it is not easily converted to a more useful form of energy (such as electricity). Because some energy in every transfer disperses as heat, and heat is not useful for doing work, we can say that the total amount of energy in the universe available for doing work is always decreasing.

Is life an exception to this inevitable flow? An organized body is hardly dispersed. Energy becomes concentrated in each new organism as the molecules of life organize into cells. Even so, living things constantly use energy—to grow, to move, to acquire nutrients, to reproduce, and so on—and some energy is lost in every one of these processes (Figure 4.3). Unless the losses are replenished with energy from another source, life's complex organization will end.

The energy that fuels most life on Earth comes from the sun. That energy flows through producers such as plants, then consumers such as animals (Figure 4.3). During this journey, the energy is transferred many times. With each transfer, some energy escapes as heat until, eventually, all of it is permanently dispersed. However, the second law of thermodynamics does not say how quickly the dispersal has to happen. Energy's spontaneous dispersal is resisted by chemical bonds. The energy in chemical bonds is a type of potential energy, which is energy stored in the position or arrangement of objects in a system (Figure 4.4). Think of the bonds in the countless molecules that make up your skin, heart, liver, fluids, and other body parts. Those bonds hold the molecules, and you, together—at least for the time being.

Take-Home Message 4.2

What is energy?

- Energy, which is the capacity to do work, cannot be created or destroyed.
- Energy disperses spontaneously.
- Energy can be transferred between systems or converted from one form to another, but some is lost (as heat, typically) during every exchange.
- Sustaining life's organization requires ongoing energy inputs to counter energy loss. Organisms stay alive by replenishing themselves with energy they harvest from someplace else.

4.3 Energy in the Molecules of Life

REMEMBER: A chemical formula (Section 2.3) indicates unvarying proportions of elements in a molecule. Reactions are processes of molecular change (2.6). Cellulose consists of chains of glucose monomers (2.7).

All cells store and retrieve energy in chemical bonds of the molecules of life, and these activities occur by way of reactions. During a reaction, one or more **reactants** (molecules that enter the reaction and become changed by it) become one or more **products** (molecules that are produced by the reaction). Intermediate molecules may form between reactants and products. We show a reaction as an equation in which an arrow points from reactants to products:

$$2H_2 \; + \; O_2 \longrightarrow 2H_2O$$
$$\text{(hydrogen)} \quad \text{(oxygen)} \quad \text{(water)}$$

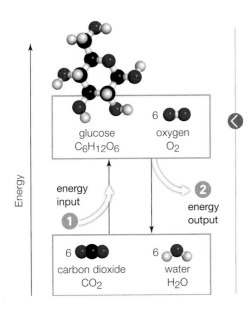

Figure 4.5 Energy inputs and outputs in chemical reactions.

1 Some reactions convert molecules with lower energy to molecules with higher energy, so they require a net energy input in order to proceed.

2 Other reactions convert molecules with higher energy to molecules with lower energy, so they end with an energy release.

Figure It Out: Which law of thermo-dynamics explains energy inputs and outputs in chemical reactions?

Answer: The first law

A. Wood continues to burn after it has been lit because the combustion reaction between cellulose molecules in wood and oxygen molecules in air releases enough energy to trigger the reaction again with other molecules. Activation energy keeps this and other energy-releasing reactions from starting without an energy input.

A number before a chemical formula in such equations indicates the number of molecules; a subscript indicates the number of atoms of that element per molecule. Note that atoms shuffle around in a reaction, but they never disappear: The same number of atoms that enter a reaction remain at the reaction's end.

Every chemical bond holds a certain amount of energy. That is the amount of energy required to break the bond, and it is also the amount of energy released when the bond forms. The particular amount of energy held by a bond depends on which elements are taking part in it. For example, two covalent bonds—one between an oxygen and a hydrogen atom in a water molecule, the other between two oxygen atoms in molecular oxygen (O_2)—both hold energy, but different amounts of it. In most reactions, the energy of the reactants differs from the energy of the products. If the reactants have less energy than the products, the reaction will not proceed without a net energy input (Figure 4.5 **1**). If the reactants have more energy than the products, the reaction will end with a net release of energy **2**.

Why Earth Does Not Go Up in Flames The molecules of life release energy when they combine with oxygen. Think of how a spark ignites wood in a camp-fire. Wood is mostly cellulose, which consists of long chains of repeating glucose monomers. A spark starts a reaction that converts cellulose (in wood) and oxygen (in air) to water and carbon dioxide. This reaction releases a lot of energy—enough to initiate the same reaction with other cellulose and oxygen molecules. That is why wood keeps burning after it has been lit (Figure 4.6A).

Earth is rich in oxygen—and in potential energy-releasing reactions. Why doesn't it burst into flames? Luckily, chemical bonds do not break without at least a small input of energy, even in an energy-releasing reaction. We call this input activation energy. **Activation energy**, the minimum amount of energy required to get a chemical reaction started, is a bit like a hill that reactants must climb before they can coast down the other side to become products (Figure 4.6B).

Both energy-requiring and energy-releasing reactions have activation energy, but the amount varies with the reaction. Consider guncotton (nitrocellulose), a highly explosive derivative of cellulose. Christian Schönbein accidentally discovered a way to manufacture it when he used his wife's cotton apron to wipe up a nitric acid spill on his kitchen table, then hung it up to dry next to the oven. The apron exploded, and being a chemist in the 1800s, Schönbein was thrilled. He immediately

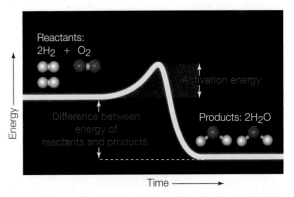

B. Most reactions will not begin without an input of acti-vation energy, which is shown as a bump in an energy hill. The graph shows an energy-releasing reaction; energy-requiring reactions also have activation energy.

Figure 4.6 Activation energy.

(A) Tero Hakala/Shutterstock.com.

activation energy Minimum amount of energy required to start a reaction.

product A molecule that is produced by a reaction.

reactant A molecule that enters a reaction and is changed by participating in it.

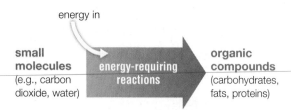

A. Cells store energy in the chemical bonds of organic compounds.

B. Cells retrieve energy stored in the chemical bonds of organic compounds.

Figure 4.7 **Cells store and retrieve energy in the chemical bonds of organic molecules.**

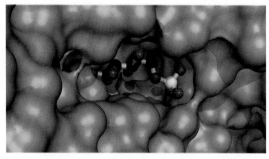

A. A glucose molecule meets up with a phosphate in the active site of a hexokinase enzyme.

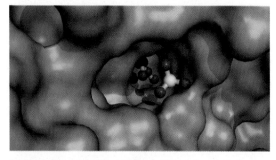

B. The reaction between glucose and phosphate produced glucose-6-phosphate, shown leaving the active site.

Figure 4.8 **Example of an active site.**
For simplicity, enzymes are often drawn as blobs or geometric shapes. This model shows the actual contours of an active site in hexokinase, an enzyme that adds a phosphate group to glucose and other six-carbon sugars.

PDB ID: 1GZX; Paoli, M., Liddington, R., Tame, J., Wilinson, A., Dodson, G.; Crystal Structure of T state hemoglobin with oxygen bound at all four haems. *J. Mol.Bio.*, v256, pp. 775–792, 1996.

tried marketing guncotton as a firearm explosive, but it was too unstable to manufacture. So little activation energy is needed to make guncotton react with oxygen that it tends to explode unexpectedly. Several manufacturing plants burned to the ground before guncotton was abandoned for use as a firearm explosive. The substitute: gunpowder, which has a higher activation energy for a reaction with oxygen.

Energy In, Energy Out Cells store energy by running energy-requiring reactions that build organic compounds (Figure 4.7A). For example, light energy drives photosynthesis, a pathway that produces glucose from carbon dioxide and water. Unlike light, glucose can be stored in a cell. Cells harvest energy by running energy-releasing reactions that break the bonds of organic compounds (Figure 4.7B). Most cells do this when they carry out aerobic respiration, a pathway that releases the energy of glucose by breaking the bonds between its carbon atoms. You will see in the next section how cells use energy released from some reactions to drive others (we return to the reactions of photosynthesis and aerobic respiration in Chapter 5).

Take-Home Message 4.3

How do cells use energy?

- Cells store and retrieve energy by making and breaking chemical bonds.
- Some reactions require a net input of energy. Others end with a net release of energy.
- Most chemical reactions require an input of activation energy to begin.

4.4 How Enzymes Work

REMEMBER: Electrons carry energy in incremental amounts (Section 2.2). The number of hydrogen ions in a fluid determines its pH (2.5). Molecules jiggle faster at higher temperatures (2.4). Enzymes speed reactions without being changed by them; metabolism includes all the enzyme-mediated reactions by which cells acquire and use energy as they build and break down organic molecules (2.6). A protein's function arises from and depends on its three-dimensional shape, which is held together by hydrogen bonds; when a protein denatures, it loses its shape and its function (2.9). ATP has an important metabolic role as an energy carrier (2.10).

The Need for Speed Metabolism requires enzymes. Why? Consider that a molecule of glucose can break down to carbon dioxide and water on its own, but the process might take decades. That same conversion takes just seconds inside your cells. Enzymes make the difference. An enzyme makes a reaction run much faster than it would on its own. The enzyme is unchanged by participating in the reaction, so it can work again and again.

Some enzymes are RNAs, but most are proteins. Each kind of enzyme interacts only with specific reactants, or **substrates**, and alters them in a specific way. Such specificity occurs because an enzyme's polypeptide (or nucleotide) chains fold to form a pocket called an **active site**, where substrates bind and a reaction proceeds (Figure 4.8). An active site is complementary in shape, size, polarity, and charge to the enzyme's substrate. That fit is the reason why each enzyme acts in a specific way on a specific substrate (Figure 4.9).

Figure 4.9 How an active site works.

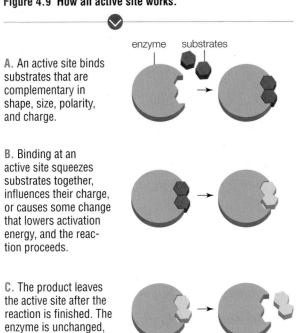

A. An active site binds substrates that are complementary in shape, size, polarity, and charge.

B. Binding at an active site squeezes substrates together, influences their charge, or causes some change that lowers activation energy, and the reaction proceeds.

C. The product leaves the active site after the reaction is finished. The enzyme is unchanged, so it can work again.

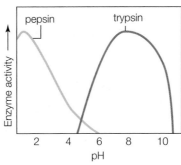

A. The pH-dependent activity of two digestive enzymes. Pepsin acts in the stomach, where the normal pH is 2. Trypsin acts in the small intestine, where the pH is normally around 7.5.

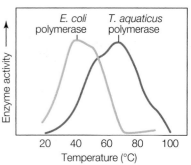

B. Temperature-dependent activity of an enzyme from two bacteria: *E. coli*, which inhabits the human gut (normally 37°C); and *Thermus aquaticus*, which lives in hot springs around 70°C.

Figure 4.10 Enzymes, temperature, and pH.
Each enzyme works best within a characteristic range of conditions—generally, the same environmental conditions in which the enzyme normally occurs.

Figure It Out: At what temperature does the *E. coli* DNA polymerase work fastest?

Answer: About 37°C

An enzyme speeds a reaction by reducing activation energy, so it lowers the barrier that prevents the reaction from proceeding. When we talk about activation energy, we are really talking about the energy required to bring reactant bonds to the breaking point. An active site can bring reactants to this state by (for example) holding them in a certain position, squeezing them, or redistributing their charge.

Factors That Influence Enzyme Activity Environmental factors such as pH, temperature, salt, and pressure influence an enzyme's shape, and so influence its function. Each enzyme works best in a particular range of conditions that reflect the environment in which it evolved.

Consider pepsin, a digestive enzyme that works best at low pH (Figure 4.10A). Pepsin begins the process of protein digestion in the very acidic environment of the stomach (pH 2). During digestion, the stomach's contents pass into the small intestine, where the pH rises to about 7.5. Pepsin denatures (unfolds) above pH 5.5, so this enzyme becomes inactive in the small intestine. Here, protein digestion continues with the assistance of trypsin, an enzyme that functions well at the higher pH.

Adding heat boosts energy, which is why the jiggling motion of atoms and molecules increases with temperature. The greater the energy of reactants, the closer they are to reaching activation energy. Thus, the rate of an enzymatic reaction typically increases with temperature—but only up to a point. An enzyme denatures above a characteristic temperature. Then, the reaction rate falls sharply as the shape of the enzyme changes and it stops working (Figure 4.10B). Body temperatures above 42°C (107.6°F) adversely affect the function of many of your enzymes, which is why such severe fevers are dangerous.

The activity of many enzymes is also influenced by the amount of salt in the surrounding fluid. Too little salt, and polar parts of the enzyme attract one another so strongly that the molecule's shape changes. Too much salt interferes with the hydrogen bonds that hold the enzyme in its characteristic shape, so the enzyme denatures.

Cells use energy released from some reactions to drive others.

active site Pocket in an enzyme where substrates bind and a reaction occurs.

substrate Of an enzyme, a reactant that is specifically acted upon by the enzyme.

Digging Into Data

One Tough Bug

The genus *Ferroplasma* consists of a few species of acid-loving archaea. One species, *F. acidarmanus*, was discovered to be the main component of slime streamers (a type of biofilm) deep inside an abandoned California copper mine (Figure 4.11A).

F. acidarmanus cells use an ancient energy-harvesting pathway that combines oxygen with iron–sulfur compounds in minerals such as pyrite. This reaction dissolves the minerals, so groundwater that seeps into the mine ends up with extremely high concentrations of metal ions such as copper, zinc, cadmium, and arsenic. The reaction also produces sulfuric acid, which lowers the pH of the water around the cells to zero.

Despite living in an environment with a composition similar to hot battery acid, *F. acidarmanus* cells maintain their internal pH at a cozy 5.0. Thus, researchers investigating *Ferroplasma* metabolic enzymes were surprised to discover that most of the cells' enzymes function best at very low pH (Figure 4.11B).

1. What does the dashed line signify?
2. Of the four enzymes profiled in the graph, how many function optimally at a pH lower than 5? How many retain significant function at pH 5?
3. What is the optimal pH for *F. acidarmanus* carboxylesterase?

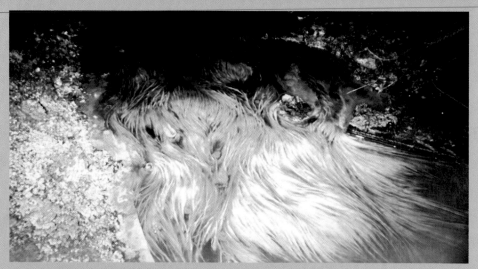

A. Deep inside one of the most toxic sites in the United States: Iron Mountain Mine, in California. The water in this stream, which is about 1 meter (3 feet) wide in this photo, is hot (around 40°C, or 104°F), heavily laden with arsenic and other toxic metals, and has a pH of zero. Slime streamers growing in it are a biofilm dominated by a species of archaea, *Ferroplasma acidarmanus*.

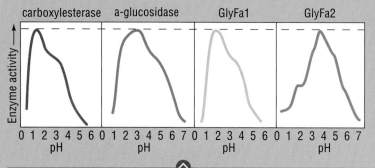

B. pH profiles of four enzymes isolated from *F. acidarmus*. Researchers had expected these enzymes to function best at the cells' cytoplasmic pH (5.0).

Figure 4.11 pH anomaly of *Ferroplasma acidarmanus* enzymes.

(A) Dr. Katrina J. Edwards; (B) From Golyshina et al., *Environmental Microbiology*, 8(3): 416–425. © 2006 John Wiley and Sons. Used with permission of the publisher.

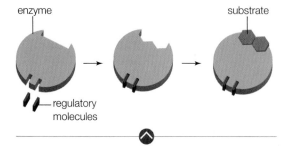

Figure 4.12 Regulatory molecule binding to enzymes. Some types of regulatory molecules (red) bind to an enzyme in a place other than the active site. This binding changes the shape of the enzyme in a way that enhances or inhibits its function.

Regulatory molecules (or ions) enhance or inhibit the activity of many enzymes. Some of these substances exert their effects by binding directly to an active site; others bind elsewhere on the enzyme. In the latter case, binding of the regulatory molecule alters the overall shape of the enzyme (Figure 4.12).

Cofactors Most enzymes (and many other proteins) can function properly only with assistance from metal ions or small organic molecules. These helpers are called **cofactors**. Many dietary vitamins and minerals are essential because they are cofactors or become converted into cofactors. **Coenzymes**, which are organic cofactors, carry chemical groups, atoms, or electrons from one reaction to another, and often into or out of organelles. In some reactions, coenzymes stay tightly bound to the enzyme. In others, they participate as separate molecules.

Unlike enzymes, many coenzymes are modified by taking part in a reaction. They are regenerated in separate reactions. Consider NAD$^+$ (nicotinamide adenine dinucleotide), a coenzyme derived from niacin (vitamin B$_3$). NAD$^+$ can accept electrons and hydrogen atoms, thereby becoming NADH. When electrons and hydrogen atoms are removed from NADH, NAD$^+$ forms again:

$$\text{NAD}^+ + \text{electrons} + \text{H}^+ \longrightarrow \boxed{\textbf{NADH}} \longrightarrow \text{NAD}^+ + \text{electrons} + \text{H}^+$$

In cells, the nucleotide ATP (adenosine triphosphate) functions as a coenzyme in many reactions. Bonds between phosphate groups hold a lot of energy, and ATP has two of these bonds holding its three phosphate groups together (Figure 4.13A). When a phosphate group is transferred to or from a nucleotide, energy is transferred along with it. Thus, the nucleotide can receive energy from an energy-releasing reaction, and it can also donate energy that contributes to the "energy in" part of an energy-requiring reaction. ATP is such an important currency in a cell's energy economy that we use a cartoon coin to symbolize it.

A reaction in which a phosphate group is transferred from one molecule to another is called a **phosphorylation**. ADP (adenosine diphosphate) forms when an enzyme transfers a phosphate group from ATP to another molecule (Figure 4.13B). Cells constantly run this reaction in order to drive a variety of energy-requiring reactions. Thus, they constantly have to replenish their stockpile of ATP—by running energy-releasing reactions that phosphorylate ADP. This cycle of using and replenishing ATP couples energy-requiring reactions with energy-releasing ones. As you will see in Chapter 5, cells harvest energy from organic compounds by running metabolic pathways that break them down. Energy that cells harvest in these pathways is not released to the environment, but rather stored in the high-energy phosphate bonds of ATP molecules and in electrons carried by coenzymes. The ATP and the coenzymes are then used to drive many of the energy-requiring reactions that a cell runs.

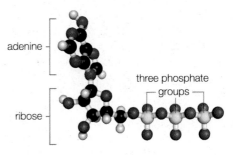

A. Bonds between phosphate groups hold a lot of energy. ATP has two of these bonds.

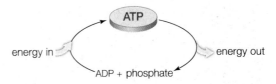

B. ADP forms in a reaction that removes a phosphate group from ATP. Energy released in this reaction drives other reactions that are the stuff of cellular work. ATP forms again in reactions that phosphorylate ADP.

Figure 4.13 ATP as an energy carrier.

Metabolic Pathways Building, rearranging, or breaking down an organic substance often occurs stepwise, in a series of enzymatic reactions called a **metabolic pathway**. Some metabolic pathways are linear, meaning that the reactions run straight from reactant to product:

$$\text{reactant} \xrightarrow{\text{enzyme 1}} \text{intermediate} \xrightarrow{\text{enzyme 2}} \text{intermediate} \xrightarrow{\text{enzyme 3}} \text{product}$$

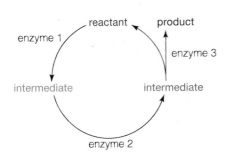

Other reactions are cyclic. In a cyclic pathway, the last step regenerates a reactant of the first step (left).

Controlling Metabolism Cells conserve energy and resources by making only what they need at any given moment—no more, no less. Several mechanisms help a cell maintain, raise, or lower its production of thousands of different substances. Consider that reactions do not only run from reactants to products. Many also run in reverse at the same time, with some of the products being converted back into reactants. The rates of the forward and reverse reactions often depend on the concentrations of reactants and products: A high concentration of reactants pushes the reaction in the forward direction; a high concentration of products pushes it in reverse.

coenzyme An organic cofactor.

cofactor A molecule or metal ion that associates with a protein and is necessary for its function.

metabolic pathway Series of enzyme-mediated reactions by which cells build, remodel, or break down an organic molecule.

phosphorylation A phosphate-group transfer.

A. Glucose reacts with oxygen (burns). Energy in the form of light and heat is released all at once as carbon dioxide and water form.

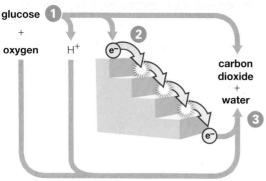

B. In cells, glucose reacts with oxygen in a stepwise fashion that involves an electron transfer chain, represented here by a staircase. Energy is released in amounts that cells are able to use.

① An input of activation energy splits glucose into carbon dioxide, electrons, and hydrogen ions (H⁺).

② Electrons lose energy ( ) as they pass from one molecule to the next in an electron transfer chain. That energy is harnessed for cellular work.

③ Electrons, hydrogen ions, and oxygen combine to form water.

Figure 4.15 Comparing uncontrolled and controlled energy release.
The overall reaction is the same in both cases.

(A) © 2004 Richard Megna–Fundamental Photographs.

Figure 4.14 Feedback inhibition.
In this example, three different enzymes act in sequence to convert a substrate to a product. The product inhibits the activity of the first enzyme.

Figure It Out: Is this a cyclic or a linear pathway?   Answer: This is a linear pathway.

Other mechanisms more actively regulate pathways. Regulatory substances govern how fast enzyme molecules are made, or influence the activity of enzymes that have already been built. Regulation of a single enzyme can affect an entire metabolic pathway. In some cases, the end product of a series of enzymatic reactions inhibits the activity of one of the enzymes in the series (Figure 4.14). This type of regulatory mechanism, in which a change that results from an activity decreases or stops the activity, is called **feedback inhibition**.

Electron Transfers The bonds of organic molecules hold a lot of energy that can be released in a reaction with oxygen. Burning is one type of reaction with oxygen, and it releases the energy of organic molecules all at once (Figure 4.15A). Cells use oxygen to break the bonds of organic molecules, but they have no way to harvest the sudden burst of energy that occurs during burning. Instead, they break the molecules apart in pathways that release the energy in small, manageable steps. Most of these steps are electron transfers, in which one molecule accepts electrons from another. Energy is harvested in these reactions. In the next chapter, you will learn about the importance of this process in electron transfer chains. An **electron transfer chain** is a series of membrane-bound enzymes and other molecules that give up and accept electrons in turn. Electrons are at a higher energy level when they enter a chain than when they leave. Energy given off by an electron as it drops to a lower energy level is harvested by molecules of the electron transfer chain to do cellular work (Figure 4.15B).

Take-Home Message 4.4

How do enzymes work in metabolic pathways?

• Binding at an enzyme's active site causes substrate bonds to reach their breaking point, and the reaction can run spontaneously to completion.
• Each enzyme works best within a characteristic range of environmental conditions. Many enzymes require the assistance of cofactors.
• ATP often couples reactions that release energy with reactions that require energy.
• A metabolic pathway is a series of enzyme-mediated reactions that builds, breaks down, or remodels an organic molecule. Some pathways involve electron transfer chains.
• Cells conserve energy and resources by producing only what they require at a given time. This metabolic control arises from mechanisms that regulate individual enzymes and often entire pathways.

4.5 **Diffusion and Membranes**

REMEMBER: An ion carries charge (Section 2.2). Atoms and molecules move faster at higher temperatures (2.4). The amount of a solute per unit volume of solution is its concentration (2.4). The number of hydrogen ions in a fluid determines its pH (2.5). Lipid bilayers are impermeable to ions and polar molecules (3.3).

gresei/Shutterstock.com.

Metabolic pathways require the participation of molecules and ions that must move across membranes and through cells. **Diffusion** is the spontaneous spreading of molecules or atoms (left), and it is an essential way in which substances move into, through, and out of cells. Diffusion occurs because an atom or molecule is always jiggling, and this internal movement causes it to randomly bounce off of nearby objects, including other atoms or molecules. Rebounds from such collisions propel solutes through a liquid, resulting in a gradual and complete mixing. How fast this occurs depends on five factors:

1. Concentration. A difference in solute concentration between adjacent regions of solution is called a concentration gradient. Solutes tend to diffuse "down" their concentration gradient, from a region of higher concentration to one of lower concentration. Why? Consider that moving objects (such as molecules) collide more often when they are more crowded. Thus, during a given interval, more molecules get bumped out of a region of higher concentration than get bumped into it.

2. Temperature. Atoms and molecules jiggle faster at higher temperature, so they collide more often. Thus, diffusion occurs more quickly at higher temperatures.

3. Size. It takes more energy to move a large object than it does to move a small one, so ions and small molecules diffuse more quickly than large molecules.

4. Charge. Each ion or charged molecule in a fluid contributes to the fluid's overall electric charge. A difference in charge between two regions of fluid can affect the rate and direction of diffusion between them. For example, positively charged substances (such as sodium ions) will tend to diffuse toward a region with an overall negative charge.

5. Pressure. Pressure squeezes objects—including atoms and molecules—closer together. Atoms and molecules that are more crowded collide and rebound more frequently. Thus, diffusion occurs faster at higher pressures.

Semipermeable Membranes Lipid bilayers are selectively permeable, which means that they are permeable to some substances and not others. Specifically, water, gases, and hydrophobic molecules can diffuse directly through a lipid bilayer; most solutes—ions and polar molecules, in particular—cannot (Figure 4.16). When a lipid bilayer separates two fluids with differing solute concentrations, water will diffuse across it. The direction of water movement depends on the relative solute concentration of the two fluids, which we describe in terms of tonicity. If the overall solute concentrations of the two fluids differ, the fluid with the lower concentration of solutes is said to be **hypotonic** (*hypo–*, under). The other one, with the higher solute concentration, is **hypertonic** (*hyper–*, over). Water diffuses from a hypotonic fluid into a hypertonic one. The diffusion will continue until the two fluids are **isotonic**, which means they have the same overall solute concentration. The movement of water across a membrane is so important in biology that it is given a special name: **osmosis** (Figure 4.17).

Figure 4.16 Selective permeability of lipid bilayers.
Hydrophobic molecules, gases, and water molecules can cross a lipid bilayer on their own. Ions in particular and most polar organic molecules such as glucose cannot.

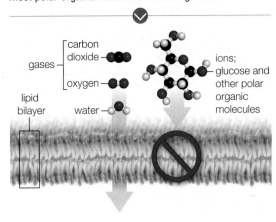

Figure 4.17 Osmosis.
Water moves across a selectively permeable membrane that separates two fluids of differing tonicity (red dots represent solutes). The fluid volume changes in the two compartments as water diffuses across the membrane.

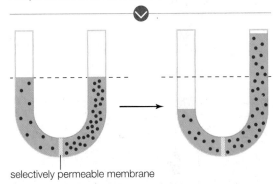

selectively permeable membrane

diffusion The spontaneous spreading of molecules or atoms.

electron transfer chain Series of enzymes and other molecules in a cell membrane that accept and give up electrons, thus releasing the energy of the electrons in steps.

feedback inhibition Mechanism by which a change that results from some activity decreases or stops the activity.

hypertonic Describes a fluid that has a high overall solute concentration relative to another fluid.

hypotonic Describes a fluid that has a low overall solute concentration relative to another fluid.

isotonic Describes two fluids with identical solute concentrations.

osmosis Diffusion of water across a selectively permeable membrane; occurs when there is a difference in solute concentration between the fluids on either side of the membrane.

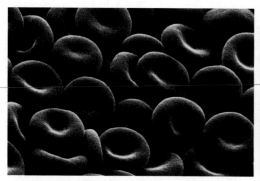
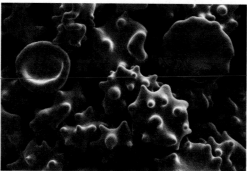

A. Red blood cells in an isotonic solution (such as the fluid portion of blood) have a normal, indented disk shape.

B. Water diffuses out of red blood cells immersed in a hypertonic solution, so they shrivel up.

C. Water diffuses into red blood cells immersed in a hypotonic solution, so they swell up. Some of these have burst.

Figure 4.18 Effects of tonicity in human red blood cells. These cells have no mechanism to compensate for differences in solute concentration between cytoplasm and extracellular fluid.

(A) Annie Cavanagh/Wellcome Images; (B, C) CMSP/Getty Images.

If a cell's cytoplasm becomes hypertonic with respect to the fluid outside of its plasma membrane, water will diffuse into the cell. If the cytoplasm becomes hypotonic, water will diffuse out. In either case, the solute concentration of the cytoplasm may change. If it changes enough, the cell's enzymes will stop working, with lethal results. Many cells have built-in mechanisms that compensate for differences in solute concentration between cytoplasm and extracellular (external) fluid. In cells with no such mechanism, the volume—and solute concentration—of cytoplasm changes when water diffuses into or out of the cell (Figure 4.18).

The rigid cell walls of plants and many protists, fungi, and bacteria can resist an increase in the volume of cytoplasm even in hypotonic environments. In the case of plant cells, cytoplasm usually contains more solutes than soil water does. Thus, water usually diffuses from soil into a plant—but only up to a point. Stiff walls keep plant cells from expanding very much, so an inflow of water causes pressure to build up inside them. Pressure that a fluid exerts against a structure that contains it is called **turgor**. When enough pressure builds up inside a plant cell, water stops diffusing into its cytoplasm. The amount of turgor that is enough to stop osmosis is called osmotic pressure.

Osmotic pressure keeps walled cells plump, just as high air pressure inside a tire keeps it inflated. A young land plant can resist gravity to stay erect because its cells are plump with cytoplasm (Figure 4.19A). When soil dries out, it loses water, so the concentration of solutes increases in it. If soil water becomes hypertonic with respect to cytoplasm, water will start diffusing out of the plant's cells, causing their cytoplasm to shrink (Figure 4.19B). As turgor inside the cells decreases, the plant wilts.

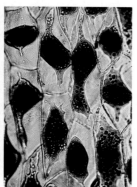

A. Osmotic pressure keeps plant parts erect. These cells in an iris petal are plump with cytoplasm.

B. Cells from a wilted iris petal. The cytoplasm shrank, and the plasma membrane is pulled away from the cell wall.

Figure 4.19 Turgor, as illustrated in cells of iris petals.

(A,B) Perennou Nuridsany/Science Source; (inset) © Evgenyi/Shutterstock.com.

Take-Home Message 4.5

What influences the movement of ions and molecules?

- Solutes tend to diffuse into an adjoining region of fluid in which they are not as concentrated. The steepness of a concentration gradient as well as temperature, molecular size, charge, and pressure affect the rate of diffusion.
- When two fluids of different solute concentration are separated by a selectively permeable membrane, water diffuses from the hypotonic to the hypertonic fluid. This movement, osmosis, is opposed by turgor.

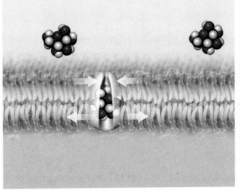

Extracellular Fluid

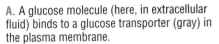 — glucose

Cytoplasm

A. A glucose molecule (here, in extracellular fluid) binds to a glucose transporter (gray) in the plasma membrane.

B. Binding causes the transport protein to change shape.

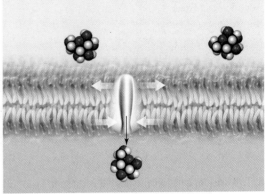

C. The transport protein releases the glucose on the other side of the membrane (in cytoplasm) and resumes its original shape.

Figure 4.20 An example of facilitated diffusion.

Figure It Out: In this example, which fluid is hypotonic: extracellular fluid or cytoplasm?

Answer: Cytoplasm

4.6 Membrane Transport Mechanisms

REMEMBER: Phospholipids swirled into water will spontaneously organize themselves into lipid bilayer sheets or bubbles; a transport protein moves specific solutes across a cell membrane (Section 3.3). Some cells use pseudopods to engulf a target (3.5).

Substances that do not diffuse directly through lipid bilayers can cross a cell membrane through transport proteins embedded in it. Each type of transport protein allows a specific substance to cross: Calcium pumps pump only calcium ions; glucose transporters transport only glucose; and so on. This specificity is an important part of homeostasis. Consider how the composition of cytoplasm depends on movement of particular solutes across the plasma membrane, which in turn depends on the transporters in it. Glucose is an important source of energy for most cells, so they normally take up as much as they can from extracellular fluid. They do so with the help of glucose transporters in the plasma membrane. As soon as a molecule of glucose enters cytoplasm, an enzyme (hexokinase, shown in Figure 4.8) phosphorylates it. Phosphorylation traps the molecule in the cell because the transporters are specific for glucose, not phosphorylated glucose. Thus, phosphorylation prevents the molecule from moving back through the transport protein and leaving the cell.

Passive Transport Osmosis is an example of **passive transport**, a membrane-crossing mechanism that requires no energy input. The diffusion of solutes through transport proteins is another example. In this case, the movement of the solute (and the direction of its movement) is driven entirely by the solute's concentration gradient. Some transport proteins form pores: permanently open channels through a membrane. Other channels are gated, which means they open and close in response to a stimulus such as a shift in electric charge or binding to a signaling molecule. With a passive transport mechanism called **facilitated diffusion**, a solute binds to a transport protein, which then changes shape so the solute is released to the other side of the membrane. A glucose transporter is an example of a transport protein that works in facilitated diffusion (Figure 4.20). This protein changes shape when it binds to a molecule of glucose. The shape change moves the glucose to the opposite side of the membrane, where it detaches from the transport protein. Then, the glucose transporter reverts to its original shape.

facilitated diffusion Passive transport mechanism in which a solute follows its concentration gradient across a membrane by moving through a transport protein.

passive transport Membrane-crossing mechanism that requires no energy input.

turgor Pressure that a fluid exerts against a membrane, wall, or other structure that contains it.

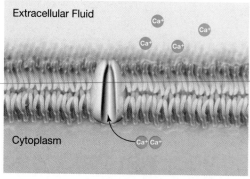
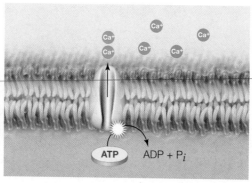
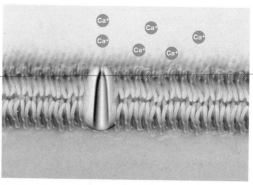

A. Two calcium ions (blue) bind to the transport protein (a calcium pump, gray).

B. A phosphate group from ATP causes the protein to change shape so that the calcium ions are ejected to the opposite side of the membrane.

C. After it loses the calcium ions, the transport protein resumes its original shape.

Figure 4.21 Active transport of calcium ions.

Active Transport Maintaining a solute's concentration often means transporting the solute against its gradient, to the side of the membrane where it is more concentrated. This takes energy. In **active transport**, a transport protein uses energy to pump a solute against its gradient across a cell membrane. Typically, an energy input (for example, in the form of a phosphate-group transfer from ATP) changes the shape of the transport protein. The shape change causes the protein to release a bound solute to the other side of the membrane.

A calcium pump moves calcium ions across cell membranes by active transport (Figure 4.21). Calcium ions act as potent messengers inside cells, and they also affect the activity of many enzymes. Thus, their concentration in cytoplasm is tightly regulated. Calcium pumps in the plasma membrane of all eukaryotic cells can keep the concentration of calcium ions in cytoplasm thousands of times lower than it is in extracellular fluid. Another example of active transport involves sodium–potassium pumps (Figure 4.22). Nearly all cells in your body have these transport proteins.

Bear in mind that the membranes of all cells, not just those of animals, have proteins that carry out active transport. In plants, for example, transport proteins in the plasma membranes of photosynthetic cells pump sucrose from cytoplasm into tubes that thread throughout the plant body.

Figure 4.22 The sodium–potassium pump.
This protein (gray) actively transports sodium ions (Na⁺) from cytoplasm to extracellular fluid, and potassium ions (K⁺) in the other direction. The transfer of a phosphate group (P) from ATP provides energy required for transporting both ions against their concentration gradient.

Membrane Trafficking Vesicles are constantly carrying materials to and from a cell's plasma membrane (Figure 4.23). Cells use vesicles to take in or expel materials in bulk (as opposed to one molecule or ion at a time via transport proteins). A cell can expel bulk materials by **exocytosis**, a pathway in which a vesicle in the cytoplasm moves to the cell's surface ❶ and fuses with the plasma membrane ❷. As the exocytic vesicle loses its identity, its contents are released to extracellular fluid ❸. The cell can take in bulk materials by **endocytosis**. In one endocytic pathway, a small patch of plasma membrane balloons into the cytoplasm, bringing with it a drop of extracellular fluid (along with solutes and particles suspended in it). As the balloon sinks into the cytoplasm, the hydrophobic tails of the lipids in the bilayer are repelled by the watery fluid on both sides. The fluid "pushes" the phospholipid tails together, which helps round off the bud as a vesicle.

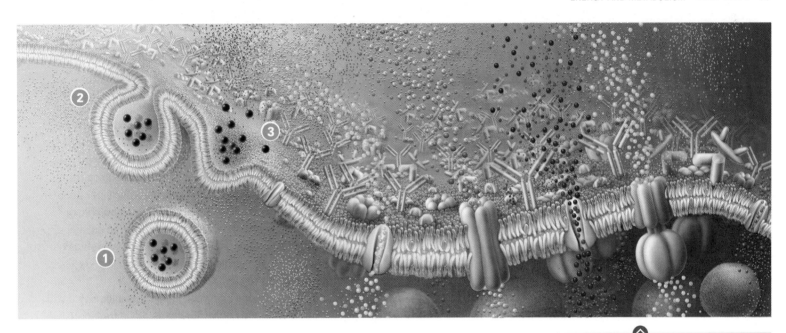

Figure 4.23 Membrane crossings.
A plasma membrane is a hub of activity: Molecules and ions (colored balls) are constantly flowing into and out of a cell via transport proteins embedded in its plasma membrane. Vesicles are also taking in or expelling bulk amounts of solutes and much larger particles.

1 Exocytosis begins as a vesicle moves to the plasma membrane.

2 The vesicle's membrane fuses with the plasma membrane.

3 As the membranes fuse, the vesicle's contents are released to the extracellular fluid.

Phagocytosis (which means "cell eating") is a type of endocytosis in which motile cells engulf microorganisms, cellular debris, or other large particles. Many

single-celled protists such as amoebas feed by phagocytosis. Some of your white blood cells use the pathway to engulf viruses and bacteria, cancerous body cells, and other threats to health. (The micrograph on the left shows a phagocytic white blood cell engulfing several tuberculosis bacteria, in red.) Phagocytosis begins when receptor proteins bind to a particular target. The binding causes microfilaments just under the plasma membrane to contract. The contraction forces a lobe of membrane-enclosed cytoplasm to bulge outward as a pseudopod. Pseudopods that merge around a target trap it inside a vesicle that sinks into the cytoplasm.

Take-Home Message 4.6

How do substances and particles that cannot diffuse through lipid bilayers cross cell membranes?

- Transport proteins allow specific ions or molecules to cross a cell membrane. The types and amounts of substances that cross a membrane depend on the transport proteins embedded in it.
- In facilitated diffusion (a type of passive transport), a solute binds to a transport protein that releases it on the opposite side of the membrane. The movement is driven by the solute's concentration gradient.
- In active transport, a transport protein pumps a solute across a membrane against its concentration gradient. The movement requires energy, as from ATP.
- Exocytosis and endocytosis move materials in bulk across a plasma membrane. Some cells can engulf large particles by phagocytosis.

active transport Energy-requiring mechanism in which a transport protein pumps a solute across a cell membrane against the solute's concentration gradient.

endocytosis Process by which a cell takes in a small amount of extracellular fluid (and its contents) by the ballooning inward of the plasma membrane.

exocytosis Process by which a cell expels a vesicle's contents to extracellular fluid.

phagocytosis "Cell eating"; an endocytic pathway by which a cell engulfs large particles such as microbes or cellular debris.

Summary

Section 4.1 Currently the most serious drug problem on college campuses is binge drinking. Drinking more alcohol than the body's enzymes can detoxify damages the body, and it can be lethal in the short term or the long term.

Section 4.2 **Energy** is the capacity to do work. Energy cannot be created or destroyed (**first law of thermodynamics**), but it can be converted from one form to another and transferred between objects or systems. Energy tends to disperse spontaneously (**second law of thermodynamics**). A bit disperses at each energy transfer, usually in the form of heat.

Living things maintain their organization only as long as they harvest energy from someplace else. Energy flows in one direction through the biosphere, starting mainly from the sun, then into and out of ecosystems. Producers and then consumers use the captured energy to assemble, rearrange, and break down organic molecules that cycle among organisms in an ecosystem.

Section 4.3 Cells store and retrieve energy by making and breaking chemical bonds in chemical reactions, in which **reactants** are converted to **products**. Some reactions require a net energy input; others end with a net energy release. **Activation energy** is the minimum energy input required to start a reaction.

Section 4.4 Enzymes greatly enhance the rate of reactions without being changed by them. Each has an **active site** that is complementary in shape, size, polarity, and charge to the enzyme's **substrate**, and each works best within a characteristic range of conditions, including temperature, salt concentration, and pH. Most enzymes require assistance from **cofactors**; organic cofactors are **coenzymes**. ATP is often used as a coenzyme to carry energy between reactions. When a phosphate group is transferred from ATP to another molecule, energy is transferred along with it. Phosphate-group transfers (**phosphorylations**) to and from ATP couple reactions that release energy with reactions that require energy.

Metabolic pathways are sequences of enzyme-mediated reactions that build, convert, and break down organic molecules. Regulating metabolic pathways allows cells to conserve energy and resources by making only what they need at a given time. The activity of many enzymes can be regulated, for example by binding of a regulatory molecule. The products of some metabolic pathways inhibit their own production, a regulatory mechanism called **feedback inhibition**. **Electron transfer chains** allow cells to harvest energy in small, manageable steps.

Section 4.5 The rate of **diffusion** is influenced by temperature, solute size, and regional differences in concentration, charge, and pressure. Gases, water, and nonpolar molecules can diffuse directly through a lipid bilayer. Ions and most polar molecules cannot. **Osmosis** is the diffusion of water across a selectively permeable membrane, from a **hypotonic** fluid toward a **hypertonic** fluid. There is no net movement of water between **isotonic** solutions. Osmotic pressure is the amount of **turgor** (fluid pressure against a cell membrane or wall) sufficient to halt osmosis.

Section 4.6 Ions and most polar molecules can cross cell membranes only with the help of a transport protein. With facilitated diffusion, a solute follows its concentration gradient across a membrane through a transport protein. **Facilitated diffusion** is a type of **passive transport** (no energy input is required). With **active transport**, a transport protein uses energy to pump a solute across a membrane against its concentration gradient. A phosphate-group transfer from ATP often supplies energy needed for active transport.

Particles and substances in bulk and large particles are moved across plasma membranes by **exocytosis** and **endocytosis**. In exocytosis, a cytoplasmic vesicle fuses with the plasma membrane, and its contents are released to the outside of the cell. In one endocytic pathway, a patch of plasma membrane balloons into the cell, taking with it a drop of extracellular fluid (along with solutes and particles in it). The balloon forms a vesicle that sinks into the cytoplasm. Some cells can engulf large particles such as other cells by the endocytic pathway of **phagocytosis**.

Self-Quiz

Answers in Appendix I

1. _____ is life's primary source of energy.
   a. Food   c. Sunlight
   b. Water   d. ATP

2. Which of the following statements is *not* correct?
   a. Energy cannot be created or destroyed.
   b. Energy cannot change from one form to another.
   c. Energy tends to disperse spontaneously.

3. If we liken a reaction to an energy hill, then a reaction that _____ is an uphill run.
   a. requires energy
   b. releases energy
   c. runs from reactants to products
   d. uses an enzyme and a cofactor

4. In an energy-requiring reaction, activation energy is a bit like _____ .
   a. a burst of speed
   b. coasting downhill
   c. a bump at the top of the hill
   d. putting on the brakes

5. _____ are always changed by participating in a reaction. (Choose all that are correct.)
   a. Enzymes
   c. Reactants
   b. Cofactors
   d. Coenzymes

6. An environmental factor that directly influences enzyme function is _____ .
   a. temperature
   c. light
   b. wind
   d. all of the above

7. A metabolic pathway _____ .
   a. may build or break down molecules
   b. generates heat
   c. can include an electron transfer chain
   d. all of the above

8. Which of the following statements is *not* correct?
   a. Some metabolic pathways are cyclic.
   b. Glucose can diffuse directly through a lipid bilayer.
   c. Feedback inhibition controls some metabolic pathways.
   d. All coenzymes are cofactors.
   e. Osmosis is a case of diffusion.

9. Ions or molecules tend to diffuse from a region where they are _____ (more/less) concentrated to another where they are _____ (more/less) concentrated.

10. _____ cannot diffuse directly across a lipid bilayer.
    a. Water
    c. Ions
    b. Gases
    d. all of the above

11. If you immerse a human red blood cell in a hypotonic solution, water will _____ .
    a. diffuse into the cell
    c. show no net movement
    b. diffuse out of the cell
    d. move in by endocytosis

12. Fluid pressure against a wall or cell membrane is called _____ .
    a. osmosis
    c. diffusion
    b. turgor
    d. osmotic pressure

13. A transport protein requires ATP to pump sodium ions across a membrane. This is a case of _____ .
    a. passive transport
    c. facilitated diffusion
    b. active transport
    d. a and c

14. Vesicles are part of _____ .
    a. endocytosis
    c. phagocytosis
    b. exocytosis
    d. all of the above

15. Match each term with its most suitable description.
    _____ reactant
    _____ phagocytosis
    _____ first law of thermodynamics
    _____ product
    _____ cofactor
    _____ concentration gradient
    _____ passive transport
    _____ active transport
    _____ ATP
    _____ cyclic pathway

    a. assists enzymes
    b. forms at reaction's end
    c. enters a reaction
    d. requires energy input
    e. one cell "eats" another
    f. energy cannot be created or destroyed
    g. basis of diffusion
    h. no energy input required
    i. goes in circles
    j. currency in a cell's energy economy

Critical Thinking

1. Beginning physics students are often taught the basic concepts of thermodynamics with two phrases: First, you can never win. Second, you can never break even. Explain.
2. Water molecules tend to diffuse in response to their own concentration gradient. How can water be more or less concentrated?
3. Dixie Bee wanted to make JELL-O shots for her next party, but felt guilty about encouraging her guests to consume alcohol. She tried to compensate for the toxicity of the alcohol by adding pieces of healthy fresh pineapple to the shots, but when she did, the JELL-O never solidified. What happened? *Hint:* JELL-O is mainly sugar and a gelatinous mixture of proteins.
4. The enzyme trypsin is sold as a dietary enzyme supplement. Explain what happens to trypsin taken with food.
5. The enzyme catalase combines two hydrogen peroxide molecules ($H_2O_2 + H_2O_2$) to make two molecules of water ($2H_2O$). A gas also forms. What is the gas?

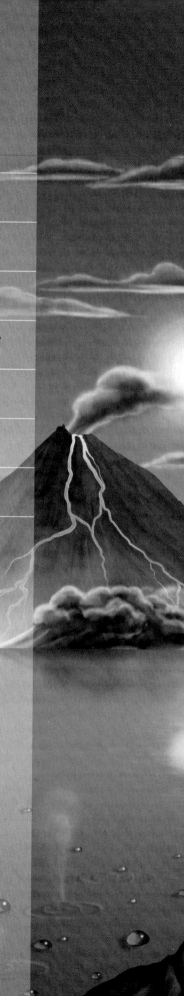

5

CAPTURING AND RELEASING ENERGY

Application ● 5.1 A Burning Concern

REMEMBER: Producers make their own food; consumers ingest tissues of other organisms (Section 1.3). Photosynthesis is the metabolic pathway (4.4) by which plants and other producers harness the energy (4.2, 4.3) in light to make sugars (2.7).

A. Air pollution—smog—blankets Lianyungang, China, in December 2013. The brown color of smog comes from one of its components, a gas (nitric oxide) that is toxic in large amounts. Invisible components include other nitrogen compounds, sulfur compounds, organic molecules, mercury and other heavy metals, and carbon dioxide.

B. Air bubbles trapped in Antarctic ice core slices such as this one are samples of Earth's atmosphere as it was when the ice formed. The deeper the slice, the older the air in the bubbles. From the deepest samples, we know that the level of carbon dioxide in the atmosphere is now higher than it has been for at least 15 million years.

Figure 5.1 Worldwide, most air pollution comes from burning fossil fuels.

(A) ChinaFotoPress/Getty Images; (B) www.photo.antarctica.ac.uk.

Your body is about 9.5 percent carbon by weight, which means that you contain an enormous number of carbon atoms. Where did they all come from? Those atoms may have passed through other consumers before you ate them, but at some point they were components of producers. The vast majority of producers get their carbon from carbon dioxide (CO_2), a gas in air. Your carbon atoms—and those of most other organisms on land—were recently part of Earth's atmosphere, in molecules of CO_2.

The main producers in the human food chain are plants. Plants make their own food by photosynthesis, a pathway that harnesses the energy of sunlight to drive the assembly of sugars from carbon dioxide and water. Photosynthesis removes carbon dioxide from the atmosphere, and fixes its carbon atoms in organic compounds. When plants and other organisms break down organic compounds for energy, carbon atoms are released in the form of CO_2, which then reenters the atmosphere. For billions of years, these two processes have constituted a more or less balanced cycle of the biosphere (you will learn more about the carbon cycle in Section 17.6). For now, know that the amount of carbon dioxide that photosynthesis removes from the atmosphere is roughly the same amount that organisms release back into it—at least it was, until humans came along. As early as 8,000 years ago, humans began burning forests to clear land for agriculture. When trees and other plants burn, most of the carbon in their tissues is released into the atmosphere as CO_2. Fires that occur naturally release carbon dioxide the same way.

Today, we burn a lot more than our ancestors ever did. In addition to wood, we are burning fossil fuels—coal, petroleum, and natural gas—to satisfy our greater and greater demands for energy. Fossil fuels are the organic remains of ancient organisms. When we burn fossil fuels, we release the carbon that has been locked in their organic molecules for hundreds of millions of years, mainly as carbon dioxide that reenters the atmosphere.

Our extensive use of fossil fuels has put Earth's atmospheric cycle of carbon dioxide out of balance: We are adding far more CO_2 to the atmosphere than photosynthetic organisms are removing from it, and the excess is fueling global climate change (we return to this topic in Section 18.5). In 2013 alone, humans released over 36 billion tons of CO_2 into the atmosphere—an increase of 61 percent over 1990 and twice as much as photosynthesis removed from the atmosphere during the same year. Most of the CO_2 that humans release comes from burning fossil fuels (Figure 5.1A). How do we know? Researchers can determine how long ago the carbon atoms in a sample of CO_2 were part of a living organism by measuring the ratio of different carbon isotopes in it (you will read more about radioisotope dating techniques in Section 11.4). The results are correlated with global statistics on fossil fuel extraction, refining, and trade.

Tiny pockets of Earth's ancient atmosphere remain in Antarctica, preserved in snow and ice that have been accumulating in layers, year after year, for millions of years (Figure 5.1B). Air and dust trapped in each layer reveal the composition of the atmosphere that prevailed when the layer formed. These layers tell us that the atmospheric CO_2 level was relatively stable for about 10,000 years before the industrial revolution began in the mid-1800s. Since then, the CO_2 level has been steadily rising. Today, the atmospheric CO_2 level is higher than it has been for *15 million years.*

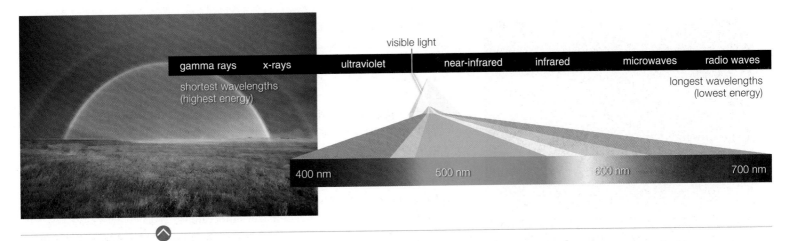

Figure 5.2 Properties of light.
Electromagnetic radiation moves through space in waves that we measure in nanometers (nm). Visible light makes up a very small part of this energy. Raindrops or a prism can separate visible light's different wavelengths, which we see as different colors. About 25 million nanometers are equal to 1 inch.

Left, © Robbie George, National Geographic Creative.

5.2 To Catch a Rainbow

REMEMBER: An electron that absorbs energy moves to a higher energy level, then emits the extra energy and moves back down (Section 2.2). Chloroplasts are organelles specialized for photosynthesis (3.5). Sustaining life's organization requires ongoing energy inputs to counter energy loss (4.2). Building, rearranging, or breaking down an organic substance often occurs stepwise, in a series of enzymatic reactions called a metabolic pathway (4.4).

Energy flow through nearly all ecosystems on Earth begins when photosynthesizers capture the energy in sunlight. Harnessing that energy is a complicated business; plants do it by converting it to chemical energy, which they and most other organisms use to drive cellular work. Understanding the conversion process requires a bit of knowledge about the nature of light.

Light is electromagnetic radiation, a type of energy that moves through space in waves, a bit like waves move across an ocean. The distance between the crests of two successive waves is called **wavelength**, and it is measured in nanometers (nm). Light that is visible to the human eye is only a tiny part of the spectrum of electromagnetic radiation emitted by the sun (Figure 5.2). This visible light travels in wavelengths between 380 and 750 nm, and it is the main form of energy that drives photosynthesis. Our eyes perceive all of these wavelengths combined as white light, and particular wavelengths in this range as different colors. White light separates into its component colors when it passes through a prism, or raindrops that act as tiny prisms. A prism bends longer wavelengths more than it bends shorter ones, so a rainbow of colors forms.

Photosynthesizers use pigments to capture the energy of visible light. A **pigment** is an organic molecule that selectively absorbs light of certain wavelengths, a bit like an antenna specialized for receiving light energy. Absorbing energy of the appropriate wavelength excites a pigment's electrons. An excited electron (one that has been boosted to a higher energy level) quickly emits its extra energy and returns to a lower energy level. As you will see, photosynthetic cells can capture energy emitted from an electron returning to a lower energy level.

pigment An organic molecule that can absorb light of certain wavelengths.

wavelength Distance between the crests of two successive waves.

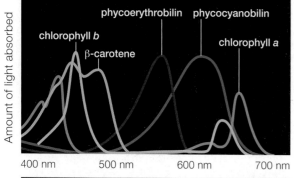

Figure 5.3 A few photosynthetic pigments.
The curves in this graph show the efficiency at which each pigment absorbs the different wavelengths of visible light. Line color indicates the pigment's characteristic color. Using a combination of pigments allows photosynthetic organisms to maximize the range of wavelengths they can capture for photosynthesis.

Top, © Photobac/Shutterstock.

All life is sustained
by inputs of energy, but
not all forms of energy
can sustain life.

Wavelengths of light that are not absorbed by a pigment are reflected, and that reflected light gives each pigment its characteristic color. **Chlorophyll a** is the most common photosynthetic pigment in plants and photosynthetic protists. It also occurs in some bacteria. Chlorophyll *a* absorbs violet, red, and orange light, and it reflects green light, so it appears green to us. Accessory pigments, including other chlorophylls, collectively harvest a wide range of additional light wavelengths for photosynthesis and for other purposes. Many accessory pigments protect cells from the damaging effects of ultraviolet (UV) light in the sun's rays. Appealing colors attract animals to ripening fruit or pollinators to flowers. You may already be familiar with some of these molecules: Carrots, for example, are orange because they contain beta-carotene (which is often abbreviated as β-carotene); roses are red and violets are blue because their cells make anthocyanins.

Most photosynthetic organisms maximize the range of wavelengths they can capture for photosynthesis by using a combination of pigments (Figure 5.3). In green plants, chlorophylls are usually so abundant that they mask the colors of the other pigments. Plants that change color during autumn are preparing for a period of dormancy; they conserve resources by moving nutrients from tender parts that would be damaged by winter cold (such as leaves) to protected parts (such as roots). Chlorophylls are not needed during dormancy, so they are disassembled and their components recycled. Yellow and orange accessory pigments are also recycled, but not as quickly as chlorophylls. Their colors begin to show as the chlorophyll content declines in leaves. Anthocyanin synthesis also increases in some plants, adding red and purple tones to turning leaf colors.

Storing Energy in Sugars All life is sustained by inputs of energy, but not all forms of energy can sustain life. Sunlight, for example, is abundant here on Earth, but it cannot be used to directly power protein synthesis or other energy-requiring reactions that all organisms must run in order to stay alive. Photosynthesis converts the energy of light into the energy of chemical bonds. Unlike light, chemical energy can power the reactions of life, and it can be stored for later use.

In plants and other photosynthetic eukaryotes, photosynthesis takes place in chloroplasts (Figure 5.4A). Plant chloroplasts have two outer membranes, and they are filled with a thick, cytoplasm-like fluid called **stroma** (Figure 5.4B). Suspended in the stroma are the chloroplast's own DNA, some ribosomes, and an inner, much-folded **thylakoid membrane**. The folds of a thylakoid membrane typically form stacks of interconnected disks called thylakoids. The space enclosed by the thylakoid membrane is a single, continuous compartment (Figure 5.4C).

Photosynthesis is often summarized by an equation:

$$CO_2 + water \xrightarrow{\text{light energy}} sugars + O_2$$

This equation means that photosynthesis converts CO_2 and water to sugars and oxygen. However, photosynthesis is not a single reaction. Rather, it is a metabolic pathway with many reactions that occur in two stages. Molecules in the thylakoid membrane carry out the reactions of the first stage, which are driven by light and thus called the light-dependent reactions. The "photo" in photosynthesis means light, and it refers to the conversion of light energy to chemical bond energy of ATP during this stage. In addition to making ATP, the main light-dependent pathway in chloroplasts splits water molecules and releases O_2. Hydrogen ions and electrons from the water molecules end up in the coenzyme NADPH (Figure 5.5A).

The "synthesis" part of photosynthesis refers to the reactions of the second stage, which build sugars from CO_2 and water. These sugar-building reactions run

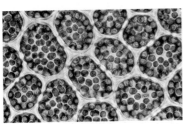

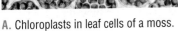

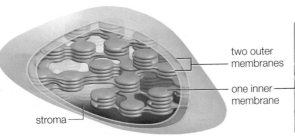

two outer
membranes

one inner
membrane

stroma

A. Chloroplasts in leaf cells of a moss.

B. Three membranes of a chloroplast.

C. Part of the thylakoid membrane, cutaway view.

Figure 5.4 Zooming in on a chloroplast.
The first stage of photosynthesis takes place at the thylakoid membrane; the second stage runs in the stroma.

(A) Michael Eichelberger/Visuals Unlimited.

in the stroma. They are collectively called the light-independent reactions because light energy does not power them. Instead, they run on energy delivered by NADPH and ATP that formed during the first stage (Figure 5.5B).

Take-Home Message 5.2

How do cells harvest light for photosynthesis?

- The sun emits light, which travels in waves. We see different wavelengths of visible light as different colors.
- Visible light is the main form of energy that drives photosynthesis. Photosynthetic species use pigments such as chlorophyll *a* to harvest this energy for photosynthesis.
- In eukaryotic cells, the first stage of photosynthesis occurs at the thylakoid membrane of chloroplasts. During these light-dependent reactions, light energy drives the formation of ATP and NADPH.
- The second stage occurs in the stroma. During these light-independent reactions, ATP and NADPH drive the synthesis of sugars from water and carbon dioxide.

ADP
NADP+
H₂O
energy

A. Light-dependent reactions of photosynthesis

ATP
NADPH
O₂

ATP
NADPH
CO₂
H₂O

B. Light-independent reactions of photosynthesis (Calvin–Benson cycle)

ADP
NADP+
sugars

Figure 5.5 Inputs and outputs of the two stages of photosynthesis.

5.3 Light-Dependent Reactions

REMEMBER: Successively higher electron energy levels can be represented as nested shells (Section 2.2). Potential energy is stored in the position or arrangement of objects in a system (4.2). Electron transfer chains can harvest the energy of electrons in small, usable increments; phosphorylation is a reaction in which a phosphate group is transferred from one molecule to another (4.4). Ions and other substances that do not diffuse directly through lipid bilayers can cross a cell membrane through transport proteins embedded in it; in active transport, a transport protein uses energy to pump a solute against its gradient across a cell membrane (4.6).

When a chlorophyll or accessory pigment absorbs light, one of its electrons jumps to a higher energy level (shell). The electron quickly drops back down to a lower shell by emitting its extra energy. In the thylakoid membrane, energy emitted by an electron is not lost to the environment. In this special membrane, photosynthetic pigments occur in clusters held together by proteins. These clusters can hold onto energy by passing it back and forth, a bit like volleyball players pass a ball among team members.

chlorophyll *a* Main photosynthetic pigment in plants.

stroma The cytoplasm-like fluid between the thylakoid membrane and the two outer membranes of a chloroplast.

thylakoid membrane A chloroplast's highly folded inner membrane system; forms a continuous compartment.

Calvin–Benson cycle Light-independent reactions of photosynthesis; cyclic carbon-fixing pathway that forms sugars from CO_2.

carbon fixation Process by which carbon from an inorganic source such as carbon dioxide gets incorporated (fixed) into an organic molecule.

electron transfer phosphorylation Process in which electron flow through electron transfer chains sets up a hydrogen ion gradient that drives ATP formation.

The reactions of photosynthesis begin when energy being passed around the thylakoid membrane reaches and becomes absorbed by a photosystem (Figure 5.6). A photosystem is a grouping of hundreds of chlorophylls, accessory pigments, and other molecules. When a photosystem absorbs light energy ❶, it releases electrons that immediately enter an electron transfer chain in the thylakoid membrane. (With this step, light energy has been converted to chemical energy.) The photosystem must replace its lost electrons, and it does so by pulling them off of water molecules in the thylakoid compartment. Water molecules do not give up electrons very easily; doing so causes them to break apart into hydrogen ions and oxygen atoms ❷. The hydrogen ions remain in the thylakoid compartment; oxygen atoms combine and diffuse out of the cell as oxygen gas (O_2).

Meanwhile, the electrons released by the photosystem have been moving through the electron transfer chain ❸. As the electrons pass from one molecule in the chain to the next, they release a bit of their extra energy. Molecules of the chain use the released energy to actively transport hydrogen ions (H^+) across the membrane, from the stroma into the thylakoid compartment ❹. Thus, the flow of electrons through the electron transfer chain sets up and maintains a hydrogen ion gradient across the thylakoid membrane.

At the end of the electron transfer chain, electrons are accepted by a second photosystem. When this photosystem absorbs light energy, it releases electrons ❺. These electrons immediately enter a second, different electron transfer chain. At the end of this chain, the coenzyme $NADP^+$ accepts the electrons along with H^+, so NADPH forms ❻:

$$NADP^+ + electrons + H^+ \longrightarrow \boxed{\textbf{NADPH}}$$

The hydrogen ion gradient that forms across the thylakoid membrane is a type of potential energy that can be tapped to make ATP. The H^+ ions want to follow their concentration gradient by moving back into the stroma, but ions cannot diffuse

Figure 5.6 Light-dependent reactions in the thylakoid membrane of chloroplasts.
ATP and oxygen gas are produced in this pathway. Electrons that travel through two different electron transfer chains end up in NADPH, which delivers them to sugar-building reactions in the stroma.

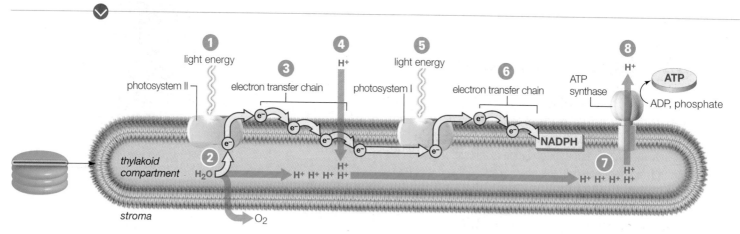

❶ A photosystem absorbs light energy and releases electrons.

❷ The photosystem pulls replacement electrons from water molecules, which then break apart into oxygen atoms and hydrogen ions. The oxygen atoms leave the cell as O_2 gas.

❸ The electrons enter an electron transfer chain in the thylakoid membrane.

❹ Energy lost by the electrons as they move through the chain is used to actively transport hydrogen ions from the stroma into the thylakoid compartment. A hydrogen ion gradient forms across the thylakoid membrane.

❺ Another photosystem absorbs light energy and releases electrons. Replacement electrons come from the first electron transfer chain.

❻ The released electrons move through a second electron transfer chain, then combine with $NADP^+$ and H^+, so NADPH forms.

❼ Hydrogen ions in the thylakoid compartment follow their gradient across the thylakoid membrane by flowing through the interior of ATP synthases.

❽ Hydrogen ion flow causes ATP synthases to phosphorylate ADP, so ATP forms in the stroma.

through the lipid bilayer. H⁺ leaves the thylakoid compartment only by flowing through proteins called ATP synthases embedded in the thylakoid membrane **7**. An ATP synthase is both a transport protein and an enzyme. When hydrogen ions flow through its interior, the protein phosphorylates ADP, so ATP forms in the stroma **8**. The process by which the flow of electrons through electron transfer chains drives ATP formation is called **electron transfer phosphorylation**.

Take-Home Message 5.3

What happens during the first stage of photosynthesis?

- Photosynthetic pigments in the thylakoid membrane transfer the energy of light to photosystems, which release electrons that enter electron transfer chains.
- The flow of electrons through electron transfer chains sets up a hydrogen ion gradient that drives ATP formation.
- Water molecules are split, oxygen is released, and electrons end up in NADPH.

5.4 Light-Independent Reactions

REMEMBER: Plant cells store sugars in the form of starch, a polymer of glucose (Section 2.7). A plant cuticle helps stems and leaves retain water (3.5). An active site is a pocket in an enzyme where substrates bind and a reaction occurs (4.4).

Energy that drives the second stage of photosynthesis is provided not by light; rather, the reactions run on the energy of electrons delivered by NADPH and phosphate-group transfers from ATP. These light-independent reactions, which are collectively called the **Calvin–Benson cycle**, produce sugars in the stroma of chloroplasts (Figure 5.7). This cyclic pathway uses carbon atoms from CO_2 to build the carbon backbones of the sugar molecules. Extracting carbon atoms from an inorganic source (such as CO_2) and incorporating them into an organic molecule is a process called **carbon fixation**.

In most plants, photosynthetic protists, and some bacteria, the enzyme rubisco fixes carbon by attaching CO_2 to a five-carbon compound called RuBP. The six-carbon intermediate that forms by this reaction is unstable, so it splits right away into two three-carbon molecules of PGA, which continue in the cycle. NADPH and ATP are used to convert these molecules to sugars.

It takes six cycles of Calvin–Benson reactions to fix the six carbon atoms necessary to make one molecule of glucose (a six-carbon sugar). Plant cells break down some of the glucose they make to access the energy stored in its bonds. However, most of the glucose they make is converted at once to sucrose or starch by other pathways that conclude the light-independent reactions. Excess glucose is stored as starch grains in chloroplast stroma. When sugars are needed in other parts of the plant, the starch is broken apart into its glucose monomers.

Alternative Carbon-Fixing Pathways The aboveground parts of most plants are covered with a cuticle that limits evaporative water loss. Gases cannot diffuse across the cuticle, but oxygen produced by the light-dependent reactions must escape

Figure 5.7 The Calvin–Benson cycle.
This sketch shows a cross-section of a chloroplast with the reactions cycling in the stroma. Six cycles of the Calvin–Benson reactions produce one six-carbon sugar.

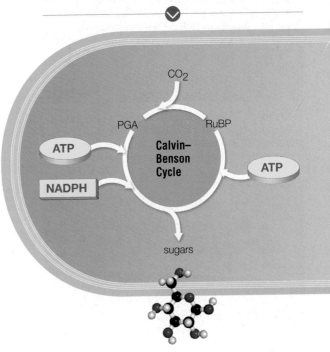

Figure 5.8 Stomata on the surface of a leaf.
When these tiny pores are open, gases can be exchanged between the plant's internal tissues and air. Gas exchange ends when stomata close to conserve water on dry days.

© D. Kucharski & K. Kucharska/Shutterstock.

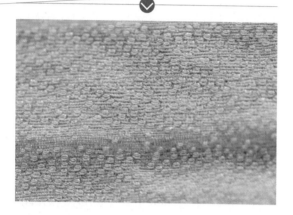

Figure 5.9 C4 and CAM plants.

(A) Image courtesy msuturfweeds.net; (B) © Tamara Kulikova/Shutterstock.

A. Crabgrass "weeds" overgrowing a lawn. Crabgrasses, which are C4 plants, thrive in hot, dry summers, easily outcompeting Kentucky bluegrass and other fine-leaved C3 grasses commonly planted in residential lawns.

B. The jade plant, *Crassula argentea*, and other CAM plants survive in hot deserts by opening stomata to fix carbon only at night. They run the Calvin–Benson cycle during the day, with stomata closed.

the plant, and carbon dioxide needed for the Calvin–Benson cycle must enter it. Thus, most leaves and stems are studded with tiny, closable gaps called **stomata** (singular, stoma; Figure 5.8). When stomata are open, CO_2 diffuses from air into photosynthetic tissues, and O_2 diffuses out of the tissues into air. Stomata close to conserve water on hot, dry days. When that happens, gas exchange comes to a halt.

Both stages of photosynthesis run during the day. With stomata closed, the O_2 level in the plant's tissues rises, and the CO_2 level declines. This outcome can reduce the efficiency of sugar production because both gases are substrates of rubisco, and they compete for its active site. Rubisco starts the Calvin–Benson cycle by attaching CO_2 to RuBP. It also attaches O_2 to RuBP. To convert the product of this alternate reaction to a substrate of the Calvin–Benson cycle, the cell uses additional ATP and NADPH, and it also loses carbon in the form of CO_2. The detrimental effects of the alternate pathway are greatest in **C3 plants**, which use only the Calvin–Benson cycle to fix carbon (they are called C3 plants because a three-carbon molecule, PGA, is the first stable intermediate to form in their light-independent reactions). In a C3 plant, sugar production becomes less and less efficient as daytime temperature rises.

An additional set of carbon-fixing reactions minimizes the effects of rubisco's inefficiency in corn, bamboo, and other **C4 plants** (so named because a four-carbon molecule is the first stable intermediate to form in their light-independent reactions). These plants also close stomata on dry days, but the efficiency of their sugar production does not decline. C4 plants fix carbon twice, in two kinds of cells. In the first cell, carbon is fixed by an enzyme that cannot use oxygen as a substrate. The resulting intermediate is transported to a second cell, where it is converted to CO_2. There, rubisco fixes carbon for the second time as the CO_2 enters the Calvin–Benson cycle. The extra C4 reactions keep the CO_2 level high near rubisco, so sugar production stays efficient even during hot, dry weather (Figure 5.9A).

Succulents, cacti, and other **CAM plants** use a carbon-fixing pathway that allows them to produce sugar efficiently even where typical daytime conditions are extremely hot and dry. CAM stands for crassulacean acid metabolism, after the Crassulaceae family of plants in which this pathway was first studied (Figure 5.9B). Like C4 plants, CAM plants fix carbon twice, but the reactions occur at different times rather than in different cells. Stomata on a CAM plant open only at night, when typically lower temperatures minimize evaporative water loss. The plant's cells use a C4 pathway to fix carbon from CO_2 in the air at this time. The product of the pathway is stored in the cell's central vacuole. When the stomata close the next day, the molecule moves out of the vacuole and becomes broken down to CO_2. Rubisco then fixes carbon for the second time as the CO_2 enters the Calvin–Benson cycle.

Take-Home Message 5.4

What happens during the second stage of photosynthesis?

- During the light-independent reactions (the second stage of photosynthesis), ATP and NADPH drive the synthesis of sugars from CO_2.
- When stomata close on hot, dry days, they also prevent the exchange of gases between plant tissues and the air. This outcome reduces the efficiency of sugar production in C3 plants.
- Plants adapted to hot, dry conditions fix carbon twice. C4 plants separate the two sets of reactions in space; CAM plants separate them in time.

5.5 **A Global Connection**

REMEMBER: Most free radicals are dangerous to life (Section 2.2). Mitochondria make ATP by aerobic respiration (3.5). Organic molecules release energy when they combine with oxygen (4.3). Most enzymes require assistance from cofactors; coenzymes carry chemical groups, atoms, or electrons between reactions; energy given off by electrons moving through an electron transfer chain drives cellular work (4.4). Ions cross a cell membrane through transport proteins; in active transport, a transport protein uses energy to move a solute across a cell membrane (4.6).

The first cells we know of appeared on Earth about 3.4 billion years ago. Like some modern prokaryotes, these ancient organisms did not tap into sunlight; rather, they extracted the energy they needed from simple molecules such as methane and hydrogen sulfide. Both gases were plentiful in the nasty brew that was Earth's early atmosphere (Figure 5.10). When photosynthesis evolved, sunlight offered cells that used it an essentially unlimited supply of energy, and they were very successful. Oxygen gas released from uncountable numbers of water molecules began seeping out of the photosynthesizers. O_2 reacts easily with metals, so at first, most of it combined with metal atoms in exposed rocks. After the exposed minerals became saturated with oxygen, the gas began to accumulate in the ocean and in the atmosphere. From that time on, the world of life would never be the same.

Before photosynthesis evolved, molecular oxygen had been a very small component of Earth's atmosphere. In what may have been the earliest case of catastrophic pollution, the new abundance of this gas exerted tremendous pressure on all life at the time. Why? Then, as now, enzymes that require metal cofactors were a critical part of metabolism. Oxygen reacts with metal cofactors, and free radicals form during those reactions. Free radicals damage biological molecules, so they are dangerous to life. Most cells had no way to cope with them, and so were wiped out everywhere except deep water, muddy sediments, and other **anaerobic** (oxygen-free) places.

By lucky circumstance, a few types of cells were already making molecules that could detoxify or prevent the formation of free radicals. Cells with these molecules were the first **aerobic** organisms—they could live in the presence of oxygen. As aerobic organisms evolved, their detoxifying molecules became incorporated into new metabolic pathways. One of these pathways put the reactive properties of oxygen to use. Today, this pathway—**aerobic respiration**—is the main ATP-producing sugar breakdown pathway in nearly all eukaryotes (including plants) and some modern bacteria. The pathway requires oxygen, and its products—carbon dioxide and water—are the raw materials used by the majority of photosynthetic organisms to build the sugars in the first place. With this connection, the cycling of carbon, hydrogen, and oxygen through living things connects full circle through the biosphere (right).

Aerobic Respiration in Mitochondria The bonds of organic molecules hold a lot of energy that can be released in a reaction with oxygen. Aerobic respiration

Figure 5.10 A view of how Earth's atmosphere was permanently changed by photosynthesis.
Earth's early atmosphere was abundant in gases such as methane, sulfur, ammonia, and chlorine. Oxygen released by early photosynthesizers changed the composition of the atmosphere. Photosynthesis is now the main pathway by which energy and carbon enter the world of life.

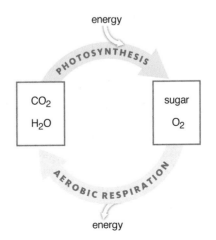

aerobic Involving or occurring in the presence of oxygen.

aerobic respiration Oxygen-requiring metabolic pathway that breaks down sugars to produce ATP.

anaerobic Occurring in the absence of oxygen.

C3 plant Type of plant that uses only the Calvin–Benson cycle to fix carbon.

C4 plant Type of plant that fixes carbon twice, in two cell types.

CAM plant Type of plant that fixes carbon twice, at different times of day.

stomata Closable gaps on aboveground plant surfaces. When open, they allow the plant to exchange gases with air. When closed, they limit water loss.

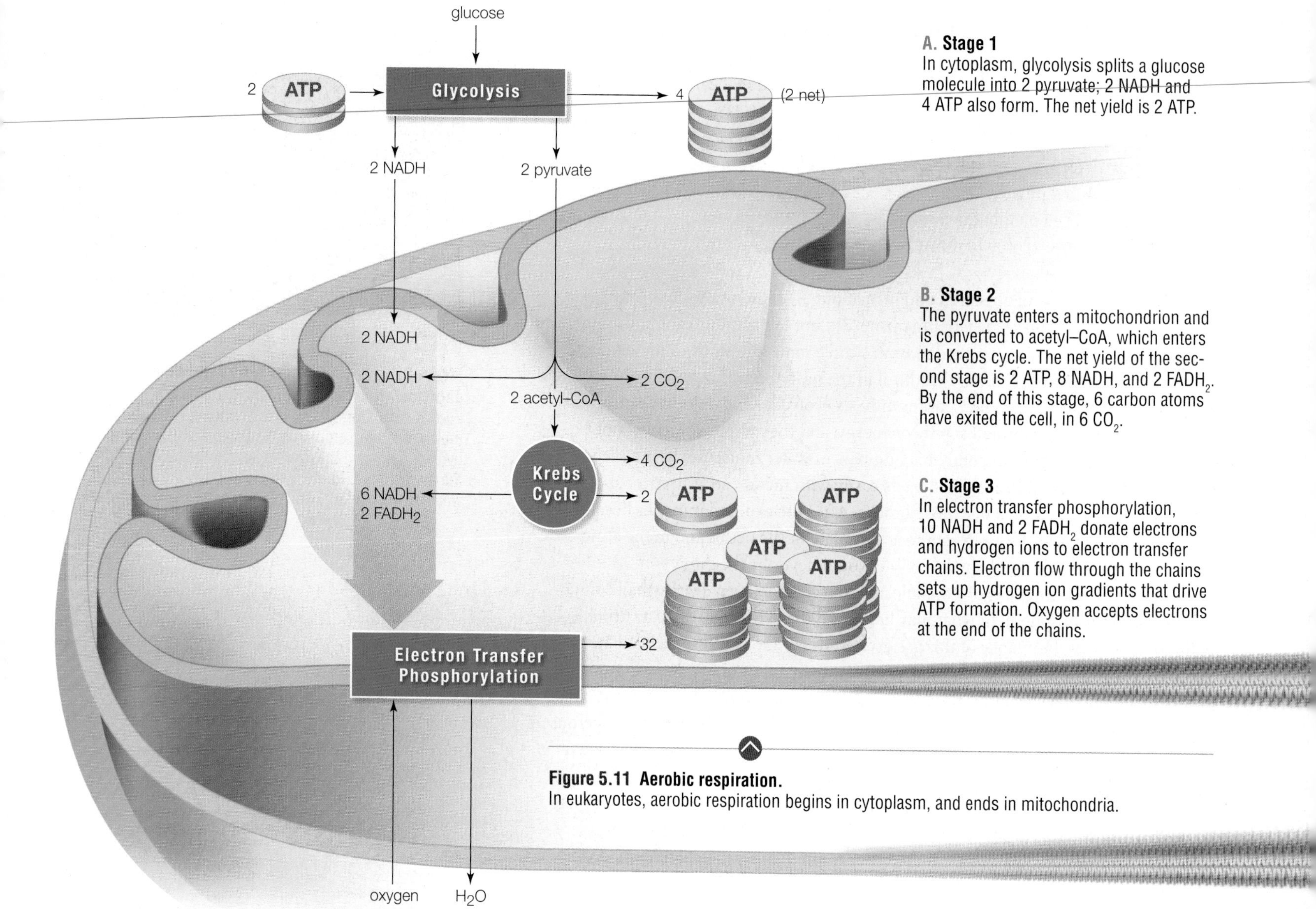

glucose

2 → ATP → **Glycolysis** → 4 ATP (2 net)

A. Stage 1
In cytoplasm, glycolysis splits a glucose molecule into 2 pyruvate; 2 NADH and 4 ATP also form. The net yield is 2 ATP.

2 NADH   2 pyruvate

2 NADH

2 NADH ← → 2 CO_2

2 acetyl–CoA

Krebs Cycle → 4 CO_2

6 NADH ←
2 $FADH_2$

Krebs Cycle → 2 ATP

B. Stage 2
The pyruvate enters a mitochondrion and is converted to acetyl–CoA, which enters the Krebs cycle. The net yield of the second stage is 2 ATP, 8 NADH, and 2 $FADH_2$. By the end of this stage, 6 carbon atoms have exited the cell, in 6 CO_2.

ATP   ATP   ATP   ATP   ATP

→ 32

C. Stage 3
In electron transfer phosphorylation, 10 NADH and 2 $FADH_2$ donate electrons and hydrogen ions to electron transfer chains. Electron flow through the chains sets up hydrogen ion gradients that drive ATP formation. Oxygen accepts electrons at the end of the chains.

Electron Transfer Phosphorylation

oxygen H_2O

Figure 5.11 Aerobic respiration.
In eukaryotes, aerobic respiration begins in cytoplasm, and ends in mitochondria.

a mitochondrion

completely breaks apart the carbon backbone of a sugar, bond by bond. Energy released as those bonds are broken drives ATP synthesis.

The reactions of aerobic respiration occur in three stages. The first stage, glycolysis, takes place in cytoplasm. **Glycolysis** is a set of reactions that convert one six-carbon sugar into two molecules of pyruvate (Figure 5.11A), an organic compound with a three-carbon backbone. The reactions occur with some variation in the cytoplasm of almost all cells; for clarity, we focus here on those that start with glucose. Two ATP are invested to begin glycolysis, but four ATP form by the end. Thus, we say that the net (overall) yield is two ATP per glucose. Electrons and hydrogen ions released by the reactions combine with the coenzyme NAD^+, so NADH also forms.

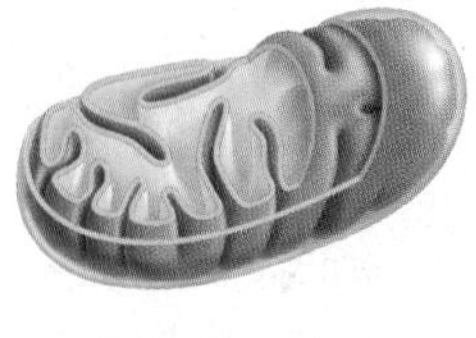

In eukaryotes, aerobic respiration continues in mitochondria. The reactions of the second stage begin when the two pyruvate molecules that formed in glycolysis enter the inner compartment of a mitochondrion (Figure 5.11B). There, each pyruvate reacts with a coenzyme named (rather unimaginatively) coenzyme A. One molecule of CO_2 forms in this reaction and then diffuses out of the cell. The reaction's other product, a molecule called acetyl–CoA, enters a pathway called the **Krebs cycle**. Each cycle of Krebs reactions releases two carbon atoms that leave the cell in molecules of CO_2.

At this point in aerobic respiration, the six-carbon backbone of one glucose molecule has been broken down completely; six carbon atoms have now exited the cell in CO_2. The two ATP that formed during the second stage add to the small net yield (two ATP) of glycolysis. However, six more NADH and two $FADH_2$ (another coenzyme) also formed. Add in the two NADH from glycolysis, and the full breakdown of one glucose molecule has a big potential payoff. Twelve coenzymes now deliver electrons—and the energy they carry—to the third and final stage of aerobic respiration, electron transfer phosphorylation (Figure 5.11C).

① NADH and $FADH_2$ deliver electrons and hydrogen ions to electron transfer chains in the inner mitochondrial membrane.

② Molecules of the electron transfer chain use energy released by the electrons to pump the hydrogen ions (H^+) across the membrane, from the matrix to the intermembrane space. This activity sets up and maintains a hydrogen ion gradient across the inner mitochondrial membrane.

③ Hydrogen ion flow back to the matrix through ATP synthases drives the formation of ATP from ADP and phosphate.

④ Electrons at the end of the electron transfer chains combine with oxygen and hydrogen ions, so water forms.

Figure 5.12 Electron transfer phosphorylation in mitochondria.
In eukaryotes, the third and final stage of aerobic respiration, electron transfer phosphorylation, occurs at the inner mitochondrial membrane.

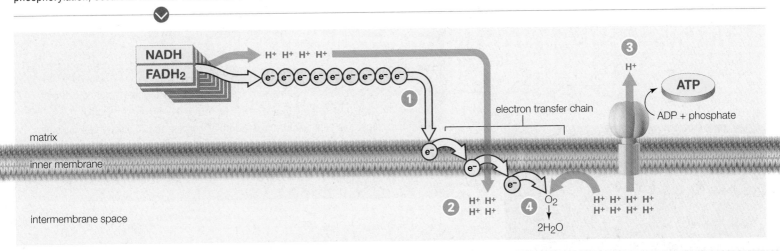

In eukaryotes, electron transfer phosphorylation occurs at the inner mitochondrial membrane (Figure 5.12). The reactions begin when NADH and $FADH_2$ donate their cargo of electrons and hydrogen ions to electron transfer chains embedded in this membrane **①**. As the electrons pass from one molecule in the chain to the next, they release a bit of their extra energy. Molecules of the chain use the released energy to actively transport hydrogen ions (H^+) across the inner mitochondrial membrane, from the matrix to the intermembrane space **②**. Thus, the flow of electrons through the electron transfer chains sets up and maintains a hydrogen ion gradient across the membrane. This gradient attracts the ions back toward the matrix, but ions cannot diffuse through a lipid bilayer. Hydrogen ions cross the inner mitochondrial membrane only by flowing through ATP synthases embedded in the membrane. The flow of hydrogen ions through ATP synthases causes these proteins to attach phosphate groups to ADP, so ATP forms **③**.

At the end of the electron transfer chains, oxygen accepts electrons and H^+ to form water, a product of the third-stage reactions **④**. The term aerobic respiration,

glycolysis Set of reactions in which glucose is broken down to two pyruvate for a net yield of two ATP.

Krebs cycle Cyclic pathway that helps break down pyruvate to carbon dioxide during aerobic respiration.

alcoholic fermentation Anaerobic pathway that breaks down sugars and produces ATP, CO_2, and ethanol.

fermentation An anaerobic pathway that breaks down sugars to produce ATP.

lactate fermentation Anaerobic pathway that breaks down sugars and produces ATP and lactate.

Aerobic respiration literally means "breathing air to live."

A. *Saccharomyces cerevisiae*, a yeast that carries out alcoholic fermentation.

B. Left, one product of *Saccharomyces* alcoholic fermentation (ethanol) makes beer alcoholic; another (CO_2) makes it bubbly. Right, holes in bread are pockets where CO_2 released by fermenting *Saccharomyces* cells accumulated in the dough.

Figure 5.13 Examples of alcoholic fermentation.

(A) © By London Scientific Films/Oxford Scientific/Getty Images; (B) left, © Elena Bosh-kovska/Shutterstock.com; right, optimarc/Shutterstock.com.

which literally means "breathing air to live," refers to this pathway's requirement for oxygen as the final acceptor of electrons. Every breath you take provides your trillions of aerobically respiring cells with a fresh supply of oxygen.

The following equation summarizes the overall pathway of aerobic respiration:

$$C_6H_{12}O_6 + O_2 \longrightarrow CO_2 + H_2O + \text{ATP}$$

glucose   oxygen   carbon dioxide   water

For each glucose molecule that enters this pathway, four ATP form in the first- and second-stage reactions. Coenzymes deliver enough H^+ and electrons to fuel the synthesis of about thirty-two additional ATP in the third stage. Thus, the breakdown of one glucose molecule typically yields thirty-six ATP. As you will see, other pathways that break down sugars also yield ATP, but not nearly as much as aerobic respiration. You and other large, multicelled eukaryotes could not survive without its higher efficiency.

Take-Home Message 5.5

What happens during aerobic respiration?

- Most cells can make ATP by breaking down sugars in the oxygen-requiring pathway of aerobic respiration.
- Aerobic respiration begins in cytoplasm with glycolysis, and, in eukaryotes, ends in the mitochondrion with electron transfer phosphorylation.
- A typical net yield of aerobic respiration is thirty-six ATP per glucose. Carbon dioxide and water also form.

5.6 Fermentation

REMEMBER: Fungi that live as single cells are called yeasts (Section 1.4). Monomers are subunits of polymers (2.6). Sugar monomers form larger carbohydrates (2.7).

Most types of eukaryotic cells use aerobic respiration exclusively, or they use it most of the time. Many bacteria, archaea, protists, and some eukaryotic cells can harvest energy from carbohydrates by fermentation. **Fermentation** refers to sugar breakdown pathways that produce ATP and do not require oxygen. Like aerobic respiration, fermentation begins with glycolysis in cytoplasm. Unlike aerobic respiration, fermentation's concluding reactions take place in cytoplasm. Electrons do not move through electron transfer chains, so no additional ATP forms, and an organic molecule (instead of oxygen) accepts electrons. In these reactions, pyruvate is converted to other molecules, but it is not fully broken down to CO_2 (as occurs in aerobic respiration). The reactions remove electrons from NADH, so NAD^+ forms; regenerating this coenzyme allows glycolysis—and the ATP it offers—to continue. Thus, the net ATP yield of fermentation consists of the two ATP that form in glycolysis (see Figure 5.11A). Fermentation is inefficient compared with aerobic respiration, but it produces enough ATP to sustain many single-celled species. It also helps cells of multicelled species under anaerobic conditions.

Some yeasts carry out **alcoholic fermentation**, a pathway that converts pyruvate to ethanol. One yeast species, *Saccharomyces cerevisiae*, helps us produce beer, wine, and bread (Figure 5.13). Beer brewers often use barley that has been germinated and dried (a process called malting) as a source of glucose for fermentation by this yeast. As the cells make ATP for themselves, they also produce ethanol (which makes the beer alcoholic) and CO_2 (which makes it bubbly). Flowers of the hop plant add flavor and help preserve the finished product. Winemakers use crushed grapes as a source of sugars for yeast fermentation. The ethanol produced by the cells makes the wine alcoholic, and the CO_2 is allowed to escape to the air.

To make bread, flour is kneaded with water, yeast, and sometimes other ingredients. Flour contains a protein (gluten) and a sugar (maltose) that consists of two glucose subunits. Kneading causes the gluten to form polymers in long, interconnected strands that make the resulting dough stretchy and resilient. The yeast cells in the dough first break down the maltose, then use the released glucose for alcoholic fermentation. The CO_2 they produce accumulates in bubbles that are trapped by the mesh of gluten strands. As the bubbles expand, they cause the dough to rise. The ethanol product of fermentation evaporates during baking.

In **lactate fermentation**, electrons and hydrogen ions carried by NADH are transferred directly to pyruvate, so NAD^+ forms. This reaction converts the pyruvate to lactic acid (also called lactate). Both molecules have three carbons: No carbons are lost, so no CO_2 is produced. We use lactate fermentation by beneficial bacteria to prepare many foods. Yogurt, for example, is made by allowing bacteria such as *Lactobacillus bulgaricus* and *Streptococcus thermophilus* to grow in milk (Figure 5.14A). Milk contains a disaccharide (lactose) and a protein (casein). The cells first break down the lactose into its monosaccharide subunits, then use the monosaccharides for lactate fermentation. The lactate they produce reduces the pH of the milk, which imparts tartness and causes the casein to form a gel.

Cells in animal skeletal muscles are fused as long fibers that carry out aerobic respiration, lactate fermentation, or both (Figure 5.14B). Red fibers have many mitochondria and produce ATP mainly by aerobic respiration. These fibers sustain prolonged activity. They are red because they contain myoglobin, a protein that stores oxygen for aerobic respiration. White muscle fibers contain few mitochondria and no myoglobin; they make most of their ATP by lactate fermentation. This pathway makes ATP quickly, so it is useful for quick, strenuous bursts of activity (Figure 5.14C). The low ATP yield does not support prolonged activity.

Most animal muscles are a mixture of white and red fibers, but the proportions vary. For example, great sprinters tend to have more white fibers in their leg muscles; great marathon runners have more red fibers. Chickens cannot fly far because their flight muscles consist mostly of white fibers (thus, the "white" breast meat). A chicken most often walks or runs. Its leg muscles consist mostly of red muscle fibers, the "dark meat." Section 20.4 returns to the structure and function of skeletal muscle fibers.

Figure 5.14 Examples of lactate fermentation.

(A) left, mexrix/Shutterstock.com; right SCIMAT/Science Source; (B) © William MacDonald, M.D.; (C) © Maxisport/Shutterstock.com.

A. Yogurt is a product of lactate fermentation by bacteria in milk. The micrograph shows *Lactobacillus bulgaricus* (red) and *Streptococcus thermophilus* (purple) in yogurt.

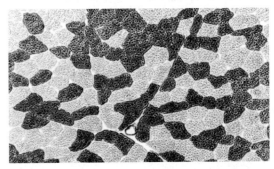

B. Lactate fermentation occurs in white muscle fibers, visible in this cross-section of human thigh muscle. The red fibers, which make ATP by aerobic respiration, sustain endurance activities.

C. Intense activity such as sprinting quickly depletes oxygen in muscles. Under the resulting anaerobic conditions, ATP is produced mainly by lactate fermentation in white muscle fibers. Fermentation does not make enough ATP to sustain this type of activity for long.

Take-Home Message 5.6

What is fermentation?

- Prokaryotes and eukaryotes use fermentation pathways, which are anaerobic, to produce ATP by breaking down carbohydrates.
- Fermentation's small ATP yield (two per molecule of glucose) occurs by glycolysis.

5.7 Food as a Source of Energy

REMEMBER: Animals store sugars in the form of glycogen (Section 2.7). Triglycerides, the main component of fats, have a glycerol head and three fatty acid tails (2.8). Proteins are polymers of amino acids (2.9).

Aerobic respiration produces a lot of ATP by fully dismantling glucose, carbon by carbon. Cells also dismantle other organic molecules to make ATP. Complex carbohydrates, fats, and proteins in food can be converted to molecules that enter glycolysis or the Krebs cycle (Figure 5.15). As in glucose metabolism, the energy of electrons that are transferred to coenzymes ultimately drives the synthesis of ATP.

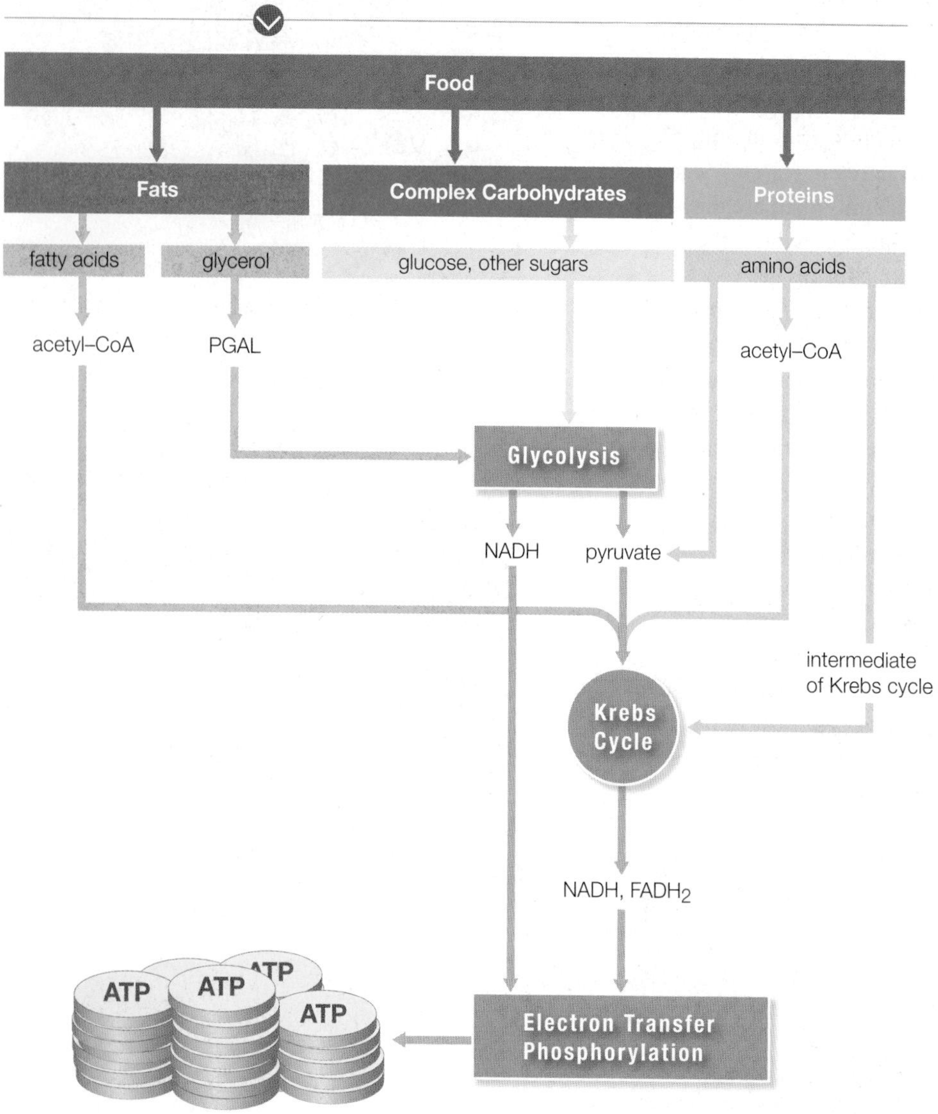

Complex Carbohydrates In humans and other mammals, the digestive system breaks down starch and other complex carbohydrates to their sugar subunits, which are quickly taken up by cells for glycolysis. When a cell produces more ATP than it uses, ATP accumulates in the cytoplasm. The high concentration of ATP causes glucose to be diverted away from glycolysis and into a pathway that builds glycogen. Liver and muscle cells especially favor the conversion of glucose to glycogen, and these cells contain the body's largest stores of it. Between meals, the liver maintains the blood glucose level by breaking down the stored glycogen to release its glucose monomers.

What happens if you eat too many carbohydrates? When the blood level of glucose gets too high, acetyl–CoA is diverted away from the Krebs cycle and into a pathway that makes fatty acids. This is why excess dietary carbohydrate ends up as fat.

Fats A triglyceride molecule has a glycerol head and three fatty acid tails. Cells dismantle triglycerides in fats by first breaking the bonds that connect fatty acid tails to the glycerol head. Nearly all cells in the body can break down the released fatty acids for energy. First, their long backbones are split into two-carbon fragments. These fragments are converted to acetyl–CoA, which can enter the Krebs cycle. The glycerol released by triglyceride breakdown is converted by liver cells to an intermediate of glycolysis.

On a per carbon basis, fats are a richer source of energy than carbohydrates. This is because typical carbohydrate backbones have many oxygen atoms bonded to them. By contrast, most fatty acid tails have no oxygen atoms bonded to them, so more reactions with oxygen are required to fully break their backbones apart. Coenzymes accept electrons in reactions that break apart carbon backbones. The more

Figure 5.15 A variety of organic compounds from food can enter the reactions of aerobic respiration.

Above, © shabaneiro/Shutterstock.com.

Food

Fats → fatty acids → acetyl–CoA

Fats → glycerol → PGAL

Complex Carbohydrates → glucose, other sugars

Proteins → amino acids → acetyl–CoA

Glycolysis → NADH

Glycolysis → pyruvate

intermediate of Krebs cycle

Krebs Cycle → NADH, FADH$_2$

Electron Transfer Phosphorylation → ATP

Digging Into Data

Biofuels

A lot of energy is locked up in the chemical bonds of molecules made by plants. That energy can fuel consumers, as when an animal cell powers ATP synthesis by aerobic respiration. It can also fuel our cars, which run on energy released by burning biofuels or fossil fuels. Both processes are fundamentally the same: They release energy by breaking the bonds of organic molecules. Both use oxygen to break those bonds, and both produce carbon dioxide. Unlike fossil fuels, biofuels are a renewable source of energy: We can always make more of them simply by growing more plants. Also unlike fossil fuels, biofuels do not contribute to global climate change, because growing plant matter for fuel recycles carbon that is already in the atmosphere.

Corn, soy, sugarcane, and other food crops are rich in oils, starches, and sugars that can be easily converted to biofuels. The starch in corn kernels, for example, can be enzymatically broken down to glucose, which is fermented to ethanol by bacteria or yeast. However, growing food crops for biofuel production typically requires a lot of energy (in the form of fossil fuels) and it damages the environment. Making biofuels from other plant matter such as weeds or agricultural waste requires additional steps, because these materials contain a higher proportion of cellulose. Breaking down this tough carbohydrate to its glucose monomers adds cost to the biofuel product.

In 2006, David Tilman and his colleagues published the results of a 10-year study comparing the net energy output of various biofuels. The researchers made biofuel from a mixture of native perennial grasses grown without irrigation, fertilizer, pesticides, or herbicides, in sandy soil that was so depleted by intensive agriculture that it had been abandoned. The energy content of this biofuel and the energy it took to produce it were measured and compared with that of biofuels made from food crops (Figure 5.16).

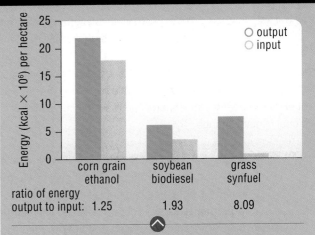

Figure 5.16 Energy inputs and outputs of various biofuels. Input: calculated energy used to grow the crop and produce the biofuel. Output: actual energy content of the biofuel. Corn and soy were grown on fertile farmland; grass, in depleted soil. One hectare is about 2.5 acres.

1. About how much energy did ethanol produced from one hectare of corn yield? How much energy did it take to grow and produce that ethanol?
2. Which of the three crops required the least amount of land to produce a given amount of biofuel energy?
3. The production of which biofuel was most efficient (which had the highest ratio of energy output to energy input)?

coenzymes that accept electrons, the more electrons can be delivered to the ATP-forming machinery of electron transfer phosphorylation.

Proteins Enzymes in the digestive system split dietary proteins into their amino acid subunits, which are absorbed into the bloodstream and used to build proteins or other molecules. When you eat more protein than your body needs for this purpose, the amino acids are broken down. The amino group is removed, and it becomes ammonia, a waste product eliminated in urine. The carbon backbone is split, and acetyl–CoA, pyruvate, or an intermediate of the Krebs cycle forms, depending on the amino acid. These molecules enter aerobic respiration's second stage.

Take-Home Message 5.7

Can the body break down molecules other than sugars for energy?

- Breaking the carbon backbone of any organic molecule releases electrons. Those electrons carry energy that can be harnessed to drive ATP formation in aerobic respiration.
- First the digestive system and then individual cells convert molecules in food (complex carbohydrates, triglycerides, and proteins) into substrates of glycolysis or aerobic respiration's second-stage reactions.

Summary

Section 5.1 By the pathway of photosynthesis, the energy of light is used to build sugars from water and carbon dioxide. Photosynthesis removes CO_2 from the atmosphere, and the metabolic activity of most organisms puts it back. This global cycle was balanced for millions of years. Human activities, especially burning fossil fuels, are currently disrupting the cycle by adding massive amounts of extra CO_2 to the atmosphere. The resulting imbalance is fueling global climate change.

Section 5.2 Light energy travels in waves. Visible light drives photosynthesis, which begins when light energy is absorbed by photosynthetic pigments. A **pigment** absorbs light of particular **wavelengths** only; wavelengths not captured are reflected as its characteristic color. The main photosynthetic pigment, **chlorophyll *a***, absorbs violet and red light, so it appears green. Accessory pigments absorb additional wavelengths.

In chloroplasts, the light-dependent reactions of photosynthesis occur at a much-folded **thylakoid membrane.** The light-independent reactions occur in the chloroplast's cytoplasm-like **stroma.** An overview is shown below:

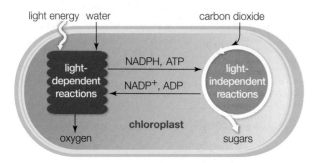

Section 5.3 Clusters of photosynthetic pigments in the thylakoid membrane absorb the energy in light and pass it to photosystems. Receiving energy causes photosystems to release electrons. The electrons flow through electron transfer chains in the thylakoid membrane, and end up in NADPH. Molecules of the electron transfer chain use energy released by the electrons to set up a hydrogen ion gradient across the thylakoid membrane. The ions move back across the membrane through ATP synthases embedded in it. The flow causes these proteins to produce ATP, a process called **electron transfer phosphorylation**.

Photosynthesis releases oxygen because a photosystem replaces lost electrons by pulling them from water molecules, which break apart into hydrogen ions and oxygen as a result.

Section 5.4 Carbon fixation occurs as part of the light-independent reactions of photosynthesis. In the stroma of chloroplasts, the **Calvin–Benson cycle** builds carbon backbones of sugars using carbon atoms from CO_2. This cyclic pathway is driven by phosphate-group transfers from ATP and electrons delivered by NADPH (both of these molecules form during light-dependent reactions).

On hot, dry days, a plant conserves water by closing **stomata**, so carbon dioxide for the light-independent reactions cannot enter it, and oxygen produced by the light-dependent reactions cannot leave. Oxygen buildup in plant tissues reduces the efficiency of sugar production in **C3 plants**. Additional **carbon fixation** reactions in **C4 plants** and **CAM plants** make sugar production more efficient on hot, dry days.

Section 5.5 When photosynthesis evolved, oxygen released by organisms that used it permanently changed the atmosphere, with profound effects on life's evolution. Organisms that could not tolerate the increased atmospheric oxygen persisted only in **anaerobic** habitats. Oxygen-detoxifying pathways allowed other organisms to thrive under **aerobic** conditions.

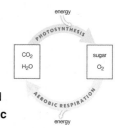

Most modern organisms convert the chemical energy of carbohydrates to the chemical energy of ATP by oxygen-requiring **aerobic respiration**. In eukaryotes, this pathway starts with **glycolysis** in cytoplasm, and ends in mitochondria. Coenzymes pick up electrons in aerobic respiration's first two stages, glycolysis and the **Krebs cycle**. The energy of those electrons drives ATP synthesis in the third stage, electron transfer phosphorylation. At the end of the electron transfer chains, oxygen accepts electrons and hydrogen ions, so water forms. Aerobic respiration yields about thirty-six ATP per glucose.

Section 5.6 Anaerobic **fermentation** pathways include **alcoholic fermentation** and **lactate fermentation**. Both begin with glycolysis, and they run in the cytoplasm. The electron acceptor at the end of these reactions is an organic molecule. The final steps produce no ATP. Thus, the breakdown of one glucose molecule yields only the two ATP from glycolysis.

Section 5.7 Organic molecules other than sugars can be broken down to make ATP. In humans and other mammals, first the digestive system and then individual cells convert fats, proteins, and complex carbohydrates in food to molecules that are substrates of glycolysis or the second-stage reactions of aerobic respiration.

Self-Quiz

1. Most of the carbon that land plants use for photosynthesis comes from _____ .
   - a. glucose
   - b. the atmosphere
   - c. water
   - d. soil

2. Plants use _____ as an energy source to drive photosynthesis.
   - a. sunlight
   - b. sugars
   - c. O_2
   - d. CO_2

3. Which of the following statements is incorrect?
   - a. Pigments absorb light of certain wavelengths only.
   - b. Many accessory pigments are multipurpose molecules.
   - c. Chlorophyll *a* is green because it absorbs green light.

4. In the light-dependent reactions, _____ .
   - a. carbon dioxide is fixed
   - b. ATP forms
   - c. CO_2 accepts electrons
   - d. sugars form

5. When a photosystem absorbs light, _____ .
   - a. sugar phosphates are produced
   - b. electrons are transferred to ATP
   - c. it ejects electrons

6. The atoms in the oxygen molecules released during photosynthesis come from _____ .
   - a. glucose
   - b. CO_2
   - c. water
   - d. O_2

7. The Calvin–Benson cycle starts when _____ .
   - a. light is available
   - b. carbon is fixed
   - c. electrons leave a photosystem

8. Closed stomata _____ .
   - a. limit gas exchange
   - b. permit water loss
   - c. restrict photosynthesis
   - d. absorb light

9. True or false? Plants make all of their ATP by photosynthesis.

10. In the third stage of aerobic respiration, _____ is the final acceptor of electrons.
   - a. water
   - b. H^+
   - c. O_2
   - d. NADH

11. In eukaryotes, the final reactions of aerobic respiration are completed in _____ .
   - a. the nucleus
   - b. mitochondria
   - c. the plasma membrane
   - d. cytoplasm

12. In eukaryotes, the final reactions of fermentation are completed in _____ .
   - a. the nucleus
   - b. mitochondria
   - c. the plasma membrane
   - d. cytoplasm

13. Your body cells can break down _____ as a source of energy to fuel ATP production.
   - a. fatty acids
   - b. glycerol
   - c. amino acids
   - d. all of the above

14. Which of the following metabolic pathways require(s) molecular oxygen (O_2)?
   - a. aerobic respiration
   - b. lactate fermentation
   - c. alcoholic fermentation
   - d. photosynthesis

15. Match the term with the best description.
   - ____ pyruvate
   - ____ anaerobic
   - ____ mitochondrion
   - ____ pigment
   - ____ carbon dioxide
   - ____ rubisco
   - ____ photosynthesis
   - a. no oxygen required
   - b. converts light to chemical energy
   - c. product of glycolysis
   - d. aerobic respiration in eukaryotes
   - e. carbon-fixing enzyme
   - f. like an antenna
   - g. big in the atmosphere

Critical Thinking

1. While looking into an aquarium, you see bubbles coming from an aquatic plant (right). What are the bubbles?

2. How is the function of the thylakoid membrane similar to that of the inner mitochondrial membrane?

3. Which of the following is *not* produced by an animal muscle cell operating under anaerobic conditions?
   - a. heat
   - b. lactate
   - c. ATP
   - d. NAD^+
   - e. pyruvate
   - f. all are produced

4. The bar-tailed godwit is a type of shorebird that makes an annual migration from Alaska to New Zealand and back. The birds make each 11,500-kilometer (7,145-mile) trip by flying over the Pacific Ocean in about nine days. One bird was observed to make the entire journey uninterrupted, a feat that is comparable to a human running a nonstop seven-day marathon at 70 kilometers (43.5 miles) per hour. Would you expect the flight (breast) muscles of bar-tailed godwits to be light or dark colored? Explain your answer.

6 DNA STRUCTURE AND FUNCTION

Application ❱ 6.1 Cloning

REMEMBER: A cell's DNA contains all of the information necessary to build a new cell and, in the case of multicelled organisms, an entire individual (Section 2.10).

On September 11, 2001, Constable James Symington drove his search dog Trakr from Nova Scotia to Manhattan. Within hours of arriving, the dog led rescuers to the area where the final survivor of the World Trade Center attacks was buried. She had been clinging to life, pinned under rubble from the building where she had worked. Symington and Trakr helped with the search and rescue efforts for three days nonstop, until Trakr collapsed from smoke and chemical inhalation, burns, and exhaustion.

Trakr survived the ordeal, but later lost the use of his limbs from a degenerative neurological disease probably linked to toxic smoke exposure at Ground Zero. The hero dog died in April 2009, but his DNA lives on in his genetic copies—his **clones**. Symington's essay about Trakr's superior nature and abilities as a search and rescue dog won the Golden Clone Giveaway, a contest to find the world's most clone-worthy dog. Trakr's DNA was shipped to Korea, where it was inserted into donor dog eggs, which were then implanted into surrogate mother dogs. Five puppies, all clones of Trakr, were delivered to Symington in July 2009 (Figure 6.1A). Today, Trakr's clones are search and rescue dogs for Team Trakr Foundation, an international humanitarian organization that Symington established in 2010.

Trakr's clones were produced by **somatic cell nuclear transfer** (**SCNT**), a laboratory procedure in which an unfertilized egg's nucleus is replaced with the nucleus of a donor's somatic cell (Figure 6.2). A somatic cell is a body cell, as opposed to a reproductive cell (*soma* is a Greek word that means body). If all goes well, the egg's cytoplasm reprograms the transplanted DNA to direct the development of an embryo, which is then implanted into a surrogate mother. The animal born to the surrogate is a clone of the donor.

SCNT is possible because all cells descended from a fertilized egg inherit the same DNA. That DNA is like a master blueprint that directs the development of the individual's body. Thus, the DNA in each body cell of a multicelled individual

> The DNA in each body cell of a multicelled individual contains all the information necessary to build the individual all over again.

A. Left, James Symington and his dog Trakr assisted in the search for victims at Ground Zero, September 2001. Symington wrote an essay about Trakr's superior abilities in search and rescue operations such as this one, and won an opportunity to have the dog cloned. Right, Symington with Trakr's clones in 2009.

B. Champion dairy cow Liz (right) and her clone, Liz II (left). Liz II, who was produced by SCNT, had already begun to win championships by the time she was one year old.

Figure 6.1 Examples of adult animal cloning.

(A) left, Splash News/Newscom; right, Ben Glass, courtesy of © BioArts International; (B) Courtesy of Cyagra, Inc.

contains all the information necessary to build the individual all over again. In fact, clones occur all the time in nature. Body cells of many plants and some animal species easily give rise to new individuals that are clones of the parent. Identical twins are the product of embryo splitting, another natural process. The first few divisions of a fertilized egg form a ball of cells that sometimes splits spontaneously. If both halves of the ball continue to develop independently, identical twins result. Humans have long exploited this phenomenon with a technique called artificial embryo splitting. A tiny ball of cells is grown from a fertilized egg in a laboratory. The ball is teased apart into two halves, each of which goes on to develop as a separate embryo. The embryos are implanted in surrogate mothers, who give birth to identical twins.

Twins produced by embryo splitting are identical to one another, but they are not identical to either parent. This is because, in humans and other animals, twins get their DNA from two parents that typically differ in their DNA sequence (Chapter 8 returns to this topic). Thus, animal breeders who want a clone of a specific individual use SCNT. Clones produced by SCNT have the same championship features as their adult donor animals (Figure 6.1B), and there are other benefits. SCNT can yield many more offspring in a given time frame than traditional breeding, and offspring can be produced from a donor animal that is castrated or even dead.

"Cloning" means making an identical copy of something, and it can refer to deliberate interventions in reproduction intended to produce an exact genetic copy of an organism. **Reproductive cloning** refers to any technology, especially SCNT, that yields animal clones. SCNT has been used since 1997, when a lamb called Dolly was cloned from a mammary cell of an adult sheep (the clone was named after voluptuous performer Dolly Parton). At first, Dolly looked and acted like a normal sheep. However, she died early, most likely because she was a clone. SCNT is technically challenging and the outcome is still unpredictable. Depending on the species, few implanted embryos may survive until birth. Until recently, most of the clones that did survive had serious health problems such as enlarged organs and obesity. Cloned mice developed lung and liver problems, and almost all died prematurely. Cloned pigs tended to limp and have heart problems; some developed without a tail or, even worse, an anus.

Why the problems? Even though a somatic cell contains all the DNA required to produce a new individual, it will not automatically start dividing and form an embryo. During early development, an embryo's cells start using different subsets of their DNA. As they do, the cells become different in form and function, a process called **differentiation**. Differentiation is usually a one-way path in animal cells, which means that once a cell has become specialized, all of its descendant cells will be specialized the same way. By the time a liver cell, muscle cell, or other differentiated cell forms, most of its DNA has been turned off, and is no longer used (Chapter 10 returns to this topic). To clone an adult, scientists must transform one of its differentiated cells into an undifferentiated cell by turning the unused DNA back on, reprogramming it to function like the DNA of an egg. Even though we are getting better at doing that, we still have quite a bit to learn. SCNT technology has improved so much in recent years that health problems are much less common in animals cloned today.

Cloning animals brings us closer to the possibility of cloning humans, both technically and ethically. The idea raises uncomfortable ethical questions. For example, if cloning a lost animal for a grieving owner is acceptable, why would it not be acceptable to clone a lost child for a grieving parent? Different people have very different answers to such questions, so controversy over cloning continues to rage even as the technique improves.

A. A cow's egg is held in place by suction through a hollow glass tube called a micropipette. DNA is identified by a purple stain.

B. Another micropipette punctures the egg and sucks out the DNA. All that remains inside the egg's plasma membrane is cytoplasm.

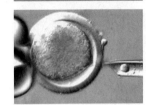

C. A new micropipette prepares to enter the egg at the puncture site. The pipette contains a cell grown from the skin of a donor animal.

D. The micropipette enters the egg and delivers the skin cell to a region between the cytoplasm and the plasma membrane.

E. After the pipette is withdrawn, the donor's skin cell is visible next to the cytoplasm of the egg. The transfer is complete.

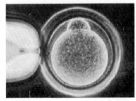

F. An electric current causes the foreign cell to fuse with and empty its nucleus into the cytoplasm of the egg. The egg begins to divide, and an embryo forms.

Figure 6.2 Somatic cell nuclear transfer (SCNT) with cells from cattle.
This series of micrographs was taken by scientists at Cyagra, a company that specializes in cloning livestock.

Courtesy of Cyagra, Inc., www.cyagra.com.

clone Genetically identical copy of an organism.
differentiation Process by which cells become specialized during development.
reproductive cloning Any of several laboratory procedures that produce genetically identical animals.
somatic cell nuclear transfer (SCNT) Reproductive cloning method in which the DNA of an adult donor's body cell is transferred into an unfertilized egg.

6.2 Fame, Glory, and DNA Structure

REMEMBER: Radioisotope tracers (Section 2.2) were used in research that led to the discovery that DNA is the hereditary material of all organisms. Protein structure varies greatly, but patterns such as helices are common (2.9). DNA is a polymer of nucleotides that have been linked into a chain (2.10).

Discovery of DNA's Function DNA, the substance (Figure 6.3), was first described in 1869 by Johannes Miescher, a chemist who extracted it from cell nuclei. Miescher determined that DNA is not a protein, and that it is rich in nitrogen and phosphorus, but he never learned its function. Sixty years later, Frederick Griffith unexpectedly found a clue. Griffith was studying pneumonia-causing bacteria in the hope of creating a vaccine. He discovered that these deadly bacteria contained a substance that could transform harmless bacteria into lethal ones. The transformation was permanent and heritable: Even after hundreds of generations, descendants of transformed cells retained the ability to kill.

What was the substance that had caused this transformation? In 1940, Oswald Avery and Maclyn McCarty set out to identify the substance, which they called the "transforming principle." The team extracted lipids, proteins, and nucleic acids from pneumonia-causing bacteria, then used a process of elimination to see which component transformed harmless bacteria into killers. Treating the extract with lipid- and protein-destroying enzymes did not destroy the transforming principle. Thus, the substance could not be lipid or protein. Avery and McCarty realized that the transforming principle must be nucleic acid—DNA or RNA. DNA-degrading enzymes destroyed the extract's ability to transform cells, but RNA-degrading enzymes did not. Thus, DNA had to be the transforming principle.

The result surprised Avery and McCarty, who, along with most other scientists, had assumed that proteins were the material of heredity. After all, traits are diverse, and proteins are the most diverse of all biological molecules. The two scientists were so skeptical that they published their results only after they had convinced themselves, by years of painstaking experimentation, that DNA was indeed hereditary material. They were also careful to point out that they had not proven DNA was the *only* hereditary material.

Avery and McCarty's tantalizing results prompted a stampede of other scientists into the field of DNA research. The resulting explosion of discovery confirmed the molecule's role as carrier of hereditary information. Key in this advance was the realization that any molecule—DNA or otherwise—had to have certain properties in order to function as hereditary material. First, a full complement of hereditary information must be transmitted along with the molecule; second, cells of a given species should contain the same amount of it; third, because the molecule functions as a genetic bridge between generations, it has to be exempt from major change; and fourth, it must be capable of encoding the almost unimaginably huge amount of information required to build a new individual.

In the late 1940s, Alfred Hershey and Martha Chase proved that DNA, and not protein, satisfies the first property of a hereditary molecule: It transmits a full complement of hereditary information. Hershey and Chase specialized in working with bacteriophages, a type of virus that infects bacteria. Like all viruses, these infectious particles carry information about how to make new viruses in their hereditary material. After a virus injects a cell with this material, the cell starts making new virus particles. Hershey and Chase carried out an elegant series of experiments proving that the material a bacteriophage injects into bacteria is DNA, not protein.

Figure 6.3 Human DNA, the substance.
DNA's role as the carrier of hereditary information was uncovered over many years, as scientists built upon one another's discoveries.

Patrick Landmann/Science Source.

Digging Into Data

The Hershey–Chase Experiments

By 1950, researchers had discovered bacteriophages, a type of virus that infects bacteria. Like all viruses, these infectious particles carry hereditary information about how to make new viruses. After a virus infects a cell, the cell starts making new virus particles. Bacteriophages inject genetic material into bacteria, but was that material DNA, protein, or both? Alfred Hershey and Martha Chase carried out experiments to determine the composition of that material. These experiments were based on the knowledge that proteins contain more sulfur (S) than phosphorus (P), and DNA contains more phosphorus than sulfur. Bacteriophage DNA and protein were labeled with radioactive tracers and allowed to infect bacteria. The virus–bacteria mixtures were then whirled in a blender to dislodge any viral components attached to the exterior of the bacteria (Figure 6.4). Afterward, radioactivity from the tracers was measured. The graph shown in Figure 6.4C, which summarizes results from these experiments, is reproduced from an original publication by Hershey and Chase.

Figure 6.4 Hershey–Chase experiments.

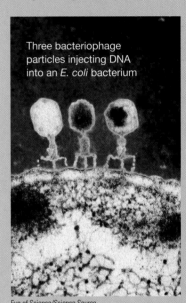

Three bacteriophage particles injecting DNA into an *E. coli* bacterium

Eye of Science/Science Source.

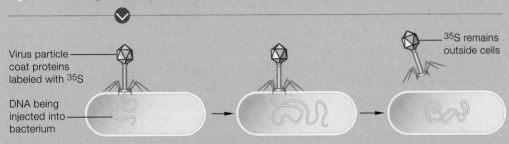

35S remains outside cells

Virus particle coat proteins labeled with 35S

DNA being injected into bacterium

A. In one experiment, bacteriophage were labeled with a radioisotope of sulfur (35S), a process that makes their protein components radioactive. The labeled viruses were mixed with bacteria long enough for infection to occur, and then the mixture was whirled in a kitchen blender. Blending dislodged viral parts that remained on the outside of the bacteria. Afterward, most of the radioactive sulfur was detected outside the bacterial cells. The viruses had not injected protein into the bacteria.

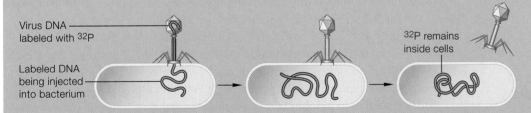

Virus DNA labeled with 32P

32P remains inside cells

Labeled DNA being injected into bacterium

B. In another experiment, bacteriophage were labeled with a radioisotope of phosphorus (32P), which makes their DNA radioactive. The labeled viruses were allowed to infect bacteria. After the external viral parts were dislodged from the bacteria, the radioactive phosphorus was detected mainly inside the bacterial cells. The viruses had injected DNA into the cells—evidence that DNA is the genetic material of this virus.

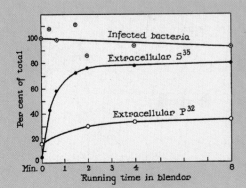

C. Detail from Alfred Hershey and Martha Chase's publication describing their experiments with bacteriophages. In this graph, "infected bacteria" refers to the percentage of bacteria that survived the blender.

Journal of General Physiology, 36(1), Sept. 20, 1952.

1. Before blending, what percentage of each isotope, 35S and 32P, was outside of the bacteria (extracellular)?

2. After 4 minutes in the blender, what percentage of each isotope was extracellular?

3. How did the researchers know that the radioisotopes in the fluid came from outside of the bacterial cells and not from bacteria that had been broken apart by whirling in the blender?

4. The extracellular concentration of which isotope increased the most with blending? Why do these results imply that bacteriophage viruses inject DNA into bacteria?

Discovery of DNA's Structure DNA is a polymer of four types of nucleotides—adenine (A), guanine (G), thymine (T), and cytosine (C). Each has a five-carbon sugar, three phosphate groups, and a nitrogen-containing base after which it is named (Figure 6.5). Just how those four nucleotides are arranged in a DNA molecule was a puzzle that took over 50 years to solve. As molecules go, DNA is gigantic, and chromosomal DNA has a complex structural organization; both factors made the molecule difficult to work with given the laboratory methods at the time.

Clues about DNA's structure started coming together around 1950, when Erwin Chargaff (one of many researchers investigating its function) made two important discoveries about the molecule. First, the amounts of thymine and adenine are identical, as are the amounts of cytosine and guanine (A = T and G = C). We call this discovery Chargaff's first rule. Chargaff's second discovery, or rule, is that the DNA of different species differs in the proportions of adenine and guanine.

Meanwhile, biologist James Watson and biophysicist Francis Crick had been sharing ideas about the structure of DNA. The helical (coiled) pattern of secondary structure that occurs in many proteins had just been discovered, and Watson and Crick suspected that the DNA molecule was also a helix. The two spent many hours arguing about the size, shape, and bonding requirements of the four DNA nucleotides. They pestered chemists to help them identify bonds they might have overlooked, fiddled with cardboard cutouts, and made models from scraps of metal connected by suitably angled "bonds" of wire.

Biochemist Rosalind Franklin (right) had also been working on the structure of DNA. Like Crick, Franklin was expert in x-ray crystallography, a technique in which x-rays are directed through a purified and crystallized substance. Atoms in the substance's molecules scatter the x-rays in a pattern that can be captured as an image. Researchers can use the pattern to calculate the size, shape, and spacing between any repeating elements of the molecules—all of which are details of molecular structure.

Franklin had been told she would be the only one in her department working on the structure of DNA, so she did not know that Maurice Wilkins was already doing the same thing just down the hall. No one had told Wilkins about Franklin's assignment; he assumed she was a technician hired to do his x-ray crystallography work. And so a clash began. Wilkins thought Franklin displayed an appalling lack of deference that technicians of the era usually accorded researchers. To Franklin, Wilkins seemed prickly and oddly overinterested in her work.

Variations in the sequence of DNA are the foundation of life's diversity.

Figure 6.5 The four nucleotides that make up DNA.
Each kind has three phosphate groups, a deoxyribose sugar (orange), and a nitrogen-containing base (blue) after which it is named.

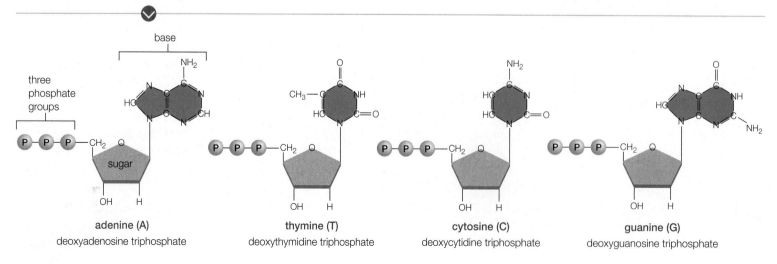

adenine (A)
deoxyadenosine triphosphate

thymine (T)
deoxythymidine triphosphate

cytosine (C)
deoxycytidine triphosphate

guanine (G)
deoxyguanosine triphosphate

Wilkins and Franklin had been given identical samples of DNA. Franklin's meticulous work with hers yielded the first clear x-ray diffraction image of DNA as it occurs in cells. She gave a presentation on this work in 1952. DNA, she said, had two chains twisted into a double helix, with a backbone containing phosphate groups on the outside, and bases arranged in an unknown way on the inside. She had calculated DNA's diameter, the distance between its chains and between its bases, the angle of the helix, and the number of bases in each coil. Crick, with his crystallography background, would have recognized the significance of the work—if he had been there. Watson was in the audience but he was not a crystallographer and did not understand the implications of Franklin's data.

Franklin started to write a research paper on her findings. Meanwhile, and perhaps without her knowledge, Watson reviewed Franklin's x-ray diffraction image with Wilkins, and Watson and Crick read a report detailing Franklin's unpublished data. Crick, who had more experience with molecular modeling than Franklin, immediately understood what the image and the data meant. Franklin's

C. Barrington Brown © 1968 J. D. Watson

data provided Watson and Crick with the last piece of the DNA puzzle. In 1953, they put together all of the clues that had been accumulating for the last fifty years and built the first accurate model of the DNA molecule (left). On April 25, 1953, Franklin's paper appeared third in a series of articles about the structure of DNA in the journal *Nature*. It supported with solid experimental evidence Watson and Crick's theoretical model, which appeared in the first article of the series.

Watson and Crick proposed that a DNA molecule consists of two chains (or strands) of nucleotides, running in opposite directions and coiled into a double helix (Figure 6.6). Covalent bonds between the sugar of one nucleotide and a phosphate of the next form the sugar–phosphate backbone of each chain. Hydrogen bonds between the internally positioned bases hold the two strands together. Only two kinds of base pairings form: A to T, and G to C, which explains the first of Chargaff's rules. Most scientists had assumed (incorrectly) that the bases had to be on the outside of the helix, because they would be more accessible to DNA-copying enzymes that way. You will see in Section 6.4 how these enzymes access the nucleotide bases on the inside of the double helix.

Dozens of scientists contributed to the discovery of DNA's structure, but only three received recognition from the general public for their work. Rosalind Franklin died in 1958. Because the Nobel Prize is not given posthumously, she did not share in the 1962 honor that went to Watson, Crick, and Wilkins for the discovery of the structure of DNA.

DNA Sequence A small piece of DNA from a tulip, a human, or any other organism might be:

one base pair

Notice how the two strands of DNA fit together. They are complementary—the base of each nucleotide on one strand pairs with a suitable partner base on the other. This base-pairing pattern (A to T, G to C) is the same in all molecules of DNA. How can just two kinds of base pairings give rise to the incredible diversity of traits we see among living things? Even though DNA is composed of only four

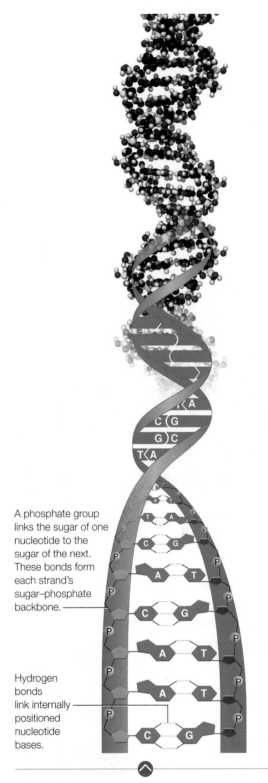

A phosphate group links the sugar of one nucleotide to the sugar of the next. These bonds form each strand's sugar–phosphate backbone.

Hydrogen bonds link internally positioned nucleotide bases.

Figure 6.6 Structure of DNA, as illustrated by a composite of three different models.
The two sugar–phosphate backbones coil in a helix around internally positioned bases.

Figure It Out: What do the yellow balls represent?
Answer: Phosphate groups

nucleotides, the *order* in which one nucleotide follows the next along a strand—the **DNA sequence**—varies tremendously among species (which explains Chargaff's second rule). DNA molecules can be hundreds of millions of nucleotides long, so their sequence can encode a massive amount of information (we return to the nature of that information in the next chapter). DNA sequence variation is the basis of traits that define species and distinguish individuals. Thus DNA, the molecule of inheritance in every cell, is the basis of life's unity. Variations in its sequence are the foundation of life's diversity.

Take-Home Message 6.2

What is DNA?

- DNA is the molecule of inheritance in all organisms.
- A DNA molecule consists of two nucleotide chains (strands) coiled into a double helix. Hydrogen bonding between internally positioned nucleotide bases (A pairs with T, and C with G) hold the two strands together.
- The sequence of bases along a DNA strand—the DNA sequence—varies among species and among individuals. This variation is the basis of life's diversity.

6.3 DNA in Chromosomes

REMEMBER: The DNA of a eukaryotic cell is contained in a nucleus (Section 3.5).

Stretched out end to end, the DNA molecules in a single human cell would be about 2 meters (6.5 feet) long. How can that much DNA cram into a nucleus that is less than 10 micrometers in diameter? Inside a cell, proteins that associate with each DNA molecule twist and pack it into a structure called a **chromosome** (Figure 6.7). In a eukaryotic cell, for example, a DNA molecule ① wraps twice at regular intervals around "spools" of proteins called **histones** ②. These DNA–histone spools look a bit like beads on a string in micrographs. Interactions among histones and other proteins twist the spooled DNA into a tight fiber ③. This fiber coils, and then it coils again into a hollow cylinder like an old-style telephone cord ④.

During most of the cell's life, each chromosome consists of one DNA molecule. When the cell prepares to divide, it duplicates its chromosomes (more about this process in the next section). After replication, each chromosome consists of two DNA molecules, or **sister chromatids**, attached to one another at a constricted region called the **centromere**:

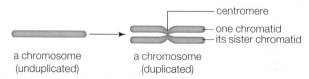

The duplicated chromosomes condense into their familiar "X" shapes ⑤ just prior to cell division.

The DNA of a eukaryotic cell is divided up among some number of chromosomes that differ in length and shape ⑥. That number is called the **chromosome number**, and it is a characteristic of the species. For example, the chromosome number of humans is 46, so our cells have 46 chromosomes.

Figure 6.7 Eukaryotic chromosome structure.
The micrograph shows a duplicated chromosome just before cell division.

Andrew Syred/Science Source.

① Two strands of DNA twist into a double helix.

② At regular intervals, the DNA (blue) wraps around a core of histone proteins (purple).

③ The DNA and proteins associated with it twist tightly into a fiber.

④ The fiber coils and then coils again to form a hollow cylinder.

⑤ At its most condensed, a duplicated chromosome has an X shape.

⑥ The DNA in the nucleus of a typical eukaryotic cell is divided into a number of chromosomes.

Figure It Out: What is the yellow structure in the upper right corner of the drawing?   Answer: A cell nucleus

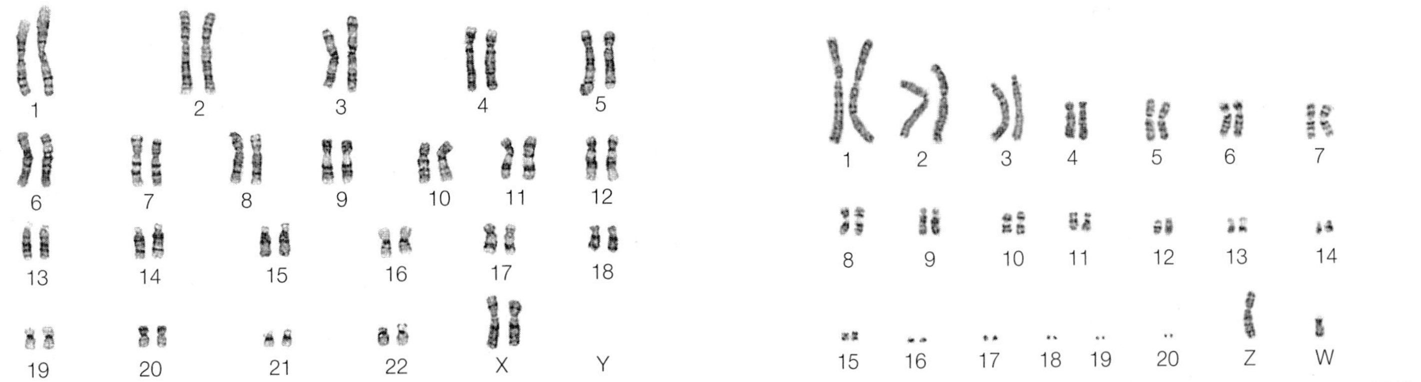

A. Karyotype of a female human, with identical sex chromosomes (XX).

B. Karyotype of a female chicken, with nonidentical sex chromosomes (ZW).

Figure 6.8 Karyotypes.
Chromosomes are numbered (as shown).

(A) © University of Washington Department of Pathology; (B) With kind permission from Springer Science+Business Media: *Chromosome Research,* Volume 17, Number 1, 99 113, DOI: 10.1007/s10577-009-9021-6; *Avian comparative genomics: reciprocal chromosome painting between domestic chicken (Gallus gallus) and the stone curlew (Burhinus oedicnemus, Charadriiformes)—An atypical species with low diploid number,* Wenhui Nie, Patricia C. M. O'Brien, Bee L. Ng, Beiyuan Fu, Vitaly Volobouev, Nigel P. Carter, Malcolm A. Ferguson-Smith, and Fengtang Yang; Figure 2a.

Actually, human body cells have two sets of 23 chromosomes—two of each type. Having two sets of chromosomes means these cells are **diploid**, or 2*n*. An image of an individual's diploid set of chromosomes is called a **karyotype** (Figure 6.8). To create a karyotype, cells taken from the individual are treated to make the chromosomes condense, and then stained so the chromosomes can be distinguished under a microscope. A micrograph of a single cell is digitally rearranged so the images of the chromosomes are lined up by centromere location, and arranged according to size, shape, and length.

In a human body cell, all but one pair of chromosomes are **autosomes**, which are the same in both females and males. The two autosomes of a pair have the same length, shape, and centromere location. They also hold information about the same traits. Think of them as two sets of books on how to build a house. Your father gave you one set. Your mother had her own ideas about wiring, plumbing, and so on. She gave you an alternate set that says slightly different things about many of those tasks.

Members of a pair of **sex chromosomes** differ between females and males, and the differences determine an individual's sex. The sex chromosomes of humans are called X and Y. The body cells of typical human females have two X chromosomes (XX, Figure 6.8A); those of typical human males have one X and one Y chromosome (XY). This pattern—XX females and XY males—is the rule among fruit flies, mammals, and many other animals, but there are other patterns (Figure 6.8B). Female butterflies, moths, birds, and certain fishes have two nonidentical sex chromosomes; the two sex chromosomes of males are identical. Environmental factors (not chromosomes) determine sex in some species of invertebrates, turtles, and frogs. As an example, the temperature of the sand in which sea turtle eggs are buried determines the sex of the hatchlings.

Take-Home Message 6.3

What is a chromosome?

- A chromosome is a molecule of DNA together with associated proteins that organize it and allow it to pack tightly.
- A eukaryotic cell's DNA is divided among a characteristic number of chromosomes, which differ in length and shape.
- Members of a pair of sex chromosomes differ between males and females. Chromosomes that are the same in both sexes are called autosomes.

autosome A chromosome that is the same in males and females.

centromere Of a duplicated eukaryotic chromosome, constricted region where sister chromatids attach to each other.

chromosome Structure that consists of DNA together with associated proteins; carries part or all of a cell's genetic information.

chromosome number The total number of chromosomes in a cell of a given species.

diploid Having two of each type of chromosome characteristic of the species (2*n*).

DNA sequence Order of nucleotides in a strand of DNA.

histone Type of protein that associates with eukaryotic DNA and structurally organizes chromosomes.

karyotype Image of an individual's complement of chromosomes arranged by size, length, shape, and centromere location.

sex chromosome Member of a pair of chromosomes that differs between males and females.

sister chromatids The two attached DNA molecules of a duplicated eukaryotic chromosome.

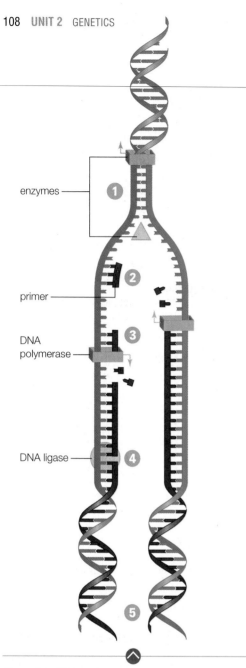

Figure 6.9 DNA replication.

1 As replication begins, enzymes begin to unwind and separate the two strands of DNA.

2 Primers base-pair with the exposed single DNA strands.

3 Starting at primers, DNA polymerases (green boxes) assemble new strands of DNA from nucleotides, using the parent strands as templates.

4 DNA ligase seals any gaps that remain between bases of the "new" DNA.

5 Each parental DNA strand (blue) serves as a template for assembly of a new strand of DNA (magenta). Both strands of the double helix serve as templates, so two double-stranded DNA molecules result. One strand of each is parental (old), and the other is new, so DNA replication is semiconservative.

6.4 DNA Replication and Repair

REMEMBER: With a few exceptions, free radicals are dangerous to life (Section 2.2). An enzyme speeds a reaction; bonds between phosphate groups hold a lot of energy (4.4). The wavelength of light is measured in nanometers (5.2).

When a cell reproduces, it divides. The two descendant cells must inherit a complete copy of genetic information or they will not function properly. Thus, in preparation for division, the cell copies its chromosomes so that it contains two sets: one for each of its future offspring. The copying process is an energy-intensive pathway called **DNA replication**. During DNA replication, the double helix of a DNA molecule is opened to expose the internally positioned bases, and an enzyme, **DNA polymerase**, links nucleotides into new strands of DNA according to the sequence of those bases. Each chromosome is copied in its entirety. Two identical molecules of DNA are the result. In eukaryotes, these molecules are sister chromatids that remain attached at the centromere until cell division occurs.

Before DNA replication, a chromosome has one molecule of DNA—one double helix (Figure 6.9). As replication begins, enzymes break the hydrogen bonds that hold the double helix together, so the two DNA strands unwind and separate **1**. Another enzyme starts making **primers**, which are short, single strands of nucleotides that serve as attachment points for DNA polymerase **2**. The nucleotide bases of a primer can form hydrogen bonds with the exposed bases of a single strand of DNA. Thus, a primer can base-pair with a complementary strand of DNA (right). The establishment of base pairing between two strands of DNA (or DNA and RNA) is called hybridization. Hybridization is spontaneous and is driven entirely by hydrogen bonding.

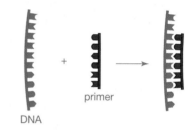

DNA polymerases attach to the hybridized primers and begin DNA synthesis. As a DNA polymerase moves along a strand, it uses the sequence of exposed nucleotide bases as a template, or guide, to assemble a new strand of DNA from free nucleotides **3**. Two of a nucleotide's three phosphate groups are removed when it is added to a DNA strand. Breaking those bonds releases enough energy to drive the attachment.

A DNA polymerase follows base-pairing rules: It adds a T to the end of the new DNA strand when it reaches an A in the template strand; it adds a G when it reaches a C; and so on. Thus, the DNA sequence of each new strand is complementary to the template (parental) strand. The enzyme DNA ligase seals any gaps, so the new DNA strands are continuous **4**. Both strands of the parent molecule are copied at the same time. As each new DNA strand lengthens, it winds up with the template strand into a double helix. So, after replication, two double-stranded molecules of DNA have formed **5**. One strand of each molecule is conserved (parental), and the other is new, so replication is said to be semiconservative. Both double-stranded molecules produced by DNA replication are duplicates of the parent molecule.

How Mutations Arise Sometimes, a new DNA strand is not exactly complementary to its parent strand. A nucleotide may get lost during DNA replication, or an extra one slips in. Occasionally, the wrong nucleotide is added. Most of these replication errors occur simply because DNA polymerases work very fast. Mistakes are inevitable, and some DNA polymerases make a lot of them. Luckily, most DNA polymerases also proofread their work. They can correct a mismatch by reversing the synthesis reaction to remove the mispaired nucleotide, then resuming synthesis in the forward direction. Replication errors also occur after the cell's DNA gets broken or otherwise damaged, because DNA polymerases do not copy damaged DNA very well. In most cases, repair enzymes and other proteins remove and replace damaged or mismatched bases in DNA before replication begins.

When proofreading and repair mechanisms fail to correct an error, it becomes a **mutation**—a permanent change in the DNA sequence of a cell's chromosome(s). Repair enzymes cannot fix a mutation after the altered DNA strand has been replicated, because they do not recognize correctly paired bases (Figure 6.10). Thus, a mutation is passed to one of the cell's offspring and all of its descendants.

Mutations alter DNA's genetic instructions, so they may have a harmful outcome. Cancer begins with mutations. Rosalind Franklin died at the age of 37, of ovarian cancer probably caused by extensive exposure to x-rays during her work. At the time, the link between x-rays, mutations, and cancer was not understood. We now know that electromagnetic energy with a wavelength shorter than 320 nanometers, including x-rays, most ultraviolet (UV) light, and gamma rays, has enough energy to knock electrons out of atoms. Exposure to such ionizing radiation damages DNA, breaking it into pieces that get lost during replication (Figure 6.11). Ionizing radiation can also cause covalent bonds to form between bases on opposite strands of the double helix, an outcome that permanently blocks DNA replication. The nucleotide bases themselves can be irreparably damaged by ionizing radiation. Repair enzymes remove bases damaged in this way, but they leave an empty space in the double helix or even a strand break. Any of these events can result in mutations.

UV light in the range of 320 to 380 nanometers does not have enough energy to knock electrons out of atoms, but it can cause a covalent bond to form between adjacent thymine or cytosine bases. The result is a nucleotide dimer that kinks the DNA strand. DNA polymerase tends to copy the kinked part incorrectly during replication, and mutations are the outcome. Exposing unprotected skin to sunlight increases the risk of cancer because the UV wavelengths cause these nucleotide dimers to form. For every second a skin cell spends in the sun, between 50 and 100 nucleotide dimers form in its DNA.

Exposure to some chemicals also causes mutations. For instance, several of the fifty-five or more cancer-causing chemicals in tobacco smoke transfer methyl groups ($—CH_3$) to the nucleotide bases in DNA. Nucleotides altered in this way do not base-pair correctly. The body converts other chemicals in the smoke to compounds that bind irreversibly to DNA. In both cases, the resulting replication errors can lead to mutation. Cigarette smoke also contains free radicals, which inflict the same damage on DNA as ionizing radiation.

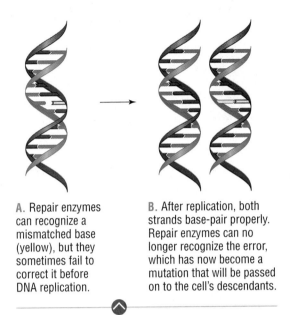

A. Repair enzymes can recognize a mismatched base (yellow), but they sometimes fail to correct it before DNA replication.

B. After replication, both strands base-pair properly. Repair enzymes can no longer recognize the error, which has now become a mutation that will be passed on to the cell's descendants.

Figure 6.10 How replication errors become mutations.

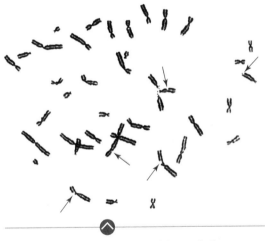

Figure 6.11 Why exposure to ionizing radiation causes mutations.
This micrograph shows major breaks (red arrows) in the chromosomes of human white blood cells that have been exposed to ionizing radiation. Pieces of broken chromosomes often become lost during DNA replication.

Olga Shovman, Andrew C. Riches, Douglas Adamson, and Peter E. Bryant. *An improved assay for radiation-induced chromatid breaks using a colcemid block and calyculin-induced PCC combination.* Mutagenesis (2008) 23(4): 267–270; first published online March 6, 2008. doi:10.1093/mutage/gen009, by permission of Oxford University Press.

Take-Home Message 6.4

What happens during DNA replication?

- When a cell copies its DNA, each strand of the double helix serves as a template for synthesis of a new, complementary strand of DNA. Two double helices result.
- Proofreading and repair mechanisms usually maintain the integrity of a cell's genetic information by correcting mispaired bases and fixing damaged DNA before replication.
- Mismatched or damaged nucleotides that are not repaired become mutations.
- DNA damage by environmental agents such as UV light, chemicals, and free radicals can result in mutations, because damaged DNA is not replicated very well.

DNA polymerase Enzyme that carries out DNA replication.

DNA replication Process by which a cell duplicates its DNA before it divides.

mutation Permanent change in DNA sequence.

primer Short, single strand of DNA that base-pairs with a specific DNA sequence.

Summary

Section 6.1 **Somatic cell nuclear transfer (SCNT)** and other types of **reproductive cloning** technologies produce genetically identical individuals (**clones**). SCNT works because the DNA in each body cell of an animal contains all the information necessary to build a new individual. However, the outcome of SCNT is still unpredictable. This is because, during development, cells of an embryo become specialized as they begin to use different subsets of their DNA (a process called **differentiation**). Reprogramming an adult cell to behave like an embryonic cell involves rewinding this developmental process, and we are still learning how to do it. Cloning animals continues to raise ethical questions, particularly about the possibility of cloning humans.

Section 6.2 It took decades of research by many scientists to determine that deoxyribonucleic acid (DNA) is the hereditary material of all life, and to unravel its structure.

Each DNA nucleotide has a five-carbon sugar, three phosphate groups, and one of four nitrogen-containing bases after which the nucleotide is named: adenine, thymine, guanine, or cytosine. DNA is a polymer that consists of two strands of these nucleotides coiled into a double helix. Hydrogen bonding between the internally positioned bases holds the strands together. The bases pair in a consistent way: adenine with thymine (A–T), and guanine with cytosine (G–C). The order of bases along a strand of DNA—the **DNA sequence**—varies among species and among individuals, and this variation is the basis of life's diversity.

Section 6.3 The DNA of eukaryotes is divided among a characteristic number of **chromosomes** that differ in length and shape. In eukaryotic chromosomes, the DNA wraps around **histones**. When duplicated, a eukaryotic chromosome has an X shape and consists of two **sister chromatids** attached at a **centromere**.

Diploid cells have two of each type of chromosome. **Chromosome number** is the sum of all chromosomes in a cell of a given species. A human body cell has twenty-three pairs of chromosomes. Members of a pair of **sex chromosomes** differ among males and females. Chromosomes that are the same in males and females are **autosomes**. Autosomes of a pair have the same length, shape, and centromere location. A **karyotype** is an image of an individual's complete set of chromosomes.

Section 6.4 Before a cell divides, it copies all of its DNA (by **DNA replication**) so both of its cellular offspring inherit a complete set of chromosomes. For each molecule of DNA that is copied, two DNA molecules are produced; each is a duplicate of the parent. DNA replication is said to be semiconservative because one strand of each molecule is new, and the other is parental.

During DNA replication, enzymes unwind the double helix. **Primers** form and base-pair (hybridize) with the exposed single strands of DNA. Starting at the primers, **DNA polymerase** enzymes use each strand as a template to assemble new, complementary strands of DNA from free nucleotides. DNA ligase seals any gaps.

Proofreading by DNA polymerases corrects most DNA replication errors as they occur. DNA damage by environmental agents, including ionizing and nonionizing radiation, free radicals, and some chemicals, can lead to replication errors because DNA polymerase does not copy damaged DNA very well.

Most DNA damage is repaired before replication begins. Uncorrected replication errors become **mutations**: permanent changes in the nucleotide sequence of a cell's DNA.

Self-Quiz

Answers in Appendix I

1. _____ is an example of reproductive cloning.
   a. Somatic cell nuclear transfer (SCNT)
   b. Multiple offspring from the same pregnancy
   c. Artificial embryo splitting
   d. a and c
   e. all of the above

2. Which is *not* a nucleotide base in DNA?
   a. adenine   c. glutamine   e. cytosine
   b. guanine   d. thymine   f. All are in DNA.

3. What are the base-pairing rules for DNA?
   a. A–G, T–C   c. A–T, G–C
   b. A–C, T–G   d. A–A, G–G, C–C, T–T

4. Variation in _____ is the basis of variation in traits.
   a. karyotype   c. the double helix
   b. DNA sequence   d. chromosome number

5. One species' DNA differs from others in its _____ .
   a. nucleotides
   b. DNA sequence
   c. double helix
   d. replication process
   e. sugar–phosphate backbone
   f. all of the above

6. In eukaryotic chromosomes, DNA wraps around _____ .
   a. histone proteins
   b. sister chromatids
   c. centromeres
   d. nucleotides

7. Chromosome number _____ .
   a. refers to a particular chromosome in a cell
   b. is a characteristic feature of a species
   c. is the number of autosomes in cells of a given type
   d. is the same in all species

8. Human body cells are diploid, which means they _____ .
   a. are complete
   b. have two sets of chromosomes
   c. contain sex chromosomes
   d. divide to form two cells

9. When DNA replication begins, _____ .
   a. the two DNA strands unwind from each other
   b. the two DNA strands condense for base transfers
   c. old strands move to find new strands
   d. mutations occur

10. DNA replication requires _____ .
    a. DNA polymerase
    b. nucleotides
    c. primers
    d. all are required

11. Energy that drives the attachment of a nucleotide to the end of a growing strand of DNA comes from _____ .
    a. phosphate-group transfers from ATP
    b. DNA polymerase
    c. the nucleotide itself
    d. a and c

12. After DNA replication, a eukaryotic chromosome _____ .
    a. consists of two sister chromatids
    b. has a characteristic X shape
    c. is constricted at the centromere
    d. all of the above

13. Exposure to _____ can lead to mutations.
    a. UV light
    b. cigarette smoke
    c. chemicals
    d. x-rays
    e. sunlight
    f. all of the above

14. All mutations _____ .
    a. arise from DNA damage
    b. lead to evolution
    c. are caused by radiation
    d. change the DNA sequence

15. Match the terms appropriately.
    _____ nucleotide
    _____ clone
    _____ autosome
    _____ DNA polymerase
    _____ mutation
    _____ bacteriophage
    _____ semiconservative replication

    a. replication enzyme
    b. does not determine sex
    c. copy of an organism
    d. nitrogen-containing base, sugar, phosphate
    e. injects DNA
    f. can cause cancer
    g. something old, something new

Critical Thinking

1. Mutations are the original source of genetic variation. How can mutations accumulate in DNA, given that cells have repair systems that fix mispaired nucleotides or breaks in DNA strands?

2. Woolly mammoths have been extinct for about 10,000 years, but we often find their well-preserved remains in Siberian permafrost. Research groups are now planning to use SCNT to resurrect these huge elephant-like mammals. No mammoth eggs have been recovered so far, so elephant eggs would be used instead. An elephant would also be the surrogate mother for the resulting embryo. The researchers may try a modified SCNT technique used to clone a mouse that had been dead and frozen for sixteen years. Ice crystals that form during freezing break up cell membranes, so cells from the frozen mouse were in bad shape. Their DNA was transferred into donor mouse eggs, and cells from the resulting embryos were fused with mouse stem cells. Four healthy clones were born from the hybrid embryos. What are some of the pros and cons of cloning an extinct animal?

Visual Question

1. Determine the complementary strand of DNA that forms on this template DNA fragment during replication:

```
G  G  T  T  T  C  T  T  C  A  A  G  A  G  A
|  |  |  |  |  |  |  |  |  |  |  |  |  |  |

_  _  _  _  _  _  _  _  _  _  _  _  _  _  _
```

GENE EXPRESSION AND CONTROL

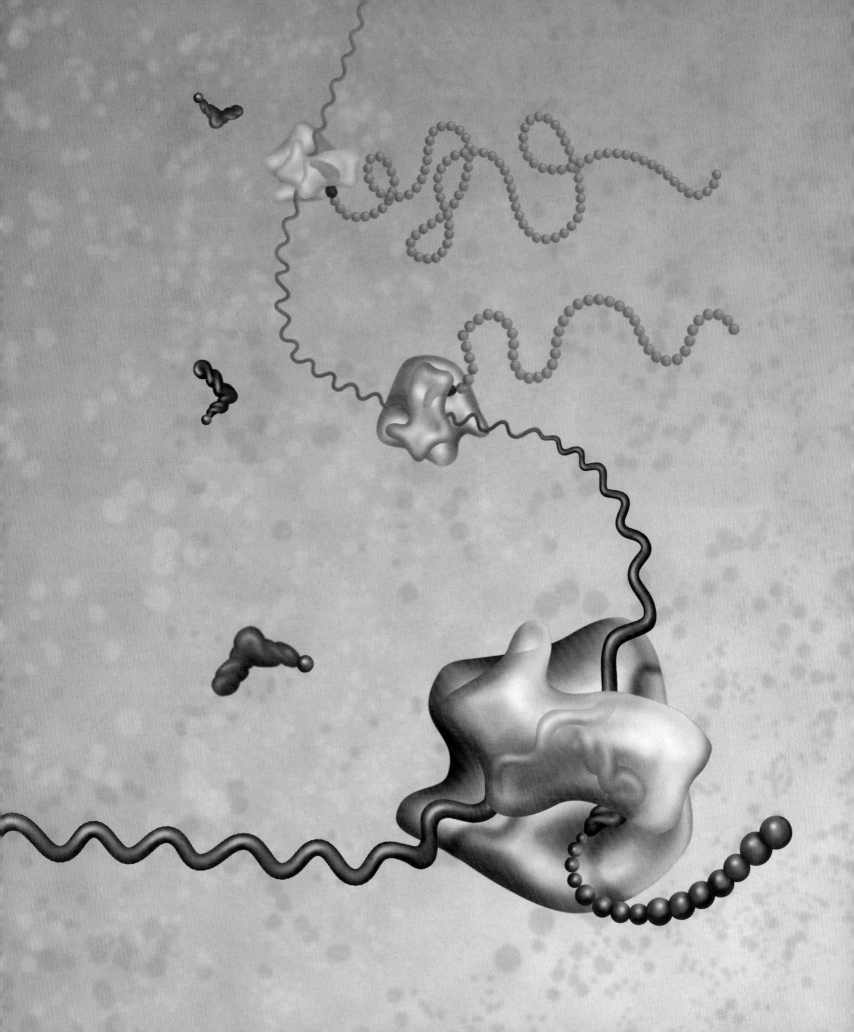

Application ❯ 7.1 **Ricin, RIP**

REMEMBER: A fat consists mainly of triglycerides (Section 2.8). A domain contributes a structural or functional property to a protein (2.9). A few strains of *E. coli* make a toxic protein that can severely damage the lining of the intestine (3.1). Ribosomes assemble polypeptides (3.4). Enzymes are unchanged by participating in a reaction (4.4). Cells can take in materials across their membrane by endocytosis (4.6).

A. Seeds of the castor-oil plant, source of ribosome-busting ricin. Eating just eight of these seeds can kill an adult human.

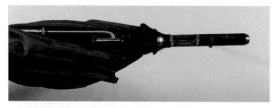

B. Bulgarian spy's weapon: an umbrella modified to fire a tiny pellet of ricin into a victim. An umbrella like this one was used to assassinate Georgi Markov on the streets of London in 1978.

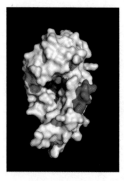

C. Researchers incorporated a peptide that binds to skin cancer cells into the enzymatic domain of Shiga-like toxin. The model shows the active site of the toxin in red; the changed bit, in blue. This engineered RIP specifically kills skin cancer cells.

Figure 7.1 RIPs, examples of source and delivery.

(A) Vaughan Fleming/SPL/Science Source; (B) Cary Wolinsky/National Geographic Creative; (C) Cheung et al. Molecular Cancer, 9:28, 2010.

Castor-oil plants grow wild in tropical regions, and they are widely cultivated for their seeds (Figure 7.1A). Castor-oil seeds are rich in ricin, a toxic protein that effectively deters beetles, birds, mammals, and other animals from eating them. They are also rich in castor oil, a fat that we extract from the seeds to use as an ingredient in plastics, cosmetics, paints, soaps, and many other items. Ricin-containing seed pulp left over from the extraction process is usually—but not always—discarded.

A dose of ricin as small as a few grains of salt can kill an adult human, and there is no antidote. Ricin's lethal effect was known as long ago as 1888, but using it as a weapon is banned by most countries under the Geneva Protocol. Controlling production of the toxin is impossible because no special skills or equipment are required to manufacture it from easily obtained raw materials. Thus, ricin appears periodically in the news as a tool of criminals. Perhaps the most famous example occurred in 1978 at the height of the Cold War, when defectors from countries under Russian control were targets for assassination. Bulgarian journalist Georgi Markov had defected to England and was working for the BBC. As he made his way to a bus stop on a London street, an assassin used a modified umbrella (Figure 7.1B) to fire a tiny pellet of ricin into Markov's leg. Markov died in agony three days later.

Ricin is called a ribosome-inactivating protein (RIP) because it inactivates ribosomes, the organelles that assemble amino acids into proteins. Other RIPs are made by some bacteria, mushrooms, algae, and many plants (including food crops such as tomatoes, barley, and spinach). Most of these proteins are not particularly toxic in humans because they do not cross intact cell membranes very well. Those that do, including ricin, have a domain that binds to molecules on our plasma membranes. Binding causes the cell to take up the RIP by endocytosis. Once inside the cell, a second domain of the RIP—an enzyme—begins to inactivate ribosomes. One molecule of ricin can inactivate more than 1,000 ribosomes per minute. If enough ribosomes are affected, protein synthesis grinds to a halt. Proteins are critical to all life processes, so cells that cannot make them die quickly.

Fortunately, few people actually encounter ricin. Contact with other toxic RIPs is much more common. Bracelets made from beautiful seeds were recalled from stores in 2011 after a botanist recognized the seeds as jequirity beans. These beans contain abrin, an RIP even more toxic than ricin. Shiga toxin, an RIP made by *Shigella dysenteriae* bacteria, causes a severe bloody diarrhea (dysentery) that can be lethal. Some strains of *E. coli* make Shiga-like toxin, an RIP that causes symptoms associated with food poisoning.

Despite their toxicity, the main function of RIPs may not be destroying ribosomes. Many are part of plant immunity, and have antiviral and anticancer activity. Plants that make toxic RIPs have been used as traditional medicines for many centuries; now, Western scientists are exploiting the unique properties of these proteins in drug design. For example, researchers have attached ricin's toxic enzyme domain to antibodies that can find specific types of cancer cells in a person's body. Other RIPs have also been modified to specifically target cancer cells (Figure 7.1C). The intent of both strategies: to assassinate the cancer cells without harming normal ones.

7.2 Gene Expression

REMEMBER: Proteins are polymers of amino acids (Section 2.9). The DNA in a body cell contains enough information to rebuild the individual (6.1). Information is encoded in the sequence of nucleotide bases in a DNA strand (6.2).

You learned in Chapter 6 that chromosomes are like a set of books that provide instructions for building and operating an individual. You already know the alphabet used to write those books: the four letters A, T, G, and C, for the four nucleotides in DNA—adenine, thymine, guanine, and cytosine. In this chapter, we investigate "words" that can be made from those letters, and "sentences" that can be made from the words.

The nature of information represented by the sequence of nucleotides in a DNA molecule occurs in hundreds or thousands of units called genes. The DNA sequence of a **gene** encodes (contains instructions for building) an RNA or protein product. Converting the information encoded by a gene into a product starts with RNA synthesis, which is called transcription. During **transcription**, enzymes use a strand of DNA as a template to assemble a strand of RNA.

Most of the RNA inside cells occurs in single strands that are similar in structure to single strands of DNA (Figure 7.2). For example, both are chains of four kinds of nucleotides. Like a DNA nucleotide, an RNA nucleotide has three phosphate groups, a sugar, and one of four bases. However, the sugar in an RNA nucleotide (ribose) is slightly different from the sugar in a DNA nucleotide (deoxyribose). Three bases (adenine, cytosine, and guanine) occur in both RNA and DNA nucleotides, but the fourth base differs between the two molecules. In DNA, the fourth base is thymine (T); in RNA, it is uracil (U). These small differences in structure give rise to very different functions. DNA's important but only role is to store a cell's genetic information. By contrast, a cell makes several kinds of RNAs on an ongoing basis, and the different types have different functions. **Messenger RNA (mRNA)** was named for its function as the "messenger" between DNA and protein. By the process of **translation**, the protein-building information in an mRNA is decoded (translated) into a sequence of amino acids. The result is a polypeptide that twists and folds into a protein.

Transcription and translation are both part of **gene expression**, the multistep process by which information in a gene guides the assembly of an RNA or protein product. Expression of a gene that encodes an RNA product (such as an mRNA) involves transcription. Expression of a gene that encodes a protein product involves both transcription and translation:

$$DNA \xrightarrow{\textit{transcription}} mRNA \xrightarrow{\textit{translation}} protein$$

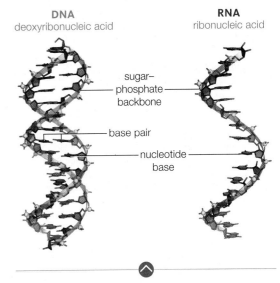

DNA
deoxyribonucleic acid

RNA
ribonucleic acid

sugar–phosphate backbone

base pair

nucleotide base

Figure 7.2 Comparing DNA and RNA.
DNA permanently stores a cell's genetic information. Cells continually make different types of RNAs that have various functions.

Take-Home Message 7.2

What is the nature of the information carried by DNA?

• Information in a DNA sequence occurs in units that are called genes.

• A cell uses the information encoded by a gene to make an RNA or protein product, a process called gene expression.

• Transcription converts information in a gene to RNA; translation converts information in an mRNA to protein.

gene A part of a chromosome that encodes an RNA or protein product in its DNA sequence.

gene expression Process by which the information in a gene guides assembly of an RNA or protein product. Includes transcription and translation.

messenger RNA (mRNA) RNA that has a protein-building message.

transcription Process by which enzymes assemble an RNA using a strand of DNA as a template.

translation Process by which the protein-building instructions in an mRNA guide the assembly of a polypeptide.

7.3 Transcription: DNA to RNA

Figure 7.3 Base pairing in DNA replication and RNA synthesis.

A. During DNA replication, a nucleotide is added to the end of a growing DNA strand only if it base-pairs with the corresponding nucleotide on the template DNA strand. Cytosine (C) pairs with guanine (G), and adenine (A) pairs with thymine (T).

B. During RNA synthesis, a nucleotide is added to the end of a growing RNA strand only if it base-pairs with the corresponding nucleotide on the template DNA strand. The same base-pairing rules apply, except that uracil in RNA base-pairs with adenine (A).

REMEMBER: Only two kinds of base pairings occur in DNA: G pairs with C, and A pairs with T (Section 6.2). Each molecule of DNA produced during DNA replication is a duplicate of the parent (6.4).

The process of transcription is similar to DNA replication. Base-pairing rules are followed, for example. In DNA replication, cytosine (C) pairs with guanine (G), and adenine (A) pairs with thymine (T). The very same base-pairing rules also govern RNA synthesis in transcription, except that RNA, remember, has uracil instead of thymine (Figure 7.3). Uracil (U) base-pairs with adenine (A). During transcription, a strand of DNA acts as a template upon which a strand of RNA is assembled from nucleotides. A nucleotide can be added to a growing RNA only if it is complementary to the corresponding nucleotide of the template DNA. As in DNA replication, each nucleotide provides the energy for its own attachment to the end of a growing strand. Transcription is also similar to DNA replication in that one strand of a nucleic acid serves as a template for synthesis of another. However, in contrast to DNA replication, only part of one DNA strand, not the whole molecule, is used as a template for transcription. The enzyme **RNA polymerase**, not DNA polymerase, adds nucleotides to the end of a growing RNA. Also, transcription produces a single strand of RNA, not two DNA double helices.

In eukaryotic cells, transcription occurs in the nucleus; in prokaryotes, it occurs in cytoplasm. In both cases, the process begins at a regulatory site called a **promoter**, a short DNA sequence that serves as a binding site for RNA polymerase. Generally, a gene's promoter is close to it and a bit upstream:

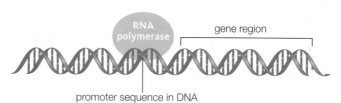

Binding at a promoter positions an RNA polymerase close to the gene that will be transcribed. The polymerase starts moving along the DNA, over the gene region.

Figure 7.4 DNA Christmas trees.
Typically, many RNA polymerases simultaneously transcribe the same gene, producing a structure called a Christmas tree after its shape. Here, several genes next to one another on the same chromosome are being transcribed simultaneously.

© O. L. Miller.

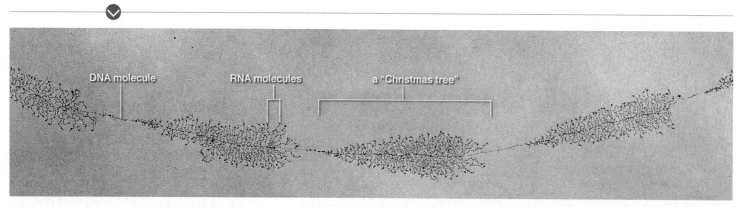

Figure It Out: Are the polymerases transcribing this DNA molecule moving from left to right or from right to left?

Answer: Left to right (the RNAs get longer as the polymerases move along the DNA)

As it moves, the polymerase unwinds the double helix just a bit so it can "read" the nucleotide sequence of the template DNA strand:

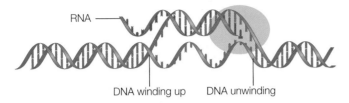

An RNA polymerase moving over a gene region joins free RNA nucleotides into a chain, in the order dictated by the DNA sequence of the gene:

direction of transcription

When the polymerase reaches the end of the gene region, it releases the DNA and the new RNA. The new RNA strand is complementary to the DNA strand from which it was transcribed. It is an RNA copy of a gene, in the same way that a paper transcript of a conversation carries the same information in a different format. Typically, many polymerases transcribe a particular gene region at the same time, so many new RNA strands can be produced very quickly (Figure 7.4).

RNA Modifications Just as a dressmaker may snip off loose threads or add bows to a dress before it leaves the shop, so do eukaryotic cells tailor their RNA before it leaves the nucleus. Consider that most eukaryotic genes contain intervening sequences called **introns**. Introns are removed in chunks from a newly transcribed RNA before it leaves the nucleus. Sequences that remain in the RNA after this process are called **exons** (Figure 7.5). Exons can be rearranged and spliced together in different combinations, so one gene may encode two or more versions of the same product. A newly transcribed RNA that will become an mRNA is further tailored after splicing. For example, a tail of 50 to 300 adenines is added to the end of a new mRNA. Among other functions, this poly-A tail is a signal that allows an mRNA to be exported from the nucleus.

An mRNA is a copy of a gene, in the same way that a paper transcript of a conversation carries the same information in a different format.

Figure 7.5 Post-transcriptional modification of RNA. Introns are removed and exons spliced together. Messenger RNAs also get a poly-A tail.

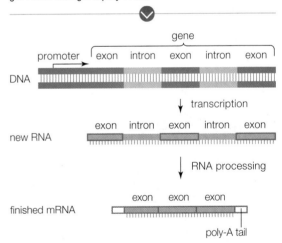

exon Nucleotide sequence that remains in an RNA after post-transcriptional modification.

intron Nucleotide sequence that intervenes between exons and is removed during post-transcriptional modification.

promoter DNA sequence that is a site where RNA polymerase binds for transcription.

RNA polymerase Enzyme that carries out transcription.

Take-Home Message 7.3

How is RNA assembled?

- During transcription, RNA polymerase uses a gene region in a chromosome as a template to assemble a strand of RNA. The new strand is an RNA copy of the gene from which it was transcribed.
- Post-transcriptional modification of RNA occurs in the nucleus of eukaryotes.

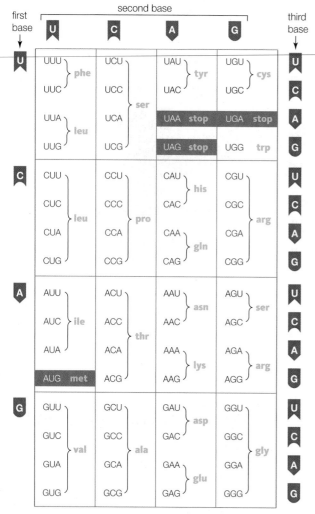

A. Codon table. Each codon in mRNA is a set of three nucleotide bases. The left column lists a codon's first base, the top row lists the second, and the right column lists the third.

Sixty-one of the triplets encode amino acids; one of those, AUG, both codes for methionine and serves as a signal to start translation. Three codons are signals that stop translation.

ala alanine (A)   gly glycine (G)   pro proline (P)
arg arginine (R)   his histidine (H)   ser serine (S)
asn asparagine (N)   ile isoleucine (I)   thr threonine (T)
asp aspartic acid (D) leu leucine (L)   trp tryptophan (W)
cys cysteine (C)   lys lysine (K)   tyr tyrosine (Y)
glu glutamic acid (E) met methionine (M)   val valine (V)
gln glutamine (Q)   phe phenylalanine (F)

B. Names and abbreviations of the 20 naturally occurring amino acids specified by the genetic code (**A**).

Figure 7.6 The genetic code.

Figure It Out: Which codons specify the amino acid lysine (lys)?

Answer: AAA and AAG

7.4 **The Genetic Code**

REMEMBER: A polypeptide is a linear sequence of amino acids joined by peptide bonds (Section 2.9).

An mRNA is essentially a disposable copy of a gene; its job is to carry the gene's protein-building information to types of RNA during translation. The protein-building message is encoded by nucleotides, "genetic words" that occur one after another along its length. Like the words of a sentence, a series of these genetic words can form a meaningful parcel of information—in this case, the sequence of amino acids of a protein.

Each of the genetic "words" carried by an mRNA is three nucleotide bases long, and each is a code—a **codon**—for a particular amino acid. The sequence of bases in a triplet determines which amino acid the codon specifies (Figure 7.6A). For instance, the codon UUU codes for the amino acid phenylalanine (phe), and UUA codes for leucine (leu).

Codons occur one after another along the length of an mRNA. When the mRNA is translated, the order of its codons determines the order of amino acids in the resulting polypeptide. Thus, the DNA sequence of a gene is transcribed into the nucleotide sequence of an mRNA, which is in turn translated into an amino acid sequence:

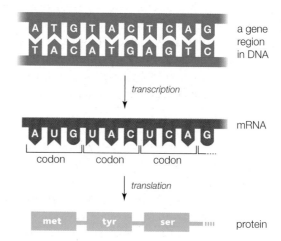

With four possible bases (G, A, U, or C) in each of the three positions of a codon, there are a total of sixty-four (or 4^3) mRNA codons. Collectively, the sixty-four codons constitute the **genetic code**. These codons specify a total of twenty naturally occurring amino acids (Figure 7.6B), so some amino acids have more than one codon. For instance, the amino acid tyrosine (tyr) is specified by two codons: UAU and UAC.

Other codons signal the beginning and end of a protein-coding sequence. In eukaryotes, the first AUG in an mRNA usually serves as the signal to start translation. AUG is the codon for methionine, so methionine is the first amino acid in new polypeptides. The codons UAA, UAG, and UGA do not specify an amino acid. These are signals that stop translation, so they are called stop codons. A stop codon marks the end of the protein-coding sequence in an mRNA.

The genetic code is highly conserved, which means that most organisms use the same code and probably always have. Bacteria, archaea, and some protists have a few codons that differ from the eukaryotic code, as do mitochondria and chloroplasts—a clue that led to a theory of how these two organelles evolved (we return to this topic in Section 13.3).

Take-Home Message 7.4

What is the genetic code?

• The genetic code consists of sixty-four codons (base triplets).

• Three codons are signals that stop translation; the remaining codons specify an amino acid. In eukaryotic mRNAs, the first occurrence of the codon that specifies methionine is a signal to begin translation.

7.5 Translation: RNA to Protein

REMEMBER: Ribosomes are organelles that carry out protein synthesis. Some enzymes are RNAs (Section 4.4).

A ribosome has two subunits, one large and one small (Figure 7.7). Each subunit consists mainly of **ribosomal RNA** (**rRNA**), with associated structural proteins. During translation, a large and a small ribosomal subunit converge as an intact ribosome on an mRNA. **Transfer RNAs** (**tRNAs**) then deliver amino acids to the intact ribosome. Ribosomal RNA is one example of RNA with enzymatic activity: During translation, the rRNA components of a ribosome (not the protein components) cause peptide bonds to form between amino acids.

Each tRNA has two attachment sites. The first is an **anticodon**, which is a triplet of nucleotides that base-pairs with an mRNA codon (Figure 7.8). The other attachment site binds to an amino acid—the one specified by the codon. Transfer RNAs with different anticodons carry different amino acids. During translation, these tRNAs deliver amino acids to a ribosome, one after another in the order specified by the codons in an mRNA. As the amino acids are delivered, the ribosome joins them via peptide bonds into a new polypeptide. Thus, the order of codons in an mRNA—DNA's protein-building message—is translated into a new protein.

Figure 7.8 tRNA structure.
Each tRNA's anticodon is complementary to an mRNA codon. Each also carries the amino acid specified by that particular codon. These models depict the tRNA that carries the amino acid tryptophan.

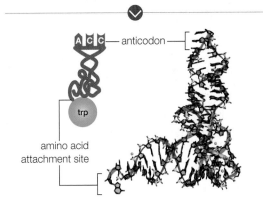

Figure 7.7 Ribosome structure.
An intact ribosome consists of a large and a small subunit. Protein components of both subunits are shown in the ribbon models in green; rRNA components, in brown.

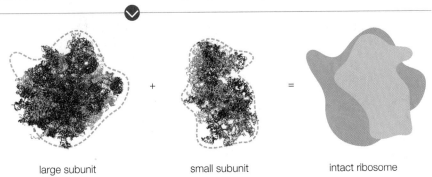

large subunit   small subunit   intact ribosome

anticodon In a tRNA, set of three nucleotides that base-pairs with an mRNA codon.

codon In an mRNA, a nucleotide base triplet that codes for an amino acid or stop signal during translation.

genetic code Complete set of sixty-four mRNA codons.

ribosomal RNA (rRNA) RNA that becomes part of ribosomes.

transfer RNA (tRNA) RNA that delivers amino acids to a ribosome during translation.

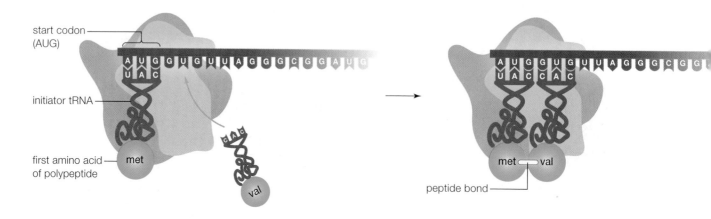

1 Transcription

2 RNA transport

ribosome
subunits

tRNA

3 Convergence of RNAs

4 Translation

mRNA

polypeptide

Figure 7.9 Overview of translation in a eukaryotic cell.
In eukaryotes, RNAs are transcribed in the nucleus, then
transported into cytoplasm. Translation begins when
ribosomal subunits and tRNA converge on an mRNA in
cytoplasm. Then, tRNAs deliver amino acids in the order
dictated by successive codons in the mRNA. The ribosome links the amino acids together as it moves along
the mRNA, so a polypeptide forms and elongates.

In all cells, translation occurs in cytoplasm. In a eukaryotic cell, RNAs that
carry out transcription are produced in the nucleus (Figure 7.9 **1**), then transported through nuclear pores into cytoplasm **2**. Translation is initiated when ribosomal subunits and tRNAs converge on an mRNA **3**. The complex of molecules is
now ready to carry out protein synthesis. The intact ribosome begins to move along
the mRNA and assemble a polypeptide **4**.

The first tRNA carries the amino acid methionine, the first amino acid of the
new polypeptide. Another tRNA joins the complex when its anticodon base-pairs
with the second codon in the mRNA. This tRNA brings with it the second amino
acid. The ribosome joins the first two amino acids by way of a peptide bond:

start codon
(AUG)

A U G G U G U U A G G G C G G A U G

U A C

initiator tRNA

first amino acid
of polypeptide

met

val

A U G G U G U U A G G G C G G

U A C C A C

met val

peptide bond

The first tRNA is released and the ribosome moves to the next codon. Another tRNA brings the third amino acid to the ribosome as its anticodon base-pairs with the third codon of the mRNA. The ribosome joins the second and third amino acids by way of a peptide bond:

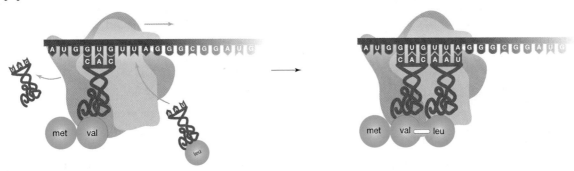

The second tRNA is released and the ribosome moves to the next codon. Another tRNA brings the fourth amino acid to the complex as its anticodon base-pairs with the fourth codon of the mRNA. The ribosome joins the third and fourth amino acids by way of a peptide bond:

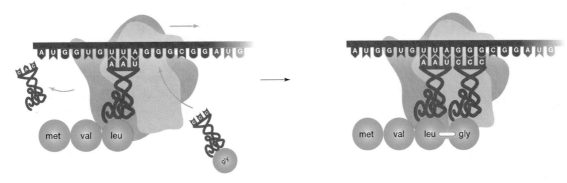

The new polypeptide continues to elongate as the ribosome joins successive amino acids delivered by tRNAs. Translation ends when the ribosome reaches a stop codon in the mRNA. The mRNA and the polypeptide detach from the ribosome, and the ribosomal subunits separate from each other.

Most of the energy that fuels translation is provided by GTP, an RNA nucleotide. Phosphate-group transfers from GTP help the ribosome move from one codon to the next along an mRNA. The RIPs you learned about in Section 7.1 are toxic because they remove a particular adenine base from one of the rRNAs in the ribosome's large subunit. The adenine is part of a binding site for proteins involved in the GTP-requiring steps of elongation. After the base has been removed, the ribosome can no longer bind to these proteins, and elongation stops.

Take-Home Message 7.5

How is mRNA translated into protein?

- Translation is an energy-requiring process in which a polypeptide is synthesized based on the sequence of codons in an mRNA.
- During elongation, amino acids are delivered to the ribosome by tRNAs in the order dictated by successive mRNA codons. As amino acids arrive, the ribosome joins each to the end of the polypeptide.

A. Hemoglobin, an oxygen-binding protein in red blood cells. This protein consists of four polypeptides: two alpha globins (blue) and two beta globins (green). Each globin forms a pocket that cradles a cofactor called a heme (red). Oxygen gas binds to the iron atom at the center of each heme.

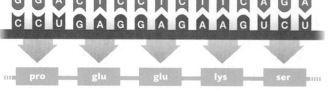

B. Part of the DNA (blue), mRNA (brown), and amino acid sequence of human beta globin. Numbers indicate nucleotide position in the mRNA.

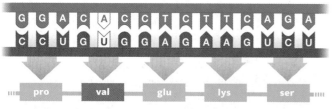

C. A base-pair substitution replaces a thymine with an adenine. When the altered mRNA is translated, valine replaces glutamic acid as the sixth amino acid. Hemoglobin with this form of beta globin is called HbS, or sickle hemoglobin.

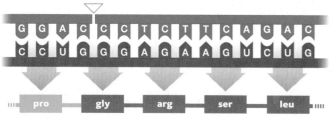

D. A base-pair deletion shifts the reading frame for the rest of the mRNA, so a completely different protein product forms. The mutation shown results in a defective beta globin. The outcome is beta thalassemia, a genetic disorder in which a person has an abnormally low amount of hemoglobin.

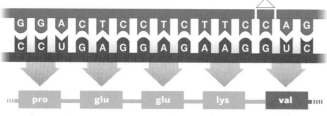

E. An insertion of one nucleotide causes the reading frame for the rest of the mRNA to shift. The protein translated from this mRNA is too short and does not assemble correctly into hemoglobin molecules. As in D, the outcome is beta thalassemia.

Figure 7.10 Examples of mutations.

(A) 1BBB, A third quaternary structure of human hemoglobin A at 1.7-A resolution. Silva, M.M., Rogers, P.H., Arnone, A., Journal: (1992) *J.Biol.Chem.* 267: 17248–17256.

7.6 Products of Mutated Genes

REMEMBER: Hydrophilic substances dissolve easily in water; hydrophobic substances do not (Section 2.4). Hemoglobin has multiple polypeptides that fold around hemes; fibrous proteins such as keratin aggregate by many thousands into much larger structures (2.9). Many proteins can function properly only with assistance from a cofactor (4.4). A diploid cell has two sets of chromosomes (6.3). A mutation is a permanent change in the DNA sequence of a cell's chromosome (6.4).

Mutations are relatively uncommon events in a normal cell. Consider that the chromosomes in a diploid human cell collectively consist of about 6.5 billion nucleotides, any of which may become mutated each time that cell divides. However, the mutation rate in human somatic cells is about 10^{-8}, which means only one nucleotide changes every 10^8 times DNA replication occurs in these cells, on average. On top of that, less than 2 percent of human DNA encodes products, so there is an extremely low probability that any mutation will occur in a coding region. When a nucleotide in a protein-coding region does change, the redundancy of the genetic code offers the cell a margin of safety. For example, a mutation that changes a codon from CCU to CCC may have no further effect, because both of these codons specify proline.

Very rarely, a mutation changes an amino acid in a protein, or results in a premature stop codon that shortens it. Such mutations can have drastic effects on an organism. Consider hemoglobin, an oxygen-transporting protein in your red blood cells. Hemoglobin's structure allows it to bind and release oxygen. In adult humans, a hemoglobin molecule consists of four polypeptides called globins: two alpha globins and two beta globins (Figure 7.10A). Each globin folds around a cofactor called heme. Oxygen molecules bind to hemoglobin at those hemes.

As red blood cells circulate through the lungs, the hemoglobin inside of them bind to oxygen molecules. The cells then travel to other regions of the body, and the hemoglobin releases its oxygen cargo wherever the oxygen level is low. When the red blood cells return to the lungs, the hemoglobin binds to more oxygen.

Mutations that alter hemoglobin's structure can greatly impact health. For example, mutations in either the alpha or beta globin genes can cause a condition called anemia, in which a person's blood is deficient in red blood cells or in hemoglobin. Both outcomes limit the blood's ability to carry oxygen, and the resulting symptoms can range from mild to life-threatening. Sickle-cell anemia arises because of a particular mutation in the beta globin gene. The mutation changes one nucleotide to another, so it is called a **base-pair substitution**. In this case, the substitution results in a version of beta globin that has valine instead of glutamic acid as its sixth amino acid (Figure 7.10B,C). Hemoglobin assembled with this altered beta globin chain is called sickle hemoglobin, or HbS.

Unlike glutamic acid, which carries a negative charge, valine carries no charge. As a result of that one base-pair substitution, a tiny patch of the beta globin polypeptide that is normally hydrophilic becomes hydrophobic. This change slightly alters the behavior of hemoglobin. Under certain conditions, HbS molecules stick together and form large, rodlike

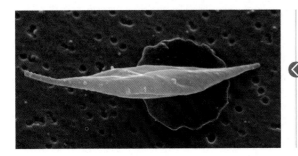

Figure 7.11 A sickled red blood cell compared with a normal one.
A single base-pair substitution gives rise to an abnormal beta globin chain that, when assembled in hemoglobin molecules, forms HbS. The sixth amino acid in these abnormal beta globin chains is valine, not glutamic acid. In the body, the difference causes HbS molecules to form rod-shaped clumps that distort normally round blood cells (red) into the sickle shape (tan) characteristic of sickle-cell anemia. Sickled cells clog small blood vessels.

EM Unit, UCL Medical School, Royal Free Campus/Wellcome Images.

clumps. Red blood cells that contain the clumps become distorted into a crescent (sickle) shape (Figure 7.11). Sickled cells clog tiny blood vessels, thus disrupting blood circulation throughout the body. Over time, repeated episodes of sickling can damage organs and eventually cause death.

A different type of anemia called beta thalassemia is caused by a **deletion**, which is a mutation in which one or more nucleotides is lost from the DNA. In this case, the twentieth nucleotide in the coding region of the beta globin gene is lost (Figure 7.10D). Like most other deletions, this one causes the reading frame of the mRNA codons to shift. A frameshift usually has drastic consequences because it garbles the genetic message, just as incorrectly grouping a series of letters garbles the meaning of a sentence:

> The fat cat ate the sad rat.
> T hef atc ata tet hes adr at.
> Th efa tca tat eth esa dra t.

The frameshift caused by the beta globin deletion results in a polypeptide that is very different from normal beta globin in amino acid sequence and in length. This outcome is the source of the anemia. Beta thalassemia can also be caused by an **insertion**, which is a mutation in which nucleotides are added to the DNA. Insertions, like deletions, often cause frameshifts (Figure 7.10E).

Some mutations can affect a gene's expression without changing any of its codons. Special nucleotide sequences in DNA affect the expression of nearby genes. A promoter is one example; an intron–exon splice site is another. Consider a mutation that causes the hairless appearance of sphynx cats (Figure 7.12). In this case, a base-pair substitution disrupts an intron–exon splice site in a gene for keratin, a fibrous protein. The intron is not correctly removed during post-transcriptional processing. The altered protein translated from the resulting mRNA cannot properly assemble into filaments that make up hair. Cats that have this mutation still make hair, but it falls out before it gets very long.

Figure 7.12 A sphynx cat.
The hairless appearance of a sphynx cat arises from a single base-pair mutation in a gene for keratin, a fibrous protein that makes up hair. The altered keratin that results from the mutation does not assemble correctly into filaments. Sphynx cats are not truly hairless; they produce hair, but it is easily dislodged.

Glennis Siverson/National Geographic Creative.

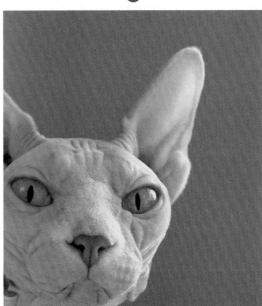

Take-Home Message 7.6

What happens after a gene becomes mutated?

- Mutations that change a protein can have drastic consequences on an organism's form and function.
- A base-pair substitution may change an amino acid in a protein, or it may introduce a premature stop codon.
- Frameshifts that occur after an insertion or deletion can change an mRNA's codon reading frame, thus garbling its protein-building instructions.

base-pair substitution Mutation in which a single base pair changes.

deletion Mutation in which one or more nucleotides are lost.

insertion Mutation in which one or more nucleotides become inserted into DNA.

Which genes a cell uses determines the molecules it will produce, which in turn determines what kind of cell it will be.

Figure 7.13 Example of a homeotic gene.
Most homeotic genes, including this one, encode transcription factors that work by binding directly to DNA.

(B) left, © Jürgen Berger, Max-Planck-Institute for Developmental Biology, Tübingen; right, Science VU/Dr. F. Rudolph Turner/Visuals Unlimited, Inc.

A. A model of the protein product (in gold) of the homeotic gene *antennapedia* attached to a promoter. Expression of *antennapedia* in embryonic tissues of the insect thorax causes legs to form.

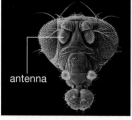 

antenna

leg

B. The head of a normal fruit fly (left) has two antennae. Right, a mutation that triggers expression of the *antennapedia* gene in embryonic tissues of the head causes legs to form instead of antennae.

7.7 Control of Gene Expression

REMEMBER: Free radicals are dangerous to life (Section 2.2). Estrogen and testosterone are steroid hormones that govern reproduction and secondary sexual traits in animals (2.8). Bacteria can produce ATP by breaking down sugars in fermentation (5.6). As cells in an embryo start using different subsets of their DNA, they become different in form and function, a process called differentiatiation (6.1). The DNA of eukaryotic cells wraps around histones; body cells of human females typically have two X chromosomes (6.3).

A typical cell in your body uses only about 10 percent of its genes at a time. Some of the active genes affect structures and metabolic pathways common to all cells; others are expressed only by certain subsets of cells. For example, most body cells express genes that encode the enzymes of glycolysis, but only immature red blood cells express globin genes. Differentiation occurs as different cell lineages begin to express different subsets of their genes during development. Which genes a cell uses determines the molecules it will produce, which in turn determines what kind of cell it will be. Thus, control over gene expression is necessary for proper development of complex, multicelled bodies. It also allows individual cells to respond appropriately to changes in their internal and external environments.

All steps of gene expression are regulated, starting with transcription and ending with delivery of an RNA or protein product to its final destination in the cell. The "switches" that turn a gene on or off are molecules or processes that affect individual steps of its expression. For example, proteins called **transcription factors** affect whether and how fast a gene is transcribed by binding directly to the DNA. Some transcription factors prevent RNA polymerase from attaching to a gene's promoter, which in turn prevents the gene's transcription. Others help RNA polymerase bind to a promoter, which speeds up transcription. The rest of this section introduces some specific examples of gene expression control in eukaryotes.

Master Genes As an animal embryo develops, its cells differentiate and form tissues, organs, and body parts. The entire process of development is driven by cascades of master gene expression. The products of **master genes** affect the expression of many other genes. Expression of a master gene causes other genes to be expressed, which in turn cause other genes to be expressed, and so on.

Gene expression that orchestrates animal development begins when different maternal mRNAs are delivered to different regions of cytoplasm in an egg as it forms. These mRNAs are translated only after the egg is fertilized. Then, their protein products diffuse away, forming gradients that span the developing embryo. The position of a cell within the embryo determines how much of these proteins it is exposed to. This in turn determines which master genes it turns on. The products of those master genes also form in gradients that diffuse away from cells expressing them. Still other master genes are transcribed depending on where a cell falls within these gradients, and so on. Eventually, the products of master genes cause undifferentiated cells to differentiate and become specialized.

A homeotic gene is a type of master gene whose expression directs the formation of a specific body part such as an eye, leg, or wing. Most homeotic genes encode transcription factors, and the function of many of them has been discovered by deliberately manipulating the genes' expression (Figure 7.13). For example, in a common experiment called a **knockout**, researchers inactivate a gene by introducing a mutation that prevents its expression, or by deleting it entirely. A knockout organism (one that has had a gene knocked out) may differ from normal individuals, and the differences are clues to the function of the missing gene product.

Homeotic genes control development by the same mechanisms in all multi-celled eukaryotes, and many are interchangeable among different species. Thus, we can infer that they evolved in the most ancient eukaryotic cells. Consider the *eyeless* gene. Eyes form in embryonic fruit flies wherever this gene is expressed, which is normally in tissues of the head only. Flies with a mutated *eyeless* gene lack eyes (Figure 7.14A). If the *eyeless* gene is expressed in another part of the developing embryo, eyes form there too (Figure 7.14B). Humans, squids, mice, fishes, and many other animals have a gene called *PAX6*, which is very similar in DNA sequence to the *eyeless* gene of flies. In humans, mutations in *PAX6* cause eye disorders such as aniridia, in which the irises are underdeveloped or missing (Figure 7.14C). If a *PAX6* gene from a human or a mouse is inserted into a fly, it has the same effect as the *eyeless* gene: An eye forms wherever it is expressed. (Because *PAX6* is just a switch, the eye that forms is a fly eye, not a human or mouse eye.) The same principle applies in reverse: The *eyeless* gene from flies switches on eye formation in frogs. Such studies are evidence of shared ancestry among these evolutionarily distant animals.

Sex Chromosome Genes

Sex Chromosome Genes In humans and other mammals, a female's cells have two X chromosomes, one inherited from her mother, the other one from her father. In each cell, one X chromosome is always tightly condensed (Figure 7.15). We call the condensed X chromosomes **Barr bodies**, after Murray Barr, who discovered them. Condensation inhibits transcription, so most of the genes on a Barr body are not expressed. This X chromosome inactivation ensures that only one of the two X chromosomes in a female's cells is active, thus equalizing expression of X chromosome genes between the sexes—a mechanism called dosage compensation. The body cells of male mammals (XY) have one set of X chromosome genes. Body cells of female mammals (XX) have two sets, but female embryos do not develop properly when both sets are expressed.

At this writing, researchers have discovered 1,805 genes on the human X chromosome. Only a few of them are associated with traits that differ between males and females; most govern nonsexual traits such as blood clotting and color perception. The Y chromosome has only 458 known genes, but one of them is *SRY*—the master gene for male sex determination in mammals. Its expression in XY embryos triggers the formation of testes, which are male reproductive organs. Some of the cells in testes make testosterone, a steroid hormone that controls the emergence of male secondary sexual traits such as facial hair, increased musculature, and deepened voice. We know that *SRY* is the master gene that controls emergence of male sexual traits because mutations in this gene cause XY individuals to develop external genitalia that appear female. An XX embryo has no Y chromosome, no *SRY* gene, and much less testosterone, so primary female reproductive organs (ovaries) form instead of testes. Ovaries make estrogens and other sex hormones that will govern the development of female secondary sexual traits, such as enlarged, functional breasts, and fat deposits around the hips and thighs.

Lactose Tolerance

Lactose Tolerance Humans and other mammals break down lactose, a carbohydrate in milk, but most do so only when young. An individual's ability to digest lactose ends at a certain age that depends on the species. In the majority of humans worldwide, this switch occurs at about age five, when transcription of the gene for lactase slows. The resulting decline in production of the enzyme causes a common condition known as lactose intolerance.

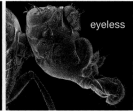

A. A normal fruit fly (left) has large, round eyes. A fruit fly with a mutation in its *eyeless* gene (right) develops without eyes.

B. Eyes form wherever the *eyeless* gene is expressed in fly embryos—here, on the head and also on the wing.

The *PAX6* gene of humans, mice, squids, and some other animals is so similar to *eyeless* that it similarly triggers eye development in fruit flies.

C. A normal human eye has a colored iris surrounding the pupil (dark area where light enters). Mutations in *PAX6* cause eyes to develop without an iris, a condition called aniridia.

Figure 7.14 Eyes and *eyeless*.

(A) David Scharf/Science Source; (B) Eye of Science/Science Source; (C) left, M. Bloch; right, Courtesy of the Aniridia Foundation International, www.aniridia.net.

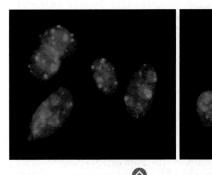

Figure 7.15 X chromosome inactivation.
Barr bodies are visible as red spots in the nucleus of the four XX cells on the left. Compare the nucleus of two XY cells to the right.

© Dr. William Strauss.

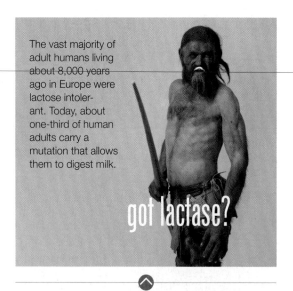

The vast majority of adult humans living about 8,000 years ago in Europe were lactose intolerant. Today, about one-third of human adults carry a mutation that allows them to digest milk.

Figure 7.16 Milk wasn't on the Stone Age menu.

South Tyrol Museaum of Archaeology/A. Ochsenreiter, as altered by Lisa Starr.

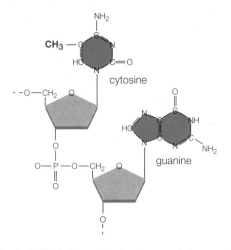

A. In the DNA of differentiated cells, a methyl group (red) is most often attached to a cytosine (C) that is followed by a guanine (G).

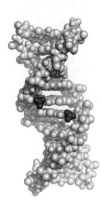

B. A model of DNA shows methyl groups (red) attached to a cytosine–guanine pair on complementary DNA strands. When the cytosine on one strand is methylated, enzymes methylate the cytosine on the other strand. This is why a methylation tends to persist in a cell's descendants.

Figure 7.17 DNA methylation.

Cells in the intestinal lining secrete lactase into the small intestine, where the enzyme cleaves lactose into its monosaccharide monomers. Monosaccharides are absorbed directly by the small intestine, but lactose is not. Thus, when lactase production slows, lactose consumed in food passes undigested through the small intestine. The lactose ends up in the large intestine, which hosts huge numbers of bacteria. These cells respond to the presence of lactose by switching on genes that break it down. Carbon dioxide and other gaseous products of their various fermentation reactions accumulate quickly in the large intestine, distending its wall and causing pain. Other products of their metabolism disrupt the solute–water balance inside the large intestine, and diarrhea results.

Not everybody is lactose intolerant. About one-third of human adults carry a mutation that allows them to digest milk; this mutation is more common in some populations than in others. Recent analyses of DNA from well-preserved skeletons show that the vast majority of adult humans living about 8,000 years ago in Europe were lactose intolerant (Figure 7.16). Around that time, a mutation appeared in the DNA of prehistoric people inhabiting a region between what is now central Europe and the Balkans. This mutation allowed its bearers to continue digesting milk as adults, and it spread rapidly to the rest of the continent along with the practice of dairy farming. Today, most adults of northern and central European ancestry are able to digest milk because they carry this mutation, a single base-pair substitution in a region of DNA that helps control the lactase gene promoter. Other mutations in the same region of DNA arose independently in North Africa, southern Asia, and the Middle East. Some people descended from these populations can continue to digest milk as adults.

DNA Methylation In a eukaryotic cell, only regions of DNA that have been unwound from histones are accessible to RNA polymerase for transcription. Modifications to histone proteins change the way they interact with DNA wrapped around them, thus affecting transcription. Some modifications make histones release their grip on DNA; others make them tighten it. For example, adding methyl groups ($-CH_3$) to a histone tightens the DNA around it, so enzymes that methylate histones can slow or shut down transcription.

Direct methylation of DNA nucleotides also suppresses transcription, often more permanently than histone modifications. Once a particular nucleotide has become methylated in a cell's DNA, it will usually stay methylated in the DNA of the cell's descendants. Methylation and other heritable modifications to DNA that affect its function but do not alter the nucleotide sequence are said to be **epigenetic**.

DNA methylation is a part of differentiation, so it begins very early in embryonic development: Genes actively expressed in a zygote (the first cell of a new individual) become silenced as their promoters get methylated. This silencing is the basis of selective gene expression that drives differentiation. During development, and also during the remainder of the individual's life, each cell's DNA continues to acquire methylations. Between 3 and 6 percent of the DNA has been methylated in a normal, differentiated body cell.

Methyl groups are usually added to a cytosine that is followed by a guanine (Figure 7.17), but which of these cytosines get methylated varies by the individual. This is because methylation is influenced by environmental factors. For instance, humans conceived during a famine end up with an unusually low number of methyl groups in the DNA of certain genes. The product of one of those genes is a hormone that promotes prenatal growth and development. The resulting increase in expression of this gene may offer a survival advantage in a poor nutritional environment.

Digging Into Data

Effect of Grandmother's Food Supply on Infant Mortality

Researchers are investigating long-reaching epigenetic effects of starvation, in part because historical data on periods of famine are widely available. Before the industrial revolution, a failed harvest in one autumn typically led to severe food shortages the following winter. A retrospective study has correlated female infant mortality at certain ages with the abundance of food during the paternal grandmother's childhood. Figure 7.18 shows some of the results of this study.

1. Compare the mortality risk of girls whose paternal grandmothers ate well at age 2 with that of girls whose grandmothers were starving at the same age. Which girl was more likely to die early? About how much more likely was she to die?

2. Children have a period of slow growth around age 9. What trend in this data can you see around that age?

3. There was no correlation between early death of a male child and eating habits of his paternal grandmother, but there was a strong correlation with the eating habits of his paternal grandfather. What does this tell you about the probable location of epigenetic changes that gave rise to these data?

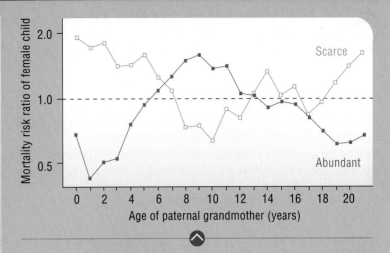

Figure 7.18   An epigenetic effect.
The graph shows relative risk of early death of a female child correlated with the age at which her paternal grandmother experienced a winter with a food supply that was scarce (blue) or abundant (red) during childhood. The dotted line represents no difference in risk of mortality. A value above the line means an increased risk; one below the line indicates a reduced risk.

Source: Pembrey et al., *European Journal of Human Genetics* (2006) 14, 159–166.

Nucleotides also become methylated by chance during DNA replication, so cells that divide a lot tend to have more methyl groups in their DNA than inactive cells. Free radicals and toxic chemicals add more methyl groups.

Factors that influence DNA methylation can have multigenerational effects. When an organism reproduces, it passes its DNA to offspring. Methylation of parental DNA is normally "reset" in the zygote, with new methyl groups being added and old ones being removed. This reprogramming does not remove all of the parental methyl groups, however, so some methylations acquired during an individual's lifetime are passed to future offspring. Inheritance of epigenetic modifications can adapt offspring to an environmental challenge much more quickly than evolution (we return to evolutionary processes in Chapter 12). Epigenetic modifications are not considered to be evolutionary because the underlying DNA sequence does not change. Even so, they may persist for generations after an environmental challenge has faded.

> Factors that influence DNA methylation can have multi-generational effects.

Take-Home Message 7.7

How is gene expression controlled?

- Gene expression can be switched on or off, or speeded or slowed, by molecules and processes that operate at each step.
- This control allows cells to respond appropriately to environmental change. It is also critical for differentiation and development in multicelled eukaryotes.
- Epigenetic modifications of DNA acquired during an individual's lifetime can be passed to offspring, so gene expression patterns can persist for generations.

epigenetic   Refers to heritable changes in gene expression that are not the result of changes in DNA sequence.

Summary

Section 7.1 The ability to make proteins is critical to all life processes. Ricin and other ribosome-inactivating proteins (RIPs) have an enzyme domain that alters ribosomes, thus destroying the cell's ability to make proteins. A cell that cannot make proteins dies quickly.

Section 7.2 Information encoded within the nucleotide sequence of DNA occurs in subsets called **genes**. **Gene expression** is the conversion of information in a gene to an RNA or protein product. RNA is produced during **transcription**. During **translation**, protein-building information in an **mRNA (messenger RNA)** is converted into a sequence of amino acids in a polypeptide:

$$\text{DNA} \xrightarrow{\ transcription\ } \text{mRNA} \xrightarrow{\ translation\ } \text{protein}$$

Section 7.3 During transcription, **RNA polymerase** binds to a **promoter** near a gene region in a chromosome, then links RNA nucleotides in the order dictated by the nucleotide base sequence of the DNA strand. The resulting RNA strand is an RNA copy of the gene. In eukaryotes, RNA is modified before it leaves the nucleus. **Intron** sequences are removed, and the remaining **exon** sequences may be rearranged and spliced in different combinations. Messenger RNAs are further modified.

Section 7.4 An mRNA's protein-building information consists of a series of **codons**, sets of three nucleotides. All sixty-four codons, most of which specify amino acids, constitute the **genetic code**.

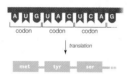

Section 7.5 Each **tRNA (transfer RNA)** has an **anticodon** that can base-pair with a codon, and it binds to the amino acid specified by that codon. Proteins and enzymatic **rRNA (ribosomal RNA)** make up the two subunits of a ribosome. During translation, codons in an mRNA direct synthesis of a polypeptide. First, the mRNA, a tRNA, and two ribosomal subunits converge. Next, successive amino acids are delivered by tRNAs in the order specified by the codons in the mRNA. The ribosome causes a peptide bond to form between the amino acids as they arrive, so a polypeptide forms. Translation ends when the ribosome encounters a stop codon in the mRNA.

Section 7.6 **Insertions**, **deletions**, and **base-pair substitutions** are mutations. A mutation that changes a gene's product may have harmful effects. In humans, an example is sickle-cell anemia, a disorder caused by a single base-pair substitution in the gene for the beta globin chain of hemoglobin.

Section 7.7 Gene expression control is the basis of cell differentiation in multicelled eukaryotes, and it also allows individual cells to respond to changes in their environment. Molecules and processes that influence gene expression operate at every step between transcription and delivery of the gene product to its final destination. Proteins called **transcription factors** influence transcription by binding directly to DNA.

All cells of an embryo share the same genes; different cells in it become specialized as they begin to use different subsets of those genes, a process called differentiation. Various **master genes** are expressed locally in different parts of the developing embryo. Their products, which diffuse through the embryo in gradients, affect expression of other master genes, which in turn affect the expression of others, and so on. Cells differentiate according to their exposure to these gradients. Eventually, master gene expression induces the expression of homeotic genes, the products of which trigger development of specific body parts. The function of many homeotic genes was revealed by **knockouts**.

In cells of female mammals, one of the two X chromosomes is condensed as a **Barr body**. The condensation makes most of this chromosome's genes permanently inaccessible. In humans, the *SRY* gene determines male sex. A mutation that abolishes control over expression of the lactase gene allows some people to continue digesting milk in adulthood.

DNA modifications (such as methylations) that affect gene expression but do not involve changes to the DNA sequence are said to be **epigenetic**. An individual acquires these modifications during its lifetime, as a consequence of cell division and also by exposure to environmental challenges. Epigenetic modifications are passed to the cell's descendants; some have multigenerational effects because they can be passed to the individual's offspring.

Self-Quiz

Answers in Appendix I

1. A chromosome contains many genes that are transcribed into different _____ .
   a. proteins
   b. polypeptides
   c. RNAs
   d. a and b

2. RNAs form by _____ ; proteins form by _____ .
   a. replication; translation
   b. translation; transcription
   c. transcription; translation
   d. replication; transcription

3. In cells, most RNA molecules are _____ , and DNA molecules are _____ .
   a. single-stranded; double-stranded
   b. double-stranded; single-stranded

4. The main function of an mRNA molecule is to _____ .
   a. store heritable information
   b. carry a translatable message
   c. form peptide bonds between amino acids

5. Where does transcription take place in a eukaryotic cell?
   a. the nucleus   c. the cytoplasm
   b. ribosomes   d. b and c are correct

6. What is the maximum number of amino acids that can be encoded by a gene with 45 bases plus a stop codon?
   a. 15   c. 90
   b. 45   d. 135

7. Most codons specify a(n) _____ .
   a. protein   c. amino acid
   b. polypeptide   d. mRNA

8. Where does translation take place in a eukaryotic cell?
   a. the nucleus   c. the cytoplasm
   b. ribosomes   d. b and c are correct

9. A mutation called a _____ often results in a frameshift that garbles the genetic message.
   a. deletion   c. base-pair substitution
   b. insertion   d. a and b

10. Muscle cells differ from bone cells because _____ .
    a. they carry different genes   c. they are eukaryotic
    b. they use different genes   d. both a and b

11. The expression of a gene may depend on _____ .
    a. the type of organism   c. the type of cell
    b. environmental conditions   d. all of the above

12. A gene that is knocked out is _____ .
    a. deleted   c. expressed
    b. inactivated   d. either a or b

13. A cell with a Barr body is _____ .
    a. prokaryotic   c. from a female mammal
    b. a sex cell   d. infected by Barr virus

14. True or false? Some gene expression patterns are heritable.

15. Put the following processes in order of their occurrence during expression of a eukaryotic gene:
    a. mRNA processing   c. transcription
    b. translation   d. RNA leaves nucleus

16. Match each term with the most suitable description.
    _____ methylation   a. cells become specialized
    _____ insertion   b. cascades of control
    _____ promoter   c. can be epigenetic
    _____ genetic message   d. assembles amino acids
    _____ differentiation   e. read in threes
    _____ *SRY*   f. makes a man out of you
    _____ master gene   g. extra nucleotides
    _____ ribosome   h. binding site for
                                  RNA polymerase

Critical Thinking

1. An anticodon has the sequence GCG. What amino acid does this tRNA carry? What would be the effect of a mutation that changed the C of the anticodon to a G?

2. Each position of a codon can be occupied by one of four (4) nucleotides. What is the minimum number of nucleotides per codon necessary to specify all 20 of the naturally occurring amino acids that are assembled into proteins?

3. Why are some genes expressed and some not?

4. Bacteria use the same stop codons as eukaryotes. However, bacterial transcription is also terminated in places where the mRNA folds back on itself to form a hairpin-looped structure like the one shown on the right. How do you think that this structure stops transcription?

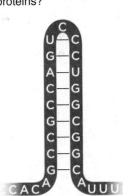

Visual Question

1. Use Figure 7.6 to translate the following sequence of bases in an mRNA into an amino acid sequence, starting at the first base. Use the one-letter abbreviations for the amino acids.
   GGUGAAAAUGAGACCAUUUGUAGU

2. Translate the base sequence in the previous question, starting at the second base.

8

HOW CELLS REPRODUCE

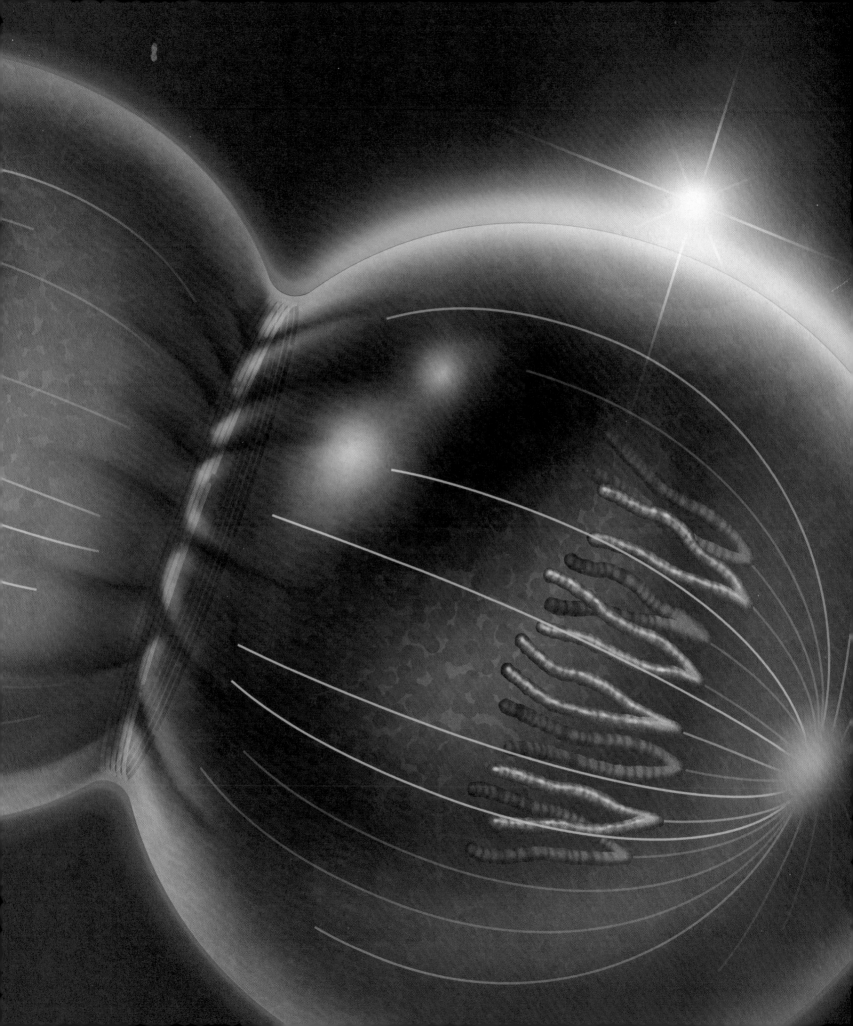

Application ❯ 8.1 Henrietta's Immortal Cells

Cells are individually alive, even as part of a multicelled body.

Each human starts out as a fertilized egg. By the time of birth, that single cell has given rise to about a trillion other cells, all organized as a human body. Even in the adult, billions of cells divide every day as new cells replace worn-out ones. However, despite the ability of human cells to continue dividing as part of a body, they tend to divide a limited number of times when grown in the laboratory. As early as the mid-1800s, researchers were trying to coax human cells to keep dividing outside of the body because they realized immortal cell lineages—cell lines—would allow them to study human diseases (and potential cures) without experimenting on people. The quest to create a human cell line continued unsuccessfully until 1951. By this time, George and Margaret Gey had been trying to culture human cells for nearly thirty years. Then their assistant, Mary Kubicek, prepared a new sample of human cancer cells. Mary named the cells HeLa, after the first and last names of the patient from whom the cells had been taken. The HeLa cells began to divide, again and again. The cells were astonishingly vigorous, quickly coating the inside of their test tube and consuming their nutrient broth. Four days later, there were so many cells that the researchers had to transfer them to more tubes. The cell populations increased at a phenomenal rate. The cells were dividing every twenty-four hours and coating the inside of the tubes within days.

Sadly, cancer cells in the patient were dividing just as fast. Only six months after she had been diagnosed with cervical cancer, malignant cells had invaded tissues throughout her body. Two months after that, Henrietta Lacks, a young woman from Baltimore, was dead.

Even after Henrietta had passed away, her cells lived on in the Geys' laboratory. The Geys discovered how to grow poliovirus in HeLa cells, a practice that enabled them to determine which strains of the virus cause polio. That work was a critical step in the development of polio vaccines, which have since saved millions of lives.

Henrietta Lacks was just thirty-one, a wife and mother of five, when runaway cell divisions of cancer killed her. Her cells, however, are still dividing, again and again, more than sixty years after she died. Frozen away in tiny tubes packed in Styrofoam boxes, HeLa cells continue to be shipped among laboratories all over the world. They are still widely used to investigate cancer (Figure 8.1), viral growth, protein synthesis, the effects of radiation, and countless other processes important in medicine and research. HeLa cells helped several researchers win Nobel Prizes, and they even traveled into space for experiments on satellites.

Understanding why cancer cells are immortal—and why we are not—begins with learning about the structures and mechanisms that cells use to divide.

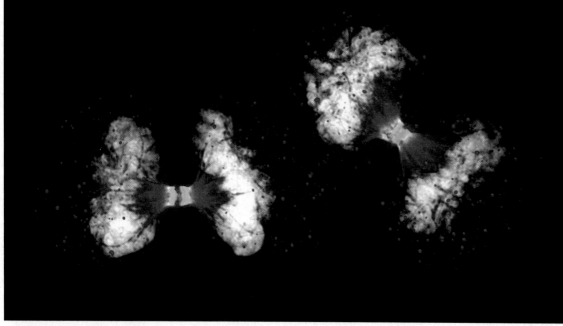

Figure 8.1 HeLa.
This micrograph shows two dividing cells of the HeLa line—cellular legacy of cancer victim Henrietta Lacks (left). Blue and green tracers identify the location of proteins that help attach chromosomes (white) to microtubules (red) during the division process. The location of the proteins is abnormal, which means the chromosomes are not properly attached to microtubules that are supposed to distribute them evenly into descendant cells. Defects in these and other proteins that orchestrate cell division result in cells with too many or too few chromosomes, an outcome associated with cancer.

Top, Dr. Paul D. Andrews/University of Dundee; inset, courtesy of the family of Henrietta Lacks.

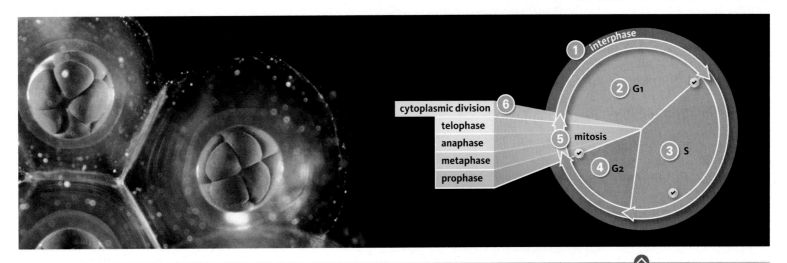

8.2 Multiplication by Division

REMEMBER: Processes by which individuals produce offspring are called reproduction (Section 1.3). Microtubules consist of tubulin subunits; microfilaments form a cell cortex that reinforces the plasma membrane; a rigid cell wall surrounds the plasma membrane of a plant cell; the motor protein myosin interacts with microfilaments to bring about contraction (3.5). When a phosphate group is transferred from ATP, energy is transferred along with it (4.4). Sister chromatids attach to one another at the centromere (6.3). A cell copies its chromosomes by DNA replication (6.4).

A life cycle is a sequence of recognizable stages that occur during an organism's lifetime, from the first cell of the new individual until its death. Multicelled organisms and free-living cells have life cycles, but what about cells that make up a multicelled body? Biologists consider such cells to be individually alive, each with its own lifetime. A cell's life passes through a series of recognizable intervals and events collectively called the **cell cycle** (Figure 8.2). A typical body cell spends most of its life in interphase ❶. During **interphase**, the cell increases its mass, roughly doubles the number of its cytoplasmic components, and copies its chromosomes in preparation for division. Interphase has three major stages: G1, S, and G2. G1 and G2 were named "Gap" phases because outwardly they seem to be periods of inactivity, but they are not. Most cells going about their metabolic business are in G1 ❷. Cells preparing to divide enter S ❸, when they undergo DNA replication. During G2 ❹, cells make the proteins that will carry out division.

The rest of the cell cycle consists of the division process itself. When the cell divides, both of its two cellular offspring end up with a blob of cytoplasm and some DNA. Each of the offspring of a eukaryotic cell inherits its DNA packaged inside a nucleus. Thus, a eukaryotic cell's nucleus has to divide before its cytoplasm does. **Mitosis** is a nuclear division mechanism that maintains the chromosome number ❺. In multicelled organisms, mitosis and cytoplasmic division ❻ are the basis of increases in body size and tissue remodeling during development, as well as ongoing replacements of damaged or dead cells in the adult. Mitosis and cytoplasmic division are also part of **asexual reproduction**, a reproductive mode by which offspring are produced by one parent only. Some multicelled eukaryotes and many single-celled ones use this mode of reproduction. (Prokaryotes do not have a nucleus and do not undergo mitosis. We discuss their reproduction in Section 13.5.)

Figure 8.2 The eukaryotic cell cycle.
The photo shows early frog embryos, each a product of three mitotic divisions of one fertilized egg.

❶ A cell spends most of its life in interphase, which includes three stages: G1, S, and G2.

❷ G1 is the phase of growth before DNA replication. The cell's chromosomes are unduplicated.

❸ S is the phase of synthesis, during which the cell makes copies of its chromosome(s) by DNA replication.

❹ G2 is the phase after DNA replication and before mitosis. The cell prepares to divide during this stage.

❺ The nucleus divides during mitosis.

❻ After mitosis, the cytoplasm may divide. Each descendant cell begins the cycle anew, in interphase.

✔ Built-in checkpoints stop the cycle from proceeding until certain conditions are met (see Section 8.3).

Figure It Out: Each of the embryos in the photo consists of how many cells?   **Answer:** Eight

asexual reproduction Reproductive mode of eukaryotes by which offspring arise from a single parent.

cell cycle The collective series of intervals and events of a cell's life, from the time it forms until it divides.

interphase In a eukaryotic cell cycle, the interval between mitotic divisions when a cell grows, roughly doubles the number of its cytoplasmic components, and replicates its DNA.

mitosis Nuclear division mechanism that maintains the chromosome number.

anaphase Stage of mitosis during which sister chromatids separate and move toward opposite spindle poles.

homologous chromosomes In the nucleus of a diploid eukaryotic cell, chromosomes with the same length, shape, and set of genes.

metaphase Stage of mitosis at which all chromosomes are aligned in the middle of the cell.

prophase Stage of mitosis during which chromosomes condense and become attached to a newly forming spindle.

spindle Temporary structure that moves chromosomes during nuclear division; consists of microtubules.

telophase Stage of mitosis during which chromosomes arrive at opposite ends of the cell. Two new nuclei form as the chromosomes loosen.

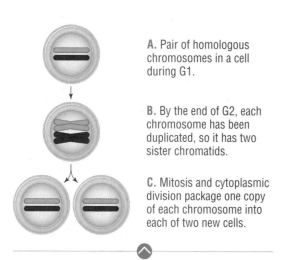

A. Pair of homologous chromosomes in a cell during G1.

B. By the end of G2, each chromosome has been duplicated, so it has two sister chromatids.

C. Mitosis and cytoplasmic division package one copy of each chromosome into each of two new cells.

Figure 8.3 How mitosis maintains the chromosome number in a diploid cell.
For clarity, only one pair of homologous chromosomes is illustrated. Pink indicates a chromosome of maternal origin; blue, a chromosome of paternal origin.

When a cell divides by mitosis, it produces two descendant cells, each with the chromosome number of the parent. However, if only the total number of chromosomes mattered, then one of the descendant cells might get, say, two pairs of chromosome 22 and no chromosome 9. A cell cannot function properly without a full complement of DNA, which means it needs to have *a copy of each* chromosome. Thus, the two cells produced by mitosis have the same number and types of chromosomes as the parent. Figure 8.3 shows how mitosis maintains the chromosome number. Consider how your body's cells are diploid, which means their nuclei contain pairs of chromosomes—two of each type. One chromosome of a pair was inherited from your father; the other, from your mother. Except for a pairing of nonidentical sex chromosomes (XY) in males, the chromosomes of each pair are homologous. **Homologous chromosomes** have the same length, shape, and genes (*hom*– means alike). When a cell is in G1, each of its chromosomes consists of one double-stranded DNA molecule. The cell replicates its DNA in S, so by G2, each of its chromosomes consists of two double-stranded DNA molecules. These molecules stay attached to one another at the centromere as sister chromatids until mitosis is almost over, and then they are pulled apart and packaged into two separate nuclei. When sister chromatids are pulled apart, each becomes an individual chromosome that consists of one double-stranded DNA molecule. Thus, each of the two new nuclei that form in mitosis contains a full complement of chromosomes. When the cytoplasm divides, these nuclei are packaged into separate cells. Each new cell has the parental chromosome number, and starts its life in G1 of interphase.

Figure 8.4 shows the details of mitosis. When a cell is in interphase, its chromosomes are loosened to allow transcription and DNA replication. Loosened chromosomes are spread out, so they are not easily visible under a light microscope ➊. In preparation for nuclear division, the chromosomes begin to pack tightly ➋. Transcription and DNA replication stop as the chromosomes condense into their most compact "X" forms. Tight condensation keeps the chromosomes from getting tangled and breaking during nuclear division. A cell reaches **prophase**, the first stage of mitosis, when its chromosomes have condensed so much that they are visible under a light microscope ➌. "Mitosis" is from *mitos*, the Greek word for thread, after the threadlike appearance of the chromosomes during nuclear division.

Also during prophase, microtubules assemble and extend from two regions on opposite sides of the cell. These microtubules form a **spindle**, a temporary structure that moves chromosomes during nuclear division. The spindle penetrates the nuclear region as the nuclear envelope breaks up. Some of the microtubules stop lengthening when they reach the middle of the cell. Others lengthen until they reach a chromosome and attach to it at the centromere. By the end of prophase, one sister chromatid of each chromosome has become attached to microtubules extending from one end of the cell, and the other sister has become attached to microtubules extending from the other end.

The opposing sets of microtubules then begin a tug-of-war by adding and losing tubulin subunits. As the microtubules lengthen and shorten, they push and pull the chromosomes. When all the microtubules are the same length, the chromosomes are aligned in the middle of the cell ➍. The alignment marks **metaphase**. During **anaphase**, the sister chromatids of each duplicated chromosome separate, so each becomes an individual, unduplicated chromosome. The spindle moves the chromosomes toward opposite sides of the cell ➎.

Telophase begins when the two clusters of chromosomes reach opposite ends of the cell ➏. Each cluster has the same number and kinds of chromosomes as the parent cell nucleus had: two of each chromosome, if the parent cell was diploid.

Mitosis in a plant cell   **Mitosis in an animal cell**

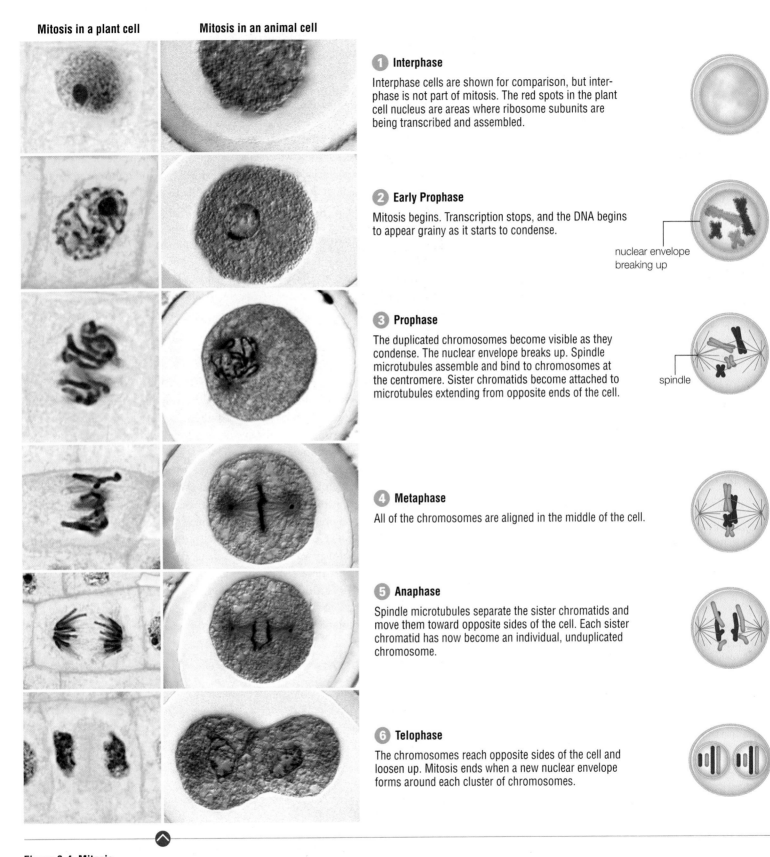

1 Interphase

Interphase cells are shown for comparison, but interphase is not part of mitosis. The red spots in the plant cell nucleus are areas where ribosome subunits are being transcribed and assembled.

2 Early Prophase

Mitosis begins. Transcription stops, and the DNA begins to appear grainy as it starts to condense.

nuclear envelope breaking up

3 Prophase

The duplicated chromosomes become visible as they condense. The nuclear envelope breaks up. Spindle microtubules assemble and bind to chromosomes at the centromere. Sister chromatids become attached to microtubules extending from opposite ends of the cell.

spindle

4 Metaphase

All of the chromosomes are aligned in the middle of the cell.

5 Anaphase

Spindle microtubules separate the sister chromatids and move them toward opposite sides of the cell. Each sister chromatid has now become an individual, unduplicated chromosome.

6 Telophase

The chromosomes reach opposite sides of the cell and loosen up. Mitosis ends when a new nuclear envelope forms around each cluster of chromosomes.

Figure 8.4 Mitosis.
Micrographs show nuclei of plant cells (onion root, left), and animal cells (fertilized egg of a roundworm, right). A diploid (2n) animal cell with two chromosome pairs is illustrated.

Plant cell nuclei, Michael Clayton/University of Wisconsin, Department of Botany; animal cell nuclei, © ISM/Phototake.

Figure 8.5 Cytoplasmic division of an animal cell.

1 In a dividing animal cell, the spindle disassembles as mitosis ends.

2 At the midpoint of the former spindle, a ring of actin and myosin filaments attached to the plasma membrane contracts.

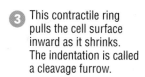

3 This contractile ring pulls the cell surface inward as it shrinks. The indentation is called a cleavage furrow.

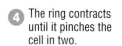

4 The ring contracts until it pinches the cell in two.

Figure 8.6 Cytoplasmic division of a plant cell.

5 Vesicles containing wall-building materials cluster at the future plane of division before mitosis ends.

6 The vesicles fuse with each other, forming a cell plate.

7 The cell plate expands to the plasma membrane. When it attaches to the membrane, it partitions the cytoplasm.

8 The cell plate matures as two new cell walls. These walls join with the parent cell wall, so each descendant cell becomes enclosed by its own wall.

A new nucleus forms around each cluster as the chromosomes loosen up again, after which telophase—and mitosis—are finished.

Cytoplasmic Division In most eukaryotes, the cell cytoplasm divides between late anaphase and the end of telophase, so two cells form, each with their own nucleus. The mechanism of cytoplasmic division differs between plants and animals.

Typical animal cells pinch themselves in two after nuclear division ends (Figure 8.5). How? The cell cortex, which is the mesh of cytoskeletal elements just under the plasma membrane, includes a band of microfilaments and motor proteins that wraps around the cell's midsection. The band is called a contractile ring because it contracts when the motor proteins are energized by phosphate-group transfers from ATP. Contraction occurs after telophase, when the spindle disassembles 1.

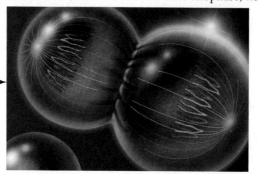

Because the contractile ring is attached to the plasma membrane, it pulls the membrane inward as it contracts 2. The sinking plasma membrane is visible on the outside of the cell as an indentation (left), which is called a **cleavage furrow** 3. The cleavage furrow advances around the cell and deepens until the cytoplasm—and the cell—is pinched in two 4. Each of the two cells formed by this division has its own nucleus and some of the parent cell's cytoplasm, and each is enclosed by a plasma membrane.

Dividing plant cells face a particular challenge because a stiff cell wall surrounds their plasma membrane. Accordingly, plant cells have their own mechanism of cytoplasmic division (Figure 8.6). By the end of anaphase, a set of short microtubules has formed on either side of the future plane of division. These microtubules now guide vesicles from Golgi bodies and the cell surface to the division plane 5. There, the vesicles and their wall-building contents start to fuse into a disk-shaped cell plate 6. The plate expands at its edges until it reaches the plasma membrane and attaches to it, thus partitioning the cytoplasm 7. In time, the cell plate will develop into two new cell walls, so each of the descendant cells will be enclosed by its own plasma membrane and wall 8.

Take-Home Message 8.2

How do eukaryotic cells reproduce?

- A eukaryotic cell reproduces by division: nucleus first, then cytoplasm.
- Mitosis (a nuclear division mechanism that maintains the chromosome number) and cytoplasmic division are part of the cell cycle, a series of events and stages through which a cell passes during its lifetime.
- DNA replication occurs before mitosis begins, so each chromosome consists of two DNA molecules attached as sister chromatids.
- During mitosis, a spindle assembles, then separates all of the sister chromatids and moves them to opposite sides of the cell. A new nuclear envelope forms around each of the two clusters of chromosomes. The two new nuclei have the same chromosome number as the parent cell.

8.3 **Mitosis and Cancer**

REMEMBER: A receptor protein triggers a change in cell activity in response to a stimulus such as binding to a certain substance; adhesion proteins help cells stick together in animal tissues (Section 3.3). Dolly the sheep died early, most likely because she was a clone (6.1). An image of a cell's chromosomes is a karyotype (6.3). Proofreading and repair mechanisms usually maintain the integrity of a cell's genetic information by correcting mispaired bases and fixing DNA damage (6.4). Gene expression can be switched on or off by molecules or processes that operate at each step; a knockout is an organism with a gene that has been deliberately inactivated (7.7).

When a cell divides—and when it does not—is determined by mechanisms of gene expression control. Like the accelerator of a car, some of these mechanisms cause the cell cycle to advance. Others are like brakes, preventing the cycle from proceeding. In the adult body, brakes on the cell cycle normally keep the vast majority of cells in G1. Most of your nerve cells, skeletal muscle cells, heart muscle cells, and fat-storing cells have been in G1 since you were born, for example.

Control over the cell cycle also ensures that a dividing cell's descendants receive intact copies of its chromosomes. Built-in checkpoints ensure the cell's DNA has been copied completely, that it is not damaged, and even that enough nutrients are available to support division (a few checkpoints are indicated in Figure 8.2). Protein products of "checkpoint genes" interact to carry out this control process. For example, a checkpoint that operates in S monitors whether the cell's chromosomes have been damaged during DNA replication. Checkpoint proteins recognize damaged DNA and bind to it. This binding puts the brakes on the cell cycle, and also enhances expression of genes involved in DNA repair. After the damage has been fixed, the brakes are lifted and the cell cycle proceeds. If the DNA remains unrepaired, other checkpoint proteins trigger events that cause the cell to self-destruct.

Cell Division Gone Wrong Sometimes a checkpoint gene mutates so that its protein product no longer works properly. In other cases, the controls that regulate its expression fail, and a cell makes too much or too little of the gene's product. When enough checkpoint mechanisms fail, a cell loses control over its cell cycle. Interphase may be skipped, so division occurs over and over with no resting period. Signaling mechanisms that cause abnormal cells to die may stop working. The problem is compounded because checkpoint malfunctions are inherited by the cell's descendants, which form a **neoplasm**—an accumulation of abnormally dividing cells.

A neoplasm that makes a lump in the body is called a **tumor**, but the two terms are sometimes used interchangeably. Once a tumor-causing mutation has occurred, the gene it affects is called an oncogene. An **oncogene** is any gene that helps transform a normal cell into a tumor cell. Oncogene mutations in reproductive cells can be passed to offspring, which is a reason that some types of tumors run in families.

Not every gene can become an oncogene; those that can are called protooncogenes. Consider how most of your body cells have receptors for growth factors, which are molecules that stimulate cell division and differentiation. When a growth factor binds to its receptor on a cell, a series of events is triggered that advances the cell cycle from interphase into mitosis. Mutations can result in a receptor that stimulates mitosis even when the growth factor is not present, and in fact most neoplasms carry mutations resulting in an overactivity or overabundance of growth factor receptors (Figure 8.7).

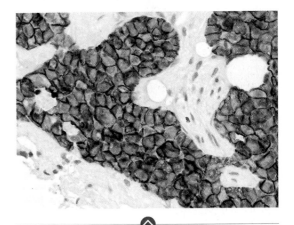

Figure 8.7 An oncogene causing a neoplasm.
In this section of human breast tissue, a brown-colored tracer shows the active form of a growth factor receptor. Normal cells are lighter in color.

The dark cells have an overactive receptor that is constantly stimulating mitosis; these cells have formed a neoplasm. Cells of most neoplasms have mutations that cause this receptor to be overproduced or overactive.

© From *Expression of the epidermal growth factor receptor (EGFR) and the phosphorylated EGFR in invasive breast carcinomas.* http://breast-cancer-research.com/content/10/3/R49.

cleavage furrow In a dividing animal cell, the indentation where cytoplasmic division will occur.

neoplasm An accumulation of abnormally dividing cells.

oncogene Gene that helps transform a normal cell into a tumor cell.

tumor A neoplasm that forms a lump.

Figure 8.8 Neoplasms and malignancy.

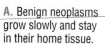

A. Benign neoplasms grow slowly and stay in their home tissue.

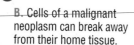

B. Cells of a malignant neoplasm can break away from their home tissue.

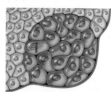

C. The malignant cells become attached to the wall of a lymph vessel or blood vessel (as shown here). They release digestive enzymes that create an opening in the wall, then enter the vessel.

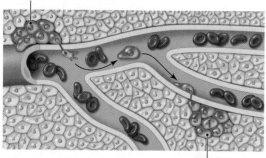

D. The cells creep or tumble along in the vessel, then exit the same way they got in. Migrating cells may start growing in other tissues, a process called metastasis.

Figure 8.9 Skin cancer can be detected and treated early with periodic screening.

(A) Dr. Allan Harris/Phototake; (B) Biophoto Associates/Science Source; (C) James Stevenson/Science Source.

A. Basal cell carcinoma is the most common type of skin cancer. This slow-growing, raised lump may be uncolored, reddish-brown, or black.

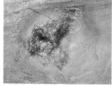

B. Squamous cell carcinoma is the second most common form of skin cancer. This pink growth, firm to the touch, grows under the skin's surface.

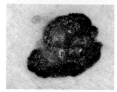

C. Melanoma spreads fastest. Cells form dark, encrusted lumps that may itch or bleed easily.

Checkpoint gene products that inhibit mitosis are called tumor suppressors because tumors form when they are missing. Two examples are *BRCA1* and *BRCA2*: Mutations in these genes give rise to neoplasms in the breast, prostate, ovary, and other tissues. The multiple functions of the *BRCA* gene products are still being unraveled, but we do know they help maintain the structure and number of chromosomes in a dividing cell, and they participate directly in DNA repair. *BRCA* gene products also bind to hormone receptors that are particularly abundant on cells of breast and ovarian tissues; the binding suppresses transcription of growth factor genes in these cells. When a mutation alters a *BRCA* gene so that its product cannot bind to these hormone receptors, the cells overproduce growth factors—an outcome associated with neoplasm formation.

Viruses such as HPV (human papillomavirus) cause a cell to make proteins that interfere with its own tumor suppressors. Infection with HPV causes skin growths called warts, and some kinds are associated with neoplasms that form on the cervix.

Cancer Benign neoplasms such as warts are not usually dangerous. They grow very slowly, and their cells retain the plasma membrane adhesion proteins that keep them properly anchored in their home tissue. A malignant neoplasm is one that gets progressively worse, and is dangerous to health. Malignant cells typically display the following three characteristics:

First, like cells of all neoplasms, malignant cells grow and divide abnormally. Controls that usually keep cells from getting overcrowded in tissues are lost in malignant cells, so their populations may reach extremely high densities with cell division occurring very rapidly. The number of small blood vessels that transport blood to the growing cell mass also increases abnormally.

Second, the cytoplasm and plasma membrane of malignant cells are altered. Both are indications of cellular malfunction. The cytoskeleton may be shrunken, disorganized, or both. Malignant cells typically have an abnormal chromosome number, with some chromosomes present in multiple copies, and others missing or damaged. The balance of metabolism is often shifted, as in an amplified reliance on ATP formation by fermentation rather than aerobic respiration.

Altered or missing proteins impair the function of the plasma membrane of malignant cells. For example, these cells do not stay anchored properly in tissues because their plasma membrane adhesion proteins are defective or missing. Malignant cells can slip easily into and out of vessels of the circulatory and lymphatic systems (Figure 8.8). By migrating through these vessels, the cells can establish neoplasms elsewhere in the body. The process in which malignant cells break loose from their home tissue and invade other parts of the body is called **metastasis**. Metastasis is the third hallmark of malignant cells.

The disease called **cancer** occurs when the abnormally dividing cells of a malignant neoplasm disrupt body tissues, both physically and metabolically. Unless chemotherapy, surgery, or another procedure eliminates malignant cells from the body, they can put an individual on a painful road to death. Each year, cancer causes 15 to 20 percent of all human deaths in developed countries. The good news is that mutations in multiple checkpoint genes are required to transform a normal cell into a malignant one, and such mutations may take a lifetime to accumulate. Lifestyle choices such as not smoking and avoiding exposure of unprotected skin to sunlight reduce one's risk of acquiring mutations in the first place. Some neoplasms can be detected with periodic screening such as gynecology or dermatology exams (Figure 8.9). If detected early enough, many types of malignant neoplasms can be removed before metastasis occurs.

Digging Into Data

HeLa Cells Are a Genetic Mess

HeLa cells can vary in chromosome number. Defects in proteins that orchestrate cell division result in descendant cells with too many or too few chromosomes, an outcome that is one of the hallmarks of cancer cells. The karyotype in Figure 8.10, originally published in 1989, shows all of the chromosomes in a single metaphase HeLa cell.

1. What is the chromosome number of this HeLa cell?
2. How many extra chromosomes does this cell have, compared with a normal human body cell?
3. Can you tell that this cell came from a female? How?

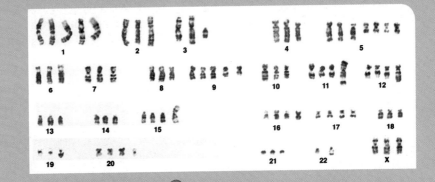

Figure 8.10 Karyotype of a HeLa cell.
© Dr. Thomas Ried, NIH and the Association for Cancer Research.

Telomeres Remember that Dolly the cloned sheep died early. The life expectancy of a sheep is normally about 10 to 12 years; by the time Dolly was five, however, she was as fat and arthritic as a twelve-year-old sheep. The following year, she contracted a lung disease typical of much older sheep and had to be euthanized.

Dolly's early demise may have been the result of abnormally short telomeres. Telomeres are noncoding DNA sequences that occur at the ends of eukaryotic chromosomes (Figure 8.11). Vertebrate telomeres consist of a short, noncoding DNA sequence repeated perhaps thousands of times. These "junk" repeats provide a buffer against the loss of more valuable DNA internal to the chromosomes.

A telomere buffer is particularly important because, under normal circumstances, a eukaryotic chromosome shortens by about 100 nucleotides with each DNA replication. When a cell's offspring receive chromosomes with too-short telomeres, checkpoint gene products halt the cell cycle, and the descendant cells die shortly thereafter. Most body cells can divide only a certain number of times before this happens. This cell division limit may be a fail-safe mechanism in case a cell loses control over the cell cycle and begins to divide again and again. A limit on the number of divisions keeps the resulting neoplasm from overrunning the body.

The cell division limit varies by species, and it may be part of the mechanism that sets an organism's life span. Dolly's DNA came from the nucleus of a mammary gland cell donated by an adult sheep. When Dolly was only two years old, her telomeres were as short as those of a six-year-old sheep—the exact age of the adult animal that had been her genetic donor.

A few normal cells in an adult retain the ability to divide indefinitely. These cells are called stem cells, and their descendants replace other cell lineages that eventually die out when they reach their division limit. Stem cells are immortal because they continue to make telomerase, an enzyme that reverses the telomere shortening associated with DNA replication (differentiated body cells produce no telomerase).

Mice that have had their telomerase enzyme knocked out age prematurely, with a life expectancy about half that of a normal mouse. When one of these knockout mice is close to the end of its shortened life span, rescuing the function of its telomerase enzyme results in lengthened telomeres. The rescued mouse also regains

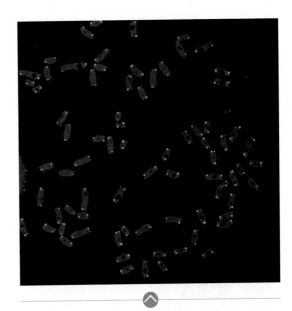

Figure 8.11 Telomeres.
The bright dots at the end of each DNA strand in these (duplicated) chromosomes are telomere sequences.
© ISM/Phototakeusa.

cancer Disease that occurs when a malignant neoplasm physically and functionally disrupts body tissues.

metastasis The process in which cells of a malignant neoplasm spread from one part of the body to another.

vitality: Worn-out tissue in the brain and other organs repairs itself and begins to function normally, and the once-geriatric individual even begins to reproduce again.

Researchers are careful to point out that shortening telomeres may be an effect of aging rather than a cause. Also, while telomerase holds therapeutic promise for rejuvenating aged tissues, it can be dangerous. Cancer cells—including the HeLa cells you learned about in Section 8.1—characteristically express high levels of this molecule, which is why, like stem cells, they can divide indefinitely.

Take-Home Message 8.3

What is cancer?

- Gene expression controls advance, delay, or block the cell cycle in response to internal and external conditions. Checkpoints built into the cell cycle allow problems to be corrected before the cycle proceeds.
- Neoplasms form when cells lose control over their cell cycle and begin dividing abnormally.
- Mutations in multiple checkpoint genes can give rise to a progressively worsening malignant neoplasm.
- Cancer is a disease that occurs when the abnormally dividing cells of a malignant neoplasm physically and metabolically disrupt body tissues.
- In most cases, lifestyle choices and early intervention can reduce one's risk of cancer.

8.4 Sex and Alleles

REMEMBER: All organisms inherit their DNA from parents by processes of reproduction (Section 1.3). Traits shared by members of a species often vary a bit among individuals, as eye color does among people (1.4). A somatic cell is a body cell (6.1). DNA sequence variation is the basis of traits that define species and distinguish individuals (6.2). The two autosomes of a pair hold information about the same traits (6.3). Mutations (6.4) that alter a gene's product can be harmful (7.6).

Your homologous chromosomes carry the same genes, but their DNA sequence may not be identical (Figure 8.12). This is because you inherited your chromosomes from two parents who are, most likely, not closely related. Unique mutations accumulated in their separate lines of descent over time. Thus, the DNA sequence of any of your genes may differ a bit from the corresponding gene on the homologous chromosome. Different forms of the same gene are called **alleles**.

Alleles may encode slightly different forms of a gene's product, and such differences influence the details of shared, inherited traits. Members of a species have the same traits because they have the same genes. However, almost every shared trait varies a bit among individuals of a sexually reproducing species. Alleles of the shared genes are the basis of this variation. Consider just one human gene, *HBB*, which encodes beta globin. *HBB* has has more than 700 alleles; a few cause sickle-cell anemia, several cause beta thalassemia, and so on. There are approximately 20,000 human genes, and most of them have multiple alleles.

On the Advantages of Sex You learned earlier in this chapter that mitosis and cytoplasmic division are part of asexual reproduction in eukaryotes, but only a few

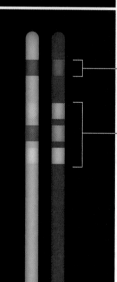

A. Corresponding colored patches in this fluorescence micrograph indicate corresponding DNA sequences in a homologous chromosome pair. These chromosomes carry the same set of genes.

Genes occur in pairs on homologous chromosomes.

The members of each pair of genes may be identical, or they may differ slightly, as alleles. (Color variations represent sequence differences.)

B. Homologous chromosomes carry the same series of genes, but the DNA sequence of any one of those genes might differ just a bit from that of its partner on the homologous chromosome.

Figure 8.12 Genes on chromosomes.
Different forms of a gene are called alleles.

(A) Courtesy of Carl Zeiss Microimaging, Thornwood, NY.

species use this reproductive mode (Figure 8.13). Most eukaryotes reproduce sexually. **Sexual reproduction** is the process in which offspring arise from two parents and inherit genes from both.

If the function of reproduction is the perpetuation of one's genes, then an asexual reproducer would seem to win the evolutionary race. When it reproduces, it passes all of its genes to every one of its offspring. Only about half of a sexual reproducer's genes are passed to each offspring. So why is sex so widespread?

All offspring of an asexual reproducer are clones: Barring new mutations, they have exactly the same alleles as their one parent. Consistency is a good evolutionary strategy in a favorable, unchanging environment; alleles that help an organism survive and reproduce in the environment do the same for its descendants. However, most environments are constantly changing, and change is not always favorable. Individuals that are identical are equally vulnerable to challenges. In a changing environment, sexual reproducers have the evolutionary edge because they carry different alleles. Sexual reproduction mixes up the genetic information of two parents that differ in the alleles they carry, so offspring typically inherit new combinations of alleles—ones that do not occur in either parent (or other offspring). Some may inherit a particular combination of alleles that suits them perfectly to a new environmental challenge. As a group, their diversity offers them a better chance of surviving environmental change than clones.

To understand why environments constantly change, think about one example: interactions between a predatory species and its prey. To the prey species, the predators are an environmental challenge. Prey individuals that can best escape predation tend to leave more offspring, so alleles that help them be best tend to become more common in the species over generations. At the same time, prey that can better escape predation are an environmental challenge to the predator species. Predator individuals best able to capture prey tend to leave more offspring, so alleles that help them be best tend to become more common in the predator species over generations (Chapter 12 returns to evolutionary processes). The two species are locked in a constant race, with each genetic improvement in one countered by a genetic improvement in the other. This idea is called the Red Queen hypothesis, a reference to Lewis Carroll's book *Through the Looking Glass*. In the book, the Queen of Hearts tells Alice, "It takes all the running you can do, to keep in the same place."

Another advantage of sexual reproduction involves the inevitable occurrence of mutations that are harmful but not lethal. A population of sexual reproducers has a better chance of weathering the effects of such mutations. With asexual reproduction, individuals bearing a harmful mutation necessarily pass it to all of their offspring. This outcome would be rare in sexual reproduction, because each offspring of a sexual union has a 50 percent chance of inheriting a parent's mutation. Thus, all else being equal, harmful mutations can spread through an asexually reproducing population more quickly than a sexually reproducing one.

Figure 8.13 Tiny New Zealand mud snails can reproduce on their own (asexually) or with a partner (sexually).
Like most sexual organisms, the sexual snails have two chromosome sets; like most animals that cannot reproduce sexually, the asexual snails have at least three. DNA has a high phosphorus content; having the extra sets of chromosomes multiplies each organism's requirement for this nutrient. Fertilizers and detergents contain phosphorus, so agricultural runoff and other types of water pollution may be fueling the gigantic populations of asexual snails currently invading ecosystems worldwide.

© Bart Zijistra.

Almost every shared trait varies a bit among individuals of a sexually reproducing species.

Take-Home Message 8.4

Why do species that reproduce sexually vary in shared traits?

- Paired genes on homologous chromosomes may vary in DNA sequence as alleles. Alleles arise by mutation.
- Alleles of shared genes are the basis of differences in traits shared by a species.
- Offspring of sexual reproducers inherit new combinations of parental alleles.

alleles Forms of a gene with slightly different DNA sequences; may encode slightly different versions of the gene's product.

sexual reproduction Reproductive mode by which offspring arise from two parents and inherit genes from both.

haploid Having one of each type of chromosome characteristic of the species.

meiosis Nuclear division process that halves the chromosome number. Basis of sexual reproduction.

8.5 Meiosis in Sexual Reproduction

Meiosis, a nuclear division mechanism that halves the chromosome number, is the process inherent to sexual reproduction that gives rise to new combinations of alleles in offspring. It parcels chromosomes into new nuclei two times (Figure 8.14 shows the stages of meiosis in a diploid (2*n*) cell). DNA replication occurs before meiosis begins, so each chromosome has two sister chromatids.

The first stage of meiosis I is prophase I ❶. During this phase, the chromosomes condense, and homologous chromosomes align tightly and swap segments (more about segment-swapping shortly). The nuclear envelope breaks up. A spindle forms, and by the end of prophase I, microtubules attach one chromosome of each homologous pair to one spindle pole, and the other to the opposite spindle pole. These microtubules grow and shrink, pushing and pulling the chromosomes as they do. At metaphase I ❷, all of the microtubules are the same length, and the chromosomes are aligned midway between the spindle poles. During anaphase I ❸, the spindle pulls the homologous chromosomes of each pair away from one another

MEIOSIS I One diploid nucleus to two haploid nuclei

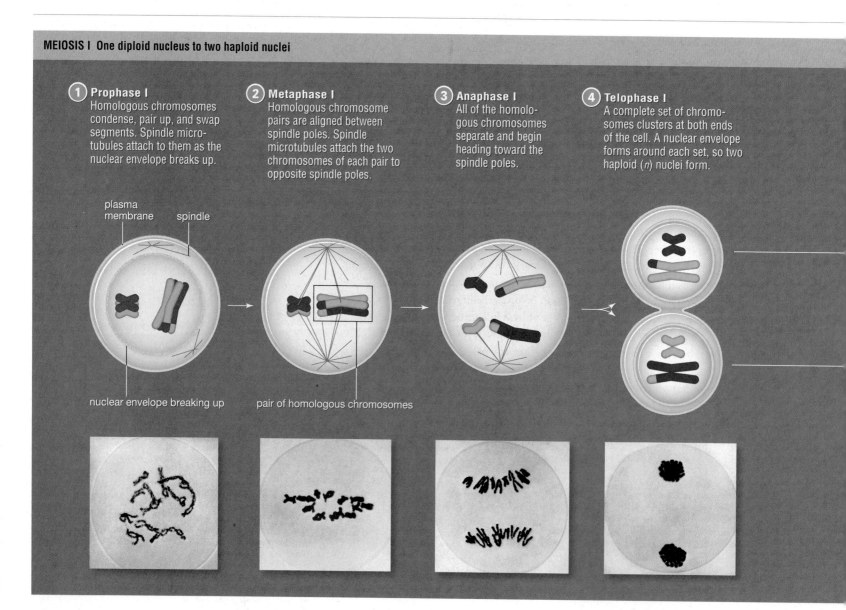

❶ **Prophase I**
Homologous chromosomes condense, pair up, and swap segments. Spindle microtubules attach to them as the nuclear envelope breaks up.

❷ **Metaphase I**
Homologous chromosome pairs are aligned between spindle poles. Spindle microtubules attach the two chromosomes of each pair to opposite spindle poles.

❸ **Anaphase I**
All of the homologous chromosomes separate and begin heading toward the spindle poles.

❹ **Telophase I**
A complete set of chromosomes clusters at both ends of the cell. A nuclear envelope forms around each set, so two haploid (*n*) nuclei form.

plasma membrane   spindle

nuclear envelope breaking up

pair of homologous chromosomes

and toward opposite spindle poles. The two sets of chromosomes reach the spindle poles during telophase I ❹ , and a new nuclear envelope forms around each cluster of chromosomes as the DNA loosens up. The two new nuclei are **haploid** (*n*); each contains one set of chromosomes—half of the diploid (2*n*) number. The cytoplasm often divides at this point. Each chromosome is still duplicated (it consists of two sister chromatids). Meiosis may pause at this point, but no DNA replication occurs before meiosis II.

Meiosis II proceeds simultaneously in both nuclei that formed in meiosis I. During prophase II ❺, the chromosomes condense and the nuclear envelope breaks up. A new spindle forms. By the end of prophase II, spindle microtubules attach each chromatid to one spindle pole, and its sister chromatid to the opposite spindle pole. These microtubules push and pull the chromosomes, aligning them midway between spindle poles at metaphase II ❻. During anaphase II ❼, the spindle microtubules pull the sister chromatids apart and toward opposite spindle poles. Each chromosome is now unduplicated (it consists of one molecule of DNA). During telophase II ❽, these chromosomes reach the spindle poles.

Figure 8.14 Meiosis.
The micrographs show meiosis in a lily cell. The illustrations show two pairs of chromosomes in a diploid (2*n*) animal cell; homologous chromosomes are indicated in blue and pink.

Bottom photos, With thanks to the John Innes Foundation Trustees, computer enhanced by Gary Head.

Figure It Out: Which phase of meiosis reduces the chromosome number?   Answer: Anaphase I

MEIOSIS II Two haploid nuclei to four haploid nuclei

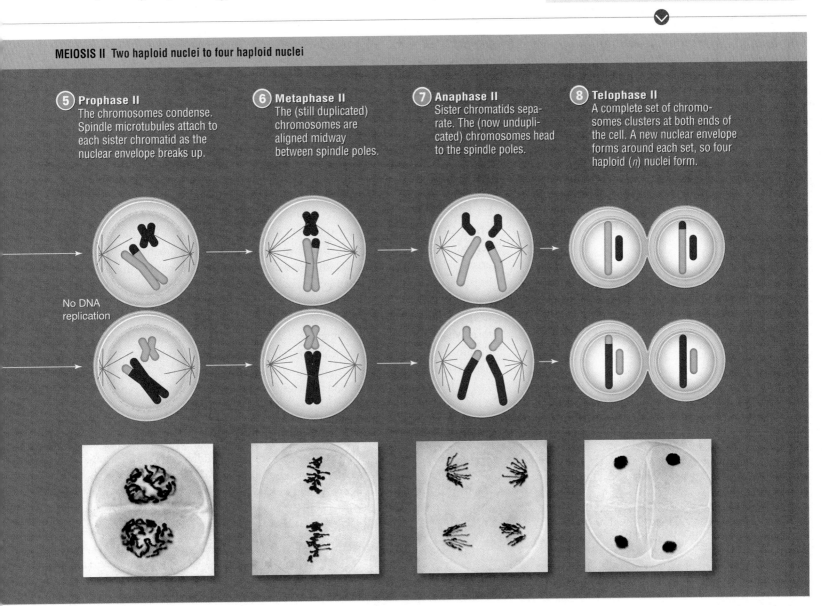

❺ **Prophase II**
The chromosomes condense. Spindle microtubules attach to each sister chromatid as the nuclear envelope breaks up.

❻ **Metaphase II**
The (still duplicated) chromosomes are aligned midway between spindle poles.

❼ **Anaphase II**
Sister chromatids separate. The (now unduplicated) chromosomes head to the spindle poles.

❽ **Telophase II**
A complete set of chromosomes clusters at both ends of the cell. A new nuclear envelope forms around each set, so four haploid (*n*) nuclei form.

No DNA replication

New nuclear envelopes form around the four clusters of chromosomes as the DNA loosens up. Each of the four nuclei that form are haploid (*n*), with one set of (unduplicated) chromosomes. The cytoplasm often divides at this point.

How Meiosis Mixes Alleles We mentioned briefly that duplicated chromosomes swap segments with their homologous partners during prophase I. As the chromosomes condense, each is drawn close to its homologous partner, so that the chromatids align along their length:

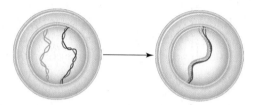

This tight, parallel orientation favors **crossing over**, a process by which a chromosome and its homologous partner exchange corresponding pieces of DNA during meiosis (Figure 8.15A–C). Homologous chromosomes may swap any segment of DNA along their length, although crossovers tend to occur more frequently in certain regions.

Swapping segments of DNA shuffles alleles between homologous chromosomes. It breaks up the particular combinations of alleles that occurred on the parental chromosomes, and makes new ones on the chromosomes that end up in offspring. Thus, crossing over introduces novel combinations of alleles—and new combinations of traits—among offspring. It is a normal and frequent process in meiosis, but the rate of crossing over varies among species and among chromosomes. In humans, between 46 and 95 crossovers occur per meiosis, so on average each chromosome crosses over at least once (Figure 8.15D).

From Gametes to Offspring Sexual reproduction involves the fusion of mature reproductive cells—**gametes**—from two parents. All gametes are haploid, and they arise by division of germ cells, which are immature reproductive cells that form in organs set aside for reproduction. Animals and plants make gametes somewhat differently. In animals, meiosis in diploid germ cells gives rise to eggs (female gametes) or sperm (male gametes). In plants, haploid germ cells form by meiosis. Gametes form when these cells divide by mitosis. We leave details of sexual reproduction in animals and plants for later chapters, but you will need to know a few concepts before you get there.

At fertilization, two haploid gametes fuse and produce a diploid **zygote**, which is the first cell of a new individual. Thus, meiosis halves the chromosome number, and fertilization restores it. If meiosis did not precede fertilization, the chromosome number would double with every generation. If the chromosome number changes, so does the individual's set of genetic instructions. An individual's set of chromosomes is like a fine-tuned blueprint that must be followed exactly, in order to build a body that functions normally. As you will see in Chapter 9, chromosome number changes can have drastic consequences for health, particularly in animals.

Fertilization also contributes to the variation that we see among offspring of sexual reproducers. Think about it in terms of human reproduction. Cells that give rise to human gametes have twenty-three pairs of homologous chromosomes. Each time a human germ cell undergoes meiosis, the four gametes that form end up with one of 8,388,608 (or 2^{23}) possible combinations of homologous chromosomes. In

Figure 8.15 Crossing over.
Blue signifies a paternal chromosome, and pink, its maternal homologue. For clarity, only one pair of homologous chromosomes is shown.

(D) James Kezer/Courtesy of Dr. Sessions.

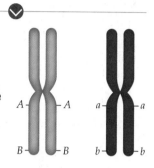

A. Here, we focus on only two of the many genes on a chromosome. In this example, one gene has alleles *A* and *a*; the other has alleles *B* and *b*.

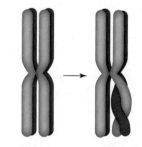

B. Close contact between homologous chromosomes promotes crossing over between nonsister chromatids. Paternal and maternal chromatids exchange corresponding pieces.

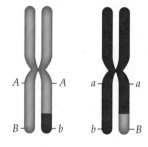

C. Crossing over mixes up paternal and maternal alleles on homologous chromosomes.

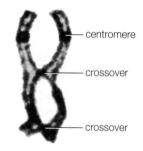

D. Each pair of homologous chromosomes can cross over multiple times. This is a normal and common process of meiosis.

centromere

crossover

crossover

Digging Into Data

BPA and Abnormal Meiosis

In 1998, researchers at Case Western University were studying meiosis in mouse oocytes when they saw an unexpected and dramatic increase in abnormal meiosis events (Figure 8.16). The improper sorting of chromosomes into gametes is one of the main causes of human genetic disorders, which we will discuss in Chapter 9.

The researchers discovered that the spike in abnormal meiosis events began immediately after the mouse facility started washing the animals' plastic cages and water bottles in a new, alkaline detergent. The detergent had damaged the plastic, which as a result was leaching bisphenol A (BPA). BPA is a synthetic chemical that mimics estrogen, the main female sex hormone in animals. Though it has been banned for use in baby bottles, BPA is still widely used to manufacture other plastic items and epoxies (such as the coating on the inside of metal cans of food). BPA-free plastics are often manufactured with a related compound, bisphenol S (BPS), that has effects similar to BPA.

1. What percentage of mouse oocytes displayed abnormalities of meiosis with no exposure to damaged caging?
2. Which group of mice had the most meiotic abnormalities?
3. What is abnormal about metaphase I as it is occurring in the oocytes shown in Figure 8.16B, C, and D?

Caging materials	Total number of oocytes	Abnormalities
Control: New cages with glass bottles	271	5 (1.8%)
Damaged cages with glass bottles		
Mild damage	401	35 (8.7%)
Severe damage	149	30 (20.1%)
Damaged bottles	197	53 (26.9%)
Damaged cages with damaged bottles	58	24 (41.4%)

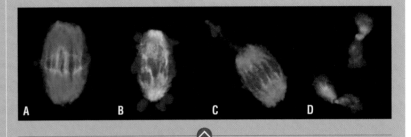

Figure 8.16 **Meiotic abnormalities associated with exposure to plastic.** Fluorescence micrographs show nuclei of single mouse oocytes (female germ cells) in metaphase I. (**A**) Normal metaphase; (**B–D**) examples of abnormal metaphase. Chromosomes are stained red; spindle fibers, green.

(A–D) Reprinted from *Current Biology*, Vol 13, (Apr 03), Authors Hunt, Koehler, Susiarjo, Hodges, Ilagan, Voight, Thomas, Thomas, and Hassold, Bisphenol A Exposure Causes Meiotic Aneuploidy in the Female Mouse, pp. 546–553, © 2003 Cell Press. Published by Elsevier Ltd. With permission from Elsevier.

addition, any number of genes may occur as different alleles on the maternal and paternal chromosomes, and crossing over makes mosaics of that genetic information. Then, out of all the male and female gametes that form, which two actually get together at fertilization is a matter of chance. Are you getting an idea of why such fascinating combinations of traits show up among the generations of your own family tree?

Meiosis halves the chromosome number, and fertilization restores it.

Take-Home Message 8.5

How does meiosis give rise to new combinations of parental alleles among the offspring of sexual reproducers?

- During meiosis, the nucleus of a diploid (2*n*) cell divides twice. DNA replication does not occur between the two divisions, so the chromosome number is reduced to the haploid number (*n*) for forthcoming gametes.
- Crossing over—recombination between nonsister chromatids of homologous chromosomes—occurs during meiosis. It makes new combinations of parental alleles.
- The union of two haploid gametes at fertilization results in a diploid zygote, the first cell of a new individual.

crossing over Process in which homologous chromosomes exchange corresponding segments during meiosis.

gamete Mature, haploid reproductive cell; e.g., an egg or a sperm.

zygote Diploid cell that forms when two gametes fuse; the first cell of a new individual.

Summary

Section 8.1 An immortal line of human cells (HeLa) is a legacy of cancer victim Henrietta Lacks. Researchers all over the world continue to work with these cells in their efforts to unravel the mechanisms of cancer.

Section 8.2 A **cell cycle** starts when a new cell forms, and ends when the cell reproduces. Most of a cell's activities occur during **interphase**. A eukaryotic cell reproduces by dividing: nucleus first, then cytoplasm. **Mitosis**, a nuclear division mechanism that maintains the chromosome number, is the basis of growth and tissue repair in multicelled species. Mitosis is also the basis of **asexual reproduction** in many species.

DNA replication occurs before mitosis begins, so each of the cell's **homologous chromosomes** consists of two molecules of DNA (sister chromatids) attached at the centromere. Mitosis occurs in four stages. During **prophase**, the chromosomes condense, the nuclear envelope breaks up, and microtubules assemble into a **spindle**. Spindle microtubules attach to the chromosomes at the centromere. At **metaphase**, all of the chromosomes are aligned in the middle of the cell. During **anaphase**, the spindle separates the sister chromatids of each chromosome and moves them toward opposite sides of the cell. During **telophase**, two new nuclei form, each with the parental chromosome number.

In most cases, nuclear division is followed by cytoplasmic division. In animal cells, a contractile ring pulls the plasma membrane inward, forming a **cleavage furrow** that pinches the cytoplasm in two. In plant cells, vesicles merge as a cell plate that expands and fuses with the plasma membrane, thus becoming a cross-wall that partitions the cytoplasm.

Section 8.3 The products of checkpoint genes work together to control the cell cycle. These molecules monitor the integrity of the cell's DNA, and can pause the cycle until breaks or other problems are fixed. When checkpoint mechanisms fail, a cell loses control over its cell cycle, and its abnormally dividing descendants form a **neoplasm**. Neoplasms may form lumps called **tumors**.

Mutations can turn some genes into tumor-causing **oncogenes**. Mutations in multiple checkpoint genes can transform benign neoplasms into malignant ones. Cells of malignant neoplasms can break loose from their home tissues and colonize other parts of the body, a process called **metastasis**. **Cancer** occurs when cells of a malignant neoplasm physically and metabolically disrupt body tissues. Telomere length limits the number of times that a normal cell can divide, a fail-safe mechanism in case the cell loses control over its cell cycle.

Section 8.4 **Sexual reproduction** mixes up the genetic information of two parents. The offspring of sexual reproducers typically vary in shared, inherited traits. Particularly in changing environments, this variation can offer an evolutionary advantage over genetically identical offspring produced by asexual reproduction.

Sexual reproduction produces offspring whose body cells contain pairs of chromosomes, one of each homologous pair from the mother and the other from the father. The two chromosomes of a homologous pair carry the same genes. Paired genes on homologous chromosomes may vary in DNA sequence, in which case they are called **alleles**. Alleles are the basis of differences in shared traits. They arise by mutation.

Section 8.5 **Meiosis**, the basis of sexual reproduction in eukaryotes, is a nuclear division mechanism that halves the chromosome number. DNA replication occurs before meiosis begins, so each chromosome consists of two molecules of DNA (sister chromatids). Two nuclear divisions occur during meiosis. In the first nuclear division (meiosis I), all of the homologous chromosomes line up and exchange corresponding segments. This **crossing over** mixes up the alleles on maternal and paternal chromosomes. The homologous chromosomes are then moved apart and packaged in separate nuclei. This stage reduces the chromosome number, from diploid ($2n$) to **haploid** (n). The second nuclear division (meiosis II) occurs in both haploid nuclei that formed in meiosis I. Sister chromatids separate in this stage, so at the end of meiosis each chromosome consists of one molecule of DNA. Four haploid nuclei typically form. Meiosis is necessary for the production of haploid **gametes**. The fusion of two gametes at fertilization restores the diploid parental chromosome number in the **zygote**, the first cell of the new individual.

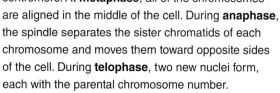

Self-Quiz

Answers in Appendix I

1. Mitosis and cytoplasmic division function in _____ .
   a. asexual reproduction of single-celled eukaryotes
   b. growth and tissue repair in multicelled species
   c. asexual reproduction in prokaryotes
   d. both a and b

2. A duplicated chromosome has how many chromatids?
   a. one   c. three
   b. two   d. four

3. A cell with two of each type of chromosome has a chromosome number that is _____ .
   a. diploid
   b. haploid
   c. tetraploid
   d. abnormal

4. Homologous chromosomes _____ .
   a. originate with two parents
   b. are sister chromatids

5. Interphase is the part of the cell cycle when _____ .
   a. a cell ceases to function
   b. the spindle forms prior to nuclear division
   c. DNA replication occurs
   d. mitosis proceeds

6. After mitosis, the chromosome number of a descendant cell is _____ the parent cell's.
   a. the same as
   b. one-half of
   c. rearranged compared to
   d. doubled compared to

7. One evolutionary advantage of sexual over asexual reproduction may be that it produces _____ .
   a. more offspring per individual
   b. more variation among offspring
   c. healthier offspring

8. Alternative forms of the same gene are _____ .
   a. gametes
   b. homologous
   c. alleles
   d. oncogenes

9. Meiosis is a necessary part of sexual reproduction because it _____ .
   a. divides two nuclei into four new nuclei
   b. reduces the chromosome number for gametes
   c. produces clones that can cross over

10. Crossing over mixes up _____ .
    a. chromosomes
    b. alleles
    c. zygotes
    d. gametes

11. Sexual reproduction in animals requires _____ .
    a. meiosis
    b. fertilization
    c. gametes
    d. all of the above

12. Which of the following is one of the very important differences between mitosis and meiosis?
    a. Chromosomes align in the middle of the cell only in meiosis.
    b. Homologous chromosomes swap segments only in meiosis.
    c. Sister chromatids separate only in meiosis.

13. The cell illustrated on the right is in anaphase I, not anaphase II. I know this because _____ .
    a. crossing over has already occurred
    b. the chromosomes are still duplicated
    c. a spindle has formed.
    d. sister chromatids have separated

14. Match each stage with the events listed.
    _____ prophase
    _____ metaphase
    _____ anaphase
    _____ telophase
    _____ interphase
    a. sister chromatids move apart
    b. chromosomes condense
    c. new nuclei form
    d. DNA replication
    e. all chromosomes are aligned in the middle of the cell

15. Match each term with the best description.
    _____ cell plate
    _____ spindle
    _____ tumor
    _____ contractile ring
    _____ gamete
    _____ cancer
    _____ zygote
    _____ prophase I
    a. lump of abnormal cells
    b. forms at fertilization
    c. divides plant cells
    d. mash-up time
    e. dangerous metastatic cells
    f. made of microtubules
    g. haploid
    h. makes an indentation

Critical Thinking

1. When a cell reproduces by mitosis and cytoplasmic division, does its life end?
3. Make a sketch of meiosis in a cell with a diploid chromosome number of 4. Now try it when the chromosome number is 3.
4. The diploid chromosome number for the body cells of a frog is 26. What would that number be after three generations if meiosis did not occur before gamete formation?
5. Which nuclear division, meiosis I or meiosis II, is most similar to mitosis?

Visual Question

The eukaryotic cell in the photo on the left is in the process of cytoplasmic division. Is this cell from a plant or an animal? How do you know?

© Michel Delarue © ISM/Phototake.

9 PATTERNS OF INHERITANCE

Application ⊙ 9.1 Menacing Mucus

REMEMBER: Polypeptide chains twist, fold, and pack into functional domains; a protein's function arises from and depends on its shape (Section 2.9). Vesicles move substances among organelles of the ER, and to and from the plasma membrane (3.5). Water moves across a membrane by osmosis, from a hypotonic to a hypertonic fluid (4.5). In active transport, a transport protein uses energy to pump a solute against its gradient across a cell membrane; in endocytosis, a small patch of plasma membrane balloons into cytoplasm and becomes a vesicle (4.6). During protein synthesis, amino acids are specified by three-nucleotide codons (7.4). A deletion is a mutation in which one or more nucleotides are lost from a chromosome (7.6).

In 1988, researchers discovered a gene that, when mutated, causes the most common fatal genetic disorder in the United States: cystic fibrosis (CF). The gene they discovered, *CFTR*, encodes an active transport protein that moves chloride ions out of epithelial cells. Sheets of epithelial cells line the passageways and ducts of the lungs, liver, pancreas, intestines, and reproductive system. When the CFTR protein pumps chloride ions out of these cells, water follows the solute by osmosis. The two-step process maintains a thin, watery film on the surface of epithelial cell sheets. Mucus slides easily over the wet sheets of cells.

The mutation most commonly associated with cystic fibrosis is a deletion that removes a single codon from the *CFTR* gene. The protein product of the mutated gene is missing one amino acid, and as a result it misfolds in a tiny region. This small defect interferes with cellular processes that would otherwise finish the protein and install it in the plasma membrane. Normally, a newly translated CFTR polypeptide is modified by the endoplasmic reticulum (ER) and exported to a Golgi body, which attaches carbohydrates to it. The finished protein is then packaged in vesicles routed to the plasma membrane. CFTR polypeptides with the missing amino acid are produced properly, but a cellular quality control mechanism recognizes the misfolded region and destroys most of them before they leave the ER. The few that make it to the plasma membrane are quickly taken back into the cell by endocytosis and destroyed.

Epithelial cell membranes that lack the CFTR protein cannot transport chloride ions. Too few chloride ions leave these cells. Not enough water leaves them either, so the surfaces of epithelial cell sheets are too dry. Mucus that normally slips and slides through the body's tubes sticks to their walls instead. Thick globs of mucus accumulate and clog passageways and ducts throughout the body. Breathing becomes difficult as the mucus obstructs the smaller airways of the lungs. Digestive problems arise as ducts that lead to the gut become clogged with mucus. Males are typically infertile because their sperm flow is hampered.

In addition to its role in chloride ion transport, the CFTR protein also helps alert the immune system to the presence of disease-causing bacteria in the lungs. It functions as a receptor by binding directly to these bacteria and causing them to be taken into the cell by endocytosis. In epithelial cells lining the respiratory tract, endocytosis of bacteria triggers an immune response. Bacteria-fighting molecules that are produced in this response keep microbial populations at bay. When the cells lack CFTR, the early alert system fails, so bacteria have time to multiply before being detected by the immune system. Thus, chronic bacterial infections of the lungs are a hallmark of cystic fibrosis. Antibiotics help control infections, but there is no cure for the disorder. Most affected people die before age thirty, when their tormented lungs fail (Figure 9.1).

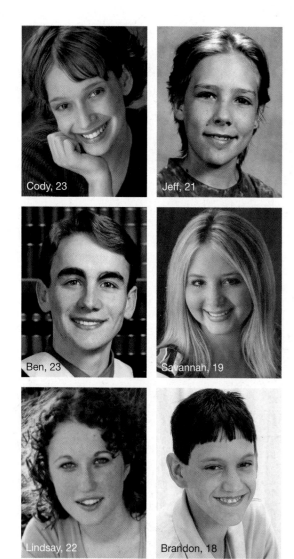

Figure 9.1 A few of the many victims of cystic fibrosis. At least one young person dies every day in the United States from complications of this disease, which occurs most often in people of northern European ancestry.

Cody, 23
Jeff, 21
Ben, 23
Savannah, 19
Lindsay, 22
Brandon, 18

Top row: left, courtesy of © The Cody Dieruf Benefit Foundation, www.breathinisbelievin.org; right, courtesy of © Bobby Brooks and The Family of Jeff Baird. Middle row: left, courtesy of The Family of Benjamin Hill, reprinted with permission of © Chappell/Marathonfoto; right, Courtesy of © The Family of Savannah Brooke Snider. Bottom row: left, courtesy of © Steve & Ellison Widener and Breathe Hope, http://breathehope.tamu.edu; right, courtesy of © The Family of Brandon Herriott.

9.2 **Tracking Traits**

REMEMBER: Inheritance is transmission of DNA to offspring (Section 1.3). DNA was proven to be hereditary material in the 1950s (6.2). Information encoded in a DNA sequence occurs in units called genes; a cell uses the information in a gene to make an RNA or protein product (7.2). Diploid body cells of sexual reproducers have two sets of chromosomes, one inherited from each of two parents; homologous chromosomes carry the same genes, which can differ as alleles; alleles may encode different forms of the gene's product; almost every shared trait varies a bit among individuals of a species, and alleles of shared genes are the basis of this variation (8.4). At fertilization, two gametes fuse to produce a zygote, the first cell of a new individual (8.5).

In the nineteenth century, people thought that hereditary material must be some type of fluid, with fluids from both parents blending at fertilization like milk into coffee. However, the idea of "blending inheritance" failed to explain what people could see with their own eyes. Children sometimes have traits such as freckles that do not appear in either parent. A cross between a black horse and a white one does not produce gray offspring. The naturalist Charles Darwin had no hypothesis to explain such phenomena, even though inheritance was central to his theory of natural selection (Chapter 11 returns to this theory). At the time, no one knew

The Moravian Museum, Brno

that hereditary information is divided into discrete units (genes), an insight that is critical to understanding how traits are inherited. However, even before Darwin presented his theory, someone had been gathering evidence that would support it. Gregor Mendel (left), an Austrian monk, had been carefully breeding thousands of pea plants. By keeping detailed records of how traits passed from one generation to the next, Mendel had been collecting evidence of how inheritance works.

Mendel's Experiments Mendel cultivated the garden pea plant (Figure 9.2). This species is naturally self-fertilizing, which means its flowers produce male and female gametes **1** that form viable embryos when they meet up. To prevent an individual pea plant from self-fertilizing, Mendel removed the pollen-bearing parts (anthers) from its flowers. He then cross-fertilized the flowers by brushing their egg-bearing parts (carpels) with pollen from another plant **2**. He collected seeds **3** from the cross-fertilized individual, and recorded the traits of the new pea plants that grew from them **4**.

Many of Mendel's experiments, which are called crosses, started with plants that "breed true" for particular traits such as white flowers or purple flowers. Breeding true for a trait means that, new mutations aside, all offspring have the same form of the trait as the parent(s), generation after generation. For example, all offspring of pea plants that breed true for white flowers also have white flowers. As you will see in the next section, Mendel discovered that crossing pea plants yields offspring with traits that often appear in predictable patterns. Mendel's meticulous work tracking pea plant traits led him to conclude (correctly) that hereditary information passes from one generation to the next in distinct units.

Inheritance in Modern Terms Mendel discovered hereditary units, which we now call genes, almost a century before the discovery of DNA. Today, we know that individuals of a species share certain traits because their chromosomes carry the

Figure 9.2 Breeding experiments with the garden pea.

Top, Valentina Razumova/Shutterstock; (#1) Jean M. Labat/ardea.com.

1 In the flowers of a garden pea plant (above), pollen grains that form in anthers (right) produce male gametes. Female gametes form in carpels.

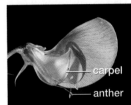
carpel
anther

2 Experimenters control the transfer of hereditary material from one pea plant to another by cutting off a flower's pollen-producing anthers (to prevent it from self-fertilizing), then brushing pollen from another flower onto its egg-producing carpel.

In this example, pollen from a plant with purple flowers is brushed onto the carpel of a white-flowered plant.

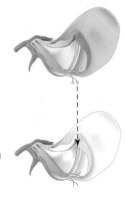

3 Later, seeds develop inside pods of the cross-fertilized plant. An embryo in each seed develops into a mature pea plant.

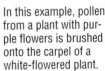

4 Every plant that arises from the cross has purple flowers. Predictable patterns such as this are evidence of how inheritance works.

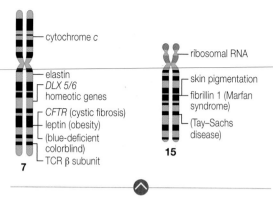

Figure 9.3 Locations of a few genes on two human chromosomes.
Genetic disorders arising from mutations in the genes are shown in parentheses. The number or letter below a chromosome is its name; staining reveals characteristic banding patterns that are illustrated here. Appendix III has a similar map of all 23 human chromosomes.

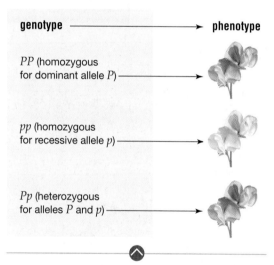

Figure 9.4 Genotype gives rise to phenotype.
In this example, the dominant allele P specifies purple flowers; the recessive allele p, white flowers.

© Tamara Kulikova/Shutterstock.

Figure It Out: Which individual is a hybrid?

Answer: The heterozygous one

same genes, and that each gene occurs at a specific location on a particular chromosome (Figure 9.3). The diploid body cells of humans and other animals contain pairs of genes, on pairs of homologous chromosomes. In most cases, both genes of a pair are expressed. Genes at the same location on a pair of homologous chromosomes may be identical, or they may vary as alleles. Organisms breed true for a trait because they carry identical alleles of genes governing that trait. An individual with two identical alleles of a gene is **homozygous** for the allele (*homo*– means the same). By contrast, an individual with two different alleles of a gene is **heterozygous** (*hetero*–, mixed). A hybrid is a heterozygous individual produced by a cross or mating between individuals that breed true for different forms of a trait. Homozygous and heterozygous are examples of **genotype**, the particular set of alleles an individual carries. Genotype is the basis of **phenotype**, which refers to the individual's observable traits. "White-flowered" and "purple-flowered" are examples of pea plant phenotypes that arise from differences in genotype.

The phenotype of a heterozygous individual depends on how the products of its two different alleles interact. In many cases, the effect of one allele influences the effect of the other, and the outcome of this interaction is reflected in the individual's phenotype. An allele is **dominant** when its effect masks that of a **recessive** allele paired with it. A dominant allele is often represented by an italic capital letter such as *A*; a recessive allele, with a lowercase italic letter such as *a*. Consider the purple- and white-flowered pea plants that Mendel studied. In these plants, the allele that specifies purple flowers (let's call it *P*) is dominant over the allele that specifies white flowers (*p*). Thus, a pea plant homozygous for the dominant allele (*PP*) has purple flowers; one homozygous for the recessive allele (*pp*) has white flowers (Figure 9.4). A heterozygous plant (*Pp*) has purple flowers.

Take-Home Message 9.2

How do alleles contribute to traits?

- Genotype (an individual's set of alleles) is the basis of phenotype (the individual's observable traits).
- A homozygous individual has two identical alleles of a gene. A heterozygous individual has two nonidentical alleles.
- A dominant allele masks the effect of a recessive allele in a heterozygous individual.

9.3 Mendelian Inheritance Patterns

REMEMBER: During the nuclear division process of meiosis, homologous chromosomes cross over and then separate; crossing over makes new combinations of parental alleles for forthcoming gametes; at fertilization, two gametes fuse and produce a zygote, which is the first cell of a new individual (Section 8.5).

When homologous chromosomes separate during meiosis, the gene pairs on those chromosomes separate too. Each gamete that forms carries only one of the two genes of a pair. Let's use our alleles for purple and white flowers in an example (Figure 9.5). Plants homozygous for the dominant allele (*PP*) can only make gametes that carry allele *P* ❶. Plants homozygous for a recessive allele (*pp*) can only make gametes that carry allele *p* ❷. If these homozygous plants are crossed (*PP* × *pp*),

Figure 9.5 Segregation of genes on homologous chromosomes into gametes.

Homologous chromosomes separate during meiosis, so the pairs of genes they carry separate too. Each of the resulting gametes carries one of the two members of each gene pair.

1 All gametes made by a parent homozygous for a dominant allele carry that allele.

2 All gametes made by a parent homozygous for a recessive allele carry that allele.

3 If these two parents are crossed, the union of any of their gametes at fertilization produces a zygote with both alleles. All offspring of this cross will be heterozygous.

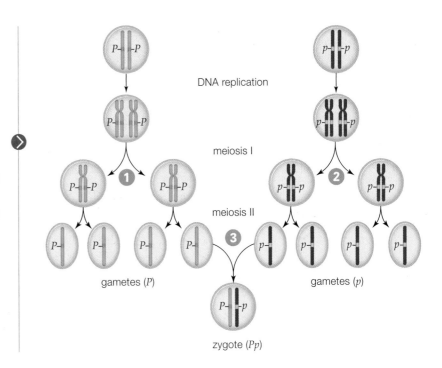

Figure 9.6 Making a Punnett square.

Parental gametes are listed in circles on the top and left sides of a grid. Each square is filled with the combination of alleles that would result if the gametes in the corresponding row and column met up.

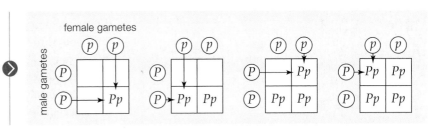

only one outcome is possible: A gamete carrying allele P meets up with a gamete carrying allele p **3**. All of the offspring of this cross will have both alleles—they will be heterozygous (Pp). Because all of the offspring will carry the dominant allele P, all will have purple flowers. A grid called a **Punnett square** is helpful for predicting the genetic and phenotypic outcomes of such crosses (Figure 9.6).

Our example illustrated a pattern so predictable that it can be used as evidence of a dominance relationship between alleles, as the following experiments illustrate.

Monohybrid Crosses Dominance relationships between alleles determine the phenotypic outcome of a **monohybrid cross**, in which individuals with the same two alleles of a gene are crossed ($Pp \times Pp$, for example). The frequency at which traits associated with the alleles appear among the offspring depends on whether one of the alleles is dominant over the other.

To make a monohybrid cross, we would start with individuals that breed true for two different forms of a trait. In garden pea plants, flower color (purple and white) is one example of a trait with two distinct forms, but there are many others. A cross between individuals that breed true for different forms of the trait yields hybrid offspring, all with the same set of alleles governing the trait. A cross between two of these F_1 (first generation) hybrids is the monohybrid cross. The frequency at which the two traits appear in the F_2 (second generation) offspring offers information about a dominance relationship between the two alleles. (F is an abbreviation for filial, which means offspring.)

dominant Refers to an allele that masks the effect of a recessive allele on the homologous chromosome.

genotype The particular set of alleles that is carried in an individual's chromosomes.

heterozygous Having two different alleles of a gene.

homozygous Having identical alleles of a gene.

monohybrid cross Cross between two individuals identically heterozygous for alleles of one gene; for example $Aa \times Aa$.

phenotype An individual's observable traits.

Punnett square Diagram used to predict the genetic and phenotypic outcomes of a cross.

recessive Refers to an allele with an effect that is masked by a dominant allele on the homologous chromosome.

codominance Inheritance pattern in which the full and separate phenotypic effects of two alleles are apparent in heterozygous individuals.

dihybrid cross Cross between two individuals identically heterozygous for alleles of two genes; for example *AaBb × AaBb*.

incomplete dominance Inheritance pattern in which one allele is not fully dominant over another, so the heterozygous phenotype is an intermediate blend between the two homozygous phenotypes.

Figure 9.7 Example of a monohybrid cross.

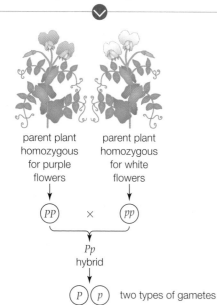

parent plant homozygous for purple flowers   parent plant homozygous for white flowers

PP × *pp*

Pp hybrid

P *p*   two types of gametes

A. All of the F₁ (first generation) offspring of a cross between two plants that breed true for different forms of a trait are identically heterozygous (*Pp*). These offspring make two types of gametes: *P* and *p*.

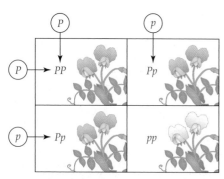

B. A monohybrid cross is a cross between the F₁ offspring. In this example, the phenotype ratio among the F₂ (second generation) offspring is 3:1 (three purple to one white).

Figure It Out: In this example, how many possible genotypes are there in the F₂ generation?

Answer: Three (PP, Pp, and pp)

A cross between two purple-flowered heterozygous plants (*Pp*) offers an example of a monohybrid cross. Each individual can make two types of gametes: ones that carry a *P* allele, and ones that carry a *p* allele (Figure 9.7A). So, in a monohybrid cross between *Pp* plants (*Pp × Pp*), the two types of gametes can meet up in four possible ways at fertilization:

Possible Event	Offspring Genotype	Resulting Phenotype
Sperm *P* meets egg *P* ⟶	zygote genotype is *PP* ⟶	individual has purple flowers
Sperm *P* meets egg *p* ⟶	zygote genotype is *Pp* ⟶	individual has purple flowers
Sperm *p* meets egg *P* ⟶	zygote genotype is *Pp* ⟶	individual has purple flowers
Sperm *p* meets egg *p* ⟶	zygote genotype is *pp* ⟶	individual has white flowers

Three of four possible outcomes of this cross include at least one copy of the dominant allele *P*. In other words, each time fertilization occurs, there are 3 chances in 4 that the resulting zygote will have a *P* allele (and the individual will make purple flowers). There is 1 chance in 4 that the zygote will have two *p* alleles (and the individual will make white flowers). Thus, the probability that a particular offspring of this cross will have purple or white flowers is 3 purple to 1 white—a ratio of 3:1 (Figure 9.7B). The 3:1 pattern is evidence that purple and white flower color are specified by alleles with a clear dominance relationship: Purple is dominant; white, recessive. If the probability of an individual inheriting a particular genotype is difficult to imagine, think about it in terms of many offspring: In this example, there will be roughly three purple-flowered plants for every white-flowered one.

Dihybrid Crosses A monohybrid cross allows us to study a dominance relationship between alleles of one gene. What about alleles of two genes? An individual heterozygous for alleles of two genes (*AaBb*, for example) is called a dihybrid, and a cross between two such individuals is a **dihybrid cross**. As with a monohybrid cross, the frequency of traits appearing among the offspring of a dihybrid cross depends on the dominance relationships between the alleles.

To make a dihybrid cross, we would start with individuals that breed true for two different traits. Let's use a gene for flower color (*P*, purple; *p*, white) and one for plant height (*T*, tall; *t*, short) in an example. Figure 9.8 shows a dihybrid cross starting with one parent plant that breeds true for purple flowers and tall stems (*PPTT*), and one that breeds true for white flowers and short stems (*pptt*). The *PPTT* plant only makes gametes with the dominant alleles (*PT*); the *pptt* plant only makes gametes with the recessive alleles (*pt*) **1**. So, all offspring from a cross between these two plants (*PPTT × pptt*) will be dihybrids (*PpTt*) with purple flowers and tall stems **2**.

Four combinations of *P* and *T* alleles are possible in the gametes of *PpTt* dihybrids **3**. If two of these plants are crossed (a dihybrid cross, *PpTt × PpTt*), the four types of gametes can combine in sixteen possible ways at fertilization **4**. Nine of the sixteen genotypes would give rise to tall plants with purple flowers; three, to short plants with purple flowers; three, to tall plants with white flowers; and one, to short plants with white flowers. Thus, the ratio of phenotypes among the offspring of this dihybrid cross would be 9:3:3:1.

How two gene pairs get sorted into gametes depends partly on whether the two genes are on the same chromosome. When homologous chromosomes separate during meiosis, either member of the pair can end up in either of the two new nuclei that form. Thus, genes on one chromosome pair assort into gametes independently of genes on the other chromosome pairs. What about genes on the same chromosome? Pea plants have seven chromosomes. Mendel studied seven pea genes, and all of them assorted into gametes independently of one another. Was he lucky enough to choose one gene on each of those chromosomes? As it turns out, some of the genes Mendel studied *are* on the same chromosome. These genes are far enough apart that crossing over occurs between them very frequently—so frequently that they tend to assort into gametes independently, just as if they were on different chromosomes. By contrast, genes that are very close together on a chromosome usually do not assort independently into gametes, because crossing over does not happen between them very often. Thus, gametes usually end up with parental combinations of alleles of these genes.

Take-Home Message 9.3

How are alleles distributed into gametes?

- Diploid cells have pairs of genes, on pairs of homologous chromosomes. The two genes of a pair (which may be identical or not) are separated from each other during meiosis, so they end up in different gametes.
- In most cases, the two genes of a pair are distributed into gametes independently of other gene pairs on other chromosomes.
- Gene pairs that are far apart on the same chromosome also tend to be distributed independently into gametes.

9.4 Beyond Simple Dominance

REMEMBER: Sickle-cell anemia is the outcome of a base-pair substitution in the beta globin gene (Section 7.6).

In the Mendelian inheritance patterns discussed in the previous section, the effect of a dominant allele on a trait fully masks that of a recessive one. Other inheritance patterns are more common, and more complex.

Incomplete Dominance In an inheritance pattern called **incomplete dominance**, one allele is not fully dominant over the other, so the heterozygous phenotype is an intermediate blend of the two homozygous phenotypes. A gene that affects flower color in snapdragon plants is an example. One allele of the gene (*R*) encodes an enzyme that makes a red pigment. The enzyme encoded by an allele with a mutation (*r*) cannot make any pigment. Plants homozygous for the *R* allele (*RR*) make a lot of red pigment, so they have red flowers. Plants homozygous for the *r* allele (*rr*) make no pigment, so their flowers are white. Heterozygous plants (*Rr*) make only enough pigment to tint their flowers pink. A cross between two pink-flowered heterozygous plants yields red-, pink-, and white-flowered offspring in a 1:2:1 ratio.

Codominance In an inheritance pattern called **codominance**, both alleles are fully expressed in heterozygous individuals; neither allele is dominant or recessive. Alleles of the *ABO* gene offer an example. This gene encodes an enzyme that modifies a carbohydrate on the surface of human red blood cells.

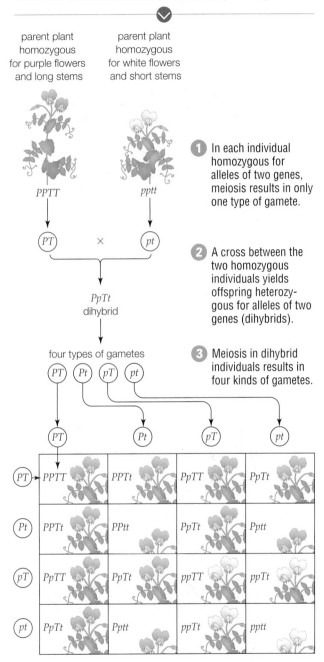

Figure 9.8 An example of a dihybrid cross.
P and *p* are dominant and recessive alleles for flower color; *T* and *t* are dominant and recessive alleles for plant height.

parent plant homozygous for purple flowers and long stems

parent plant homozygous for white flowers and short stems

PPTT

pptt

PT × *pt*

1 In each individual homozygous for alleles of two genes, meiosis results in only one type of gamete.

PpTt dihybrid

2 A cross between the two homozygous individuals yields offspring heterozygous for alleles of two genes (dihybrids).

four types of gametes
PT *Pt* *pT* *pt*

3 Meiosis in dihybrid individuals results in four kinds of gametes.

	PT	*Pt*	*pT*	*pt*
PT	*PPTT*	*PPTt*	*PpTT*	*PpTt*
Pt	*PPTt*	*PPtt*	*PpTt*	*Pptt*
pT	*PpTT*	*PpTt*	*ppTT*	*ppTt*
pt	*PpTt*	*Pptt*	*ppTt*	*pptt*

4 If two of the dihybrid individuals are crossed, the four types of gametes can meet up in 16 possible ways. Of 16 possible offspring genotypes, 9 will result in plants that are purple-flowered and tall; 3, purple-flowered and short; 3, white-flowered and tall; and 1, white-flowered and short. Thus, the ratio of phenotypes is 9:3:3:1.

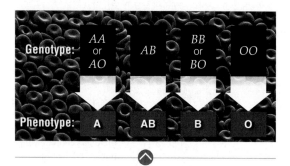

Figure 9.9 Combinations of alleles that are the basis of blood type.

Photo, Annie Cavanagh/Wellcome Images.

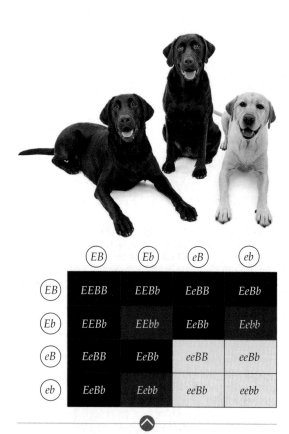

Figure 9.10 An example of epistasis.
Interactions among products of two gene pairs affect fur color in Labrador retrievers. Dogs with alleles *E* and *B* have black fur. Those with an *E* and two recessive *b* alleles have brown fur. Dogs homozygous for the recessive *e* allele have yellow fur.

Top, Susan Schmitz/Shutterstock.

Two alleles, *A* and *B*, encode slightly different versions of this enzyme, which in turn modify the carbohydrate differently. A third allele, *O*, has a mutation that prevents the enzyme product from becoming active, so the carbohydrate remains unmodified.

The alleles of the *ABO* gene that you carry determine the form of the carbohydrate on your blood cells, so they are the basis of your ABO blood type (Figure 9.9). The *A* and the *B* allele are codominant when paired. If your genotype is *AB*, then you have both versions of the enzyme, and your blood type is AB. The *O* allele is recessive when paired with either the *A* or *B* allele. If your genotype is *AA* or *AO*, your blood type is A. If your genotype is *BB* or *BO*, it is type B. If you are *OO*, it is type O.

The immune system attacks any cell bearing molecules that do not occur in one's own body, so receiving incompatible blood in a transfusion can be dangerous. An immune attack causes red blood cells to clump or burst, with potentially fatal results. Almost everyone makes the red blood cell carbohydrate (which is later modified in people with an *A* or *B* allele), so type O blood does not usually trigger an immune response in transfusion recipients. People with type O blood are called universal donors because they can donate blood to anyone. However, because their body is unfamiliar with the modified form of the carbohydrate made by people with type A or B blood, they can receive type O blood only. People with type AB blood can receive a transfusion of any ABO blood type, so they are called universal recipients.

Pleiotropy and Epistasis In an inheritance pattern called **pleiotropy**, a single gene influences multiple traits. Mutations that affect the gene's product or its expression affect all of the traits. Many complex genetic disorders, including sickle-cell anemia and cystic fibrosis, are caused by mutations in single genes. Marfan syndrome, another example, is a result of mutations that affect fibrillin. Long fibers of this protein are part of elastic tissues that make up the heart, skin, blood vessels, tendons, and other body parts. Mutations can cause tissues to form with defective fibrillin or none at all. The largest blood vessel leading from the heart, the aorta, is particularly affected. Without a proper scaffold of fibrillin, the aorta's thick wall is not as elastic as it should be, and it eventually stretches and becomes leaky. Thinned and weakened, the aorta can rupture during exercise—an abruptly fatal outcome. About 1 in 5,000 people have Marfan syndrome, and there is no cure. Its effects—and risks—are manageable with early diagnosis, but symptoms are easily missed. Affected people are tall and loose-jointed, but there are plenty of tall, loose-jointed people who do not have the syndrome. Thus, people with Marfan syndrome may die suddenly and early without ever knowing they had the disorder.

In a common inheritance pattern called polygenic inheritance or **epistasis**, one trait is affected by multiple gene products. Consider how fur color in dogs and other animals arises from pigments called melanins. A dark brown form of melanin gives rise to brown or black fur; a reddish melanin, to yellow fur. The products of several genes interact to carry out melanin synthesis and deposition in fur. In Labrador retriever dogs, alleles of two of these genes determine whether the individual has black, brown, or yellow fur (Figure 9.10).

Human skin color offers another example of epistasis. At least 100 gene products affect this trait, which begins with melanosomes (organelles that make melanins). Most people have about the same number of melanosomes in their skin cells. Variations in skin color arise from differences in the size, shape, and cellular distribution of melanosomes in the skin, as well as in the kinds and amounts of melanins they make. These variations have a genetic basis. Consider one gene that encodes a transport protein in melanosome membranes. Nearly all people of African, Native American, or east Asian descent carry the same allele of this gene. A mutation that

Digging Into Data

The Cystic Fibrosis Mutation and Typhoid Fever

The CF allele that causes most cases of cystic fibrosis is at least 50,000 years old and very common: 1 in 25 people carry it in some populations. Why is this allele so common if it is so dangerous? Consider that the CF allele is eventually lethal in homozygous individuals, but not in those who are heterozygous. This allele is codominant with the normal allele. Heterozygous individuals typically have no symptoms of cystic fibrosis because their cells make enough of the normal CFTR protein to have normal chloride ion transport.

Researchers think the CF allele has persisted because it offers heterozygous individuals an advantage in surviving certain deadly infectious diseases. The unmutated CFTR protein triggers endocytosis when it binds to bacteria. This process is an essential part of the body's immune response to bacteria in the respiratory tract. However, the same function of CFTR allows bacteria to enter cells of the gastrointestinal tract, where they can be deadly. For example, internalization of *Salmonella typhi* bacteria by epithelial cells in the gut causes a common worldwide disease called typhoid fever. Symptoms include extreme fever and diarrhea, and the resulting dehydration causes delirium that may last several weeks. If untreated, it kills up to 30 percent of those infected. Around 600,000 people, most of whom are children, die from the disease each year.

In 1998, Gerald Pier and his colleagues compared the uptake of *S. typhi* by different types of epithelial cells: those heterozygous for the CF allele, and those homozygous for the normal allele. (Cells homozygous for the CF allele do not take up any *S. typhi* bacteria.) Some of the results are shown in Figure 9.11.

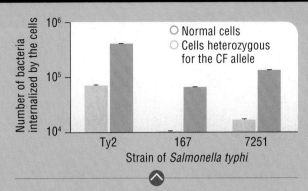

Figure 9.11 **Effect of the CF mutation on uptake of three strains of *Salmonella typhi* bacteria by epithelial cells.**

1. Regarding the Ty2 strain of *S. typhi*, about how many more bacteria were able to enter normal cells (those expressing unmutated *CFTR*) than cells expressing the gene with the CF mutation?
2. Which strain of bacteria entered normal epithelial cells most easily?
3. Entry of all three *S. typhi* strains into the heterozygous epithelial cells was inhibited. Is it possible to tell from this graph which strain was most inhibited?

occurred between 6,000 and 12,000 years ago gave rise to a different allele. This mutation, a single base-pair substitution, changed the 111th amino acid of the transport protein from alanine to threonine. The change results in less melanin—and lighter skin color—than the original African allele does. Today, nearly all people of European descent are homozygous for the mutated allele.

A person of mixed ethnicity may make gametes that contain different combinations of alleles for dark and light skin. It is fairly rare that one of those gametes contains all of the alleles for dark skin, or all of the alleles for light skin, but it happens (Figure 9.12). Skin color is only one of many human traits that vary as a result of single nucleotide mutations. The small scale of such changes offers a reminder that all of us share the genetic legacy of common ancestry.

Figure 9.12 Variation in human skin color begins with differences in alleles.
Twins Kian and Remee are shown with their parents. Both of the children's grandmothers are of European descent, and have pale skin. Both of their grandfathers are of African descent, and have dark skin. The twins inherited different alleles of some genes that affect skin color from their parents, who, given the appearance of their children, must be heterozygous for those alleles.

© Gary Roberts/worldwidefeatures.com.

Take-Home Message 9.4

Do all traits appear in a Mendelian inheritance pattern?

- Some alleles are not dominant or recessive when paired. The heterozygous phenotype may include both homozygous phenotypes (codominance), or it may be a blend of the two homozygous phenotypes (incomplete dominance).
- In some cases, one gene influences multiple traits. In other cases, multiple genes influence the same trait.

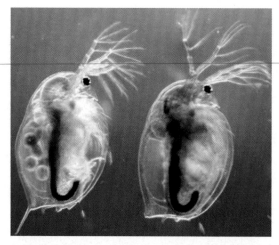

A. Under low-oxygen conditions, a water flea switches on genes involved in producing hemoglobin. Producing this red protein enhances the individual's ability to take up oxygen from water. The flea on the left has been living in water with a normal oxygen content; the one on the right, in water with a low oxygen content.

B. The color of the snowshoe hare's fur varies by season. In summer, the fur is brown (left); in winter, white (right). Both forms offer seasonally appropriate camouflage from predators.

C. The height of a mature yarrow plant depends on the elevation at which it grows.

Figure 9.13 Some environmental effects on phenotype.

(A) From *Science* 4 February 2011: Vol 331 no. 6017 pp. 555–561, Reprinted with permission from AAAS; (B) left, Jupiter Images Corporation; right, © age fotostock/SuperStock; (C) photo, Igor Sokolov (breeze)/Shutterstock.com.

9.5 Complex Variation in Traits

REMEMBER: Mutations are uncommon in a normal cell (Section 7.6). Expression of a homeotic gene directs formation of a specific body part; environmentally triggered changes in DNA methylation can be permanent and heritable (7.7). Growth factors stimulate cell division and differentiation (8.3). Genetic diversity offers sexual reproducers an advantage in a changing environment (8.4).

The pea plant phenotypes that Mendel studied appeared in two or three forms, which made them easy to track through generations. However, many other traits do not appear in distinct forms. Such traits are often the result of complex genetic interactions—multiple genes, multiple alleles, or both—with added environmental influences. Tracking traits with complex variation presents a special challenge, which is why the genetic basis of many of them has not yet been completely unraveled.

The phrase "nature versus nurture" refers to a centuries-old debate about whether human behavioral traits arise from one's genetics (nature) or from environmental factors (nurture). We now know that both play a substantial role. The environment affects the expression of many genes, which in turn affects phenotype—including behavioral traits. We can summarize this thinking with an equation:

$$\text{genotype} \;+\; \textbf{environment} \;\longrightarrow\; \text{phenotype}$$

Epigenetics research is revealing that the environment has an even greater contribution to this equation than most biologists had suspected. Environmental cues trigger internal pathways that methylate particular regions of DNA, so they suppress gene expression in those regions. In humans and other animals, DNA methylation patterns can be permanently and heritably affected by diet, stress, and exercise, and also by exposure to drugs and toxins such as tobacco, alcohol, arsenic, and asbestos.

Mechanisms that adjust phenotype in response to external cues are part of an individual's normal ability to adapt to environmental change. Consider the water flea, a tiny aquatic animal that lives in standing pools of fresh water such as seasonal ponds and ditches. In these pools, water conditions such as temperature, oxygen content, and salinity vary dramatically over time and between different parts of the pool. Water fleas have a lot of genes—many more than humans and other animals. Environmental cues trigger adjustments in gene expression that change a flea's form and function to suit its current environment. For example, a water flea that swims to the bottom of a pond can survive the low oxygen conditions there by turning on expression of seven genes involved in the production of hemoglobin—and turning red (Figure 9.13A). The newly produced hemoglobin improves the individual's ability to absorb oxygen from the water. In the presence of insect predators, a water flea makes a protective pointy helmet, a long tail spine, and teeth on its neck. Water fleas can also switch between asexual and sexual modes of reproduction. During early spring, food and space are typically abundant, and competition for these resources is scarce. Under these conditions, water fleas reproduce rapidly by asexual means, giving birth to large numbers of female offspring that quickly fill the pool. Later in the season, competition intensifies as the pool's water becomes warmer, saltier, and more crowded. Then, some of the water fleas start giving birth to males, and the population begins to reproduce sexually. The increased genetic diversity of sexually produced offspring may offer an advantage in the more challenging environment.

In many mammals, seasonal changes in temperature and the length of day affect the production of melanin and other pigments that color skin and fur. These species have different color phases in different seasons (Figure 9.13B). Hormonal signals triggered by the seasonal changes cause fur to be shed, and new fur grows back with different types and amounts of pigments deposited in it. The resulting change in phenotype provides these animals with seasonally appropriate camouflage from predators.

In plants, a flexible phenotype gives immobile individuals an ability to thrive in diverse habitats. For example, genetically identical yarrow plants grow to different heights at different altitudes (Figure 9.13C). More challenging temperature, soil, and water conditions are typically encountered at higher altitudes. Differences in altitude are also correlated with changes in the reproductive mode of yarrow: Plants at higher altitude tend to reproduce asexually, and those at lower altitude tend to reproduce sexually.

Researchers recently discovered several mutations associated with five human psychiatric disorders: autism, depression, schizophrenia, bipolar disorder, and attention deficit hyperactivity disorder (ADHD). However, there must be environmental components to these disorders too, because the majority of people who carry these mutations never end up with a psychiatric disorder. Moreover, one person with the mutations might get one type of disorder, while a relative with the same mutations might get another: two different results from the same genetic underpinnings.

Animal models are helping us unravel some of the mechanisms by which environment can influence mental state. For example, we have discovered that learning and memory are associated with dynamic and rapid DNA modifications in animal brain cells. Mood is, too. Stress-induced depression causes methylation-based silencing of a particular nerve growth factor gene; some antidepressants work by reversing this methylation. As another example, rats whose mothers are not very nurturing end up anxious and having a reduced resilience for stress as adults. The difference between these rats and ones who had nurturing maternal care is traceable to epigenetic DNA modifications that result in a lower than normal level of another nerve growth factor. Drugs can reverse these modifications—and their effects. We do not yet know all of the genes that influence human mental state, but the implication of such research is that future treatments for many psychiatric disorders will involve deliberate modification of methylation patterns in an individual's DNA.

Continuous Variation Some traits occur in a range of small differences that is called **continuous variation**. Continuous variation can be an outcome of epistasis, in which multiple genes affect a single trait. The more genes and environmental factors that influence a trait, the more continuous is its variation. Traits that arise from genes with a lot of alleles may also vary continuously. Some genes have regions of DNA in which a series of 2 to 6 nucleotides is repeated many times in a row. These **short tandem repeats** can spontaneously expand or contract very quickly compared with the typical rate of mutation, and the resulting changes in the gene's DNA sequence may be preserved as alleles. For example, in dogs, short tandem repeats have given rise to 12 alleles of a homeotic gene that influences the length of the face, with longer repeats associated with longer faces (Figure 9.14A).

Human skin color varies continuously, as does human eye color (Figure 9.14B). The colored part of the eye is a doughnut-shaped structure called the iris. Iris color, like skin color, is the result of interactions among gene products that make and distribute melanins. The more melanin deposited in the iris, the less light is reflected from it. Dark irises have dense melanin deposits that absorb almost all light, and reflect almost none. Green and blue eyes have the least amount of melanin, so they reflect the most light.

> The environment affects the expression of many genes, which in turn affects phenotype.

A. Face length is a trait that varies continuously in dogs. A gene with 12 alleles influences this trait; all arose by the spontaneous insertion of short tandem repeats. The longer the alleles, the longer the face.

B. Iris color is a trait that varies continuously in humans.

Figure 9.14 Examples of continuous variation.

(A) WilleeCole/Shutterstock.com; (B) from left: first row, © szefel/Shutterstock.com; © Aaron Amat/Shutterstock.com; © Villedieu Christophe/Shutterstock.com; David Goodin; second row, jayfish/Shutterstock.com; © J. Helgason/Shutterstock.com; © Tischenko Irina/ Shutterstock.com; © Tatiana Makotra/Shutterstock.com; third row, © Vaaka/Shutterstock.com; © evantravels/Shutterstock.com; © rawcaptured/Shutterstock.com; Anemone/Shutterstock .com; fourth row, © Andrey Armyagov/Shutterstock.com; © Tatiana Makatra/Shutterstock.com; © lightpoet/Shutterstock.com; © Anemone/Shutterstock.com.

continuous variation A range of small differences in a shared trait.

short tandem repeat In chromosomal DNA, a sequence of a few nucleotides repeated multiple times in a row.

bell curve Bell-shaped curve; typically results from graphing frequency versus distribution for a trait that varies continuously.

pedigree Chart of family connections that shows the appearance of a trait through generations.

A. To see if human height varies continuously, male biology students at the University of Florida were divided into categories of one-inch increments in height and counted.

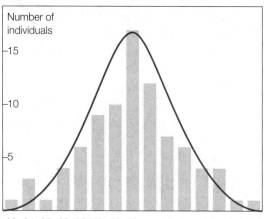

Number of individuals

Measured values

B. Graphing the data that resulted from the experiment in **A** produces a bell-shaped curve, which is evidence that height does vary continuously in humans.

Figure 9.15 How to determine whether a particular trait varies continuously.

(A) Courtesy of Ray Carson, University of Florida News and Public Affairs.

How do we know if a particular trait varies continuously? Let's use another human trait, height, in an example. First, the total range of phenotypes is divided into measurable categories—inches, in this case (Figure 9.15A). The individuals in each category are counted; these counts reveal the relative frequencies of phenotypes across the range of values. Finally, this data is plotted as a bar chart (Figure 9.15B). A graph line around the top of the bars shows the distribution of values for the trait. If the line is a bell-shaped curve, or **bell curve**, then the trait varies continuously.

Take-Home Message 9.5

Do all traits occur in distinct forms?

• Many traits do not occur in distinct forms. The more genes and other factors that influence a trait, the more continuous is its range of variation.

• The environment influences gene expression, and therefore can alter phenotype.

9.6 Human Genetic Analysis

REMEMBER: Sampling error is a difference between results obtained from a subset, and results from the whole (Section 1.6). Environmentally driven epigenetic modifications of DNA—and the resulting gene expression patterns—can be inherited (7.7).

Some organisms, including pea plants and fruit flies, are ideal for genetic studies. They have relatively few chromosomes, they reproduce quickly under controlled conditions, and breeding them in controlled settings poses few ethical problems. It does not take long to follow a trait through many generations. Humans, however, are a different story. We live under variable conditions, in different places, and we live as long as the geneticists who study our inheritance patterns. Most of us select our own mates and reproduce if and when we want to. Our families tend to be on the small side, so sampling error is a major factor in studying them. Because of these and other challenges, geneticists often use historical records to track traits through many generations of a family. They make and study **pedigrees**, standardized charts that illustrate the phenotypes of family members and genetic connections among them (Figure 9.16). Analysis of a pedigree can reveal whether a trait is associated with a dominant or recessive allele, and whether the allele is on an autosome or a sex chromosome. Pedigree analysis also allows geneticists to determine the probability that a trait will reappear in future generations of a family or a population.

Types of Genetic Variation Few easily observed human traits follow Mendelian inheritance patterns. Like the flower color of Mendel's pea plants, these traits arise from a single gene with alleles that have a clear dominance relationship. Consider the gene called *MC1R*. In humans, dogs, and other animals, *MC1R* encodes a protein that triggers production of the brownish melanin. Mutations can result in a defective protein; an allele with one of these mutations is recessive when paired with an unmutated allele. A person who is homozygous for a mutated allele does not make the brownish melanin—only the reddish type—so this individual has red hair. Most other human traits are polygenic, and many have epigenetic contributions that can originate in parents or even grandparents. Environmental effects make these traits even harder to study.

Most of what we know about human genetics comes from research on inherited abnormalities and disorders, because this information helps us develop treatments for affected people. A genetic abnormality is a rare or uncommon version of a trait, such as having six fingers on a hand or a web between two toes. Such abnormalities are not inherently life-threatening, and how you view them is a matter of opinion. By contrast, a genetic disorder sooner or later causes medical problems that may be severe. A genetic disorder is often characterized by a specific set of symptoms (a syndrome). The next section discusses a few genetic disorders and abnormalities that are caused by mutations in single genes. Alleles that give rise to severe genetic disorders are generally rare in populations because they compromise the health and reproductive ability of their bearers. Why do they persist? Mutations periodically reintroduce them. In some cases, a codominant allele offers a survival advantage in a particular environment. You already learned about one example, the CF allele that causes cystic fibrosis. You will see additional examples in later chapters. Keep in mind that single-gene disorders are the least common kind. Far more people are affected by disorders that arise from a complex interplay of multiple genes and environmental factors. Diabetes, asthma, obesity, cancer, heart disease, and multiple sclerosis are like this. Despite intense research, our understanding of these disorders remains incomplete.

Take-Home Message 9.6

How do we study inheritance patterns in humans?

- Human inheritance patterns are often studied by tracking genetic abnormalities or disorders through family trees.
- A genetic disorder is an inherited condition that causes medical problems.
- A genetic abnormality is a rare but harmless version of an inherited trait.
- Only a few genetic disorders are governed by single genes and inherited in a Mendelian pattern. Most human traits are polygenic, and many have epigenetic contributions.

9.7 Human Genetic Disorders

REMEMBER: Thousands of tiny pores span the nuclear envelope; intermediate filaments of lamins support the nuclear envelope, and also help regulate processes inside the nucleus such as DNA replication; basement membrane supports and organizes animal tissues (Section 3.5). Autosomes are the same in both females and males; sex chromosomes differ between the sexes (6.3). An insertion is a mutation in which nucleotides are added to DNA; anemia is a deficiency in red blood cells or hemoglobin (7.6). One of the two X chromosomes in each cell of a female is an inactivated Barr body (7.7). Growth factors stimulate cell division (8.3).

Human genetic disorders inherited in a Mendelian pattern are typically categorized by the chromosome of origin (autosome or sex chromosome) and whether alleles associated with them are dominant or recessive.

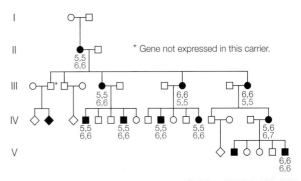

A. Standard symbols used in pedigrees.

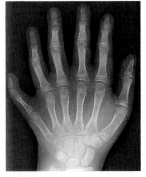

B. Above, a pedigree for polydactyly, a genetic abnormality in which a person has extra fingers (right), toes, or both. The number of fingers on each hand is indicated in black; toes on each foot, in red. Polydactyly that appears on its own is typically inherited in an autosomal dominant pattern. When part of a syndrome (such as Ellis–van Creveld syndrome), it can be inherited in an autosomal recessive pattern.

C. For more than 30 years, researcher Nancy Wexler has studied the genetic basis of Huntington's disease, an inherited disorder that causes progressive degeneration of the nervous system. Wexler and her team constructed an extended family tree for nearly 10,000 Venezuelans. Their pedigree analysis of relationships among unaffected and affected individuals revealed that a dominant allele on human chromosome 4 is the culprit. Wexler has a personal interest in the disorder: It runs in her family.

Figure 9.16 Pedigrees.

(B) Courtesy of Irving Buchbinder, DPM, DABPS, Community Health Services, Hartford CT; (C) Acey Harper/ Time & Life Pictures/Getty Images.

Table 9.1 Some Autosomal Dominant Traits in Humans

Disorder/Abnormality	Main Symptoms
Achondroplasia	One form of dwarfism
Aniridia	Defects of the eyes
Camptodactyly	Rigid, bent fingers
Familial hypercholesterolemia	High cholesterol, clogged arteries
Huntington's disease	Degeneration of the nervous system
Marfan syndrome	Abnormal or missing connective tissue
Polydactyly	Extra fingers, toes, or both
Progeria	Drastic premature aging
Neurofibromatosis	Tumors of nervous system, skin

The Autosomal Dominant Pattern A trait associated with a dominant allele on an autosome appears in people who are heterozygous for it as well as those who are homozygous. Table 9.1 lists a few examples. Such traits appear in every generation of a family, and they occur with equal frequency in both sexes. When one parent is heterozygous, and the other is homozygous for the recessive allele, each of their children has a 50 percent chance of inheriting the dominant allele and having the associated trait (Figure 9.17A).

A form of hereditary dwarfism called achondroplasia offers an example of an autosomal dominant disorder (one caused by a dominant allele on an autosome). Mutations associated with achondroplasia occur in a gene for a growth factor receptor. The mutations cause the receptor, which normally slows bone development, to be overly active. About 1 in 10,000 people is heterozygous for one of these mutations. As adults, affected people are, on average, about 4 feet 4 inches (1.3 meters) tall, with arms and legs that are short relative to torso size (Figure 9.17B). An allele that causes achondroplasia can be passed to children because its expression does not interfere with reproduction, at least in heterozygous people. The homozygous condition results in severe skeletal malformations that cause early death.

Huntington's disease is also inherited in an autosomal dominant pattern. Mutations that cause this disorder alter a gene for a cytoplasmic protein whose function is still unknown. The mutations are insertions that occur when a three-nucleotide short tandem repeat expands spontaneously. The oversized protein product of the altered gene gets chopped into pieces inside nerve cells of the brain. The pieces accumulate in cytoplasm as large clumps that eventually prevent the cells from functioning properly. Brain cells involved in movement, thinking, and emotion are particularly affected. Dramatic, involuntary jerking and writhing movements that are symptoms of the most common form of Huntington's appear after age thirty. Affected people die during their forties or fifties. With this and other late-onset disorders, people may reproduce before symptoms appear, so the allele can be passed unknowingly to children.

Figure 9.17 Autosomal dominant inheritance.

(B) © Newcastle Photos and Ivy & Violet Broadhead and family; (C) Photo courtesy of The Progeria Research Foundation and John Hurley.

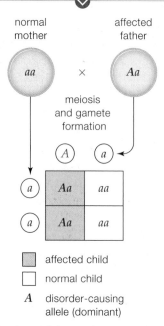

A. A dominant allele on an autosome (red) is fully expressed in heterozygous people.

B. Achondroplasia affects Ivy Broadhead (left), her brother, father, and grandfather.

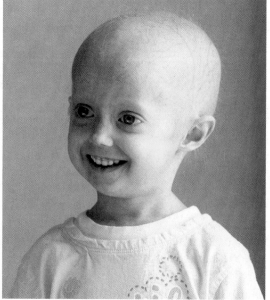

C. Symptoms of Hutchinson–Gilford progeria are already evident in Megan Nighbor at age 5.

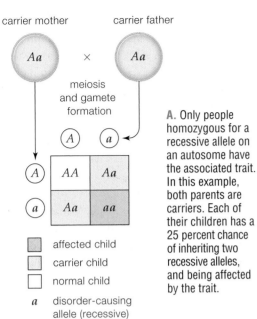

carrier mother   carrier father

Aa   ×   *Aa*

meiosis and gamete formation

A   *a*

	A	*a*
A	*AA*	*Aa*
a	*Aa*	*aa*

◼ affected child
◻ carrier child
◻ normal child

a disorder-causing allele (recessive)

A. Only people homozygous for a recessive allele on an autosome have the associated trait. In this example, both parents are carriers. Each of their children has a 25 percent chance of inheriting two recessive alleles, and being affected by the trait.

B. The albino phenotype is associated with autosomal recessive alleles that cause a deficiency in melanin.

Figure 9.18 Autosomal recessive inheritance.

(B) © Rick Guidotti, Positive Exposure.

Hutchinson–Gilford progeria is an autosomal dominant disorder characterized by drastically accelerated aging. It is usually caused by a mutation that affects lamin A, a protein component of intermediate filaments that support the nuclear envelope. In cells that carry this mutation, the nucleus is grossly abnormal, with improperly assembled nuclear pore complexes and membrane proteins on the wrong side of the nuclear envelope. The function of the nucleus as protector of chromosomes and gateway for transcription is severely impaired, and DNA damage accumulates quickly. The effects are pleiotropic. Outward symptoms begin to appear before age two, as skin that should be plump and resilient starts to thin, muscles weaken, and bones soften. Premature baldness is inevitable (Figure 9.17C). Most people with the disorder die in their early teens as a result of a stroke or heart attack brought on by hardened arteries, a condition typical of advanced age. Progeria does not run in families because affected people do not live long enough to reproduce.

The Autosomal Recessive Pattern A recessive allele on an autosome is expressed only in homozygous individuals, so traits associated with the allele tend to skip generations. They also appear in both sexes at equal frequency. Table 9.2 lists a few examples. Heterozygous individuals are called carriers because they have the allele but not the trait. Any child of two carriers has a 25 percent chance of inheriting the allele from both parents—and developing the trait (Figure 9.18A).

Albinism, a phenotype characterized by an abnormally low level of melanin, is inherited in an autosomal recessive pattern. Mutations associated with albinism affect proteins involved in melanin synthesis. Skin, hair, or eye pigmentation may be reduced or missing. In the most dramatic form, the skin is very white and does not tan, and the hair is white (Figure 9.18B). The irises of the eyes appear red because the lack of pigment allows underlying blood vessels to show through. Melanin also plays a role in the retina, so vision problems are typical. In skin, melanin acts as a sunscreen; without it, the skin is defenseless against UV radiation. Thus, people with the albino phenotype have a very high risk of skin cancer.

With late-onset disorders, people may reproduce before symptoms appear, so the allele can be passed unknowingly to children.

Table 9.2 Some Autosomal Recessive Traits in Humans

Trait	Description
Albinism	Absence of pigmentation
Cystic fibrosis	Abnormally thick mucus damages tissues and organs
Ellis–van Creveld syndrome	Dwarfism, heart defects, polydactyly
Friedreich's ataxia	Progressive loss of motor and sensory function
Hereditary methemoglobinemia	Blue skin coloration
Phenylketonuria (PKU)	Mental impairment
Sickle-cell anemia	Red blood cells can sickle and disrupt circulation
Tay–Sachs disease	Deterioration of mental and physical abilities; early death

polyploid Having three or more of each type of chromosome.

Figure 9.19 Tay–Sachs disease.
Conner Hopf was diagnosed with Tay–Sachs when he was 7 months old. He died before his second birthday.

Courtesy of © Conner's Way Foundation, www.connersway.com.

Inheriting more than two full sets of chromosomes is fatal in humans.

Table 9.3 Some X-Linked Traits in Humans

Disorder or Abnormality	Main Symptoms
Androgen insensitivity syndrome	XY individual has the traits of a female; sterility
Red–green color blindness	Inability to distinguish red from green
Fragile X syndrome	Intellectual, emotional disability
Hemophilia	Impaired blood clotting
Incontinentia pigmenti	Abnormalities of skin, hair, teeth, nails, eyes
Muscular dystrophies	Progressive loss of muscle function
SCID-X1	Severe immune system deficiency
X-linked anhidrotic dysplasia	Mosaic skin; other ill effects

Alleles associated with Tay–Sachs disease are inherited in an autosomal recessive pattern. In the general population, about 1 in 300 people is a carrier for one of these alleles, but the incidence is ten times higher in some groups, such as Jews of eastern European descent. The gene altered in Tay–Sachs encodes a lysosomal enzyme responsible for breaking down a particular type of lipid. Mutations result in an enzyme that misfolds and becomes destroyed, so cells make the lipid but cannot break it down. Typically, newborns homozygous for a Tay–Sachs allele seem normal, but within three to six months they become irritable, listless, and may have seizures as the lipid accumulates in their nerve cells. Blindness, deafness, and paralysis follow. Affected children usually die by age five (Figure 9.19).

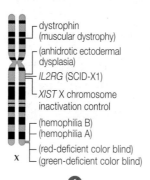

Figure 9.20 The human X chromosome.
This chromosome carries about 1,900 genes—almost 10 percent of the total. Most X chromosome alleles that cause genetic disorders are inherited in a recessive pattern. A few disorders are listed (in parentheses).

dystrophin (muscular dystrophy)
(anhidrotic ectodermal dysplasia)
IL2RG (SCID-X1)
XIST X chromosome inactivation control
(hemophilia B)
(hemophilia A)
(red-deficient color blind)
(green-deficient color blind)
X

The X-Linked Recessive Pattern Many genetic disorders are associated with alleles on the X chromosome (Figure 9.20 and Table 9.3). Almost all of them are inherited in a recessive pattern, probably because those caused by dominant X chromosome alleles tend to be lethal in male embryos.

A recessive allele on an X chromosome leaves two clues when it causes a genetic disorder. First, an affected father never passes the disorder to a son, because all children who inherit their father's X chromosome are female (Figure 9.21A). Thus, a heterozygous female is always the bridge between an affected male and his affected grandson. Second, the disorder appears in males more often than in females. This is because all males who carry the allele have the disorder, but not all heterozygous females do. One of the two X chromosomes in each cell of a female is an inactive Barr body, so only about half of a heterozygous female's cells express the recessive allele. The other half of her cells express the dominant, normal allele that she carries on her other X chromosome, and this expression can mask the phenotypic effects of the recessive allele.

Duchenne muscular dystrophy (DMD) is a genetic disorder characterized by progressive muscle degeneration. It is caused by mutations in the X chromosome gene for dystrophin, a protein that links actin microfilaments in cytoplasm to a complex of proteins in the plasma membrane. This complex structurally and functionally links the cell to basement membrane. When dystrophin is absent, the entire protein complex is unstable. Muscle cells, which are subject to stretching, are particularly affected. Their plasma membrane is easily damaged, and they become flooded with calcium ions. Eventually, the cells die and become replaced by fat cells and connective tissue. DMD affects about 1 in 3,500 people; almost all are boys. Symptoms begin between ages three and seven. Anti-inflammatory drugs can slow the progression of the disorder, but there is no cure. When an affected boy is about twelve years old, he will begin to use a wheelchair and his heart will start to fail. Even with the best care, he will probably die before the age of thirty, from a heart disorder or respiratory failure (suffocation).

Hemophilias are genetic disorders in which the blood does not clot properly. Most of us have a blood clotting mechanism that quickly stops bleeding from minor injuries. That mechanism involves two proteins, clotting factor VIII and IX, both products of X chromosome genes. Mutations in these two genes cause two types of hemophilia (A and B, respectively). Males who carry one of these mutations have prolonged bleeding, as do homozygous females (heterozygous females make enough clotting factor to have a clotting time that is close to normal). Affected people bruise easily, but internal bleeding is their most serious problem. Repeated bleeding inside the joints disfigures them and causes chronic arthritis. Today, about 1 in 7,500 people in the general population is affected. That number may be rising because the disorder is now treatable, and more affected people now live long enough to transmit a mutated allele to children.

Color blindness refers to a range of conditions in which an individual cannot distinguish among colors in the spectrum of visible light. These conditions are typically inherited in an X-linked recessive pattern, because most of the genes involved in color vision are on the X chromosome. Humans can sense the differences among 150 colors, and this perception depends on pigment-containing receptors in the eyes. Mutations that result in altered or missing receptors affect color vision. For example, people who have red–green color blindness see fewer than 25 colors because receptors that respond to the red and green wavelengths of light are weakened or absent (Figure 9.21B,C). Some people with red–green color blindness confuse red and green, and others see green as gray.

Take-Home Message 9.7

How do we know whether a trait is associated with a dominant or recessive allele on an autosome or sex chromosome?

- With an autosomal dominant inheritance pattern, anyone with the allele (whether homozygous or heterozygous) has the associated trait. The trait characteristically appears in every generation.
- With an autosomal recessive inheritance pattern, only persons who are homozygous for an allele have the associated trait. The trait tends to skip generations.
- Men who have an X-linked allele have the associated trait, but not all heterozygous women do. Thus, the trait appears more often in men. Men transmit an X-linked allele to their daughters, but not to their sons.

9.8 Chromosome Number Changes

REMEMBER: Researchers risk interpreting their results in terms of what they want to find out; sampling error can be a substantial problem when testing a small group (Section 1.6). Amyloid fibrils form insoluble deposits that disrupt brain function (2.9). Cells that are diploid (2*n*) have two sets of chromosomes (6.3). Meiosis reduces the chromosome number during gamete formation; the fusion of two haploid (*n*) gametes at fertilization restores the diploid chromosome number in the resulting zygote (8.5).

A **polyploid** individual has three or more complete sets of chromosomes. About 70 percent of flowering plant species are polyploid, as are some insects, fishes, and other animals—but not humans. In our species, inheriting more than two full sets of chromosomes is invariably fatal, although some somatic cells are normally polyploid in adult tissues. Aneuploidy—having too many or too few copies of a particular

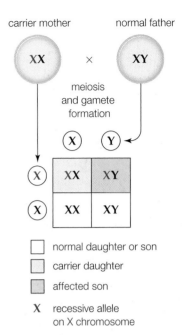

A. In this example of X-linked inheritance, the mother carries a recessive allele on one of her two X chromosomes (red).

B. A view of color blindness. The image on the left shows how a person with red–green color blindness sees the image on the right. The perception of blues and yellows is normal; red and green appear similar.

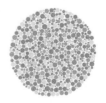

C. Part of a standardized test for color blindness. A set of 38 of these circles is commonly used to diagnose deficiencies in color perception. You may have one form of red–green color blindness if you see a 7 instead of a 29 in the circle on the left. You may have another form of red–green color blindness if you see a 3 instead of an 8 in the circle on the right.

Figure 9.21 X-linked recessive inheritance.

(B) Photos by Gary L. Friedman, www.FriedmanArchives.com; (C) Life Nature Library, The Primates, 1965, Sarel Eimerl and Irven DeVore.

Figure 9.22 An example of nondisjunction during meiosis.
Of the two pairs of homologous chromosomes shown here, one fails to separate during anaphase I. The chromosome number is altered in the resulting gametes.

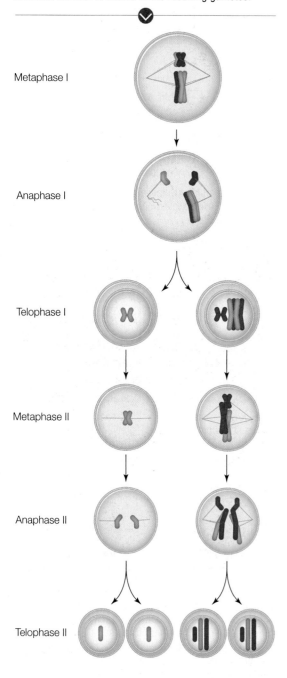

Metaphase I

Anaphase I

Telophase I

Metaphase II

Anaphase II

Telophase II

Figure It Out: During which stage of meiosis does nonjunction occur in this example?

Answer: Anaphase I

chromosome—does occur in humans. A few babies (less than 1 percent) are born with a diploid chromosome number that differs from the normal 46. Changes in chromosome number are usually an outcome of **nondisjunction**, the failure of chromosomes to separate properly during mitosis or meiosis. Nondisjunction during meiosis (Figure 9.22) can affect chromosome number at fertilization. For example, if a normal gamete (n) fuses with a gamete that has an extra chromosome ($n+1$), the resulting zygote will have three copies of one type of chromosome and two of every other type ($2n+1$), an aneuploid condition called trisomy. If a normal gamete (n) fuses with a gamete missing a chromosome ($n-1$), the new individual will have one copy of one chromosome and two of every other type ($2n-1$), an aneuploid condition called monosomy. A few disorders associated with aneuploidy are listed in Table 9.4.

Autosomal Change and Down Syndrome In most cases, inheriting the wrong number of autosomes is fatal in humans before birth or shortly thereafter. An important exception is trisomy 21. A person born with three chromosomes 21 will have Down syndrome and a high likelihood of surviving infancy. Mild to moderate mental impairment and health problems such as heart disease are hallmarks of this syndrome. Other effects may include a somewhat flattened facial profile, a fold of skin that starts at the inner corner of each eyelid, white spots on the iris (Figure 9.23), and one deep crease (instead of two shallow creases) across each palm. The skeleton grows and develops abnormally, so older children have short body parts, loose joints, and misaligned bones of the fingers, toes, and hips. Muscles and reflexes are weak, and motor skills such as speech develop slowly. Early training can help these individuals learn to care for themselves and to take part in normal activities. By age 40, all persons with Down syndrome will begin to have symptoms of Alzheimer's disease, a condition of progressive mental deterioration. A gene on chromosome 21 is the culprit. Having three of these chromosomes, a person with Down syndrome makes an excess of the gene's product, a protein that forms the main component of amyloid fibrils characteristic of Alzheimer's. Down syndrome occurs in about 1 of 700 live births, and the risk increases with maternal age.

Change in the Sex Chromosome Number About 1 in 400 human babies is born with an atypical number of sex chromosomes. Most often, such alterations lead to mild difficulties in learning and impaired motor skills such as a speech delay, but these problems may be very subtle.

About 1 in 2,500 girls is born with Turner syndrome, an outcome of having an X chromosome and no corresponding X or Y chromosome (XO). This aneuploid condition is thought to arise most frequently with an unstable Y chromosome inherited from the father. The zygote starts out being genetically male, with an X and a Y chromosome. The Y chromosome breaks up and is lost during early development, so the embryo continues to develop as a female. Affected individuals grow up well proportioned but a bit short, with an average height of 4 feet 8 inches (1.4 meters). Extra folds of skin on the neck, mild skeletal abnormalities, and heart defects are common. The ovaries do not develop properly, so these individuals do not make enough sex hormones to become sexually mature or develop secondary sexual traits such as enlarged breasts. Hormone therapy can trigger sexual development, but even with treatment most girls affected by the syndrome are infertile.

A female may inherit multiple X chromosomes, a condition called triple X syndrome or trisomy X. This syndrome occurs in about 1 of 1,000 births; as with Down syndrome, risk increases with maternal age. Only one X chromosome is typically

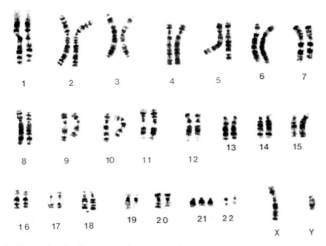

A. Example of a Down syndrome genotype.

B. Example of a Down syndrome phenotype. Excess tissue deposits on the iris give rise to a ring of starlike white speckles, a lovely effect of the chromosome number change that causes Down syndrome.

Figure 9.23 Down syndrome, genotype and phenotype.

(A) L. Willatt, East Anglian Regional Genetics Service/Science Source; (B) Ciarra, photo by © Michelle Harmon.

active in female cells, so having extra X chromosomes usually does not cause physical or medical problems, but mild mental impairment may occur.

About 1 out of every 500 males has two or more X chromosomes (XXY, XXXY, and so on). The resulting disorder, Klinefelter syndrome, develops at puberty. As adults, affected males tend to be overweight and tall, with small testes. Underproduction of the hormone testosterone interferes with sexual development and can result in sparse facial and body hair, a high-pitched voice, enlarged breasts, and infertility. Testosterone injections during puberty can minimize some of these traits.

About 1 in 1,000 males is born with an extra Y chromosome (XYY), a result of nondisjunction of the Y chromosome during sperm formation. Adults affected by the resulting XYY syndrome tend to be taller than average, but are within a normal range of phenotype. Sexual development occurs normally, and fertility is normal. Having XYY syndrome was once thought to predispose an individual to a life of crime. This misguided view was based on sampling error (too few cases in narrowly chosen groups such as prison inmates) and bias (the researchers who gathered the karyotypes also took the personal histories of the participants). That view has since been disproven: Men with XYY syndrome are only slightly more likely to be convicted for crimes than other men. Researchers believe this slight increase can be explained by poor socioeconomic conditions related to mild mental impairment of individuals with the syndrome.

Table 9.4 Some Traits Associated With Chromosome Number Changes

Disorder or Abnormality	Main Symptoms
Down syndrome	Mental impairment; heart defects
Turner syndrome (XO)	Sterility; abnormal ovaries, abnormal sexual traits
Klinefelter syndrome (XXY)	Sterility, mental impairment
XXX syndrome	Minimal abnormalities
XYY syndrome	Mild mental impairment

Take-Home Message 9.8

What are the effects of chromosome number changes in humans?

- Polyploidy is fatal in humans, but not in some other organisms.
- Aneuploidy can arise from nondisjunction during meiosis. In humans, most cases of aneuploidy are associated with some degree of mental impairment.

nondisjunction Failure of chromosomes to separate properly during nuclear division.

9.9 Genetic Screening

REMEMBER: Cells make the thousands of different proteins they need from only twenty kinds of amino acid monomers (Section 2.9). During early development, an embryo's cells start using different subsets of their DNA, thus becoming different in form and function—a process called differentiation (6.1).

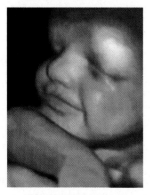

A. Conventional ultrasound.

B. 4D ultrasound.

C. Fetoscopy.

Figure 9.24 Three ways of imaging a human fetus.

(A) Mediscan/Corbis; (B) Dr. Benoit/Mona Lisa/LooksatSciences/Phototake; (C) © Neil Bromhall/Science Source.

Studying human inheritance patterns has given us many insights into how genetic disorders arise and progress, and how to treat them. Surgery, prescription drugs, hormone replacement therapy, and dietary controls can minimize and in some cases eliminate the symptoms of a genetic disorder. Some disorders can be detected early enough to start countermeasures before symptoms develop. For these reasons, most hospitals in the United States now screen newborns for mutations that cause phenylketonuria, or PKU. The mutations affect an enzyme that converts one amino acid (phenylalanine) to another (tyrosine). Without this enzyme, the body becomes deficient in tyrosine, and phenylalanine accumulates to high levels. The imbalance inhibits protein synthesis in the brain, which in turn results in severe neurological symptoms. Restricting all intake of phenylalanine can slow the progression of PKU, so routine early screening has resulted in fewer individuals suffering from the symptoms of the disorder.

Parents can also benefit from human genetics studies. The probability that a future child will inherit a genetic disorder can be estimated by testing prospective parents for alleles known to be associated with genetic disorders. Karyotypes and pedigrees are also useful in this type of screening, which can help people make informed decisions about family planning.

Genetic screening is also done post-conception, in which case it is called prenatal diagnosis (prenatal means before birth). Prenatal diagnosis checks an embryo or fetus for physical and genetic abnormalities. Early diagnosis of these conditions gives parents time to prepare for the birth of an affected child, and an opportunity to decide whether to continue with the pregnancy or terminate it. Dozens of conditions are detectable prenatally, including aneuploidy, hemophilia, Tay–Sachs disease, sickle-cell anemia, muscular dystrophy, and cystic fibrosis. If a disorder is treatable, early detection can allow the newborn to receive prompt and appropriate treatment. A few conditions are even surgically correctable before birth.

As an example of how prenatal diagnosis works, consider a woman who becomes pregnant at age thirty-five. Her doctor will probably perform a procedure called obstetric sonography, in which ultrasound waves directed across the woman's abdomen form images of the fetus's limbs and internal organs (Figure 9.24A,B). If the images reveal a physical defect that may be the result of a genetic disorder, a more invasive technique such as fetoscopy would be recommended for further diagnosis. With fetoscopy, sound waves pulsed from inside the mother's uterus yield images much higher in resolution than ultrasound (Figure 9.24C). Samples of tissue or blood are often taken at the same time, and some corrective surgeries can be performed.

Human genetics studies show that our thirty-five-year-old woman has about a 1 in 80 chance that her baby will be born with a chromosomal abnormality, a risk more than six times greater than when she was twenty years old. Thus, even if no abnormalities are detected by ultrasound, she probably will be offered an additional diagnostic procedure, amniocentesis, in which a small sample of fluid is drawn from the amniotic sac enclosing the fetus (Figure 9.25A). The fluid contains cells shed by the fetus, and those cells can be tested for genetic disorders. Chorionic villus sampling (CVS) can be performed earlier than amniocentesis. With this technique, a few cells from the chorion are removed and tested (the chorion is a membrane that surrounds the amniotic sac and helps form the placenta, an organ that allows substances to be exchanged between mother and embryo).

An invasive procedure often carries a risk to the fetus. The risks vary by the procedure. Amniocentesis has improved so much that, in the hands of a skilled physician, it no longer increases the risk of miscarriage. CVS occasionally disrupts the placenta's development and thus causes underdeveloped or missing fingers and toes in 0.3 percent of newborns. Fetoscopy raises the miscarriage risk by a whopping 2 to 10 percent, so it is rarely performed unless surgery or another medical procedure is required before the baby is born.

Couples who discover they are at high risk of having a child with a genetic disorder may opt for reproductive interventions such as *in vitro* fertilization. With this procedure, sperm and eggs taken from the prospective parents are mixed in a test tube. If an egg becomes fertilized, the resulting zygote will begin to divide. In about forty-eight hours, it will have become an embryo that consists of a ball of eight cells (Figure 9.25B). All of the cells in this ball have the same genes, but none has yet committed to being specialized one way or another. Doctors can remove one of these undifferentiated cells and analyze its genes, a procedure called preimplantation diagnosis. The withdrawn cell will not be missed. If the embryo has no detectable genetic defects, it is inserted into the woman's uterus to develop. Most of the resulting "test-tube babies" are born in good health.

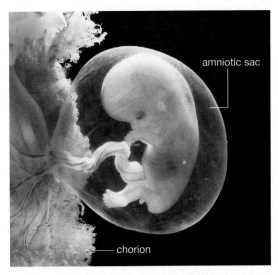

A. With amniocentesis, a tiny bit of the fluid inside the amniotic sac is removed, and fetal cells that have been shed into the fluid are tested for genetic disorders. Chorionic villus sampling tests cells of the chorion, which is part of the placenta.

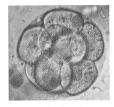

B. About 48 hours after *in vitro* fertilization, a human embryo is a ball of eight identical cells. If one cell is removed for genetic analysis, the remaining seven can continue development.

Figure 9.25 Cells tested for prenatal and preimplantation diagnosis.

(A) © Lennart Nilsson/Bonnierforlagen AB; (B) Fran Heyl Associates © Jacques Cohen, computer-enhanced by © Pix Elation.

Take-Home Message 9.9

How do we use what we know about human inheritance?

- Studying inheritance patterns for genetic disorders has helped researchers develop treatments for some of them.
- Genetic testing can provide prospective parents with information about the health of their future children.

Summary

Section 9.1 Symptoms of cystic fibrosis are pleiotropic effects of mutations in one gene. The allele associated with most cases persists at high frequency despite its devastating effects in homozygous people. Carrying the allele may offer heterozygous individuals protection from dangerous gastrointestinal tract infections.

Section 9.2 Gregor Mendel indirectly discovered the role of genes and alleles in inheritance by breeding pea plants and carefully tracking traits of the offspring over many generations. Each gene occurs at a particular location on a chromosome. Individuals with

identical alleles are **homozygous** for the allele. **Heterozygous** individuals have two nonidentical alleles.

A **dominant** allele masks the effect of a **recessive** allele partnered with it on the homologous chromosome. **Genotype** (an individual's particular set of alleles) gives rise to **phenotype** (the individual's observable traits).

Section 9.3 Diploid cells have pairs of genes on homologous chromosomes. The two genes of a pair (which may differ as alleles) are separated from each other during meiosis, so they end up in different gametes. Crossing individuals that breed

true for two forms of a trait yields offspring that are identically heterozygous for alleles governing the trait. A cross between such offspring is a **monohybrid cross**. The frequency at which the two forms of the trait appear among the offspring of a monohybrid cross can reveal a dominance relationship between the alleles.

In most cases, the two genes of a pair on homologous chromosomes are distributed into gametes independently of other gene pairs on other chromosomes. Crossing individuals that breed true for two forms of two traits yields offspring that are identically heterozygous for alleles governing those traits. A cross between such offspring is a **dihybrid cross**. The frequency at which the two forms of the two traits appear among the offspring of such crosses can reveal dominance relationships between the alleles.

A **Punnett square** can be useful for determining the probability that certain genotypes (and phenotypes) will appear among the offspring of a monohybrid cross or a dihybrid cross.

Section 9.4 Not all traits have a Mendelian inheritance pattern. With **incomplete dominance**, the phenotype of heterozygous individuals is an intermediate blend of the two homozygous phenotypes. With **codominance**, heterozygous individuals have both homozygous phenotypes. With **epistasis**, two or more genes affect the same trait. With **pleiotropy**, one gene affects two or more traits.

Section 9.5 Environmental factors can influence phenotype by altering gene expression. A trait that is influenced by multiple genes often occurs in a range of phenotype called **continuous variation**. Continuous variation typically occurs as a **bell curve** in the range of values. Multiple alleles such as those that arise in regions of **short tandem repeats** can give rise to continuous variation.

Section 9.6 Geneticists study inheritance patterns in humans by tracking genetic disorders and abnormalities through generations of families. A genetic abnormality is an uncommon version of a heritable trait that does not result in medical problems. A genetic disorder sooner or later causes mild or severe medical problems (which often occur in a syndrome). **Pedigrees** can reveal inheritance patterns for alleles that can be predictably associated with specific phenotypes.

Section 9.7 An allele is inherited in an autosomal dominant pattern if the trait it specifies appears in everyone who carries it, and both sexes are affected with equal frequency. Such traits appear in every generation of families that have the allele. An allele is inherited in an autosomal recessive pattern if the trait it specifies

appears only in homozygous people. Such traits also appear in both sexes equally, but they can skip generations.

An allele is inherited in an X-linked pattern when it occurs on the X chromosome. Most X-linked disorders are inherited in a recessive pattern, and these tend to appear in men more often than in women. Heterozygous women have a dominant, normal allele that can mask the effects of the recessive one; men do not. Men can transmit an X-linked allele to their daughters, but not to their sons. Only a woman can pass an X-linked allele to a son.

Section 9.8 Occasionally, new individuals end up with the wrong chromosome number. Consequences of such changes range from minor to lethal alterations in form and function.

Chromosome number change is usually an outcome of **nondisjunction**, in which chromosomes fail to separate properly during nuclear division. **Polyploid** individuals have three or more of each type of chromosome. Polyploidy is lethal in humans, but not in flowering plants and some insects, fishes, and other animals.

In humans, most cases of autosomal aneuploidy are lethal. Trisomy 21, which causes Down syndrome, is an exception. Some changes in the number of sex chromosomes result in an impairment in learning and motor skills.

Section 9.9 Prospective parents can use genetic screening to estimate their risk of transmitting a harmful allele to offspring. The procedure involves analysis of parental pedigrees and genotype by a genetic counselor. Amniocentesis and other methods of prenatal testing can reveal a genetic disorder before birth.

Self-Quiz

Answers in Appendix I

1. A heterozygous individual has a _____ for a trait being studied.
   a. pair of identical alleles
   b. pair of nonidentical alleles
   c. haploid condition, in genetic terms

2. An organism's observable traits constitute its _____ .
   a. phenotype   c. genotype
   b. variation   d. pedigree

3. The offspring of the cross $AA \times aa$ are _____ .
   a. all AA
   b. all aa
   c. all Aa
   d. 1/2 AA and 1/2 aa

4. The probability of a crossover occurring between two genes on the same chromosome _____ .
   a. is unrelated to the distance between them
   b. decreases with the distance between them
   c. increases with the distance between them

5. If one parent is heterozygous for a dominant allele on an autosome and the other parent does not carry the allele, any child of theirs has a _____ chance of being heterozygous.
   a. 25 percent
   b. 50 percent
   c. 75 percent

6. True or false? All traits are inherited in a Mendelian pattern.

7. One gene that affects three traits is an example of _____ .
   a. dominance
   b. codominance
   c. pleiotropy
   d. epistasis

8. _____ in a trait is indicated by a bell curve.
   a. An epigenetic effect
   b. Nondisjunction
   c. Incomplete dominance
   d. Continuous variation

9. Pedigree analysis is necessary when studying human inheritance patterns because _____ .
   a. humans have approximately 20,000 genes
   b. of ethical problems with experimenting on humans
   c. inheritance in humans is more complicated than it is in other organisms
   d. genetic disorders occur only in humans
   e. all of the above

10. A female child inherits one X chromosome from her mother and one from her father. What sex chromosome does a male child inherit from each of his parents?

11. Nondisjunction at meiosis can result in _____ .
    a. base-pair substitutions
    b. aneuploidy
    c. crossing over
    d. pleiotropy

12. True or false? An individual with three or more complete sets of chromosomes is polyploid.

13. Klinefelter syndrome (XXY) is most easily diagnosed by _____ .
    a. pedigree analysis
    b. aneuploidy
    c. karyotyping
    d. a Punnett square

14. Match each example with the best description.
    _____ dihybrid cross
    _____ monohybrid cross
    _____ homozygous
    _____ heterozygous
    a. bb
    b. $AaBb \times AaBb$
    c. Aa
    d. $Aa \times Aa$

15. Match the terms appropriately.
    _____ polyploid
    _____ syndrome
    _____ aneuploidy
    _____ Mendelian
    _____ genotype
    _____ Huntington's disease
    a. symptoms of a genetic disorder
    b. extra sets of chromosomes
    c. caused by a short tandem repeat
    d. one extra chromosome
    e. dominant > recessive
    f. an individual's alleles

Critical Thinking

1. Mendel crossed a true-breeding pea plant with green pods and a true-breeding pea plant with yellow pods. All offspring had green pods. Which color is recessive?

2. Assuming that independent assortment occurs during meiosis, what type(s) of gametes will form in individuals with the following genotypes?
   a. $AABB$
   b. $AaBB$
   c. $Aabb$
   d. $AaBb$

3. Refer to problem 2. Determine the frequencies of each genotype among offspring from an $AABB \times aaBB$ cross.

4. Duchenne muscular dystrophy (DMD), which is inherited in an X-linked recessive pattern, occurs almost exclusively in males. Suggest why.

5. Heterozygous individuals perpetuate some alleles that have lethal effects in homozygous individuals. A mutated allele (M^L) associated with taillessness in Manx cats is an example (left). Cats homozygous for this allele (M^LM^L) typically die before birth due to severe spinal cord defects. In a case of incomplete dominance, cats heterozygous for the M^L allele and the normal, unmutated allele (M) have a short, stumpy tail or none at all. Two M^LM cats mate. What is the probability that any one of their surviving kittens will be heterozygous?

Leslie Faltersek/Clacritter Manx.

10

BIOTECHNOLOGY

Application ❯ 10.1 Personal Genetic Testing

Personalized genetic testing is already beginning to revolutionize medicine.

REMEMBER: Some lipoproteins consist of variable amounts and types of proteins and lipids (Section 2.9). A base-pair substitution is a mutation in which one nucleotide replaces another (7.6). *BRCA1* mutations give rise to neoplasms (8.3). Alleles may encode slightly different forms of a gene's product, and such differences influence the details of traits shared by a species (8.4). A heterozygous individual has two nonidentical alleles of a gene (9.2). A person who is homozygous for a mutated allele of the *MC1R* gene makes only the reddish type of melanin, so this individual has red hair; most other human traits are polygenic, and many have epigenetic contributions (9.6). Alzheimer's disease is a condition of progressive mental deterioration (9.8).

About 99 percent of your DNA is exactly the same as everyone else's. The shared part is what makes you human; the differences make you a unique member of the species. If you compared your DNA with your neighbor's, about 2.97 billion nucleotides of the two sequences would be identical; the remaining 30 million nonidentical nucleotides are sprinkled throughout your chromosomes. The sprinkling is not entirely random because some regions of DNA vary less than others; these conserved regions are of particular interest to researchers because they are most likely to have an essential function. If a conserved sequence does vary among people, the variation tends to be in single nucleotides at a particular location. A base-pair substitution that is carried by a measurable percentage of a population, usually above 1 percent, is called a **single-nucleotide polymorphism**, or **SNP** (pronounced "snip").

Alleles of most genes differ by single nucleotides, and differences in alleles are the basis of the variation in human traits that makes each individual unique. Thus, SNPs account for many of the differences in the way humans look, and they also have a lot to do with differences in the way our bodies work—how we age, respond to drugs, weather assaults by pathogens and toxins, and so on. Finding out which ones you carry has never been easier. Genetic testing companies can extract your DNA from a few drops of spit or a cheek swab, then analyze it using a tiny glass plate called a SNP-chip (Figure 10.1A). Results may include the predicted likelihood of having traits associated with your particular SNPs. For example, the test will probably determine whether you are homozygous for one allele of the *MC1R* gene. If you are, then you have red hair. Few SNPs have such a clear effect, however. Most human traits arise from a complex interplay of genes and environmental factors that we are still unraveling. Thus, although a DNA test can reliably determine an individual's SNPs, it cannot reliably predict the effect of most of those SNPs on the individual.

Consider the lipoprotein particles that carry fats and cholesterol through our bloodstreams. These particles consist of variable amounts and types of lipids and proteins, one of which is specified by the gene *APOE*. About one in four people carries an allele of this gene, *E4*, that increases the risk of developing Alzheimer's disease later in life. If you are heterozygous for this allele, a DNA testing company

Figure 10.1 Personal genetic tests.

(A) Image courtesy of Illumina, Inc.; (B) © Oli Scarff/Getty Images.

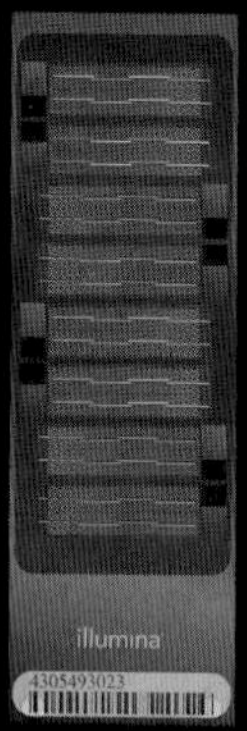

A. Only about 1 percent of the 3 billion bases in a person's DNA are unique to the individual. Personal genetic testing companies use chips like this one to analyze their customers' chromosomes for SNPs. This chip reveals which versions of 4,301,331 SNPs occur in the DNA of four individuals at a time.

B. Celebrity Angelina Jolie chose preventive treatment after genetic testing showed she had a very high risk of breast cancer. She carries a *BRCA1* mutation associated with an 87% lifetime risk of developing breast cancer. Even though Jolie did not yet have cancer, she underwent a double mastectomy, thereby reducing her risk of breast cancer to 5%.

cannot tell you whether you will develop Alzheimer's. However, it may report your lifetime risk of developing the disease, which is about 30 percent, as compared with about 14 percent for someone who has no *E4* allele.

What, exactly, does a 30 percent lifetime risk of Alzheimer's disease mean? The number is a probability statistic; it means, on average, 3 of every 10 people who have one *E4* allele eventually get the disease. However, a risk is just that. Not everyone who has an *E4* allele develops Alzheimer's, and not everyone who develops the disease has the allele. Other unknown factors, including epigenetic modifications of DNA, contribute to the disease. We still have a limited understanding of how genes contribute to many health conditions, particularly age-related ones such as Alzheimer's disease. Geneticists believe that it will be at least five to ten more years before genotyping can be used to accurately predict an individual's future health problems. Nonetheless, we are at a tipping point; personalized genetic testing is already beginning to revolutionize medicine. Cancer treatments are now being tailored to fit the genetic makeup of individual patients. People who discover they carry alleles associated with a heightened risk of a medical condition are being encouraged to make lifestyle changes that could delay the condition's onset or prevent it entirely. Preventive treatments based on personal genetics are becoming more common—and more mainstream (Figure 10.1B).

10.2 Finding Needles in Haystacks

REMEMBER: A tracer has a detectable component (Section 2.2). Typical prokaryotes have plasmids that carry a few genes (3.4). Cloning technologies produce identical copies of an organism (6.1). A bacteriophage infects a bacterium by injecting DNA into it; hydrogen bonds between bases hold the two strands of DNA together in a double helix (6.2). Hybridization is the spontaneous establishment of base-pairing between nucleic acid strands; primers are short, single strands of nucleotides; DNA ligase seals gaps in DNA strands during DNA replication (6.4).

Cutting and Pasting DNA In the 1950s, excitement over the discovery of DNA's structure gave way to frustration: No one could determine the order of nucleotides in a molecule of DNA. Identifying a single nucleotide among thousands or millions of others turned out to be a huge technical hurdle. Research in a seemingly unrelated field yielded a solution when Werner Arber, Hamilton Smith, and their coworkers discovered how some bacteria resist infection by bacteriophages. These bacteria have enzymes that chop up any injected viral DNA before it has a chance to integrate into the bacterial chromosome. The enzymes restrict viral growth; hence their name, restriction enzymes. A **restriction enzyme** cuts DNA wherever a specific nucleotide sequence occurs (Figure 10.2). For example, the enzyme *Eco*RI (named after *E. coli*, the bacteria from which it was isolated) cuts DNA at the nucleotide sequence GAATTC . Other restriction enzymes cut at different sequences.

The discovery of restriction enzymes allowed researchers to cut chromosomal DNA into manageable chunks. It also allowed them to combine DNA fragments from different organisms. How? Many restriction enzymes, including *Eco*RI, leave single-stranded tails on DNA fragments ❷. Researchers realized that complementary tails will base-pair, regardless of the source of the DNA ❸. The tails are called "sticky ends," because two DNA fragments stick together when their matching tails base-pair. DNA ligase can be used to seal the gaps between base-paired sticky ends, so continuous DNA strands form ❹. Thus, using appropriate restriction enzymes

❶ The restriction enzyme *Eco*RI recognizes the nucleotide sequence GAATTC in DNA from two sources.

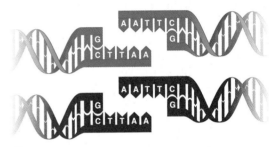

❷ The enzyme cuts the DNA into fragments. *Eco*RI leaves single-stranded tails ("sticky ends").

❸ When the DNA fragments from the two sources are mixed together, matching sticky ends base-pair.

❹ DNA ligase joins the base-paired DNA fragments to produce molecules of recombinant DNA.

Figure 10.2 Making recombinant DNA.

restriction enzyme Type of enzyme that cuts DNA at a specific nucleotide sequence.

single-nucleotide polymorphism (SNP) A one-nucleotide DNA sequence variation carried by a measurable percentage of a population.

cloning vector A DNA molecule that can accept foreign DNA and be replicated inside a host cell.

DNA cloning Set of methods that uses living cells to mass-produce targeted DNA fragments.

DNA library Collection of cells that host different fragments of foreign DNA, often representing an organism's entire genome.

genome An organism's complete set of genetic material.

PCR Polymerase chain reaction. Method that rapidly generates many copies of a specific section of DNA.

probe Fragment of DNA or RNA labeled with a tracer; can hybridize with a nucleotide sequence of interest.

recombinant DNA A DNA molecule that contains genetic material from more than one organism.

and DNA ligase, researchers can cut and paste DNA from different sources. The result, a hybrid molecule that consists of genetic material from two or more organisms, is called **recombinant DNA**.

Making recombinant DNA is the first step in **DNA cloning**, a set of laboratory methods that uses living cells to mass-produce specific DNA fragments. Researchers clone a fragment of DNA by inserting it into a **cloning vector**, which is a molecule that can carry foreign DNA into host cells. Bacterial plasmids may be used as cloning vectors, for example (Figure 10.3). When a bacterium reproduces, its offspring inherit a full complement of genetic information—one chromosome plus plasmids. If a plasmid carries a fragment of foreign DNA, that fragment gets copied and distributed to descendant cells along with the plasmid DNA. A host cell into which a recombinant cloning vector has been inserted can be grown in the laboratory (cultured) to yield a huge population of genetically identical cells. Each of these clones contains a copy of the vector and the inserted DNA fragment. The hosted DNA fragment can be harvested in large quantities from the clones.

DNA Libraries The entire set of genetic material—the **genome**—of most organisms consists of thousands of genes. To study or manipulate a single gene, researchers first find it, and then separate it from all of the other genes in a genome. They often begin by cutting an organism's DNA into fragments, and then cloning all the fragments. The result is a set of clones that collectively contain all of the DNA in a genome. A set of cells that hosts various cloned DNA fragments is a **DNA library**.

In DNA libraries, a cell that contains a particular DNA fragment of interest is mixed up with thousands or millions of others that do not—a needle in a genetic haystack. One way to find that one clone among the others involves the use of a **probe**, a fragment of DNA or RNA labeled with a tracer. For example, to find a targeted gene, researchers may use radioactive nucleotides to synthesize a short strand of DNA complementary in sequence to a similar gene. Because the nucleotide sequences of the probe and the gene are complementary, the two can hybridize. When the probe is mixed with DNA from a library, it will hybridize with the gene, but not with other DNA. Researchers can pinpoint a cell that hosts the gene by detecting the label on the probe. That cell is isolated and cultured, and DNA can be extracted in bulk from the cultured cells for research or other purposes.

Figure 10.3 An example of cloning.
Here, a fragment of chromosomal DNA is inserted into a plasmid.

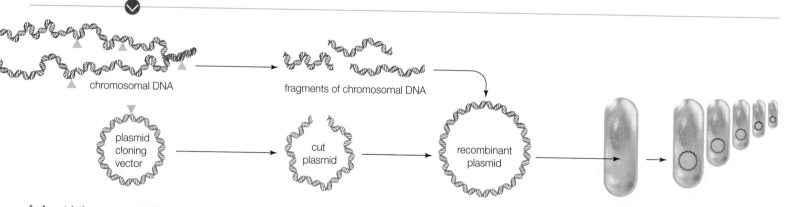

chromosomal DNA

fragments of chromosomal DNA

plasmid cloning vector

cut plasmid

recombinant plasmid

A. A restriction enzyme (gold triangles) cuts a specific nucleotide sequence in chromosomal DNA and also in a plasmid cloning vector.

B. A fragment of chromosomal DNA and the cut plasmid base-pair at their sticky ends. DNA ligase joins the two pieces of DNA, so a recombinant plasmid forms.

C. The recombinant plasmid is inserted into a host bacterial cell. When the cell reproduces, it copies the plasmid along with its chromosome. Each descendant cell receives a plasmid.

Figure 10.4 Two rounds of PCR.
Each cycle of this reaction can double the number of copies of a targeted sequence of DNA. Thirty cycles can make a billion copies.

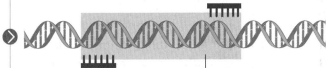

targeted section

1 DNA (blue) with a targeted sequence is mixed with primers (pink), nucleotides, and heat-tolerant *Taq* DNA polymerase.

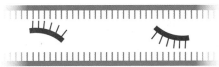

2 When the mixture is heated, the double-stranded DNA separates into single strands. When the mixture is cooled, some of the primers base-pair with the DNA at opposite ends of the targeted sequence.

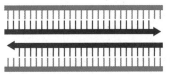

3 *Taq* polymerase begins DNA synthesis at the primers, so it produces complementary strands of the targeted DNA sequence.

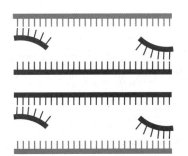

4 The mixture is heated again, so all double-stranded DNA separates into single strands. When it is cooled, primers base-pair with the targeted sequence in the original template DNA and in the new DNA strands.

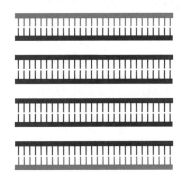

5 Each cycle of heating and cooling can double the number of copies of the targeted DNA section.

PCR The polymerase chain reaction (**PCR**) is a technique used to mass-produce copies of a particular section of DNA without having to clone it in living cells (Figure 10.4). The reaction can transform a needle in a haystack—that one-in-a-million fragment of DNA—into a huge stack of needles with a little hay in it.

The starting material for PCR is any sample of DNA with at least one molecule of a targeted sequence. It might be extracted from a mixture of 10 million different clones, a sperm, a hair left at a crime scene, or a mummy—essentially any sample that has DNA in it.

The PCR reaction is similar to DNA replication. It requires two synthetic DNA primers, each designed to base-pair with one end of the section of DNA to be amplified, or mass-produced **1**. Researchers mix these primers with the starting (template) DNA, nucleotides, and DNA polymerase, then expose the reaction mixture to repeated cycles of high and low temperatures. A few seconds at high temperature disrupts the hydrogen bonds that hold two strands of DNA together, so every molecule of DNA unwinds and becomes single-stranded. As the temperature of the reaction mixture is lowered, the single DNA strands hybridize with the primers **2**.

The DNA polymerases of most organisms denature at the high temperature required to separate DNA strands. The kind that is used in PCR reactions, *Taq* polymerase, is from *Thermus aquaticus*. This bacterial species lives in hot springs and hydrothermal vents, so its DNA polymerase necessarily tolerates heat. *Taq* polymerase, like other DNA polymerases, recognizes hybridized primers as places to start DNA synthesis **3**. Synthesis proceeds along the template strand until the temperature rises and the DNA separates into single strands **4**. The newly synthesized DNA is a copy of the targeted section. When the mixture is cooled, the primers rehybridize, and DNA synthesis begins again. Each cycle of heating and cooling takes only a few minutes, but it can double the number of copies of the targeted section of DNA **5**. Thirty PCR cycles may amplify that number a billionfold.

Take-Home Message 10.2

What techniques allow researchers to study DNA?

- DNA cloning uses living cells to mass-produce particular DNA fragments. Restriction enzymes cut DNA into fragments, then DNA ligase seals the fragments into cloning vectors. Recombinant DNA molecules result.
- A cloning vector that holds foreign DNA can be introduced into a living cell. When the host cell divides, it gives rise to huge populations of genetically identical cells (clones), each of which contains a copy of the foreign DNA.
- Researchers can isolate one gene from the many others in a genome by making a DNA library. A probe can be used to identify one clone that hosts a targeted DNA fragment among many other clones in the library.
- PCR quickly mass-produces copies of a targeted section of DNA.

10.3 Studying DNA

REMEMBER: A fat is a substance that consists mainly of triglycerides (Section 2.8). James Watson and Francis Crick built the first accurate model of the DNA molecule in 1953 (6.2). In DNA replication, DNA polymerases begin assembling new strands of DNA at hybridized primers; for each molecule of DNA that is copied, two DNA molecules are produced, each a duplicate of the parent—one strand is new, and the other is parental (6.4). A knockout is an organism in which a gene has been deliberately inactivated (7.7). A short tandem repeat is a series of a few nucleotides repeated several times in a row in chromosomal DNA (9.5).

Sequencing the Human Genome Once a fragment of DNA has been isolated (for example by cloning or PCR), researchers can use a technique called **sequencing** to determine the order of nucleotides in it. The most common method uses DNA polymerase. This enzyme is mixed with a primer, nucleotides, and the DNA to be sequenced (the template). Starting at the primer, the polymerase joins the nucleotides into a new strand of DNA, in the order dictated by the sequence of the template. The DNA fragments are then separated by length. In a technique called **electrophoresis**, an electric field pulls the fragments through a semisolid gel. Fragments of different sizes move through the gel at different rates. The shorter the fragment, the faster it moves, because shorter fragments slip through the tangled molecules of the gel faster than longer fragments do. All fragments of the same length move through the gel at the same speed, so they gather into bands. The order of the bands in the gel reflects the sequence of the template DNA (Figure 10.5A).

The sequencing method we have just described was invented in 1975. Ten years later, it had become so routine that scientists began to consider sequencing the entire human genome—all 3 billion nucleotides. Proponents of the idea said it could provide huge payoffs for medicine and research. Opponents said this daunting task would divert attention and funding from more urgent research. It would require 50 years to sequence the human genome given the techniques of the time. However, the techniques continued to improve rapidly, and with each improvement more nucleotides could be sequenced in less time. Automated (robotic) DNA sequencing and PCR had just been invented. Both were still too cumbersome and expensive to be useful in routine applications, but they would not be so for long. Waiting for faster, cheaper technologies seemed the most efficient way to sequence the genome, but just how fast did they need to be before the project should begin?

A few privately owned companies decided not to wait, and started sequencing. One of them intended to determine the genome sequence in order to patent it. The idea of patenting the human genome provoked widespread outrage, but it also spurred commitments in the public sector. In 1988, the National Institutes of Health (NIH) essentially took over the project by hiring James Watson (of DNA structure fame) to head an official Human Genome Project, and providing $200 million per year to fund it. A partnership formed between the NIH and international institutions that were sequencing different parts of the genome. Watson set aside 3 percent of the funding for studies of ethical and social issues arising from the work. He later resigned over a patent disagreement, and geneticist Francis Collins took his place.

Amid ongoing squabbles over patent issues, Celera Genomics formed in 1998. With biologist Craig Venter at its helm, the company intended to commercialize human genetic information. Celera invented faster techniques for sequencing genomic DNA because the first to have the complete sequence had a legal basis for patenting it. The competition motivated the international partnership to accelerate

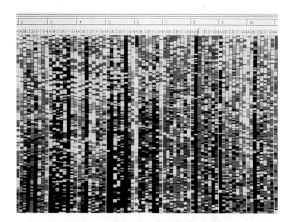

A. Human DNA sequence data. On this computer screen, the four nucleotides (adenine, thymine, guanine, and cytosine) are color-coded green, red, yellow, and blue. The order of colored bands in each vertical "lane" represents part of the sequence of the template DNA.

B. Today's automated DNA sequencing machines can sequence an individual's entire genome in 2–4 hours, for a cost of about $1,000.

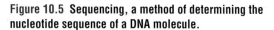

Figure 10.5 Sequencing, a method of determining the nucleotide sequence of a DNA molecule.

(A) Patrick Landmann/Science Source; (B) © Michelle McLoughlin/Reuters/Corbis.

```
758 GATAATCCTGTTTTGAACAAAAGGTCAAATTGCTGAATAGAAA-GTCTTGATTAACTAAAAGATGTACAAAGTGGAATTA 836  Human
752 GATAATCCTGTTTTGAACAAAAGGTCAAATTGCTGAATAGAAA-GTCTTGATTAACTAAAAGATGTACAAAGTGGAATTA 830  Mouse
751 GATAATCCTGTTTTGAACAAAAGGTCAAATTGCTGAATAGAAA-GTCTTGATTAACTAAAAGATGTACAAAGTGGAATTA 829  Rat
754 GATAATCCTGTTTTGAACAAAAGGTCAAATTGCTGAATAGAAA-GTCTTGATTAACTAAAAGATGTACAAAGTGGAATTA 832  Dog
782 GATAATCCTGTTTTGAACAAAAGGTCAAATTGCTGAATAGAAA-GTCTTGATTAACTAAAAGATGTACAAAGTGGAATTA 860  Chicken
758 GATAATCCTGTTTTGAACAAAAGGTCAAATTGCTGAATAGAAA-GTCTTGATTAAGTAAAAGATGTACAAAGTGGAATTA 836  Frog
823 GATAATCCTGTTTTGAACAAAAGGTCAGATTGCTGAATAGAAAAGGCTTGATTAAAGCAGAGATGTACAAAGTGGACGCA 902  Zebrafish
763 GATAATCCTGTTTTGAACAAAAGGTCAAATTGTTGAATAGAGACGCTTTGATAAAGCGGAGGAGGTACAAAGTGGGACC- 841  Pufferfish
```

Figure 10.6 Alignment of a section of genomic DNA from various species.
This is a region of the gene for a DNA polymerase. Nucleotides that differ from those in the human sequence are highlighted. The chance that any two of these sequences would randomly match is 1 in 10^{46}.

its efforts. Then, in 2000, U.S. President Bill Clinton and British Prime Minister Tony Blair jointly declared that the sequence of the human genome could not be patented. Celera kept sequencing anyway, and, in 2001, the competing governmental and corporate teams published about 90 percent of the sequence. In 2003, fifty years after the discovery of the structure of DNA, the sequence of the human genome was officially completed.

Genomics It took 15 years to sequence the human genome for the first time, but the technology has improved so much that sequencing an entire genome now takes a few hours (Figure 10.5B). Anyone can now pay to have their genome sequenced. However, despite our ability to determine the sequence of an individual's genome, it will be a long time before we understand all the information coded within that sequence. The human genome contains a massive amount of seemingly cryptic data. We can decipher some of this data by comparing genomes of different species, the premise being that all organisms are descended from shared ancestors, so all genomes are related to some extent. We see evidence of such genetic relationships simply by comparing the raw sequence data, which, in some regions of DNA, is extremely similar across many species (Figure 10.6).

The study of genomes is called **genomics**, a broad field that encompasses whole-genome comparisons, structural analysis of gene products, and surveys of small-scale variations in sequence. Genomics is providing powerful insights into evolution, and it has many medical benefits. We have learned the function of many human genes by studying their counterparts in other species. For instance, researchers comparing human and mouse genomes discovered a human version of a mouse gene, *APOA5*, that encodes a lipoprotein. Mice with an *APOA5* knockout have four times the normal level of triglycerides in their blood. The researchers then looked for—and found—a correlation between *APOA5* mutations and high triglyceride levels in humans. High triglycerides are a risk factor for coronary artery disease.

DNA Profiling As you learned in Section 10.1, about 99 percent of the human genome sequence is identical in every member of the species. The differences you carry in your DNA make you unique. In fact, those differences are so unique that they can be used to identify you. Identifying an individual by his or her DNA is a method called **DNA profiling**. One DNA profiling method uses SNP-chips (an example is shown in Figure 10.1). A SNP-chip is a tiny glass plate with microscopic spots of DNA stamped on it. The DNA sample in each spot is a short, synthetic single strand with a unique SNP sequence. When an individual's genomic DNA is washed over a SNP-chip, it hybridizes only with DNA spots that have a matching SNP sequence. Probes reveal where the genomic DNA has hybridized—and which of the SNPs are carried by the individual (Figure 10.7).

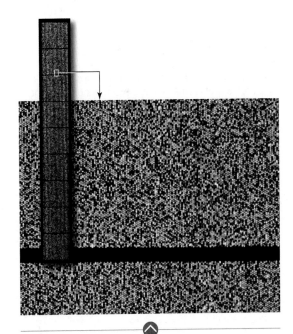

Figure 10.7 A SNP-chip analysis.
This SNP chip (inset), shown actual size, tests 550,000 SNPs. The small white box indicates the magnified portion shown. Each spot is a region where the individual's genomic DNA has hybridized with one SNP sequence. A red or green dot means that the individual is homozygous for a SNP; a combined signal (yellow dot) indicates heterozygosity.

genetic engineering Process by which deliberate changes are introduced into an individual's genome.

genetically modified organism (GMO) Organism whose genome has been modified by genetic engineering.

transgenic Refers to a genetically modified organism that carries a gene from a different species.

Another method of DNA profiling involves analysis of short tandem repeats in an individual's chromosomes. Short tandem repeats usually occur in the same location in human chromosomes, but the number of times a sequence is repeated in each location differs among individuals. For example, one person's DNA may have fifteen repeats of the nucleotides TTTTC at a certain spot on one chromosome. Another person's DNA may have four repeats of this sequence in the same location. Short tandem repeats slip spontaneously into DNA during replication, and their numbers grow or shrink over generations. Unless two people are identical twins, the chance that they have identical short tandem repeats in even three regions of DNA is 1 in a quintillion (10^{18}), which is far more than the number of people who have ever lived. Thus, an individual's array of short tandem repeats is, for all practical purposes, unique.

Analyzing a person's short tandem repeats begins with PCR, which is used to copy ten to thirteen particular regions of chromosomal DNA known to have repeats. The lengths of the copied DNA fragments differ among most individuals, because the number of tandem repeats in those regions also differs. Thus, electrophoresis can be used to reveal an individual's unique array of short tandem repeats (Figure 10.8).

A. Gray boxes indicate the regions of the individual's DNA that were tested.

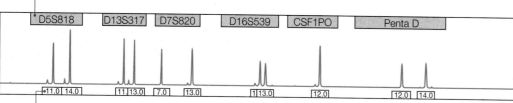

| D5S818 | D13S317 | D7S820 | D16S539 | CSF1PO | Penta D |

11.0 14.0   11 13.0   7.0   13.0   1 13.0   12.0   12.0 14.0

B. The number of repeats is shown in a box below each peak. A peak's location on the x-axis corresponds to the length of the DNA fragment amplified (a measure of the number of repeats). Peak size reflects amount of DNA.

Figure 10.8 An individual's (partial) short tandem repeat profile.
Remember, human body cells are diploid. Double peaks appear on a profile when the two members of a chromosome pair carry a different number of repeats.

Figure It Out: How many repeats does this individual have at the Penta D region?

Answer: 12 on one chromosome, and 14 on the other

Short tandem repeat analysis will soon be replaced by full genome sequencing, but for now it continues to be a common DNA profiling method. Geneticists compare short tandem repeats on Y chromosomes to determine relationships among male relatives, and to trace an individual's ethnic heritage. They also track mutations that accumulate in populations over time by comparing DNA profiles of living humans with those of ancient ones. Such studies are allowing us to reconstruct population dispersals that happened in the ancient past.

Short tandem repeat profiles are routinely used to resolve kinship disputes, and as evidence in criminal cases. Within the context of a criminal or forensic investigation, DNA profiling is called DNA fingerprinting. As of January 2014, the database of DNA fingerprints maintained by the Federal Bureau of Investigation (the FBI) contained the short tandem repeat profiles of 10.7 million convicted offenders, and had been used in more than 200,000 criminal investigations. DNA fingerprinting is also used to identify human remains, including the individuals who died in the World Trade Center on September 11, 2001.

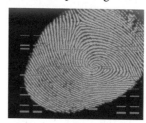

Take-Home Message 10.3

How do we use what researchers discover about DNA?

- Improvements in DNA sequencing techniques allowed the human genome sequence to be determined. The technology has improved so much that an individual's genome can now be sequenced in a few hours.
- Analysis of the human genome sequence is yielding new information about our genes and how they work.
- DNA profiling identifies individuals by the unique parts of their DNA.

10.4 Genetic Engineering

REMEMBER: Carrots are orange because they contain the photosynthetic accessory pigment β-carotene (Section 5.2). The DNA sequence of a gene encodes an RNA or protein product; gene expression is the process by which information in a gene guides the assembly of an RNA or protein product (7.2).

Traditional cross-breeding methods can alter genomes, but only if individuals with the desired traits will interbreed. Genetic engineering takes gene-swapping to an entirely different level. **Genetic engineering** is a process by which an individual's genome is deliberately modified. A gene from one species may be transferred to another to produce an organism that is **transgenic**, or a gene may be altered and reinserted into an individual of the same species. Both methods yield a **genetically modified organism**, or **GMO**.

Genetically Modified Microorganisms Most genetic engineering involves yeast and bacteria (Figure 10.9). Both types of cells have the metabolic machinery to make complex organic molecules, and they are easily engineered to produce, for example, medically important proteins. People with diabetes were among the first beneficiaries of such organisms. Insulin for their injections was once extracted from animals, but it provoked an allergic reaction in some people. Human insulin, which does not provoke allergic reactions, has been produced by transgenic *E. coli* since 1982. Slight modifications of the gene have yielded fast-acting and slow-release forms of human insulin.

Genetically engineered microorganisms also make proteins used in foods. For example, enzymes produced by modified microorganisms improve the taste and clarity of beer and fruit juice, slow bread staling, or modify certain fats. Cheese is traditionally made with an enzyme, chymosin, extracted from calf stomachs. Today, almost all cheese is made with calf chymosin produced by transgenic yeast.

Designer Plants As crop production expands to keep pace with human population growth, it places unavoidable pressure on ecosystems everywhere. Irrigation leaves mineral and salt residues in soils. Tilled soil erodes, taking topsoil with it. Runoff clogs rivers, and fertilizer in it causes algae to grow so fast that fish suffocate. Pesticides can be harmful to humans and other animals, including beneficial insects such as bees.

Pressured to produce more food at lower cost and with less damage to the environment, many farmers have begun to rely on genetically engineered crop plants. Genes can be introduced into plant cells by way of electric or chemical shocks, by blasting them with microscopic DNA-coated pellets, or by using *Agrobacterium tumefaciens* bacteria. *A. tumefaciens* carries a plasmid with genes that cause tumors to form on infected plants; hence the name Ti plasmid (for Tumor-inducing). Researchers replace the tumor-inducing genes with foreign or engineered genes, then use the plasmid as a vector to deliver the desired genes into plant cells. Whole plants can be grown from plant cells that integrate a recombinant plasmid into their chromosomes (Figure 10.10).

Many genetically modified crops carry genes that impart resistance to devastating plant diseases and pests. GMO crops such as Bt corn and soy help farmers use smaller amounts of toxic pesticides. Organic farmers often spray their crops with spores of Bt (*Bacillus thuringiensis*), a bacterial species that makes a protein toxic only to some insect larvae. Researchers transferred the gene encoding the Bt protein

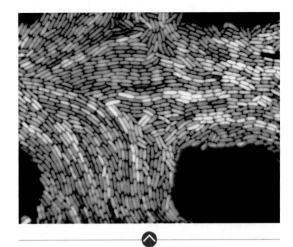

Figure 10.9 Genetically modified bacteria.
These *E. coli* bacteria are transgenic for a fluorescent jellyfish protein. The cells are genetically identical, so the visible variation in fluorescence among them reveals differences in gene expression. Such differences may help us discover why some bacteria of a population become dangerously resistant to antibiotics, and others do not.

Courtesy of Systems Biodynamics Lab, P. I. Jeff Hasty, UCSD Department of Bioengineering, and Scott Cookson.

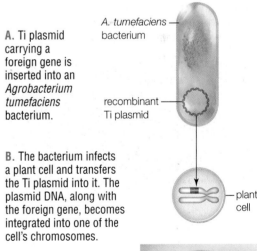

A. Ti plasmid carrying a foreign gene is inserted into an *Agrobacterium tumefaciens* bacterium.

A. tumefaciens bacterium

recombinant Ti plasmid

B. The bacterium infects a plant cell and transfers the Ti plasmid into it. The plasmid DNA, along with the foreign gene, becomes integrated into one of the cell's chromosomes.

plant cell

C. The infected plant cell divides, and its descendants form an embryo, then a plant. Cells of the transgenic plant carry and express the foreign gene.

Figure 10.10 How to make a transgenic plant.

(C) Pascal Goetgheluck/Science Source.

A. The genetically modified plants that produced the corn on the left carry a gene from the bacteria *Bacillus thuringiensis* (Bt) that conferred insect resistance. Compare the corn from unmodified plants, right. No pesticides were used on either crop.

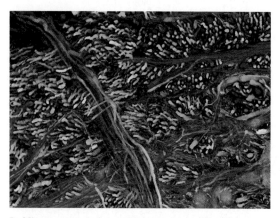

B. Mice transgenic for multiple pigments ("brainbow mice") are allowing researchers to map the complex neural circuitry of the brain. Individual nerve cells in the brain stem of a brainbow mouse are visible in this fluorescence micrograph.

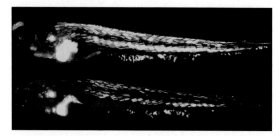

C. Zebrafish genetically modified to glow in places where BPA, an endocrine-disrupting chemical, is present. The fish are literally illuminating where this pollutant acts in the body—and helping researchers discover what it does when it gets there.

Figure 10.11 Some useful GMOs.

(A) The Bt and Non-Bt corn photos were taken as part of field trial conducted on the main campus of Tennessee State University at the Institute of Agricultural and Environmental Research. The work was supported by a competitive grant from the CSREES, USDA titled *Southern Agricultural Biotechnology Consortium for Underserved Communities*, (2000–2005). Dr. Fisseha Tegegne and Dr. Ahmad Aziz served as Principal and Co-principal Investigators respectively to conduct the portion of the study in the State of Tennessee; (B) Courtesy of © Dr. Jean Levit. The Brainbow technique was developed in the laboratories of Jeff W. Lichtman and Joshua R. Sanes at Harvard University. This image has received the Bioscape imaging competition 2007 prize; (C) © Charles Taylor/University of Exeter.

into plants. The engineered plants produce the Bt protein, but otherwise they are essentially identical with unmodified plants. Larvae die shortly after eating their first and only GMO meal. Farmers can use much less pesticide on crops that make their own (Figure 10.11A).

Genetic modifications can make food plants more nutritious. For example, rice plants have been engineered to make β-carotene, an orange photosynthetic pigment that is remodeled by cells of the small intestine into vitamin A. These rice plants carry two genes in the β-carotene synthesis pathway: one from corn, the other from bacteria. One cup of their seeds—grains of Golden Rice—has enough β-carotene to satisfy a child's daily need for vitamin A.

The USDA Animal and Plant Health Inspection Service (APHIS) regulates the introduction of GMOs into the environment. At this writing, APHIS has deregulated ninety-two crop plants, which means the plants are approved for unrestricted use in the United States. Worldwide, more than 330 million acres are currently planted in GMO crops, the majority of which are corn, sorghum, cotton, soy, canola, and alfalfa genetically engineered for resistance to the herbicide glyphosate. Rather than tilling the soil to control weeds, farmers can spray their fields with glyphosate, which kills the weeds but not the GMO crops.

Crops genetically engineered to resist glyphosate have been used in conjunction with the herbicide since the mid-1970s. Genes that confer glyphosate resistance are now appearing in weeds and other wild plants, as well as in unmodified crops—which means that recombinant DNA can (and does) escape into the environment. Glyphosate resistance genes are probably being transferred from transgenic plants to nontransgenic ones via pollen carried by wind or insects.

Biotech Barnyards Genetically modified animals can be produced by injecting DNA into a zygote, and then implanting the resulting embryo in a surrogate to complete its development. Genetically modified mice are invaluable in research (Figure 10.11B). We have discovered the function of many human genes (including the *APOA5* gene discussed in Section 10.3) by inactivating their counterparts in mice. Genetically modified mice are also used as models of human diseases. For example, researchers inactivated the molecules involved in the control of glucose metabolism, one by one. Studying the effects of the knockouts in mice has resulted in much of our current understanding of how diabetes works in humans.

Genetically engineered animals other than mice are also useful in research (Figure 10.11C), and some make molecules that have medical and industrial applications. Various transgenic goats produce proteins used to treat cystic fibrosis, heart attacks, blood clotting disorders, and even nerve gas exposure. Milk from goats transgenic for lysozyme, an antibacterial protein in human milk, may protect infants and children in developing countries from acute diarrheal disease. Goats transgenic for a spider silk gene produce the silk protein in their milk; researchers can spin this protein into nanofibers that have medical and electronics applications. Rabbits make human interleukin-2, a protein that triggers immune cells to divide and is used as a cancer drug. We also engineer food animals. Genetic engineering has given us pigs with heart-healthy fat and environmentally friendly low-phosphate feces, muscle-bound trout, chickens that do not transmit bird flu, and cows that do not get mad cow disease, among other examples.

Many people think that genetically engineering livestock is unconscionable. Others see it as an extension of thousands of years of acceptable animal husbandry practices. The techniques have changed, but not the intent: We humans continue to have an interest in improving our livestock. Either way, tinkering with the genes of

Digging Into Data

Enhanced Spatial Learning Ability in Mice With an Autism Mutation

Autism is a neurobiological disorder with symptoms that include impaired social interactions and stereotyped patterns of behavior. Around 10 percent of autistic people also have an extraordinary skill or talent such as greatly enhanced memory. Mutations in the gene for neuroligin 3, an adhesion protein that connects brain cells to one another, have been associated with autism. One of these mutations is called *R451C* because the altered gene encodes a protein with an amino acid substitution: a cysteine (C) instead of an arginine (R) in position 451.

In 2007, Katsuhiko Tabuchi and his colleagues introduced the *R451C* mutation into the neuroligin 3 gene of mice. The researchers discovered that the genetically modified mice had impaired social behavior, and also that spacial learning ability was affected.

Spatial learning in mice is tested with a water maze, which consists of a small platform submerged a bit below the surface of a pool of water so it is invisible to swimming mice. Mice do not particularly enjoy swimming, so they try to locate the hidden platform as quickly as they can. When tested again, they remember the platform's location by checking visual cues around the edge of the pool. How quickly they remember is a measure of their spatial learning ability. Figure 10.12 shows some of Tabuchi's results.

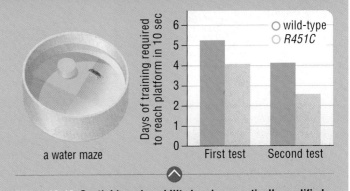
a water maze

Figure 10.12 Spatial learning ability in mice genetically modified to have a mutation associated with autism.
Performance of genetically modified mice (*R451C*) in a water maze was compared with that of unmodified (wild-type) mice.

1. In the first test, how many days did unmodified mice need to learn to find the location of a hidden platform within 10 seconds?
2. Did the modified or the unmodified mice learn the location of the platform faster in the first test?
3. Which mice learned faster the second time around?
4. Which mice had the greatest improvement in memory?

animals raises a host of ethical dilemmas. Consider animals genetically engineered to carry mutations associated with human diseases—multiple sclerosis, cystic fibrosis, diabetes, cancer, or Huntington's disease, for example. Researchers study these animals in order to understand the diseases, and to test potential treatments, without experimenting on humans. However, the modified animals often suffer the same terrible symptoms of the condition as humans do.

Some worry that our ability to tinker with genetics has surpassed our ability to understand the impact of the tinkering. Controversy raised by GMO use invites you to read the research and form your own opinions. The alternative is to be swayed by media hype (the term "Frankenfood," for instance), or by reports from possibly biased sources (such as herbicide manufacturers).

> Some worry that our ability to tinker with genetics has surpassed our ability to understand the impact of the tinkering.

Take-Home Message 10.4

What is genetic engineering?

- Genetic engineering is the directed alteration of an individual's genome, and it results in a genetically modified organism (GMO).
- A transgenic organism carries a gene from a different species. Transgenic organisms, including bacteria and yeast, are used in research, medicine, and industry.
- Genetically modified crop plants can help farmers be more productive. However, widespread use of these crops is having unintended environmental effects.
- Genetic engineering creates animals that would be impossible to produce using traditional cross-breeding methods.

10.5 Modifying Humans

REMEMBER: A gene that helps transform a normal cell into a tumor cell is an onco-gene; a proto-oncogene can become an oncogene (Section 8.3). An X-linked genetic disorder is associated with an allele on the X chromosome (9.7).

Gene Therapy We know of more than 15,000 serious genetic disorders. Collectively, they cause 20 to 30 percent of infant deaths each year, and account for half of all mentally impaired patients and a fourth of all hospital admissions. They also contribute to many age-related disorders, including cancer, Parkinson's disease, and diabetes. Drugs and other treatments can minimize the symptoms of some genetic disorders, but gene therapy is the only cure. **Gene therapy** is the transfer of recombinant DNA into an individual's body cells, with the intent to correct a genetic disorder or treat a disease. Typically, the transfer inserts an unmutated gene into the individual's chromosomes.

DNA can be introduced into human cells in many ways, for example by direct injection, electrical pulses, lipid clusters, nanoparticles, or genetically engineered viruses. Viruses have molecular machinery that delivers their genomes into cells they infect. Those used as vectors have DNA that splices itself into the infected cells' chromosomes, along with foreign DNA that has been inserted into it.

Human gene therapy is a compelling reason to embrace genetic engineering research. It is now being tested as a treatment for AIDS, muscular dystrophy, heart attack, sickle-cell anemia, cystic fibrosis, hemophilia A, Parkinson's disease, Alzheimer's disease, several types of cancer, and inherited diseases of the eye, the ear, and the immune system.

People have already benefited from gene therapy. Consider SCID-X1, a severe X-linked genetic disorder that stems from a mutated allele of the *IL2RG* gene. The gene encodes a receptor for an immune signaling molecule. Without treatment, people affected by this disorder can survive only in germ-free isolation tents because they cannot fight infections (a diminished life in a sterile isolation tent was the source of the term "bubble boy"). In the late 1990s, researchers used a genetically engineered virus to insert unmutated copies of *IL2RG* into cells taken from the bone marrow of twenty boys with SCID-X1. Each child's modified cells were infused back into his bone marrow. Within months of their treatment, eighteen of the boys left their isolation tents (Figure 10.13A). Gene therapy had repaired their immune systems.

Gene therapy is now being tested as a treatment for acute lymphocytic leukemia, a typically fatal cancer of bone marrow cells. A viral vector is used to insert a gene into immune cells extracted from patients. When the resulting genetically modified cells are reintroduced into the patients' bodies, the inserted gene directs the destruction of cancer cells (Figure 10.13B). The therapy seems to work astonishingly well: In one patient, all traces of the leukemia vanished in eight days.

Despite the successes, manipulating a gene within the context of a living individual is unpredictable even when we know its sequence and location on a chromosome. No one, for example, can predict with absolute certainty where a virus-injected gene will become integrated into a chromosome. Its insertion might disrupt other genes. If it interrupts a gene that is part of the controls over cell division, then cancer might be the outcome. Consider that five of the twenty boys first treated with gene therapy for SCID-X1 developed leukemia, and one of them died. Developers of the gene therapy had wrongly predicted that cancer related to it would be rare. Research now implicates the very gene targeted for repair, especially

A. Rhys Evans, shown here at age 10, was born with SCID-X1. His immune system has been permanently repaired by gene therapy.

B. Six-year-old Emily Whitehead was just days from death when she underwent an experimental gene therapy for acute lympho-cytic leukemia. Two years later, she remains cancer-free.

Figure 10.13 Gene therapy: success stories.

(A) © Huw Evans Agency, Cardiff; (B) © Kari Whitehead Photography.

when combined with the virus that delivered it. The viral DNA preferentially inserted itself into the children's chromosomes at a site near a proto-oncogene. The insertion activated the gene by triggering its transcription, and that is how the leukemia began. Since that time, researchers have used PCR to detect viral integration sites and improve the design of the vector. The development of more efficient and specific viral vectors has reduced the risk associated with all types of gene therapy.

Eugenics Eliminating undesirable human traits is part of **eugenics**, the philosophy of deliberately improving the genetic qualities of the human race. Eugenics has been used as a justification for some of the most horrific episodes in human history, including the genocide of 6 million Jews during World War II; thus, it continues to be a hotly debated social issue. For example, using gene therapy to cure genetic disorders seems like an acceptable goal to most people, but imagine taking this idea a bit further. Would it also be acceptable to engineer the genome of an individual who is within a normal range of phenotype in order to modify a particular trait? Researchers have already produced mice that have improved memory, enhanced learning ability, bigger muscles, and longer lives. Why not people?

Given the pace of genetics research, the debate is no longer about how we would engineer desirable traits, but how we would choose the traits that are desirable. Realistically, cures for many severe but rare genetic disorders will not be found, because the financial return would not cover the cost of the research. Eugenics, however, may be profitable. How much would potential parents pay to be sure that their child will be tall or blue-eyed, with breathtaking strength or intelligence? What about a treatment that can help you lose that extra weight, and keep it off permanently? The gray area between interesting and abhorrent can be very different depending on who is asked. In a survey conducted in the United States, more than 40 percent of those interviewed said they would be fine with using gene therapy to make smarter and cuter babies. In one poll of British parents, 10 percent would use it to keep a child from growing up to be homosexual, and 18 percent would be willing to use it to keep a child from being aggressive.

Some people are concerned that gene therapy puts us on a slippery slope that may result in irreversible damage to ourselves and to the biosphere. We as a society may not have the wisdom to know how to stop once we set foot on that slope; one is reminded of our peculiar human tendency to leap before we look. And yet, something about the human experience allows us to dream of such things as wings of our own making, a capacity that carried us into space. In this brave new world, the questions before you are these: What do we stand to lose if serious risks are not taken? And, do we have the right to impose the potential consequences on people who would choose not to take those risks?

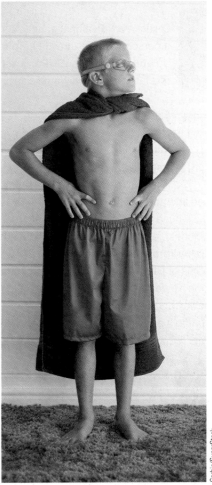

Corbis/SuperStock

The debate is no longer about how we would engineer desirable traits, but how we would choose the traits that are desirable.

Take-Home Message 10.5

Can people be genetically modified?

- Genes can be transferred into a person's cells to correct a genetic disorder or treat a disease. However, the outcome of altering a person's genome remains unpredictable given our current understanding of how the genome works.
- We as a society continue to work our way through the ethical implications of applying DNA technologies.

eugenics Idea of deliberately improving the genetic qualities of the human race.

gene therapy Treating a genetic defect or disorder by transferring a normal or modified gene into the affected individual.

Summary

Section 10.1 Personal genetic testing, which reveals a person's unique array of **single-nucleotide polymorphisms**, or **SNPs**, is beginning to revolutionize the way medicine is practiced.

Section 10.2 In **DNA cloning**, researchers use **restriction enzymes** to cut a sample of DNA into fragments, and then use DNA ligase to splice the fragments into plasmids or other **cloning vectors**. The resulting molecules of **recombinant DNA** are inserted into host cells such as bacteria. Division of host cells produces huge populations of genetically identical descendant cells (clones), each with a copy of the cloned DNA fragment.

A **DNA library** is a collection of cells that contain different fragments of DNA, often representing an organism's entire **genome**. Researchers can use **probes** to identify cells that carry a specific fragment of DNA. The polymerase chain reaction (**PCR**) uses primers and a heat-resistant DNA polymerase to rapidly increase the number of copies of a targeted section of DNA.

Section 10.3 Advances in **sequencing**, which reveals the order of nucleotides in DNA, allowed the DNA sequence of the entire human genome to be determined. DNA polymerase is used to partially replicate a DNA template. The reaction produces a mixture of DNA fragments of all different lengths; **electrophoresis** separates the fragments by length into bands.

Genomics gives us insights into the function of the human genome. Similarities between genomes of different organisms are evidence of evolutionary relationships, and can be used as a predictive tool in research. **DNA profiling** identifies a person by the unique parts of his or her DNA. An example is the determination of an individual's unique array of SNPs or short tandem repeats. Within the context of a criminal investigation, a DNA profile is called a DNA fingerprint.

Section 10.4 Recombinant DNA technology is the basis of **genetic engineering**, the directed modification of an organism's genetic makeup with the intent to modify its phenotype. A gene from one species is inserted into an individual of a different species to make a **transgenic** organism, or a gene is modified and reinserted into an individual of the same species. The result of either process is a **genetically modified organism (GMO)**.

Bacteria and yeast, the most common genetically engineered organisms, produce proteins that have medical value. The majority of the animals that are being created by genetic engineering are used for medical applications or research. Most transgenic crop plants, which are now in widespread use worldwide, were created to help farmers produce food more efficiently. Some crops are genetically modified to have enhanced nutritional value.

Section 10.5 With **gene therapy**, a gene is transferred into body cells to correct a genetic defect or treat a disease. Potential benefits of genetically modifying humans must be weighed against potential risks. The practice raises ethical issues such as whether **eugenics** would be desirable in some circumstances.

Self-Quiz

Answers in Appendix I

1. _____ cut(s) DNA molecules at specific sites.
   a. DNA polymerase   c. Restriction enzymes
   b. DNA probes   d. DNA ligase

2. A _____ is a molecule that can be used to carry a fragment of DNA into a host organism.
   a. cloning vector   c. GMO
   b. chromosome   d. cDNA

3. For each species, all _____ in the complete set of chromosomes is/are the _____ .
   a. genomes; genotype   c. SNPs; genome
   b. DNA; genome   d. transgenics; GMOs

4. A set of cells that host various DNA fragments collectively representing an organism's entire set of genetic information is called a _____ .
   a. genome   d. GMO
   b. DNA library   e. single-nucleotide
   c. clone   polymorphism

5. PCR can be used _____ .
   a. to increase the number of specific DNA fragments
   b. in DNA profiling
   c. to modify a human genome
   d. as a cloning vector
   e. a and b are correct
   f. all of the above

6. Fragments of DNA can be separated by electrophoresis according to _____ .
   a. sequence   c. species
   b. length   d. SNPs

7. *Taq* polymerase is used for PCR because it _____ .
   a. tolerates the high temperature needed to separate DNA strands
   b. is an enzyme from a bacterium
   c. does not require primers
   d. is genetically modified

8. _____ is a technique to determine the order of nucleotides in a fragment of DNA.
   a. PCR
   b. Sequencing
   c. Electrophoresis
   d. Nucleic acid hybridization

9. Which of the following can be used to carry foreign DNA into host cells? Choose all of the correct answers.
   a. RNA
   b. viruses
   c. PCR
   d. plasmids
   e. lipid clusters
   f. blasts of microscopic pellets
   g. a DNA library
   h. sequencing

10. Put the following tasks in the order they would occur during a cloning experiment.
    a. using DNA ligase to seal DNA fragments into vectors
    b. using a probe to identify a clone in the library
    c. using DNA polymerase to sequence the DNA of the clone
    d. making a DNA library of clones
    e. cutting genomic DNA with restriction enzymes

11. A transgenic organism _____ .
    a. carries a gene from another species
    b. has been genetically modified
    c. both a and b

12. True or false? A transgenic organism can pass its foreign gene to offspring.

13. _____ can correct a genetic defect in an individual.
    a. Cloning vectors
    b. Gene therapy
    c. Xenotransplantation
    d. a and b

14. True or false? Some humans are genetically modified.

15. Match the terms with the best description.
    _____ DNA profile
    _____ Ti plasmid
    _____ eugenics
    _____ SNP
    _____ transgenic
    _____ GMO

    a. GMO with a foreign gene
    b. alleles commonly have them
    c. a person's unique collection of short tandem repeats
    d. selecting "desirable" traits
    e. genetically modified
    f. used in plant gene transfers

Critical Thinking

1. Restriction enzymes in bacterial cytoplasm cut injected bacteriophage DNA wherever certain sequences occur. Why do you think these enzymes do not chop up the bacterial chromosome, which is exposed to the enzymes in cytoplasm?

2. In 1918, an influenza pandemic that originated with avian flu killed 50 million people. Researchers isolated samples of that virus from bodies of infected people preserved in Alaskan permafrost since 1918. From the samples, they sequenced the viral genome, then reconstructed the virus. The reconstructed virus is 39,000 times more infectious than modern influenza strains, and 100 percent lethal in mice.

   Understanding how this virus works can help us defend ourselves against other strains that may arise. For example, discovering what makes it so infectious and deadly would help us design more effective vaccines. Critics of the research are concerned: If the virus escapes the containment facilities (even though it has not done so yet), it might cause another pandemic. Worse, the published DNA sequence and methods to make the virus could be used for criminal purposes. Do you think this research makes us more or less safe?

Visual Question

Table 10.1 shows the results of a paternity test using short tandem repeats. Who's the daddy? How sure are you?

Table 10.1 Paternity Test Results

Marker	DNA Samples Tested			
	Mother	Baby	Alleged Father #1	Alleged Father #2
CSF1PO	15, 17	17, 23	23, 27	17, 15
FGA	9, 9	9, 9	9, 12	9, 12
THO1	29, 29	29, 27	27, 28	29, 28
TPOX	14, 18	18, 20	15, 20	17, 22
VWA	14, 14	14, 14	14, 14	14, 16
D3S1358	11, 14	14, 16	12, 16	14, 20
D5S818	11, 13	10, 13	8, 10	18, 18
D7S820	7, 13	13, 13	13, 19	13, 13
D8S1179	13, 13	13, 15	12, 15	10, 12
D13S317	12, 12	10, 12	8, 10	12, 17
D16S539	12, 14	14, 12	14, 14	18, 25
D18S51	5, 6	6, 22	22, 6	5, 22
D21S11	15, 17	17, 22	15, 22	22, 22

Application ● 11.1   **Reflections of a Distant Past**

How do you think about time? Perhaps you can conceive of a few hundred years of human events, maybe a few thousand, but how about a few million? Envisioning the very distant past requires an intellectual leap from the familiar to the unknown. One way to make that leap involves, surprisingly, asteroids. Asteroids are small planets hurtling through space. They range in size from 1 to 1,500 kilometers (roughly 0.5 to 1,000 miles) across. Millions of them orbit the sun between Mars and Jupiter—cold, stony leftovers from the formation of our solar system. Asteroids do not emit light, so they are difficult to see even with the best telescopes. Thus, asteroids can pass very close to Earth undetected, and many are discovered as they do. Some have not passed us by at all.

Consider the mile-wide Barringer Crater in Arizona (Figure 11.1A). An asteroid 45 meters (150 feet) wide made this impressive pockmark in the desert sandstone when it slammed into Earth 50,000 years ago, its impact 150 times more powerful than the bomb that leveled Hiroshima. No humans were in North America at the time of the impact. If there were no witnesses, how is it possible to know anything about what happened? We often reconstruct history by studying physical evidence of past events. Geologists were able to infer the most probable cause of the Barringer Crater by analyzing tons of meteorites, melted sand, and other rocky clues at the site.

Similar evidence points to even larger impacts in the more distant past. For example, fossil hunters have long known about a mass extinction, or permanent loss of major groups of organisms, that occurred 66 million years ago. The event is marked by an unusual, worldwide formation of sedimentary rock (Figure 11.1B) called the K–Pg boundary sequence (it was formerly known as the K–T bound-

A. What made the Barringer Crater? Rocky evidence points to a 300,000-ton asteroid that collided with Earth 50,000 years ago.

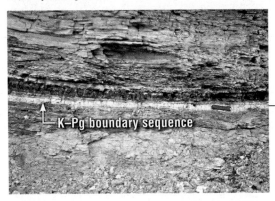
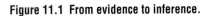

ary). There are plenty of dinosaur fossils below this formation. Above it, in layers of rock that were deposited more recently, there are no dinosaur fossils, anywhere. The formation consists of an unusual clay that is rich in iridium, an element rare on Earth's surface but common in asteroids. It also contains shocked quartz (left) and small glass spheres called tektites— rocks that form when quartz or sand (respectively) undergoes a sudden, violent application of extreme pressure. The only

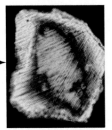

processes on Earth known to produce these minerals are nuclear explosions and asteroid impacts.

Geologists concluded that the K–Pg boundary layer must have originated with extraterrestrial material, and began looking for evidence of an asteroid that hit Earth 66 million years ago—one big enough to cover the entire planet with its debris. Twenty years later, they found it: an impact crater the size of Ireland off the coast of the Yucatán Peninsula. To make a crater this big, an asteroid 20 km (12 miles) wide would have slammed into Earth with a force *40 million times* more powerful than the one that made the Barringer Crater—enough to cause an ecological disaster of sufficient scale to wipe out almost all life on Earth.

You are about to make an intellectual leap through time, to places that were not even known a few centuries ago. We invite you to launch yourself from this premise: Natural phenomena that occurred in the past can be explained by the same physical, chemical, and biological processes that operate today. That premise is the foundation for scientific research into the history of life. The research represents a shift from experience to inference—from the known to what can only be surmised—and it gives us astonishing glimpses into the distant past.

B. The K–Pg boundary sequence is an unusual, world-wide formation of sedimentary rock that formed 66 million years ago. This rock formation marks an abrupt transition in the fossil record that implies a mass extinction of the dinosaurs. It also contains materials consistent with an asteroid impact. The impact of an asteroid big enough to blanket the Earth with its debris would have had catastrophic aftereffects on life at the time. The red pocketknife gives an idea of scale.

Figure 11.1   From evidence to inference.

A. Ostrich, native to Africa.

B. Rhea, native to South America.

C. Emu, native to Australia.

Figure 11.2 Similar-looking, related species native to distant geographic realms.
These birds are unlike most others in several unusual features, including long, muscular legs and an inability to fly. All are native to open grassland regions about the same distance from the equator.

(A) Rebecca Yale/Getty Images; (B) © Nico Stengert/Novarc Images/Alamy; (C) © Earl & Nazima Kowall/Corbis.

11.2 Confusing Discoveries

REMEMBER: Naming species in a consistent way became a priority in the 18th century (Section 1.4). Scientific discoveries may provoke controversy when a society's moral standards are interwoven with its understanding of nature (1.6).

About 2,300 years ago, the Greek philosopher Aristotle described nature as a continuum of organization, from lifeless matter through complex plants and animals. Aristotle's work greatly influenced later European thinkers, who adopted his view of nature and modified it in light of their own beliefs. By the fourteenth century, Europeans generally believed that a "great chain of being" extended from the lowest form of life (plants), up through animals, humans, and spiritual beings. Each link in the chain was a species, and each was said to have been forged at the same time, in one place, and in a perfect state. The chain was complete. Because everything that needed to exist already did, there was no room for change.

In the 1800s, European naturalists embarked on globe-spanning survey expeditions and brought back tens of thousands of plants and animals from Asia, Africa, North and South America, and the Pacific Islands. Each newly discovered species was carefully catalogued as another link in the chain of being. The explorers began to see patterns in where species live and similarities in body plans, and had started to think about natural forces that shape life. These explorers were pioneers in **biogeography**, the study of patterns in the geographic distribution of species and communities. Some of the patterns raised questions that could not be answered within the framework of prevailing belief systems. For example, globe-trotting explorers had discovered plants and animals living in extremely isolated places. The isolated species looked suspiciously similar to species living on the other side of impassable mountain ranges, or across vast expanses of open ocean. Consider the emu, rhea, and ostrich, three types of bird native to three different continents. These birds share a set of unusual features (Figure 11.2). Alfred Wallace, an explorer particularly interested in the geographical distribution of animals, thought that the shared traits might mean that the birds descended from a common ancestor (and he was correct), but he had no idea how they could have ended up on different continents.

Naturalists of the time also had trouble classifying organisms that are very similar in some features, but different in others. For example, both plants shown in Figure 11.3 live in hot deserts where water is seasonally scarce. Both have rows of sharp spines that deter herbivores, and both store water in their thick, fleshy stems.

Figure 11.3 Similar-looking, unrelated species.
On the left, saguaro cactus (*Carnegiea gigantea*), native to the Sonoran Desert of Arizona. On the right, an African milk barrel cactus (*Euphorbia horrida*), native to the Great Karoo desert of South Africa.

Left, © Marka/SuperStock; right, © Richard J. Hodgkiss, www.succulent-plant.com.

biogeography Study of patterns in the geographic distribution of species and communities.

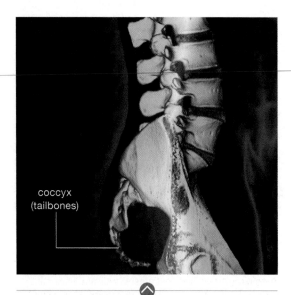

coccyx
(tailbones)

Figure 11.4 A vestigial structure: human tailbones.
Nineteenth-century naturalists were well aware of—but had trouble explaining—body structures such as human tailbones that had apparently lost most or all function.

© Zephyr/Science Photo Library/Science Source.

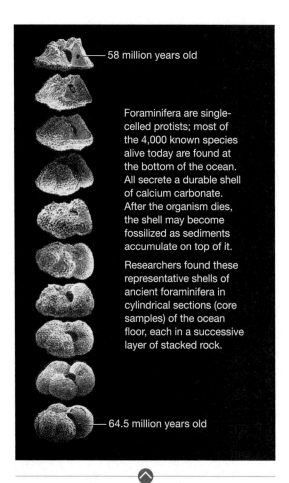

— 58 million years old

Foraminifera are single-celled protists; most of the 4,000 known species alive today are found at the bottom of the ocean. All secrete a durable shell of calcium carbonate. After the organism dies, the shell may become fossilized as sediments accumulate on top of it.

Researchers found these representative shells of ancient foraminifera in cylindrical sections (core samples) of the ocean floor, each in a successive layer of stacked rock.

— 64.5 million years old

Figure 11.5 Sequence of ten fossil foraminifera.
Courtesy of Daniel C. Kelley, Anthony J. Arnold, and William C. Parker, Florida State University Department of Geological Science.

However, their reproductive parts are very different, so these plants cannot be (and are not) as closely related as their outward appearance might suggest.

Observations such as these are examples of **comparative morphology**, the study of anatomical patterns: similarities and differences among the body plans of organisms. Today, comparative morphology is only one branch of taxonomy, but in the nineteenth century it was the only way to distinguish species. In some cases, comparative morphology revealed anatomical details (body parts with no apparent function, for example) that added to the mounting confusion. If every species had been created in a perfect state, then why were there useless parts such as wings in birds that do not fly, eyes in moles that are blind, or remnants of a tail in humans (Figure 11.4)?

Fossils were puzzling too. A **fossil** is physical evidence—remains or traces—of an organism that lived in the ancient past. Geologists mapping rock formations exposed by erosion or quarrying had discovered identical sequences of rock layers in different parts of the world. Deeper layers held fossils of simple marine life. Layers above those held similar but more complex fossils (Figure 11.5). In higher layers, fossils that were similar but even more complex resembled modern species. What did these sequences mean? Fossils of many animals unlike any living ones were also being unearthed. If these animals had been perfect at the time of creation, then why had they become extinct?

Taken as a whole, the accumulating discoveries from biogeography, comparative morphology, and geology did not fit with prevailing beliefs of the nineteenth century. If species had not been created in a perfect state (and extinct species, fossil sequences, and "useless" body parts implied that they had not), then perhaps species had indeed changed over time.

Take-Home Message 11.2

Why did observations of nature change our thinking in the nineteenth century?

- Increasingly extensive observations of nature in the nineteenth century did not fit with prevailing belief systems.
- Cumulative findings from biogeography, comparative morphology, and geology led naturalists to question traditional ways of interpreting the natural world.

11.3 A Flurry of New Ideas

REMEMBER: Epigenetic modifications of DNA that alter gene expression may persist for generations (Section 7.7). Paired genes on homologous chromosomes may vary as alleles; alleles are the basis of differences in traits shared by a species (8.4). The environment influences gene expression, and therefore affects phenotype (9.5).

Squeezing New Evidence Into Old Beliefs In the nineteenth century, naturalists were faced with increasing evidence that life on Earth, and even Earth itself, had changed over time. Around 1800, Georges Cuvier (left), an expert in zoology and paleontology, was trying to make sense of the new information. He knew that many fossil species seemed to have no living counterparts. Given this evidence, he proposed an idea

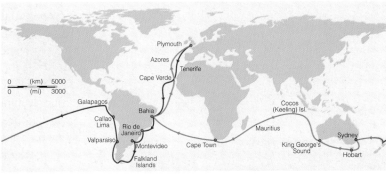

Figure 11.6 Charles Darwin and the *Beagle*.
With Darwin (left) aboard as ship's naturalist, the *Beagle* (middle) originally set sail to map the coast of South America, but ended up circumnavigating the globe. Right, the path of the voyage is shown from red to blue.

Darwin's detailed observations of the geology, fossils, plants, and animals he encountered on this expedition changed the way he thought about evolution.

Left, painting by George Richmond; middle, © Gordon Chancellor.

startling for the time: Many species that had once existed were now extinct. Cuvier also knew about evidence that Earth's surface had changed. For example, he had seen fossilized seashells in rocks at the tops of mountains far from modern seas. Like most others of his time, he assumed Earth's age to be in the thousands, not billions, of years. He reasoned that geologic forces unlike any known at the time would have been necessary to raise seafloors to mountaintops in such a short time span. These catastrophic geological events would have caused extinctions, after which surviving species repopulated the planet. Cuvier's idea came to be known as catastrophism. We now know that catastrophism is incorrect; geologic processes have not changed over time.

Another naturalist, Jean-Baptiste Lamarck, was thinking about processes that might drive **evolution**, or change in a line of descent. A line of descent is also called

a **lineage**. Lamarck (left) thought that a species gradually improved over generations because of an inherent drive toward perfection, up the chain of being. The drive directed an unknown "fluida" into body parts needing change. By Lamarck's hypothesis, environmental pressures cause an internal requirement for change in an individual's body, and the resulting change is inherited by offspring. Imagine using Lamarck's hypothesis to explain why a giraffe's neck is very long. You might predict that some short-necked ancestor of the modern giraffe stretched its neck to browse on leaves beyond the reach of other animals. The stretches may have even made its neck a bit longer. By Lamarck's hypothesis, the animal's offspring would inherit a longer neck. The modern giraffe would have been the result of many generations that strained to reach ever loftier leaves. Lamarck was correct in thinking that environmental factors affect traits, but his understanding of how inheritance works was incomplete.

Darwin and the HMS *Beagle* Lamarck's ideas about evolution influenced the thinking of Charles Darwin, who, at the age of 22, joined a survey expedition to South America on a ship called *Beagle*. Since he was eight years old, Darwin had wanted to hunt, fish, collect shells, or watch insects and birds—anything but sit in school. After a failed attempt to study medicine in college, he earned a degree in theology from Cambridge. During his studies, Darwin had spent most of his time with faculty members and other students who embraced natural history.

The *Beagle* set sail for South America in December 1831 (Figure 11.6). The young man who had hated school and had no formal training in science quickly became an enthusiastic naturalist. During the *Beagle*'s five-year voyage, Darwin found many unusual fossils, and saw diverse species living in environments that ranged from the sandy shores of remote islands to plains high in the Andes. Along

Accumulating discoveries in the nineteenth century did not fit with prevailing beliefs.

comparative morphology The scientific study of similarities and differences in body plans.

evolution Change in a line of descent.

fossil Physical evidence of an organism that lived in the ancient past.

lineage Line of descent.

the way, he read the first volume of a new and popular book, Charles Lyell's *Principles of Geology*. Lyell (left) was a proponent of what became known as the theory of uniformity, the idea that gradual, repetitive change had shaped Earth. For many years, geologists had been chipping away at the sandstones, limestones, and other types of rocks that form from accumulated sediments at the bottom of lakes, rivers, and oceans. These rocks held evidence that the gradual processes of geologic change operating in the present were the same ones that operated in the distant past.

By the theory of uniformity, strange catastrophes were not necessary to explain Earth's surface. Gradual, everyday geologic processes such as erosion by wind and water could have sculpted Earth's current landscape over great spans of time. This theory challenged the prevailing belief that Earth was 6,000 years old. According to traditional scholars, people had recorded everything that happened in those 6,000 years—and in all that time, no one had mentioned seeing a species evolve. However, by Lyell's calculations, it must have taken millions of years to sculpt Earth's surface. Darwin's exposure to Lyell's ideas gave him insights into the geologic history of the regions he would encounter on his journey. Was millions of years enough time for species to evolve? Darwin thought that it was.

A Key Insight—Variation in Traits Among the thousands of specimens Darwin collected on his voyage and sent to England were fossil glyptodons from Argentina. These armored mammals are extinct, but they have many traits in common with modern armadillos. Like glyptodons, armadillos have helmets and protective shells that consist of unusual bony plates (Figure 11.7). Armadillos also live only in places where glyptodons once lived. Could the shared traits and distribution mean that glyptodons were ancient relatives of armadillos? If so, perhaps traits of their common ancestor had changed in the line of descent that led to armadillos. But why would such changes have occurred?

After Darwin returned to England, he pondered his notes and fossils, and read

an essay by one of his contemporaries, economist Thomas Malthus. Malthus (left) had correlated increases in the size of human populations with episodes of famine, disease, and war. He proposed the idea that humans run out of food, living space, and other resources because they tend to reproduce beyond the capacity of their environment to sustain them. When that happens, the individuals of a population must either compete with one another for the limited resources, or develop new technologies to increase productivity. Darwin realized that Malthus's ideas had wider application: All populations, not just human ones, must have the capacity to produce more individuals than their environment can support.

Reflecting on his journey, Darwin started thinking about how individuals of a species often vary a bit in the details of shared traits such as size, coloration, and so on. He saw such variation among finch species on isolated islands of the Galápagos archipelago. This island chain is separated from South America by 900 kilometers (550 miles) of open ocean, so most species living on the islands did not have the opportunity for interbreeding with mainland populations. The Galápagos island birds resembled finch species in South America, but many had unique traits that suited their particular island habitat.

Darwin was familiar with dramatic variations in traits that selective breeding could produce in pigeons, dogs, and horses. He recognized that a natural environment could similarly select traits that make individuals of a population suited to it. It dawned on Darwin that having a particular form of a shared trait might give an

A. A modern armadillo, about a foot long.

B. Fossil of a glyptodon, an automobile-sized mammal that existed from 2 million to 15,000 years ago.

Figure 11.7 Ancient relatives: armadillo and glyptodon.
Even though these animals are widely separated in time, they share a restricted distribution and unusual traits, including a shell and helmet of keratin-covered bony plates—a material similar to crocodile and lizard skin. (The fossil in B is missing its helmet.) Their similarities were a clue that helped Darwin develop the theory of evolution by natural selection.

(A) © John White; (B) 2004 Arent.

individual an advantage over competing members of its species. In any population, some individuals have forms of shared traits that make them better suited to their environment than others. In other words, individuals of a natural population vary in fitness. Today, we define **fitness** as the degree of adaptation to a specific environment, and measure it by relative genetic contribution to future generations. A form of a heritable trait that enhances an individual's fitness is called an **adaptive trait**, or evolutionary **adaptation**.

Over many generations, individuals that have adaptive traits tend to survive longer and reproduce more than their less fit rivals. Darwin understood that this process, which he called **natural selection**, could be a mechanism by which evolution occurs. If an individual has a form of a trait that makes it better suited to an environment, then it is better able to survive. If an individual is better able to survive, then it has a better chance of living long enough to produce offspring. If individuals with an adaptive form of a trait produce more offspring than those that do not, then the frequency of that form will tend to increase in the population over successive generations. Table 11.1 summarizes this reasoning in modern terms.

Great Minds Think Alike Darwin wrote out his ideas about natural selection,

The Natural History Museum/Alamy.
but let ten years pass without publishing them. In the meantime, Alfred Wallace (left), who had been studying wildlife in the Amazon basin and the Malay Archipelago, wrote an essay and sent it to Darwin for advice. Wallace's essay outlined evolution by natural selection—the very same theory as Darwin's. Wallace had written earlier letters to Darwin and Lyell about patterns in the geographic distribution of species, and had come to the same conclusion.

In 1858, just weeks after Darwin received Wallace's essay, the theory of evolution by natural selection was presented at a scientific meeting. Both Darwin and Wallace were credited as authors. Wallace was still in the field and knew nothing about the meeting, which Darwin did not attend. The next year, Darwin published *On the Origin of Species*, which laid out detailed evidence in support of the theory. Many people had already accepted the idea of descent with modification (evolution). However, there was a fierce debate over the idea that natural selection drives evolution. Decades would pass before experimental evidence from the field of genetics led to its widespread acceptance as a theory by the scientific community. As you will see in the remainder of this unit, the theory of evolution by natural selection is supported by and helps explain the fossil record as well as similarities and differences in the form, function, and biochemistry of living things.

Observations about populations
- Natural populations have an inherent capacity to increase in size over time.
- As a population expands, resources that are used by its individuals (such as food and living space) eventually become limited.
- When resources are limited, the individuals of a population compete for them.

Observations about genetics
- Individuals of a species share certain traits.
- Individuals of a natural population vary in the details of those shared traits.
- Shared traits have a heritable basis, in genes. Slightly different forms of those genes (alleles) give rise to variation in shared traits.

Inferences
- A certain form of a shared trait may make its bearer better able to survive.
- Individuals of a population that are better able to survive tend to leave more offspring.
- Thus, an allele associated with an adaptive trait tends to become more common in a population over time.

Table 11.1 Principles of Natural Selection

Take-Home Message 11.3

How did new evidence change the way people in the 19th century thought about the history of life?

- Evidence found in the 1800s led to the idea that Earth and the species on it had changed over very long spans of time. This idea set the stage for Darwin's theory of evolution by natural selection.
- Natural selection is a process that drives evolutionary change: Individuals of a population survive and reproduce with differing success depending on the details of their shared, heritable traits.
- Adaptive traits enhance fitness.

adaptive trait (adaptation) A form of a heritable trait that enhances an individual's fitness.

fitness Degree of adaptation to an environment, as measured by an individual's relative genetic contribution to future generations.

natural selection Differential survival and reproduction of individuals of a population based on differences in shared, heritable traits.

A. Fossil skeleton of an ichthyosaur that lived about 200 million years ago. These marine reptiles were about the same size as modern porpoises, breathed air like them, and probably swam as fast, but the two groups are not closely related.

B. Extinct wasp encased in amber, which is ancient tree sap. This 9-mm-long insect lived about 20 million years ago.

C. Fossilized leaf from a 260-million-year-old *Glossopteris*, a type of plant called a seed fern.

D. Ancient footprints of a theropod, a type of carnivorous dinosaur that arose 250 million years ago. *Tyrannosaurus rex* was a theropod.

E. Coprolite (fossilized feces). Fossilized food remains and parasitic worms in coprolites offer clues about the diet and health of extinct species. A foxlike animal excreted this one.

Figure 11.8 Examples of fossils.

(A) Jonathan Blair; (B) © Dr. Michael Engel, University of Kansas; (C) Martin Land/Science Source; (D) © Pixtal/SuperStock; (E) Courtesy of Stan Celestian/Glendale Community College Earth Science Image Archive; (in text) Natural History Museum, London/Science Photo Library/Science Source.

11.4 **Fossil Evidence**

REMEMBER: Atoms of a radioisotope have an unstable nucleus that breaks up spontaneously, emitting radiation in a process called radioactive decay (Section 2.2). Molecular oxygen had been a very small component of Earth's atmosphere before photosynthesis evolved (5.5).

Even before Darwin's time, fossils were recognized as stone-hard evidence of earlier forms of life. Most fossils consist of mineralized bones, teeth, shells, seeds, spores, or other durable body parts (Figure 11.8). Trace fossils such as footprints and other body impressions, nests, burrows, trails, eggshells, or feces, are evidence of an ancient organism's activities.

The process of fossilization typically begins when an organism or its traces become covered by sediments, mud, or ash. Groundwater then seeps into the remains, filling spaces around and inside of them. Minerals dissolved in the water gradually replace minerals in bones and other hard tissues. Mineral particles that settle out of the groundwater and crystallize inside cavities and impressions form detailed imprints of internal and external structures. Sediments that slowly accumulate on top of the site exert increasing pressure, and, after a very long time, extreme pressure transforms the mineralized remains into rock.

Most fossils are found in layers of sedimentary rock (Figure 11.9). Sedimentary rocks form as rivers wash silt, sand, volcanic ash, and other materials from land to sea. Mineral particles in the materials settle on seafloors in horizontal layers that often vary in thickness and composition. After hundreds of millions of years, the layers of sediments become compacted into layers of rock. Even though most sedimentary rock forms at the bottom of a sea, geologic processes can tilt the rock and lift it far above sea level, where the layers may become exposed by the erosive forces of water and wind.

Biologists study sedimentary rock formations in order to understand the historical context of ancient life. Usually, the deepest layers in a particular formation of rock were the first to form, and those closest to the surface formed most recently. Thus, in general, the deeper the layer, the older the fossils it contains. Features of the formations can provide information about local and global events that were occur-

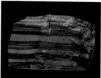

ring as they formed. Consider banded iron, a unique formation named after its distinctive striped appearance (left). Huge deposits of this sedimentary rock are the source of most iron we mine for steel today, but they also hold a record of how the evolution of photosynthesis changed the chemistry of Earth.

Banded iron started forming about 2.4 billion years ago, right after photosynthesis evolved. At that time, Earth's atmosphere and ocean contained very little oxygen, and ocean water contained a lot of dissolved iron. Oxygen released into the ocean by early photosynthetic bacteria quickly combined with the dissolved iron. The resulting iron compounds are completely insoluble in water, and they began to rain down on the ocean floor in massive quantities, forming layers of sediment that compacted into banded iron formations. This process continued for about 600 million years, until ocean water no longer contained very much dissolved iron.

The Fossil Record We have fossils for more than 250,000 known species. Considering the current range of biodiversity, there must have been many millions more, but we will never know all of them. Why not? The odds are against finding evidence of an extinct species, because fossils are relatively rare. Typically, when an organism dies, its body is quickly eaten by predators, scavengers, or both. Unconsumed remains decompose in the presence of moisture and oxygen, so they can only endure if they dry out, freeze, or become encased in an air-excluding material such as sap, tar, or mud. Fossils that do form are often crushed or scattered by erosion and other geologic assaults. In order for us to know about an extinct species that existed long ago, we have to find a fossil of it. At least one specimen had to escape being eaten, and be preserved before it decomposed. Its burial site also had to persist intact, and end up in a place that we can find today.

Most ancient species had no hard parts to fossilize, so we do not find much evidence of them. For example, there are many more fossils of bony fishes and hard-shelled mollusks than fossils of soft-bodied animals (such as jellyfishes and worms) that were probably much more numerous in life. Also think about relative numbers of organisms. Fungal spores and pollen grains are typically released by the billions. By contrast, the earliest humans lived in small bands and few of their offspring survived. The odds of finding even one fossilized human bone are much smaller than the odds of finding a fossilized fungal spore. Finally, imagine two species, one that existed only briefly and the other for billions of years. Which is more likely to be represented in the fossil record?

Despite these challenges, the fossil record is substantial enough to help us reconstruct large-scale patterns in the history of life.

Radiometric Dating Atoms of a radioisotope become atoms of other elements—daughter elements—as their nucleus disintegrates. This radioactive decay is not influenced by temperature, pressure, chemical bonding state, or moisture; it is influenced only by time. Thus, like the ticking of a perfect clock, each type of

Figure 11.9 Fossil hunting.
Fossil hunters found this fossilized trilobite (an ancient marine relative of centipedes) in a shale formation in Yoho National Park, British Columbia. Fossils are most often found in layered sedimentary rock. This type of rock forms over hundreds of millions of years, often at the bottom of a sea. Geologic processes can tilt the rock and lift it far above sea level, where the layers become exposed by the erosive forces of water and wind.

© Michael Melford/National Geographic Creative.

Digging Into Data

Abundance of Iridium in the K–Pg Boundary Layer

In the late 1970s, geologist Walter Alvarez was investigating the composition of the K–Pg boundary sequence in different parts of the world. He asked his father, Nobel Prize–winning physicist Luis Alvarez, to help him analyze the elemental composition of the layer.

The Alvarezes and their colleagues tested samples of the layer taken from formations in Italy and Denmark. The researchers discovered that the K–Pg boundary sequence contains a much higher iridium content than the surrounding rock layers. Some of their results are shown in Figure 11.10.

Iridium belongs to a group of elements that are much more abundant in asteroids and other solar system materials than they are in Earth's crust, so the Alvarez group concluded that the K–Pg boundary sequence must have originated with extraterrestrial material. They calculated that an asteroid about 14 kilometers (8.7 miles) in diameter would contain enough iridium to account for the extra iridium in the K–Pg boundary sequence.

1. What was the iridium content of the K–Pg boundary layer?
2. What was the difference in iridium content between the boundary layer and the sample taken 0.7 meters above the layer?

Sample Depth	Average Abundance of Iridium (ppb)*
+ 2.7 m	< 0.3
+ 1.2 m	< 0.3
+ 0.7 m	0.36
boundary layer	41.6
– 0.5 m	0.25
– 5.4 m	0.30

Figure 11.10 Abundance of iridium in and near the K–Pg boundary sequence.
Left, Luis and Walter Alvarez with a section of the boundary. The researchers tested the iridium content of many rock samples above, below, and at this boundary in different parts of the world. Right, some of the Alvarezes' results from Stevns Klint, Denmark. Sample depths are given as meters above or below the boundary.
* ppb, parts per billion.

The iridium content of an average Earth rock is 0.4 ppb. The average meteorite contains about 550 ppb iridium.

Lawrence Berkeley National Laboratory.

half-life Characteristic time it takes for half of a quantity of a radioisotope to decay.

radiometric dating Method of estimating the age of a rock or fossil by measuring the content and proportions of a radioisotope and its daughter elements.

radioisotope decays at a constant rate. The time it takes for half of the atoms in a sample of radioisotope to decay is called **half-life** (Figure 11.11). Half-life is a characteristic of each radioisotope. For example, radioactive uranium 238 decays into thorium 234, which decays into something else, and so on until it becomes lead 206. The half-life of the decay of uranium 238 to lead 206 is 4.5 billion years.

The predictability of radioactive decay can be used to find the age of a volcanic rock (the date it solidified). Rock forms from magma, which is a hot, molten material deep under Earth's surface. Atoms swirl and mix in this material. When magma cools, for example after reaching the surface as lava, it hardens and becomes rock. Different kinds of minerals crystallize in the rock as it hardens, each with a characteristic structure and composition. For example, the mineral called zircon consists mainly of orderly arrays of zirconium silicate molecules ($ZrSiO_4$). Some of the molecules in a newly formed zircon crystal have uranium atoms substituted for zirconium atoms, but never lead atoms. However, uranium decays into lead at a predictable rate. Thus, over time, uranium atoms disappear from a zircon crystal, and lead atoms accumulate in it. The ratio of uranium atoms to lead atoms in a zircon crystal can be measured precisely, and that ratio can be used to calculate how long ago the crystal formed (its age).

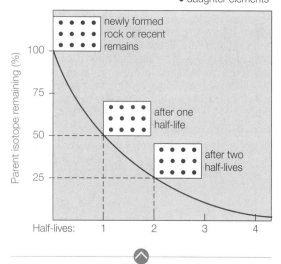

Figure 11.11 Half-life.

Figure It Out: How much of any radioisotope remains after two half-lives have passed? **Answer: 25 percent.**

Figure 11.12 The oldest Earth rock.
Dr. Simon Wilde holds a 4.4 billion-year-old speck of zircon embedded in protective plastic.

AP Images/Andy Manis.

Figure 11.13 Example of how radiometric dating can be used to find the age of a carbon-containing fossil.
Carbon 14 (^{14}C) is a radioisotope of carbon that decays into nitrogen. It forms in the atmosphere and combines with oxygen to become CO_2, which enters food chains by way of photosynthesis.

(A) © PhotoDisc/Getty Images.

A. Long ago, ^{14}C and ^{12}C were incorporated into the tissues of a nautilus. Both carbon isotopes were part of organic molecules in the animal's food. ^{12}C is stable and ^{14}C decays, but the proportion of the two isotopes in the nautilus's tissues remained the same. Why? The nautilus continued to gain both types of carbon atoms in the same proportions from its food.

B. The nautilus stopped eating when it died, so its body stopped gaining carbon. The ^{12}C atoms in its tissues were stable, but the ^{14}C atoms (represented as red dots) were decaying into nitrogen atoms. Thus, over time, the amount of ^{14}C decreased relative to the amount of ^{12}C. After 5,730 years, half of the ^{14}C had decayed; after another 5,730 years, half of what was left had decayed, and so on.

C. Fossil hunters discover the fossil and measure its content of ^{14}C and ^{12}C. They use the ratio of these isotopes to calculate how many half-lives have passed since the organism died. For example, if its ^{14}C to ^{12}C ratio is one-eighth of the ratio in living organisms, then three half-lives $(1/2)^3$ must have passed since it died. Three half-lives of ^{14}C is 17,190 years.

We have just described **radiometric dating**, a method that can reveal the age of a material by measuring its content of a radioisotope and daughter elements. The oldest known terrestrial rock, a tiny zircon crystal found in Australia, formed 4.4 billion years ago (Figure 11.12).

Fossils that still contain carbon can be radiometrically dated by measuring the ratio of carbon isotopes in them (Figure 11.13). Most of the ^{14}C in a fossil will have decayed after about 60,000 years. The age of fossils older than that can be estimated by dating volcanic rock formations above and below the fossil-containing layer.

Missing Links The discovery of intermediate forms of cetaceans (an order of animals that includes whales, dolphins, and porpoises) offers an example of how fossil finds and radiometric dating can be used to reconstruct evolutionary history. For some time, evolutionary biologists had thought that the ancestors of modern cetaceans walked on land, then took up life in the water. Evidence in support of this line of thinking includes a set of distinctive features of the skull and lower jaw that cetaceans share with some kinds of ancient carnivorous land animals. DNA sequence comparisons indicate that the ancient land animals were probably artiodactyls, hooved mammals with an even number of toes (two or four) on each foot (Figure 11.14A). Modern representatives of the artiodactyl lineage include camels, hippopotamuses, pigs, deer, sheep, and cows.

Until recently, no one had discovered fossils demonstrating gradual changes in skeletal features that would have accompanied a transition of whale lineages from terrestrial to aquatic life. These intermediate forms had to exist, because a representative fossil skull of an ancient whalelike animal had been discovered, but without a complete skeleton the rest of the story remained speculative.

Then, in 2000, Philip Gingerich and his colleagues unearthed complete fossilized skeletons of two ancient whales. They found *Rodhocetus kasrani* (Figure 11.14B) embedded in a 47-million-year-old rock formation in Pakistan, and *Dorudon atrox* (Figure 11.14C), in a 37-million-year-old formation in Egypt. Both fossil skeletons had whalelike skull bones, as well as intact ankle bones. The ankle bones of both fossils have distinctive features in common with those of extinct and modern artiodactyls. Modern cetaceans do not have even a remnant of an ankle bone (Figure 11.14D).

The proportions of limbs, skull, neck, and thorax indicate *Rodhocetus* swam with its feet, not its tail. Like modern whales, *Dorudon* was clearly a fully aquatic tail-swimmer: The entire hind limb was only about 12 centimeters (5 inches) long, much too small to have supported the animal's 5-meter (16-foot) body out of water.

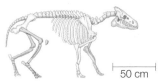

A. *Elomeryx*, a small terrestrial animal that lived about 30 million years ago. This is a member of the same artiodactyl group (even-toed hooved mammals) that gave rise to modern representatives, including hippopotamuses. *Elomeryx* is thought to resemble a 60-million-year-old ancestor that it shares with whales.

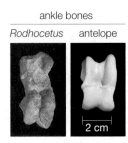

B. *Rodhocetus kasrani*, an ancient whale that lived about 47 million years ago. Its distinctive ankle bones are evidence of a close evolutionary connection to artiodactyls. Artiodactyls are defined by the unique "double-pulley" shape of the bone (right) that forms the lower part of their ankle joint.

ankle bones
Rodhocetus   antelope

C. *Dorudon atrox*, an ancient whale that lived about 37 million years ago. Its tiny, artiodactyl-like ankle bones were much too small to have supported the weight of its body on land, so this animal had to be fully aquatic.

D. Modern cetaceans such as the sperm whale have remnants of a pelvis and leg, but no ankle bones.

Figure 11.14 Comparison of cetacean skeletons.
The ancestor of whales was an artiodactyl that walked on land. Over millions of years, the lineage transitioned from life on land to life in water, and as it did, bones of the hind limb (highlighted in blue) became smaller.

(B left and right) © Phillip Gingerich, University of Michigan.

Take-Home Message 11.4

What do rocks and fossils have to do with biology?

- Sedimentary rock holds evidence of the historical context of fossils embedded in it.
- Fossils are a stone-hard historical record of ancient life. The fossil record will never be complete, but even so it is substantial enough to help us reconstruct patterns and trends in the history of life.
- The predictability of radioisotope decay can be used to estimate the age of rock layers and fossils in them. Radiometric dating helps evolutionary biologists retrace changes in ancient lineages.

geologic time scale Chronology of Earth's history.

Gondwana Supercontinent that existed before Pangea, more than 500 million years ago.

Pangea Supercontinent that began to form about 300 million years ago; broke up 100 million years later.

plate tectonics theory Theory that Earth's outermost layer of rock is cracked into plates, the slow movement of which conveys continents to new locations over geologic time.

11.5 Drifting Continents

REMEMBER: A hypothesis is a testable explanation for a natural phenomenon (Section 1.5). The scientific community consists of critically thinking people trying to poke holes in one another's ideas (1.6).

Wind, water, and other erosive forces continuously sculpt Earth's surface, but they are only part of a much bigger picture of geological change. Earth itself also changes dramatically. Consider that all continents on Earth today were once part of a bigger supercontinent—**Pangea**—that split into fragments and drifted apart. The idea that continents move around, originally called continental drift, was proposed in the early 1900s to explain why the Atlantic coasts of South America and Africa seem to "fit" like jigsaw puzzle pieces, and why the same types of fossils occur in identical rock formations on both sides of the Atlantic Ocean. It also explained why the magnetic poles of gigantic rock formations point in different directions on different continents. As magma solidifies into rock, some iron-rich minerals in it become magnetic, and their magnetic poles align with Earth's poles when they do. If the continents never moved, then all of these ancient rocky magnets should be aligned

Figure 11.15 Plate tectonics.
Huge pieces of Earth's outer layer of rock slowly drift apart and collide. As these plates move, they convey continents around the globe.

Right, © Kevin Schafer/Corbis.

1 At oceanic ridges, magma (red) welling up from Earth's interior drives the movement of tectonic plates. New crust spreads outward as it forms on the surface, forcing adjacent tectonic plates away from the ridge and into trenches elsewhere.

2 At trenches, the advancing edge of one plate plows under an adjacent plate and buckles it.

3 Faults are ruptures in Earth's crust where plates meet. The diagram shows a rift fault, in which two plates move apart. The aerial photo on the right shows a strike-slip fault, in which two abutting plates slip against one another in opposite directions.

4 Magma ruptures a tectonic plate at what are called "hot spots." The Hawaiian Islands have been forming from magma that continues to erupt at a hot spot under the Pacific Plate.

The San Andreas Fault, which extends 800 miles through California, marks the boundary between two tectonic plates.

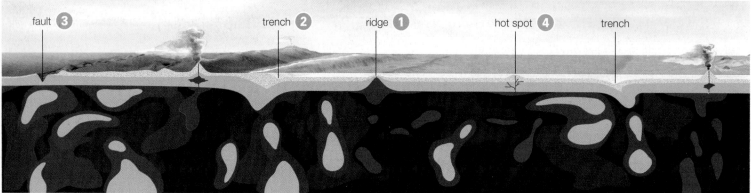

fault **3**   trench **2**   ridge **1**   hot spot **4**   trench

north-to-south, like compass needles. Indeed, the magnetic poles of rocks in each formation are aligned with one another, but the alignment is not always north-to-south. Either Earth's magnetic poles veer dramatically from their north–south axis, or the continents wander.

Continental drift was initially greeted with intense skepticism because there was no known mechanism for continents to move. Then, in the late 1950s, deep-sea explorers found immense ridges and trenches stretching thousands of kilometers across the seafloor. The discovery led to the **plate tectonics theory**, which explains how continents move (Figure 11.15). By this theory, Earth's outer layer of rock is cracked into huge plates, like a gigantic cracked eggshell. Magma welling up at an undersea ridge **1** or continental rift at one edge of a plate pushes old rock at the opposite edge into a trench **2**. The movement is like that of a colossal conveyor belt that slowly transports continents on top of it to new locations. The plates move no more than 10 centimeters (4 inches) a year—about half as fast as your toenails grow—but it is enough to carry a continent all the way around the world after 40 million years or so.

Evidence of tectonic movement is all around us, in faults **3** and other geological features of our landscapes. For example, volcanic island chains (archipelagos) form as a plate moves across an undersea hot spot. These hot spots are places where magma ruptures a tectonic plate **4**.

The fossil record also provides evidence in support of plate tectonics. Consider an unusual geological formation that occurs in a belt across Africa. The sequence of rock layers in this formation is so complex that it is quite unlikely to have formed more than once, but identical sequences also occur in huge belts that span India, South America, Madagascar, Australia, and Antarctica. Across all of these continents, the layers are the same ages. They also hold fossils found nowhere else, including remains of the seed fern *Glossopteris* (pictured in Figure 11.8C), which lived 299–252 million years ago, and an early reptile called *Lystrosaurus* that existed 270–225 million years ago. The unusual layered rock formation that contains fossils of these organisms probably formed in one long belt on a single continent that later broke up.

We have evidence of at least five supercontinents that formed and split up again since Earth's outer layer of rock solidified 4.55 billion years ago. One of them, a supercontinent called **Gondwana**, formed about 600 million years ago. Over the next 300 million years, Gondwana wandered across the South Pole, then drifted north until it merged with another supercontinent to form Pangea 300 million years ago (Figure 11.16). Most of the landmasses currently in the Southern Hemisphere as well as India and Arabia were once part of Gondwana. Some modern species, including the birds pictured in Figure 11.2, live only in these places.

Geologic changes brought on by plate tectonics would have had a profound impact on life. For example, when two continents collided into one, they brought together organisms that had been living apart on the separate landmasses, and physically separated organisms living in an ocean. When one continent broke up, organisms living on it would have been separated, and those living in different parts of the ocean would have come together. Events like these have been a major driving force of evolution, as you will see in the next chapter.

Putting Time Into Perspective Similar sequences of sedimentary rock layers occur around the world. Transitions between the layers mark boundaries between great intervals of time in the **geologic time scale**, which is a chronology of Earth's history. Each layer's composition offers clues about conditions on Earth during the

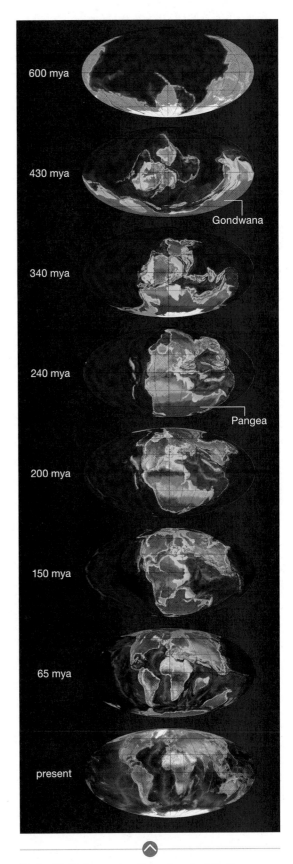

Figure 11.16 Reconstructions of ancient Earth.
mya: million years ago.

Eon	Era	Period	Epoch	mya*	Major Geologic and Biological Events
Phanerozoic	Cenozoic	Quaternary	Holocene	0.01	Modern humans evolve. Major extinction event is now under way.
			Pleistocene	2.6	
		Neogene	Pliocene	5.3	Tropics, subtropics extend poleward. Climate cools; dry woodlands and grasslands emerge. Adaptive radiations of mammals, insects, birds.
			Miocene	23.0	
		Paleogene	Oligocene	33.9	
			Eocene	56.0	
			Paleocene	66.0 ◄	**Major extinction event**
	Mesozoic	Cretaceous	Upper		Flowering plants diversify; sharks evolve. All dinosaurs and many marine organisms disappear at the end of this epoch.
				100.5	
			Lower		Climate very warm. Dinosaurs continue to dominate. Important modern insect groups appear (bees, butterflies, termites, ants, and herbivorous insects including aphids and grasshoppers). Flowering plants originate and become dominant land plants.
				145.0	
		Jurassic			Age of dinosaurs. Lush vegetation; abundant gymnosperms and ferns. Birds appear. Pangea breaks up.
				201.3 ◄	**Major extinction event**
		Triassic			Recovery from the major extinction at end of Permian. Many new groups appear, including turtles, dinosaurs, pterosaurs, and mammals.
				252 ◄	**Major extinction event**
	Paleozoic	Permian			Supercontinent Pangea and world ocean form. Adaptive radiation of conifers. Cycads and ginkgos appear. Relatively dry climate leads to drought-adapted gymnosperms and insects such as beetles and flies.
				299	
		Carboniferous			High atmospheric oxygen level fosters giant arthropods. Spore-releasing plants dominate. Age of great lycophyte trees; vast coal forests form. Ears evolve in amphibians; penises evolve in early reptiles (vaginas evolve later, in mammals only).
				359 ◄	**Major extinction event**
		Devonian			Land tetrapods appear. Explosion of plant diversity leads to tree forms, forests, and many new plant groups including lycophytes, ferns with complex leaves, seed plants.
				419	
		Silurian			Radiations of marine invertebrates. First appearances of land fungi, vascular plants, bony fishes, and perhaps terrestrial animals (millipedes, spiders).
				443 ◄	**Major extinction event**
		Ordovician			Major period for first appearances. The first land plants, fishes, and reef-forming corals appear. Gondwana moves toward the South Pole and becomes frigid.
				485	
		Cambrian			Earth thaws. Explosion of animal diversity. Most major groups of animals appear (in the oceans). Trilobites and shelled organisms evolve.
				541	
Precambrian	Proterozoic				Oxygen accumulates in atmosphere. Origin of aerobic metabolism. Origin of eukaryotic cells, then protists, fungi, plants, animals. Evidence that Earth mostly freezes over in a series of global ice ages between 750 and 600 mya.
				2,500	
	Archean and earlier				3,800–2,500 mya. Origin of bacteria and archaea.
					4,600–3,800 mya. Origin of Earth's crust, first atmosphere, first seas. Chemical, molecular evolution leads to origin of life (from protocells to anaerobic single cells).

Figure 11.17 The geologic time scale (above) correlated with sedimentary rock exposed by erosion in the Grand Canyon (opposite).
Red triangles mark times of great mass extinctions. "First appearance" refers to appearance in the fossil record, not necessarily first appearance on Earth.
* mya: million years ago. Dates are from the International Commission on Stratigraphy, 2014.

Photo, © Michael Pancier.

Figure It Out: Which formation is marked with the in the photo?
Answer: Tapeats sandstone

time the layer was deposited. Fossils in the layers are a record of life during that period of time (Figure 11.17).

Take-Home Message 11.5

How has Earth changed over geological time?

• Over geologic time, movements of Earth's crust have caused dramatic changes in continents and oceans. The changes profoundly influenced the course of life's evolution.

Ka Kaibab Limestone

To Toroweap Formation

Permian

Co Coconino Sandstone

He Hermit Shale

Es Esplanade Sandstone

We Wescogame Formation

Carboniferous

Ma Manakacha Formation

Wa Watahomigi Formation

Re Redwall Limestone

Te Temple Butte Formation

Mu Muav Limestone

Cambrian

Br Bright Angel Shale

Ta Tapeats Sandstone

Proterozoic

*Chuar Group

Nankoweap Formation

*Unkar Group

Vi Vishnu Basement Rocks

Layers not visible in this view of the Grand Canyon

Each rock layer has a composition and set of fossils that reflect events during its deposition. For example, Coconino Sandstone, which stretches from California to Montana, is mainly weathered sand. Ripple marks and reptile tracks are the only fossils in it. Many think it is the remains of a vast sand desert, similar to the modern Sahara.

11.6 Evidence in Form

Evolutionary biologists are a bit like detectives, using clues to piece together history that no human witnessed. Fossils provide some clues. The body form and function of organisms that are alive today provide others.

Morphological Divergence Comparative morphology can be used to unravel evolutionary relationships in many cases. Body parts that appear similar in separate lineages because they evolved in a common ancestor are called **homologous structures** (*hom–* means "the same"). Homologous structures may be used for different purposes in different groups, but the same genes direct their development.

A body part that outwardly appears very different in separate lineages may be homologous in underlying form. Vertebrate forelimbs, for instance, vary in size, shape, and function. However, they are alike in the structure and positioning of internal elements such as bones, nerves, blood vessels, and muscles.

Populations that are not interbreeding tend to diverge genetically, and in time these genetic divergences give rise to changes in body form. Change from the body form of a common ancestor is an evolutionary pattern called **morphological divergence**. Consider the limb bones of modern vertebrate animals. Fossil evidence suggests that many vertebrates are descended from a family of ancient "stem reptiles" that crouched low to the ground on five-toed limbs. Descendants of this ancestral group diversified over millions of years, and eventually gave rise to modern reptiles,

Figure 11.18 Morphological divergence among vertebrate forelimbs.
The number and position of many skeletal elements were preserved when these diverse forms evolved from a stem reptile; notice the bones of the forearms. Certain bones were lost over time in some of the lineages (compare the digits numbered 1 through 5). Drawings are not to scale.

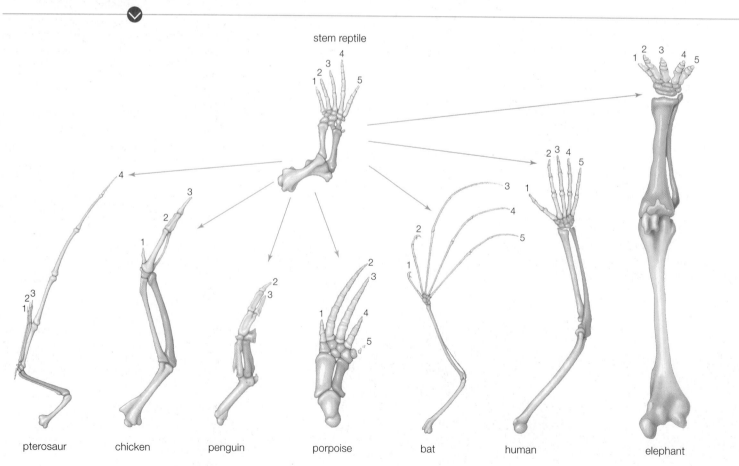

birds, and mammals. As you learned in Section 11.4, a few lineages even returned to life in the seas. As these lineages diversified, their five-toed limbs became adapted for appropriate purposes (Figure 11.18). The limbs became modified for flight in extinct reptiles called pterosaurs and in bats and most birds. In penguins and cetaceans, they are now flippers useful for swimming. Human forelimbs are arms and hands with four fingers and an opposable thumb. Elephant limbs are strong and pillarlike, capable of supporting a great deal of weight. The five-toed limb degenerated to nubs in pythons and boa constrictors, and disappeared entirely in other snakes.

Morphological Convergence Body parts that appear similar in different species are not always homologous; they sometimes evolve independently in lineages subject to the same environmental pressures. The independent evolution of similar body parts in different lineages is an evolutionary pattern called **morphological convergence**. Structures that are similar as a result of morphological convergence are **analogous structures**, which look alike but did not evolve in a shared ancestor; they evolved independently after the lineages diverged. For example, bird, bat, and insect wings all perform the same function, which is flight. However, several clues tell us that the wing surfaces are not homologous. All of the wings are adapted to the same physical constraints that govern flight, but each is adapted in a different way. In the case of birds and bats, the limbs themselves are homologous, but the adaptations that make those limbs useful for flight differ. The surface of a bat wing is a thin, membranous extension of the animal's skin. By contrast, the surface of a bird wing is a sweep of feathers, which are specialized structures derived from skin. Insect wings differ even more. An insect wing forms as a saclike extension of the body wall. Except at forked veins, the sac flattens and fuses into a thin membrane. The sturdy, chitin-reinforced veins structurally support the wing. Unique adaptations for flight are evidence that wing surfaces of birds, bats, and insects are analogous structures that evolved after the ancestors of these modern groups diverged (Figure 11.19A).

As another example of morphological convergence, consider the similar external structures of the African euphorbia and American cactus shown in Figure 11.3. These structures adapt the plants to similarly harsh desert environments where rain is scarce. Accordion-like pleats allow the plant body to swell with water when rain does come; water stored in the plants' tissues allows them to survive long dry periods. As the stored water is used, the plant body shrinks, and the folded pleats provide some shade in an environment that typically has none. Despite these similarities, a closer look reveals differences that indicate the two types of plants are not closely related (Figure 11.19B). For example, cactus spines have a simple fibrous structure; they are modified leaves that arise from dimples on the plant's surface. Euphorbia spines project smoothly from the plant surface, and they are not modified leaves: In many species the spines are dried flower stalks.

A. The flight surfaces of an insect wing, a bat wing, and a bird wing are analogous structures. The diagram shows how the evolution of wings (red dots) occurred independently in the three separate lineages. You will read more about diagrams that show evolutionary relationships in Section 12.8.

B. Spines of a saguaro cactus (left) differ from those of an African milk barrel cactus (a type of *Euphorbia*, right). This and other differences indicate the two plants are not closely related. Their similar appearances (see Figure 11.3) are an outcome of morphological convergence.

Figure 11.19 Examples of morphological convergence.

Take-Home Message 11.6

What evidence does evolution leave in body form?

• Body parts are often modified differently in different lines of descent.

• Body parts that appear alike may have evolved independently in lineages that have faced similar environmental pressures.

11.7 Evidence in Function

REMEMBER: A statistically significant result is very unlikely to have occurred by chance alone (Section 1.6). Mitochondria have their own DNA and divide independently of the cell (3.5). In eukaryotes, the third stage of aerobic respiration—electron transfer phosphorylation—occurs at the inner mitochondrial membrane (5.5). Most codons specify an amino acid (7.4). Some mutations have no effect; those that alter a gene product can have drastic consequences (7.6). Embryonic development is orchestrated by layers of master gene expression; expression of a homeotic gene directs the formation of a specific body part (7.7). Members of a species have the same traits because they have the same genes (8.4). All organisms are descended from shared ancestors, so all genomes (10.2) are related to some extent (10.3).

The genomes of closely related species tend to be more similar than those of distantly related ones.

Over time, inevitable mutations change a genome's DNA sequence. Most of these mutations are neutral—they have no effect on an individual's survival or reproduction—and they alter the DNA of each lineage independently of all other lineages. The more recently two lineages diverged, the less time there has been for unique mutations to accumulate in the DNA of each one. That is why the genomes of closely related species tend to be more similar than those of distantly related ones— a general rule that can be used to estimate relative times of divergence. Thus, similarities in the nucleotide sequence of a shared gene (or in the amino acid sequence of a shared protein) are often used as evidence of an evolutionary relationship. Biochemical comparisons like these are often combined with morphological comparisons, in order to provide data for hypotheses about shared ancestry.

Two species with very few similar proteins probably have not shared an ancestor for a long time—long enough for many mutations to have accumulated in the DNA of their separate lineages. Evolutionary biologists often compare a protein's sequence among several species, and use the number of amino acid differences as a measure of relative relatedness (Figure 11.20).

Among species that diverged relatively recently, many proteins have identical amino acid sequences. Nucleotide sequence differences may be instructive in such cases. Even if the amino acid sequence of a protein is identical among species, the nucleotide sequence of the gene that encodes the protein may differ because of redundancies in the genetic code. The DNA from nuclei, mitochondria, or chloroplasts can be used in nucleotide comparisons. Mitochondrial DNA accumulates mutations faster than nuclear DNA, so it can even be used to compare different individuals of the same sexually reproducing animal species. In most animals, mitochondria are inherited intact from a single parent (usually the mother). Thus, in most cases, differences in mitochondrial DNA sequences between maternally related individuals are due to mutations.

Figure 11.20 Example of a protein comparison. Here, part of the amino acid sequence of the same protein from 19 species is aligned. This protein, cytochrome *b*, is a component of mitochondrial electron transfer chains. The honeycreeper sequence is identical in ten species; amino acids that differ in the other species are shown in red. Dashes are gaps in the alignment.

Figure It Out: By this comparison, which species is most closely related to honeycreepers?

Answer: The song sparrow

honeycreepers (10)	...CRDVQFGWLIRNLHANGASFFFICIYLHIGRGIYYGSYLNK--ETWNIGVILLLTLMATAFVGYVLPWGQMSFWG...
song sparrow	...CRDVQFGWLIRNLHANGASFFFICIYLHIGRGIYYGSYLNK--ETWNVGIILLLALMATAFVGYVLPWGQMSFWG...
Gough Island finch	...CRDVQFGWLIRNIHANGASFFFICIYLHIGRGLYYGSYLYK--ETWNVGVILLLTLMATAFVGYVLPWGQMSFWG...
deer mouse	...CRDVNYGWLIRYMHANGASMFFICLFLHVGRGMYYGSYTFT--ETWNIGIVLLFAVMATAFMGYVLPWGQMSFWG...
Asiatic black bear	...CRDVHYGWIIRYMHANGASMFFICLFMHVGRGLYYGSYLLS--ETWNIGIILLFTVMATAFMGYVLPWGQMSFWG...
bogue (a fish)	...CRDVNYGWIIRNLHANGASFFFICIYLHIGRGLYYGSYLYK--ETWNIGVVLLLLVMGTAFVGYVLPWGQMSFWG...
human	...TRDVNYGWIIRYLHANGASMFFICLFLHIGRGLYYGSFLYS--ETWNIGIILLLATMATAFMGYVLPWGQMSFWG...
thale cress (a plant)	...MRDVEGGWLLRYMHANGASMFLIVVYLHIFRGLYHASYSSPREFVWCLGVVIFLLMIVTAFIGYVLPWGQMSFWG...
baboon louse	...ETDVMNGWMVRSIHANGASWFFIMLYSHIFRGLWVSSFTQP--LVWLSGVIILFLSMATAFLGYVLPWGQMSFWG...
baker's yeast	...MRDVHNGYILRYLHANGASFFFMVMFMHMAKGLYYGSYRSPRVTLWNVGVIIFTLTIATAFLGYCCVYGQMSHWG...

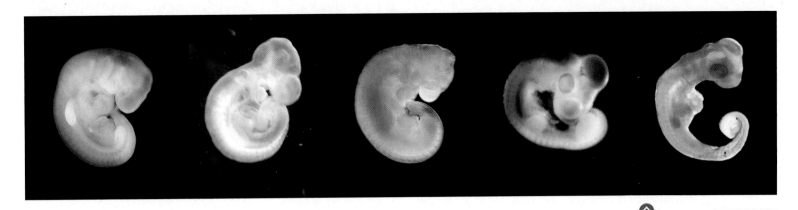

Figure 11.21 Comparing vertebrate embryos.
All vertebrates go through an embryonic stage in which they have four limb buds, a tail, and divisions called somites along their back. From left to right: human, mouse, bat, chicken, alligator.

From left, © Lennart Nilsson/Bonnierforlagen AB; Courtesy of Anna Bigas, IDIBELL-Institut de Recerca Oncologica, Spain; From *Embryonic staging system for the short-tailed fruit bat, Carollia perspicillata, a model organism for the mammalian order Chiroptera, based upon timed pregnancies in captive-bred animals.* C.J. Cretekos et al., Developmental Dynamics Volume 233, Issue 3, July 2005, Pages: 721–738. Reprinted with permission of Wiley-Liss, Inc. a subsidiary of John Wiley & Sons, Inc.; Courtesy of Prof. Dr. G. Elisabeth Pollerberg, Institut für Zoologie, Universität Heidelberg, Germany; USGS.

Getting useful information from comparing DNA requires a lot more data than comparing proteins. This is because coincidental homologies are statistically more likely to occur with DNA comparisons—there are only four nucleotides in DNA versus twenty amino acids in proteins. However, DNA sequencing has become so fast that there is a lot of data available to compare. Genomics studies with such data have shown us (for example) that about 88 percent of the mouse genome sequence is identical with the human genome, as is 73 percent of the zebrafish genome, 47 percent of the fruit fly genome, and 25 percent of the rice genome.

Patterns in Animal Development In general, the more closely related animals are, the more similar is their development. For example, all vertebrates go through a stage during which a developing embryo has four limb buds, a tail, and a series of [   ]—divisions of the body that give rise to the backbone and associated skin [mus]cle (Figure 11.21). Animals have similar patterns of embryonic develop[ment bec]ause the very same master genes direct the process. Because a mutation in [a master g]ene can unravel development completely, these genes tend to be highly [conserved.] Even among lineages that diverged a very long time ago, many master [genes retain] similar sequences and functions.

[If the s]ame genes direct development in all vertebrate lineages, how do the adult [forms end u]p so different? Part of the answer is that there are differences in the tim[ing of the s]teps in development. These differences are brought about by variations in master gene expression patterns. Consider homeotic genes called *Hox*, which, like other homeotic genes, help sculpt details of the body's form during embryonic development. Insects have a *Hox* gene called *antennapedia* that determines the identity of the thorax (the body part with legs). Humans and other vertebrates have a version of the same gene, *Hoxc6*, which determines the identity of the back (as opposed to the neck or tail). Expression of the *Hoxc6* gene in a developing embryo triggers the formation of ribs on a vertebra (Figure 11.22).

Figure 11.22 How differences in body form arise from differences in master gene expression.
Expression of the *Hoxc6* gene is indicated by purple stain in two vertebrate embryos, chicken (left) and garter snake (right). Expression of this homeotic gene causes a vertebra to develop ribs as part of the back. Chickens have 7 vertebrae in their back and 14 to 17 vertebrae in their neck; snakes have upwards of 450 back vertebrae and essentially no neck. *Hoxc6* is a vertebrate version of the *antennapedia* gene in insects.

Courtesy of Ann C. Burke, Wesleyan University.

Take-Home Message 11.7

What evidence does evolution leave in body function?

- Lineages that diverged long ago generally have more differences between their DNA (and their proteins) than do lineages that diverged more recently.
- Similarities in patterns of embryonic development are the result of master genes that have been conserved over evolutionary time.

Summary

Section 11.1 Events of the ancient past can be explained by the same physical, chemical, and biological processes that operate today. An asteroid impact 66 million years ago may have wiped out the dinosaurs and most other life on Earth.

Section 11.2 Expeditions by nineteenth-century explorers yielded increasingly detailed observations of nature. Geology, **biogeography**, and **comparative morphology** of organisms and their **fossils** led to new ways of thinking about the natural world.

Section 11.3 In the 19th century, attempts to reconcile traditional beliefs with physical evidence of **evolution**, which is change in a **lineage** over time, led to new ways of thinking about the natural world. Charles Darwin and Alfred Wallace came up with a theory of how environments select traits. A population tends to grow until it exhausts environmental resources. As that happens, competition for those resources intensifies among the population's members. Individuals with forms of shared, heritable traits that make them more competitive for the resources tend to produce more offspring. Thus, **adaptive traits** (**adaptations**) imparting greater **fitness** tend to become more common in the population over generations. The process in which environmental pressures result in the differential survival and reproduction of individuals of a population is called **natural selection**. It is one of the processes that drives evolution.

Section 11.4 Fossils are typically found in stacked layers of sedimentary rock. Fossils of many organisms are relatively scarce, so the fossil record will always be incomplete. A radioisotope's characteristic **half-life** can be used to determine the age of rocks and fossils. This technique, **radiometric dating**, helps us understand the ancient history of many lineages.

Section 11.5 According to the **plate tectonics theory**, Earth's crust is cracked into giant plates that carry landmasses to new positions as they move. Earth's landmasses have periodically converged as supercontinents such as **Gondwana** and **Pangea**. Transitions in the fossil record are the boundaries of great intervals of the **geologic time scale**.

Section 11.6 Comparative morphology can reveal evidence of evolutionary connections among lineages. **Homologous structures** are similar body parts that, by **morphological divergence**, became modified differently in different lineages. Such parts are evidence of a common ancestor. **Analogous structures** are body parts that look alike in different lineages but did not evolve in a common ancestor. By the process of **morphological convergence**, they evolved separately after the lineages diverged.

Section 11.7 We can discover and clarify evolutionary relationships by comparing amino acid sequences of proteins, or DNA sequences, from different organisms. In general, these sequences are more similar among lineages that diverged more recently. Master genes that affect development tend to be highly conserved, so similarities in patterns of embryonic development reflect shared ancestry that can be evolutionarily ancient.

Self-Quiz

Answers in Appendix I

1. The number of species on an island usually depends on the size of the island and its distance from a mainland. This statement would most likely be made by _____ .
   - a. an explorer
   - b. a biogeographer
   - c. a geologist
   - d. a philosopher

2. The bones of a bird's wing are similar to the bones in a bat's wing. This observation is an example of _____ .
   - a. uniformity
   - b. evolution
   - c. comparative morphology
   - d. a lineage

3. Evolution _____ .
   - a. is natural selection
   - b. is change in a line of descent
   - c. can occur by natural selection
   - d. b and c are correct

4. A trait is adaptive if it _____ .
   - a. arises by mutation
   - b. increases fitness
   - c. is passed to offspring
   - d. occurs in fossils

5. In which type of rock are you more likely to find a fossil?
   - a. basalt, a dark, fine-grained volcanic rock
   - b. limestone, composed of sedimented calcium carbonate
   - c. slate, a volcanically melted and cooled shale
   - d. granite, which forms by crystallization of molten rock below Earth's surface

6. True or false? Wrinkly textures in rock that formed from ancient biofilms living in marine sediments are considered trace fossils.

7. If the half-life of a radioisotope is 20,000 years, then a sample in which three-quarters of that radioisotope has decayed is _____ years old.
   a. 15,000
   b. 26,667
   c. 30,000
   d. 40,000

8. Forces that cause geologic change include _____ (select all that are correct).
   a. water movement
   b. natural selection
   c. volcanic activity
   d. tectonic plate movement
   e. wind
   f. asteroid impacts

9. Did Pangea or Gondwana form first?

10. The dinosaurs disappeared about _____ million years ago.
    a. 10
    b. 16.5
    c. 66
    d. 200

11. Through _____ , a body part of an ancestor is modified differently in different lines of descent.
    a. homologous evolution
    b. morphological convergence
    c. adaptive divergence
    d. morphological divergence

12. Homologous structures among major groups of organisms may differ in _____ .
    a. size
    b. shape
    c. function
    d. all of the above

13. A mutation that alters the embryonic expression pattern of a _____ may lead to major differences in the adult form.
    a. derived trait
    b. master gene
    c. homologous structure
    d. all of the above

14. Match each term with the most suitable description.
    _____ fitness
    _____ fossils
    _____ natural selection
    _____ homeotic genes
    _____ half-life
    _____ catastrophism
    _____ uniformity
    _____ analogous structures
    _____ homologous structures
    _____ sedimentary rock
    _____ neutral mutation

    a. does not affect fitness
    b. geological change occurs continuously
    c. geological change occurs in unusual major events
    d. good for finding fossils
    e. survival of the fittest
    f. characteristic of a radioisotope
    g. insect wing and bird wing
    h. human arm and bird wing
    i. evidence of life in distant past
    j. measured by reproductive success
    k. similar across diverse taxa

15. All of the following data types can be used as evidence of shared ancestry except similarities in _____ .
    a. amino acid sequences
    b. DNA sequences
    c. fossil morphologies
    d. embryonic development
    e. form due to convergence
    f. all are appropriate

Critical Thinking

1. In the late 1800s, a biologist studying animal embryos coined the phrase "ontogeny recapitulates phylogeny," meaning that the physical development of an animal embryo (ontogeny) seemed to retrace the changing form of the species during its evolutionary history (phylogeny). Why would embryonic development retrace evolutionary steps?

2. Radiometric dating does not measure the age of an individual atom. It is a measure of the age of a quantity of atoms— a statistic. As with any statistical measure, its values may deviate around an average (see sampling error, Section 1.6). Imagine that one sample of rock is dated ten different ways. Nine of the tests yield an age close to 225,000 years. One test yields an age of 3.2 million years. Do the nine consistent results imply that the one that deviates is incorrect, or does the one odd result invalidate the nine that are consistent?

Visual Question

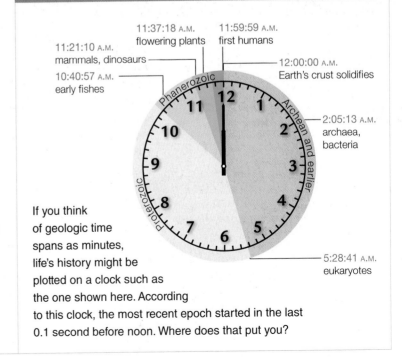

11:21:10 A.M. mammals, dinosaurs
11:37:18 A.M. flowering plants
11:59:59 A.M. first humans
10:40:57 A.M. early fishes
12:00:00 A.M. Earth's crust solidifies
2:05:13 A.M. archaea, bacteria
5:28:41 A.M. eukaryotes

If you think of geologic time spans as minutes, life's history might be plotted on a clock such as the one shown here. According to this clock, the most recent epoch started in the last 0.1 second before noon. Where does that put you?

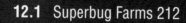

12

PROCESSES OF EVOLUTION

Application ❯ 12.1 Superbug Farms

REMEMBER: Contamination of food with disease-causing bacteria can result in food poisoning that is sometimes fatal (Section 3.1). Mutations can be introduced during DNA replication (6.4). Genetic diversity in a population is an advantage in a changing environment (8.4). An allele associated with an adaptive trait tends to become more common in a population over time (11.3).

Scarlet fever, tuberculosis, and pneumonia once caused one-fourth of the annual deaths in the United States. Since the 1940s, we have been relying on antibiotics to fight these and other dangerous bacterial diseases. We have also been using them in other, less dire circumstances. For an unknown reason, antibiotics promote growth in cattle, pigs, poultry, and even fish. The agricultural industry uses a lot of antibiotics, mainly for this purpose. In 2011, 13.7 million kilograms (about 30 million pounds) of antibiotics were used for agriculture in the U.S.—more than four times the amount used to treat people in the same year.

A natural population of bacteria is diverse, and it can evolve astonishingly fast. Consider how each cell division is an opportunity for mutation. The common intestinal bacteria *E. coli* can divide every 17 minutes, so even if a population starts out as clones, its cells diversify quickly. Bacteria share DNA even among distantly related species, and this adds even more genetic diversity to their populations. When a natural population of bacteria is exposed to a selection pressure such as an antibiotic, some cells in the population are likely to survive because they carry an allele that offers an advantage—antibiotic resistance, in this case. As susceptible cells die and the survivors reproduce, the frequency of antibiotic-resistance alleles in the population increases. A typical two-week course of treatment with antibiotics can exert selection pressure on over a thousand generations of bacteria. The pressure drives genetic change in bacterial populations so they become composed mainly of antibiotic-resistant cells. Thus, using antibiotics on an ongoing basis effectively guarantees the production of antibiotic-resistant bacterial populations (Figure 12.1).

Farms where antibiotics are used to promote growth are hot spots for the evolution of antibiotic-resistant bacteria and their spread to humans. Veterinarians and other people who work with the animals on these farms tend to carry more antibiotic-resistant bacteria in their bodies. So do neighbors who live within a mile. The bacteria spread much farther than the farm, however. Bacteria on an animal's skin or in its digestive tract can easily contaminate its meat during slaughter, and contaminated meat ends up in restaurant and home kitchens. A 2013 investigation found "worrisome" amounts of bacteria in 97% of the chicken meat in stores across the United States. About half of the samples tested were contaminated with superbugs—bacteria that are resistant to multiple antibiotics—and one in ten contained multiple types of superbugs. An earlier study found antibiotic-resistant bacteria in more than half of supermarket ground beef and pork chops, and in over 80 percent of ground turkey. Bacteria can be killed by the heat of cooking, but it is almost impossible to prevent them from spreading from contaminated meat to kitchen surfaces—and to people—during the process.

We have only a limited number of antibiotic drugs, and developing new ones is much slower than bacterial evolution. As resistant bacteria become more common, the number of antibiotics that can be used to effectively treat infections in humans dwindles. Using a particular antibiotic only in animals, or only in humans, is not a solution to this problem because there are only a few mechanisms by which these drugs kill bacteria; resistance to one antibiotic often confers resistance to others.

Figure 12.1 A breeding ground for antibiotic-resistant bacteria.
The vast majority of chickens raised for meat in the United States spend their lives in gigantic flocks that crowd huge buildings like this one. Growth-promoting antibiotics are given to the entire flock in food, a practice that pressures normal bacterial populations to become antibiotic resistant.

Bob Nichols/USDA photo.

For example, bacteria that become resistant to Flavomycin® (an antibiotic used only in animals) also resist vancomycin (an antibiotic used only in humans). Superbugs that are resistant to most currently available antibiotics are turning up at a very alarming rate.

All of this amounts to bad news. An infection with antibiotic-resistant bacteria tends to be longer, more severe, and more likely to be deadly than one more easily treatable with antibiotics. Superbugs cause more than 2 million cases of serious illness each year in the United States alone; they outright kill 23,000 of these people. Many, many more die because the infection complicates another, preexisting illness.

12.2 Alleles in Populations

REMEMBER: A population is a group of interbreeding individuals of the same species living in a given area (Section 1.2). Individuals of a species share a unique set of inherited traits (1.4) because they have the same genes; and alleles of the shared genes are the basis of variation in shared traits (8.4). Sexual reproduction mixes up genetic information from two parents, giving rise to offspring that vary in shared traits (8.5). In a Mendelian inheritance pattern, the phenotypic effect of a dominant allele fully masks that of a recessive allele (9.3). The more genes and environmental factors that influence a trait, the more continuous is its range of variation (9.4–9.6). Neutral mutations have no effect on an individual's survival or reproduction (11.7).

The individuals of a population (and a species) share certain features. Humans, for example, normally have a short neck, a thumb on each hand, and so on. These are examples of morphological traits (*morpho–* means form). Individuals of a species also share physiological traits, such as details of metabolism. They also respond the same way to certain stimuli, as when hungry humans eat food. These responses are behavioral traits.

Some traits appear in two distinct forms, or morphs, in which case the forms are called a dimorphism (*di–* means two). Flower color in the pea plants that Gregor Mendel studied is a dimorphic trait. The interaction of two alleles that have a clear dominance relationship gives rise to the dimorphism—purple or white flowers—in these plants. Other traits appear in three or more distinct forms, in which case the forms are called polymorphisms (*poly–*, many). ABO blood type in humans, which is determined by the codominant alleles of the *ABO* gene, is an example. Most other traits are complex (Figure 12.2), as is their genetic basis. Any or all of the genes that influence such traits may have multiple alleles.

In earlier chapters, you learned that alleles arise by mutation, and other events shuffle them among individuals of a population (Table 12.1). To understand the potential scope of variation that results from these events, consider alleles in our species. There are more than $10^{100,000,000}$ potential combinations of human alleles at fertilization. Not even 10^{10} people are living today. Unless you have an identical twin, it is extremely unlikely that another person with your precise genetic makeup has ever lived, or ever will.

An Evolutionary View of Mutations Being the original source of new alleles, mutations are worth another look, this time in the context of populations. We cannot predict when or in which individual a particular gene will mutate. We can, however, predict the average mutation rate of a species, which is the probability that

Figure 12.2 Sampling morphological variation among zigzag snails (top) and humans (bottom).
Variation in shared traits is an outcome of differences in alleles that influence those traits.

Top, © David McIntyre/Photographer's Direct; bottom, clockwise from top left, © Roderick Hulsbergen/http://www.photography.euweb.nl; Olga Reutska/Shutterstock; NinaMalyna/Shutterstock; Lane Oatey/Blue Jean Images/Getty Images; Djomas/Shutterstock; Paul Matthew Photography/Shutterstock.

Table 12.1 Some Sources of Variation in Shared Traits

Genetic Event	Effect
Mutation	Original source of new alleles
Crossing over at meiosis I	Introduces new combinations of alleles into chromosomes
Independent assortment at meiosis I	Mixes maternal and paternal chromosomes
Fertilization	Combines alleles from two parents

allele frequency Abundance of a particular allele among members of a population.

directional selection Mode of natural selection in which a phenotype at one end of a range of variation is favored.

disruptive selection Mode of natural selection in which traits at the extremes of a range of variation are adaptive, and intermediate forms are not.

gene pool All the alleles of all the genes in a population; a pool of genetic resources.

microevolution Change in allele frequency.

stabilizing selection Mode of natural selection in which an intermediate form of a trait is favored over extreme forms.

> Evolution is not purposeful. It simply fills the nooks and crannies of opportunity.

a mutation will occur in a given interval. In the human species, that rate is about 2.2×10^{-9} mutations per base pair per year. In other words, about 70 nucleotides in the human genome sequence change every decade.

In humans at least, most mutations are neutral. For instance, a mutation that results in your earlobes being attached to your head instead of swinging freely should not in itself stop you from surviving and reproducing as well as anybody else. So, natural selection would not affect the frequency of this mutation in the human population. Other mutations give rise to structural, functional, or behavioral alterations that reduce an individual's chances of surviving and reproducing. Consider collagen, a protein component of the skin, bones, tendons, lungs, blood vessels, and other vertebrate organs. If one of the genes for collagen mutates in a way that changes the protein's function, the entire body may be affected. A mutation such as this can change phenotype so drastically that it results in death, in which case it is called a lethal mutation.

Occasionally, a change in the environment favors a mutation that had previously been neutral or even somewhat harmful. Even if a beneficial mutation bestows only a slight advantage, its frequency tends to increase in a population over time. This is because natural selection operates on traits with a genetic basis. With natural selection, remember, environmental pressures result in an increase in the frequency of an adaptive form of a trait in a population over generations. Mutations have been altering genomes for billions of years, and they continue to do so. Cumulatively, mutations have given rise to Earth's staggering biodiversity. Think about it: The reason you do not look like an avocado or an earthworm or even your next-door neighbor began with mutations that occurred in different lines of descent.

Allele Frequency Together, all the alleles of all the genes of a population constitute a pool of genetic resources—a **gene pool**. Members of a population breed with one another more often than they breed with members of other populations, so their gene pool is more or less isolated. The abundance of a particular allele in a gene pool is called **allele frequency**, and it is expressed as a percentage. For example, if a particular allele occurs on half of the chromosomes carried by the population, then the frequency of that allele is 50 percent.

Allele frequency can change, and this change is called **microevolution**. Microevolution is always occurring in natural populations because, as you will see in the next sections, processes that drive it—mutation, natural selection, and genetic drift—are always operating in nature. As you learn about these patterns, remember an important point: Evolution is not purposeful; it simply fills the nooks and crannies of opportunity.

Take-Home Message 12.2

What is microevolution?

• Individuals of a natural population share morphological, physiological, and behavioral traits characteristic of the species. Alleles are the basis of differences in the details of those shared traits.

• All alleles of all individuals in a population make up the population's gene pool. An allele's abundance in the gene pool is called its allele frequency.

• Microevolution is change in allele frequency. It is always occurring in natural populations because processes that drive it are always operating.

12.3 Modes of Natural Selection

REMEMBER: Many vitamins and minerals are essential in the diet because they are cofactors or become converted to them; coenzymes modified in reactions are regenerated in separate reactions (Section 4.4). A homozygous individual has two identical alleles of a gene, and a heterozygous individual has two nonidentical alleles; the effect of a dominant allele masks that of a recessive allele paired with it (9.2). Blood clotting in humans involves proteins called clotting factors (9.7). With natural selection, individuals of a population survive and reproduce with differing success depending on the details of their shared, heritable traits (11.3).

Natural selection influences allele frequency by operating on phenotypes that have a genetic basis. How phenotype is affected depends on the species and the selection pressures in the environment. With **directional selection**, forms at one end of a range of phenotypic variation become more common over time (Figure 12.3A, B). With **stabilizing selection**, an intermediate form of a trait is favored, and extreme forms are selected against (Figure 12.3C). With **disruptive selection**, forms of a trait at both ends of a range of variation are favored, and intermediate forms are selected against (Figure 12.3D).

Directional Selection Antibiotic use that fosters resistant bacterial populations is one example of directional selection; another example involves rats. Rats thrive in urban centers, where garbage is plentiful and natural predators are not. Part of their success stems from an ability to reproduce very quickly: Rat populations can expand within weeks to match the amount of garbage available for them to eat. For decades, people have been fighting rats with poisons. Baits laced with warfarin, an organic compound that interferes with blood clotting, became popular in the 1950s. Rats that ate the poisoned baits died within days after bleeding internally or losing blood through cuts or scrapes. Warfarin was extremely effective, and its impact on harmless species was much lower than that of other rat poisons. It quickly became the rat poison of choice. By 1980, however, about 10 percent of rats in urban areas were resistant to warfarin. What happened?

Warfarin interferes with blood clotting because it inhibits the function of an enzyme called VKORC1. This enzyme regenerates vitamin K, which functions as a coenzyme in the production of blood clotting factors. When vitamin K is not regenerated, the clotting factors are not properly produced, and clotting cannot occur. Rats resistant to warfarin have a mutated version of the *VKORC1* gene. The enzyme encoded by this allele is insensitive to warfarin. "What happened" was evolution by natural selection. Rats with the normal allele died after eating warfarin; the lucky ones with a mutated allele survived and passed it to offspring. The rat populations recovered quickly, and a higher proportion of individuals in the next generation carried a mutation. With each onslaught of warfarin, the frequency of the mutation in rat populations increased. Exposure to warfarin had exerted directional selection.

The mutation that results in warfarin resistance also reduces the activity of the VKORC1 enzyme, so rats that have it require a lot of extra vitamin K. However, being vitamin K deficient is not so bad when compared with being dead from rat poison. In the absence of warfarin, though, rats with the allele are at a serious disadvantage because they cannot easily obtain enough vitamin K from their diet to sustain normal blood clotting and bone formation. Thus, the frequency of a warfarin resistance allele in a rat population declines quickly after warfarin exposure ends—another example of directional selection.

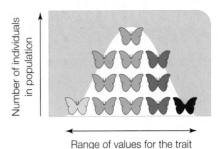

A. Population before selection occurs.

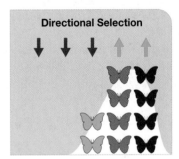

B. With directional selection, forms of a trait at one end of a range of variation is adaptive.

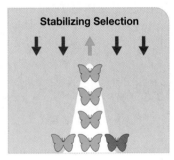

C. With stabilizing selection, extreme forms of a trait are eliminated, and an intermediate form is maintained.

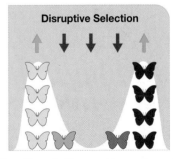

D. With disruptive selection, a midrange form of a trait is eliminated, and extreme forms are maintained.

Figure 12.3 Comparing three modes of natural selection.
In these examples, red arrows indicate which forms are being selected against; blue, forms that are adaptive.

Digging Into Data

Resistance to Rodenticides in Wild Rat Populations

Beginning in 1990, rat infestations in northwestern Germany started to intensify despite continuing use of rat poisons. In 2000, Michael H. Kohn and his colleagues tested wild rat populations around Münster. For part of their research, they trapped wild rats in five towns, and tested those rats for resistance to warfarin and the more recently developed poison bromadiolone. The results are shown in Figure 12.4.

1. In which of the five towns were most of the rats susceptible to warfarin?
2. Which town had the highest percentage of poison-resistant wild rats?
3. What percentage of rats in Olfen were warfarin resistant?
4. In which town do you think the application of bromadiolone was most intensive?

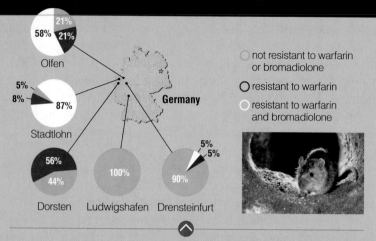

○ not resistant to warfarin or bromadiolone
◉ resistant to warfarin
○ resistant to warfarin and bromadiolone

Figure 12.4 Poison resistance in wild rats in Germany, 2000.
Photo, © Rollin Verlinde/Vilda.

Directional selection also affects the color of rock pocket mice in Arizona's Sonoran Desert. Rock pocket mice are small mammals that spend the day sleeping in underground burrows, emerging at night to forage for seeds. Light brown granite dominates their environment, but there are also patches of dark basalt: the remains of ancient lava flows. Most of the mice in populations that inhabit the dark rock have dark coats. Most of the mice in populations that inhabit the light brown rock have light brown coats. The difference arises because mice that match the rock color in each habitat are camouflaged from their natural predators. Night-flying owls more easily see mice that do not match the rocks, and they preferentially eliminate these mice from each population. Thus, in both habitats, selective predation has resulted in a directional shift in the frequency of alleles that affect coat color.

Another well-documented case of directional selection involves coloration changes in peppered moths. These moths feed and mate at night, then rest on trees during the day. In preindustrial England, the vast majority of peppered moths were white with black speckles, and a small number were much darker. At the time, the air was clean, and light-gray lichens grew on the trunks and branches of most trees. Light-colored moths that rested on lichen-covered trees were well camouflaged, but darker moths were not (Figure 12.5A). By the 1850s, the industrial revolution had begun, and smoke emitted by coal-burning factories was killing the lichens. Dark moths, which were better camouflaged on lichen-free, soot-darkened trees, had become more common (Figure 12.5B).

Figure 12.5 Adaptive value of two color forms of the peppered moth.

J. A. Bishop, L. M. Cook.

A. Light-colored moths on a nonsooty tree trunk (left) are hidden from predators; dark ones (right) stand out.

B. Where soot darkens tree trunks, the dark color (left) provides more camouflage than the light color (right).

Scientists suspected that predation by birds was the selective pressure that shaped moth coloration, and in the 1950s, H. B. Kettlewell set out to test this hypothesis. He bred dark and light moths in captivity, marked them for easy identification, then released them in several areas. His team recaptured more of the dark moths in the polluted areas and more light ones in the less polluted ones. The researchers also observed predatory birds eating more light-colored moths in soot-darkened forests, and more dark-colored moths in cleaner, lichen-rich forests. Dark-colored moths were clearly at a selective advantage in industrialized areas.

Pollution controls went into effect in 1952. As a result of improved environmental standards, tree trunks gradually became free of soot, and lichens made a comeback. Kettlewell observed that moth phenotypes shifted too: Wherever pollution decreased, the frequency of dark moths decreased as well. Recent research has confirmed Kettlewell's results implicating birds as selective agents of peppered moth coloration, and also that this selection causes a shift in the frequency of alleles underlying the coloration. Peppered moth color is determined by a single gene; individuals with a dominant allele of this gene are dark, and those homozygous for a recessive allele are light.

Stabilizing Selection With stabilizing selection, an intermediate form of a trait is favored, and extreme forms are selected against. Consider how environmental pressures maintain an intermediate body mass in populations of sociable weavers (Figure 12.6). These birds live in the African savanna, and their body mass has a genetic basis. Between 1993 and 2000, Rita Covas and her colleagues investigated selection pressures that operate on sociable weaver body mass by capturing and weighing thousands of birds before and after the breeding seasons. The results of this study indicated that optimal body mass in sociable weavers is a trade-off between the risks of starvation and predation. Birds that carry less fat are more likely to starve than fatter birds. However, birds that carry more fat spend more time eating, which in this species means foraging in open areas where they are easily accessible to predators. Fatter birds are also more attractive to predators, and not as agile when escaping. Thus, predators are agents of selection that eliminate the fattest individuals. Birds of intermediate weight have the selective advantage, and they make up the bulk of sociable weaver populations.

Disruptive Selection With disruptive selection, forms of a trait at both ends of a range of variation are favored, and intermediate forms are selected against. Disruptive selection maintains a dimorphism in black-bellied seedcrackers. These birds are native to Cameroon, Africa, and the size of their beaks (bills) has a genetic basis. The bill of a typical black-bellied seedcracker, male or female, is either 12 millimeters wide, or wider than 15 millimeters (Figure 12.7). Birds with a bill size between 12 and 15 millimeters are uncommon. It is as if every human adult were 4 feet or 6 feet tall, with no one of intermediate height.

Seedcrackers with the large and small bill forms inhabit the same geographic range, and they breed randomly with respect to bill size. The dimorphism arises from (and is maintained by) environmental factors that affect feeding performance. The finches feed mainly on the seeds of two types of sedge, which is a grasslike plant. One sedge produces hard seeds; the other produces soft seeds. Small-billed birds are better at opening the soft seeds, but large-billed birds are better at cracking the hard ones. Both hard and soft sedge seeds are abundant during Cameroon's semiannual wet seasons. At these times, all seedcrackers feed on both seed types. The seeds become scarce during the region's dry seasons. As competition for food

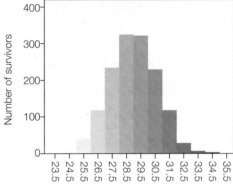

Figure 12.6 Stabilizing selection in sociable weavers. The graph shows the number of birds (out of 977) that survived a breeding season. Compare Figure 12.3C.

Top, Peter Chadwick/Science Source.

Figure It Out: According to these data, what is the optimal weight of a sociable weaver?

Answer: About 29 grams

lower bill 12 mm wide   lower bill 15 mm wide

Figure 12.7 Disruptive selection in African seedcracker populations maintains a dimorphism in bill size. Competition for scarce food during dry seasons favors birds with bills that are either 12 millimeters wide (left) or 15 to 20 millimeters wide (right). Birds with bills of intermediate size are selected against.

Thomas Bates Smith.

intensifies, each bird focuses on eating the seeds that it opens most efficiently: Small-billed birds feed mainly on soft seeds, and large-billed birds feed mainly on hard seeds. Birds with intermediate-sized bills cannot open either type of seed as efficiently as the other birds, so they are less likely to survive the dry seasons.

A. Male elephant seals engaged in combat. Males of this species typically compete for access to clusters of females.

B. A male bird of paradise engaged in a courtship display has caught the eye (and, perhaps, the sexual interest) of a female.

C. Female stalk-eyed flies prefer to mate with males that have the longest eyestalks. In this species, long eyestalks provide no known survival advantage.

Figure 12.8 Sexual selection in action.

(A) © Ingo Arndt/Nature Picture Library; (B) Tim Laman/National Geographic Creative; (C) Minden Pictures/SuperStock.

Take-Home Message 12.3

How does natural selection drive evolution?

- Natural selection influences allele frequency by operating on traits that have a genetic basis.
- With directional selection, forms at one end of a range of variation in a trait are favored.
- With stabilizing selection, an intermediate form of a trait is favored, and extreme forms are selected against.
- With disruptive selection, an intermediate form of a trait is selected against, and extreme forms are favored.

12.4 Natural Selection and Diversity

REMEMBER: A particular mutation in the beta globin gene gives rise to sickle-cell anemia (Section 7.6). A homozygous individual has two identical alleles of a gene, and a heterozygous individual has two nonidentical alleles; the effect of a dominant allele masks that of a recessive allele paired with it (9.2). In some cases, a codominant allele offers a survival advantage in a particular environment (9.6).

Survival of the Sexiest Not all evolution is driven by selection for traits that enhance survival. Competition for mates is another selective pressure that can shape form and behavior. Consider how individuals of many sexually reproducing species have a distinct male or female phenotype (a trait that differs between males and females is called a sexual dimorphism). Individuals of one sex are more colorful, larger, or more aggressive than individuals of the other sex. These traits can seem puzzling because they take energy and time away from activities that enhance survival, and some actually hinder an individual's ability to survive. Why, then, do they persist? The answer is **sexual selection**, in which the evolutionary winners outreproduce others of a population because they are better at securing mates. With this mode of natural selection, the most adaptive forms of a trait are those that help individuals defeat rivals for mates, or are most attractive to the opposite sex.

For example, the females of some species cluster in defensible groups when they are sexually receptive, and males compete for sole access to the groups. Competition for the ready-made harems favors brawny, combative males (Figure 12.8A).

Males or females that are choosy about mates act as selective agents on their own species. The females of some species shop for a mate among males that display species-specific cues such as a highly specialized appearance or courtship behavior (Figure 12.8B). The cues often include flashy body parts or movements, traits that tend to attract predators and in some cases are a physical hindrance. However, to a female member of the species, a flashy male's survival despite his obvious handicap may imply health and vigor, two traits that are likely to improve her chances of bearing healthy, vigorous offspring. Selected males pass alleles for their attractive traits to the next generation of males, and females pass alleles that influence mate preference to the next generation of females. Highly exaggerated traits can be an evolutionary outcome (Figure 12.8C).

Maintaining Multiple Alleles Any mode of natural selection may keep two or more alleles circulating at relatively high frequency in a population's gene pool, a state called balanced polymorphism. For example, sexual selection maintains multiple alleles that govern eye color in populations of *Drosophila* fruit flies. Female flies prefer to mate with rare white-eyed males, until the white-eyed males become more common than red-eyed males, at which point the red-eyed flies are again preferred.

Balanced polymorphism can also arise in environments that favor heterozygous individuals. Consider the gene that encodes the beta globin chain of hemoglobin. *HbA* is the normal allele; the codominant *HbS* allele carries a mutation that causes sickle-cell anemia. Even with medical care, about 15 percent of individuals homozygous for the *HbS* allele die by age 18 from complications of the disorder. Despite being so harmful, the *HbS* allele persists at very high frequency among the human populations in tropical and subtropical regions of Asia, Africa, and the Middle East (Figure 12.9A). Why? Populations with the highest frequency of the *HbS* allele also have the highest incidence of malaria (Figure 12.9B). Mosquitoes transmit *Plasmodium*, the parasitic protist that causes malaria, to human hosts. *Plasmodium* multiplies in the liver and then in red blood cells. The cells rupture and release new parasites during recurring bouts of severe illness. People who make both normal and sickle hemoglobin are more likely to survive malaria than people who make only normal hemoglobin. In *HbA/HbS* heterozygous individuals, *Plasmodium*-infected red blood cells sometimes sickle. The abnormal shape brings the cells to the attention of the immune system, which destroys them along with the parasites they harbor. By contrast, *Plasmodium*-infected red blood cells of individuals homozygous for the *HbA* allele do not sickle, so the parasite may remain hidden from the immune system.

In areas where malaria is common, the persistence of the *HbS* allele is a matter of relative evils. Malaria and sickle-cell anemia are both potentially deadly. Heterozygous individuals may not be completely healthy, but they do have a better chance of surviving malaria than people homozygous for the normal allele (*HbA/HbA*). With or without malaria, people who have both alleles (*HbA/HbS*) are more likely to live long enough to reproduce than individuals homozygous for the sickle allele (*HbS/HbS*). The result is that nearly one-third of people living in the most malaria-ridden regions of the world carry the *HbS* allele.

Figure 12.9 Malaria and sickle-cell anemia.
After Ayala and others.

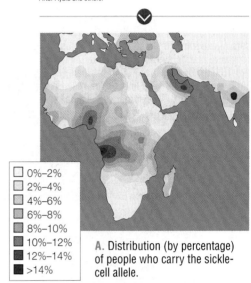

- ☐ 0%–2%
- ☐ 2%–4%
- ☐ 4%–6%
- ☐ 6%–8%
- ☐ 8%–10%
- ■ 10%–12%
- ■ 12%–14%
- ■ >14%

A. Distribution (by percentage) of people who carry the sickle-cell allele.

B. Distribution of malaria cases (orange) in Africa, Asia, and the Middle East in the 1920s, before the start of programs to control mosquitoes, which transmit the parasitic protist that causes the disease. Notice the correlation with the distribution of the sickle-cell allele in **A**.

Take-Home Message 12.4

How does natural selection maintain diversity?

- With sexual selection, adaptive forms of a trait are those that give an individual an advantage in securing mates.
- Sexual selection can reinforce phenotypic differences between males and females, and sometimes it results in exaggerated traits.
- Balanced polymorphism can be an outcome of sexual selection, or of environmental pressures that favor heterozygous individuals.

sexual selection Mode of natural selection in which some individuals outreproduce others of a population because they are better at securing mates.

Figure 12.10 Genetic drift in flour beetles.
A flour beetle is shown at above on a flake of cereal. In two sets of experiments, beetles heterozygous for alleles *b*+ and *b* were maintained for 20 generations in populations of 10 individuals (**A**) or 100 individuals (**B**). Graph lines in **B** are smoother than in **A**, which means that less genetic drift occurred in the larger populations.

Allele *b*+ was lost in one population (one graph line ends at 0). Notice that the average frequency of allele *b*+ rose at the same rate in both groups, an indication that natural selection was at work too: Allele *b*+ was weakly favored.

(A,B) Adapted from S. S. Rich, A. E. Bell, and S. P. Wilson, "Genetic drift in small populations of Tribolium," *Evolution* 33:579–584, Fig. 1, p. 580, © 1979 by John Wiley and Sons. Used by permission of the publisher; above, Photo by Peggy Greb/USDA.

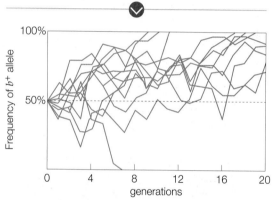

A. In these experiments, population size was maintained at 10 beetles. Several populations were tested; notice that allele *b*+ was lost in one (one graph line ends at 0).

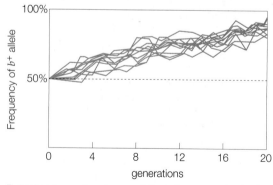

B. In these experiments, population size was maintained at 100 beetles. Genetic drift was less pronounced in these populations than in the 10-beetle populations in **A**.

Figure It Out: In how many populations did allele *b*+ become fixed?

Answer: Six

12.5 Genetic Drift and Gene Flow

REMEMBER: Members of a sexually reproducing population vary in the details of their shared traits because they carry different alleles; this diversity offers an evolutionary advantage in a changing environment (Section 8.4). Codominant alleles of the *ABO* gene are the basis of ABO blood type in humans (9.4). Engineered genes are being transferred in pollen from genetically modified plants to wild plants (10.4).

In a natural population, individuals reproduce with differing success, and the difference is not always an outcome of natural selection. By chance, a perfectly fit and healthy individual may not pass its alleles to offspring, for example by dying in a random event before it has a chance to reproduce. Such events can change a population's allele frequencies. Change in allele frequency brought about by chance alone is called **genetic drift**.

Genetic drift occurs in all natural populations of sexual reproducers, but its effects are more pronounced in small ones (Figure 12.10). For example, genetic drift makes small populations particularly vulnerable to the loss of genetic diversity. To understand why, imagine a hypothetical gene with two alleles, neither of which confers a selective advantage. These alleles (let's call them *A* and *a*) occur in a population's gene pool at a frequency of 95 percent and 5 percent, respectively. If the population consists of 10 members, then one individual is heterozygous (*Aa*) and the remaining nine are homozygous (*AA*). A random event that eliminates the heterozygous individual from the population before it reproduces also eliminates allele *a* from the population's gene pool. If all individuals of a population are homozygous for an allele (*A*, in this example), we say that the allele is **fixed**. The frequency of a fixed allele will not change unless a new mutation occurs, or an individual bearing another allele enters the population. Now imagine that our population with alleles *A* and *a* consists of 100 members instead of 10. Five individuals in this larger population would be heterozygous (*Aa*). In order for allele *a* to be lost from the population's gene pool, all five would have to be eliminated before they reproduce. The chance of random events eliminating all heterozygous individuals is smaller in the larger population. This is a simplified example of a general effect: The loss of genetic diversity is possible in all populations, but it is more likely to occur in small ones.

Bottlenecks and the Founder Effect A drastic reduction in population size, which is called a **bottleneck**, can greatly reduce diversity. Consider the northern elephant seal (pictured in Figure 12.8A). Overhunting during the late 1890s left only about twenty individuals of this species alive. Since then, hunting restrictions have allowed the population to recover, but genetic diversity among its members has been greatly reduced. The bottleneck and subsequent genetic drift eliminated many alleles that had previously been present in the population.

A loss of genetic diversity can also occur when a small group of individuals establishes a new population. If the founding group is not representative of the original population in terms of allele frequencies, then the new population will not be representative of it either. This outcome is called the **founder effect** (Figure 12.11A). Consider that all three *ABO* alleles for blood type are common in most human populations. Native Americans are an exception, with the majority of individuals being homozygous for the *O* allele. Native Americans are descendants of early humans that migrated from Asia between 14,000 and 21,000 years ago, across a narrow land bridge that once connected Siberia and Alaska. Analysis of DNA from ancient skeletal remains reveals that most early Americans were also homozygous for the *O* allele. Modern Siberians have all three alleles. Thus, the humans who first populated the Americas were probably members of a small group that had reduced genetic diversity compared with the general population.

Founding populations are often necessarily inbred. **Inbreeding** is nonrandom breeding or mating between close relatives. Closely related individuals tend to share more alleles than nonrelatives do, so inbred populations often have unusually high numbers of individuals homozygous for recessive alleles, some of which are harmful. This outcome is minimized in human populations that discourage or forbid incest (mating between parents and children or between siblings).

The Old Order Amish in Lancaster County, Pennsylvania, offer an example of the effects of inbreeding. Amish people marry only within their community. Intermarriage with other groups is not permitted, and no "outsiders" are allowed to join the community. As a result, Amish populations are moderately inbred, and many of their individuals are homozygous for harmful recessive alleles. The Lancaster community has an unusually high frequency of a recessive allele that causes Ellis–van Creveld syndrome, a genetic disorder characterized by dwarfism, heart defects, and polydactyly (extra fingers or toes), among other symptoms. This allele has been traced to a man and his wife, two of a group of 400 Amish people who immigrated to the United States in the mid-1700s. As a result of the founder effect and inbreeding since then, about 1 of 8 people in the Lancaster community is now heterozygous for the allele, and 1 in 200 is homozygous for it (Figure 12.11B).

Gene Flow Individuals tend to mate or breed most frequently with other members of their own population. However, not all populations of a species are completely isolated from one another, and nearby populations may occasionally interbreed. Also, individuals sometimes leave one population and join another. **Gene flow**, the movement of alleles between populations, occurs in both cases, and it can change or stabilize allele frequencies.

Gene flow is common among populations of animals, but it also occurs in less mobile organisms. Consider the acorns that jays disperse when they gather nuts for

the winter (left). Every fall, these birds visit acorn-bearing oak trees repeatedly, then bury the acorns in the soil of territories as much as a mile away. The jays transfer acorns (and the alleles carried by these seeds) among populations of oak trees that may otherwise be genetically isolated. Gene flow also occurs when wind or an animal transfers pollen from one plant to another, often over great distances. Many opponents of genetic engineering cite the movement of engineered genes from transgenic crop plants into wild populations via pollen.

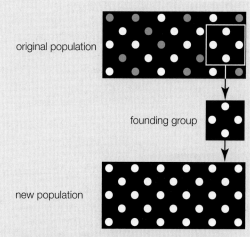

A. The founder effect: a group that founds a new population is not genetically representative of the original population, so allele frequencies differ between the new and the old populations.

B. An allele that causes Ellis–van Creveld syndrome occurs at high frequency in the gene pool of the Lancaster Amish—a result of the founder effect. Outward indications of the disorder (polydactyly and dwarfism) appear in this Amish baby.

Figure 12.11 The founder effect and one outcome.
(B) © Dr. Victor A. McKusick.

Take-Home Message 12.5
What mechanisms other than natural selection affect allele frequencies?
• Genetic drift, or change in allele frequency that occurs by chance alone, makes small populations particularly vulnerable to losing genetic diversity.
• Loss of genetic diversity can occur after a bottleneck or because of the founder effect.
• Gene flow, which is the movement of alleles between populations, can stabilize or change allele frequencies.

bottleneck Reduction in population size so severe that it reduces genetic diversity.

fixed Refers to an allele for which all members of a population are homozygous.

founder effect After a small group of individuals found a new population, allele frequencies in the new population differ from those in the original population.

gene flow The movement of alleles between populations.

genetic drift Change in allele frequency due to chance alone.

inbreeding Mating among close relatives.

12.6 Speciation

REMEMBER: A "species" is a convenient but artificial construct of the human mind (Section 1.4). Mitochondria have their own DNA and divide independently of the cell (3.5). Wavelengths of light that are not absorbed by a pigment are reflected, and that reflected light gives each pigment its characteristic color (5.2). A somatic cell is a body cell (6.1). Mitosis, a nuclear division mechanism that maintains the chromosome number, is the basis of growth and tissue repair in multicelled organisms (8.2). Homologous chromosomes undergo crossing over during meiosis; at fertilization in sexual reproducers, an egg (a female gamete) fuses with a sperm (a male gamete) to produce a zygote, the first cell of a new individual (8.5). A polyploid individual has three or more complete sets of chromosomes (9.8). The slow movement of Earth's crustal plates conveys continents to new locations over geologic time (11.5). Lineages that diverged long ago generally have more differences between their DNA (and their proteins) than do lineages that diverged more recently (11.7).

When two populations of a species do not interbreed, the number of genetic differences between them increases because mutation, natural selection, and genetic drift occur independently in each one. Over time, the populations may become so different that we consider them to be different species. Evolutionary processes in which new species arise are called **speciation**.

Evolution is a dynamic, extravagant, messy, and ongoing process that can be challenging for people who like neat categories. Speciation offers a perfect example, because it rarely occurs at a precise moment in time: Individuals often continue to interbreed even as populations are diverging, and populations that have already diverged may come together and interbreed again.

Reproductive Isolation Every time speciation happens, it happens in a unique way, which means that each species is a product of its own unique evolutionary history. However, there are recurring patterns. For example, **reproductive isolation**, the end of gene flow between populations, is always part of the process by which sexually reproducing species attain and keep their separate identities. Several mechanisms of reproductive isolation prevent successful interbreeding, and thus reinforce differences between diverging populations (Figure 12.12). For example, some closely related species cannot interbreed because the timing of their reproduction differs **1**. Consider the periodical cicada (right). Larvae of these insects feed on roots as they mature underground, then the adults emerge to reproduce. Three cicada species reproduce every 17 years. Each has a sibling species with nearly identical form and behavior, except that the siblings emerge on a 13-year cycle instead of a 17-year cycle. Sibling species have the potential to interbreed, but they can only get together once every 221 years!

Alvin E. Staffan/Science Source.

Adaptation to different environmental conditions may prevent closely related species from interbreeding **2**. For example, two species of manzanita, a plant native to the Sierra Nevada mountain range, rarely hybridize. One species that lives on dry, rocky hillsides is better adapted for conserving water. The other species, which requires more water, lives on lower slopes where water stress is not as intense. The physical separation makes cross-pollination unlikely.

Differences in behavior can prevent mating between related animal species **3**. For example, males and females of many animal species engage in courtship displays

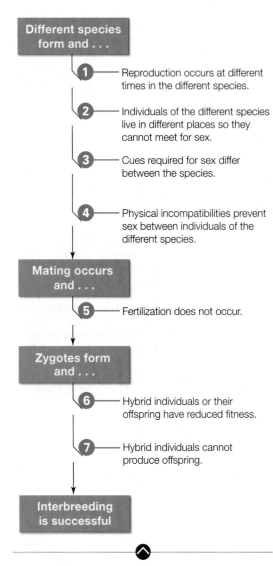

Different species form and . . .

1 — Reproduction occurs at different times in the different species.

2 — Individuals of the different species live in different places so they cannot meet for sex.

3 — Cues required for sex differ between the species.

4 — Physical incompatibilities prevent sex between individuals of the different species.

Mating occurs and . . .

5 — Fertilization does not occur.

Zygotes form and . . .

6 — Hybrid individuals or their offspring have reduced fitness.

7 — Hybrid individuals cannot produce offspring.

Interbreeding is successful

Figure 12.12 How reproductive isolation prevents interbreeding.

Figure 12.13 An example of reproductive isolation: courtship displays in peacock spiders. This male peacock spider is signaling his intent to mate with the female by raising and waving colorful flaps, and gesturing his legs in time with abdominal vibrations. If his species-specific courtship display fails to impress her, she will kill him.

© Jürgen Otto.

before sex (Figure 12.13). In a typical pattern, the female recognizes the sounds and movements of a male of her species as an overture to sex; females of different species do not.

The size or shape of an individual's reproductive parts may prevent it from mating with members of closely related species ❹. For example, plants called black sage and white sage grow in the same areas, but hybrids rarely form because the flowers of these two related species have become specialized for different pollinators (Figure 12.14).

Even if gametes of different species do meet up, they often have molecular incompatibilities that prevent a zygote from forming ❺. For example, the molecular signals that trigger pollen germination in flowering plants are species-specific (we return to pollen germination and other aspects of flowering plant reproduction in Section 28.2). Gamete incompatibility may be the primary speciation route among animals that release their eggs and free-swimming sperm into water.

Genetic changes are the basis of divergences in form, function, and behavior. Even chromosomes of species that diverged relatively recently may be different enough that a hybrid zygote ends up with extra or missing genes, or genes with incompatible products. Such outcomes typically disrupt embryonic development ❻. Hybrid individuals that do survive embryonic development often have reduced fitness. For example, hybrid offspring of lions and tigers have more health problems and a shorter life expectancy than individuals of either parent species. If hybrids live long enough to reproduce, their offspring often have lower and lower fitness with each successive generation. Incompatible nuclear and mitochondrial DNA may be the cause (mitochondrial DNA is inherited from the mother only).

Some interspecies crosses produce robust but sterile hybrid offspring ❼. For example, mating between a female horse (64 chromosomes) and a male donkey (62 chromosomes) produces a mule (63 chromosomes: 32 from the horse, and 31 from the donkey). Mules are healthy but their chromosomes cannot pair up properly for crossing over, so this animal makes few viable gametes.

A. Black sage is pollinated mainly by honeybees and other small insects.

B. The flowers of black sage are too delicate to support larger insects. Big insects, including this carpenter bee, access the nectar of small sage flowers only by piercing from the outside. When they do so, they avoid touching the flower's reproductive parts.

anthers

stigma

C. The reproductive parts (anthers and stigma) of white sage flowers are too far away from the petals to be brushed by honeybees, so honeybees are not efficient pollinators of this species. White sage is pollinated mainly by larger bees and hawkmoths, which brush the flower's stigma and anthers as they pry apart the petals to access nectar.

Figure 12.14 An example of reproductive isolation: pollinator specialization in sage.

(A) Courtesy of Dr. James French; (B) Courtesy of © Ron Brinkmann, www.flickr.com/photos/ronbrinkmann; (C) © David Goodin.

allopatric speciation Speciation pattern in which a physical barrier arises and ends gene flow between populations.

sympatric speciation Divergence within a population leads to speciation; occurs in the absence of a physical barrier to gene flow.

Allopatric Speciation Genetic changes that lead to a new species can begin with physical separation between populations. With **allopatric speciation**, a physical barrier arises and separates two populations, ending gene flow between them (*allo–* means different; *patria*, fatherland). Then, reproductive isolating mechanisms evolve that prevent interbreeding even if the diverging populations meet again.

Gene flow between populations separated by distance is often inconsistent. Whether a geographic barrier can completely block that gene flow depends on how the species travels (such as by swimming, walking, or flying), and how it reproduces (for example, by internal fertilization or by pollen dispersal).

A geographic barrier can arise in an instant, or over an eon. The Great Wall of China is an example of a barrier that arose relatively quickly. As it was being built, the wall interrupted gene flow among nearby populations of insect-pollinated plants; DNA sequence comparisons show that trees, shrubs, and herbs on either side of the wall are diverging genetically. Geographic barriers usually arise much more slowly. For example, it took millions of years of tectonic plate movements to bring the two continents of North and South America close enough to collide. The land bridge where the two continents now connect is called the Isthmus of Panama. When this isthmus formed about 4 million years ago, it cut off the flow of water— and gene flow among populations of aquatic organisms—as it separated one large ocean into what are now the Pacific and Atlantic Oceans (Figure 12.15).

Sympatric Speciation In **sympatric speciation**, populations in the same geographic region speciate in the absence of a physical barrier between them (*sym–* means together). Sympatric speciation can occur in a single generation when the chromosome number multiplies. Polyploidy typically arises when an abnormal nuclear division during meiosis or mitosis doubles the chromosome number. For example, if the nucleus of a somatic cell in a flowering plant fails to divide during mitosis, the resulting cell—which is polyploid—may proliferate and give rise to shoots and flowers. If the flowers can self-fertilize, a new polyploid species may be the result. Common bread wheat originated after related species hybridized, and then the chromosome number of the hybrid offspring doubled (Figure 12.16).

Sympatric speciation can also occur with no change in chromosome number. The sage plants you just learned about speciated with no physical barrier to gene flow. As another example, more than 500 species of cichlid fishes arose by sympatric

Figure 12.15 Example of allopatric speciation.
When the Isthmus of Panama formed 4 million years ago, it cut off gene flow among ocean-dwelling populations of snapping shrimp. Today, shrimp species on opposite sides of the isthmus are so similar that they might interbreed, but they are reproductively isolated. Instead of mating when they are brought together, they snap their claws at one another aggressively. The photos show two of the many closely related species that live on opposite sides of the isthmus.

Right, © Arthur Anker.

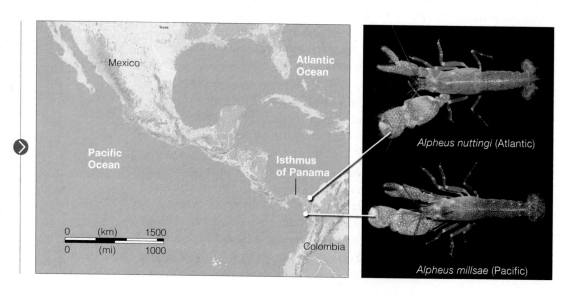

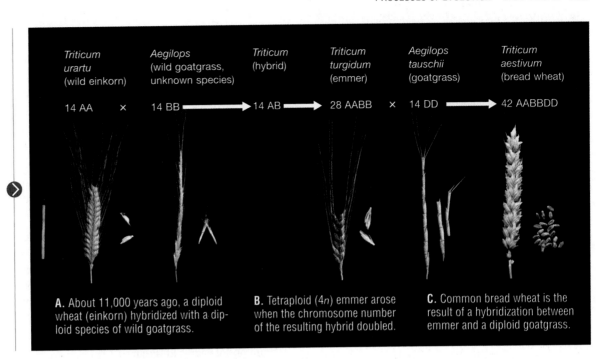

Figure 12.16 Sympatric speciation in wheat.
The wheat genome, which consists of seven chromosomes, occurs in slightly different forms called A, B, C, D, and so on. Many wheat species are polyploid, carrying more than two copies of the genome. For example, modern bread wheat (*Triticum aestivum*) is hexaploid, with six copies of the wheat genome: two each of genomes A, B, and D (or 42 AABBDD).

Photos by © J. Honegger, courtesy of S. Stamp, E. Merz, www.sortengarten/ethz.ch.

Triticum urartu (wild einkorn) — *Aegilops* (wild goatgrass, unknown species) — *Triticum* (hybrid) — *Triticum turgidum* (emmer) — *Aegilops tauschii* (goatgrass) — *Triticum aestivum* (bread wheat)

14 AA × 14 BB → 14 AB → 28 AABB × 14 DD → 42 AABBDD

A. About 11,000 years ago, a diploid wheat (einkorn) hybridized with a diploid species of wild goatgrass.

B. Tetraploid (4*n*) emmer arose when the chromosome number of the resulting hybrid doubled.

C. Common bread wheat is the result of a hybridization between emmer and a diploid goatgrass.

speciation in the shallow waters of Lake Victoria. This large freshwater lake sits isolated from river inflow on an elevated plain in Africa's Great Rift Valley. Since Lake Victoria formed about 400,000 years ago, it has dried up three times. DNA sequence comparisons indicate that almost all of the cichlid species in this lake arose since the last dry spell, which was 12,400 years ago. How could hundreds of species arise so quickly? In this case, the answer begins with differences in the color of ambient light in different parts of the lake. The light in the lake's shallower, clear water is mainly blue; light that penetrates the deeper, muddier water is mainly red. The cichlid species vary in color (Figure 12.17), and female cichlids prefer to mate with brightly colored males. Their preference has a genetic basis, in alleles that encode light-sensitive pigments of the retina (part of the eye). Retinal pigments made by species that live mainly in shallow areas of the lake are more sensitive to blue light. The males of these species are also the bluest. Retinal pigments made by species that prefer deeper areas of the lake are more sensitive to red light. Males of these species are redder. In other words, the colors that a female cichlid sees best are the same colors displayed by males of her species. Thus, mutations that affect color perception are likely to affect a female's choice of mates. Such mutations are probably the way sympatric speciation occurs in these fishes.

Figure 12.17 Males of four closely related species of cichlid native to Lake Victoria, Africa.
Hundreds of cichlid species arose by sympatric speciation in this lake. Mutations that affect female cichlids' perception of the color of ambient light in deeper or shallower regions of the lake also affect their choice of mates. Female cichlids prefer to mate with the most brightly colored males.

Kevin Bauman, www.african-cichlid.com

Figure It Out: Which form of natural selection is driving sympatric speciation in these cichlids?

Answer: Sexual selection

Take-Home Message 12.6

How do species attain and maintain separate identities?

- Speciation is an evolutionary process by which new species form. It varies in its details and duration, but always includes reproductive isolation.

- A physical barrier that intervenes between populations of a species prevents gene flow among them. When gene flow ends, genetic divergences give rise to new species.

- Divergence within a population can lead to new species that inhabit the same geographical area, with no physical barrier to gene flow.

12.7 Macroevolution

REMEMBER: The scientific community consists of critically thinking people trying to poke holes in one another's ideas (Section 1.6). Predator and prey are locked in a constant race, with each genetic improvement in one species countered by a genetic improvement in the other (8.4). The dinosaurs died out in a mass extinction that occurred 66 million years ago (11.1). Transitions in the fossil record are correlated with boundaries between rock layers in the geologic time scale (11.5).

Microevolution is change in allele frequency within a single species or population. We also see evolutionary patterns on a larger scale than microevolution, and these large-scale patterns are called **macroevolution**. Macroevolution includes trends such as land plants evolving from green algae, all dinosaurs disappearing in a mass extinction, a burst of divergences from a single species, and so on.

Very little change may occur over a very long period of time. Consider coelacanths, an order of ancient lobe-finned fish that had been assumed extinct for at least 70 million years until a fisherman caught one in 1938. In its unique form and other traits, modern coelacanth species are similar to fossil specimens hundreds of millions of years old (Figure 12.18).

Major evolutionary novelties often stem from the adaptation of an existing structure for a completely new purpose. For example, the feathers that allow modern birds to fly are derived from feathers that first evolved in some dinosaurs. Those dinosaurs could not have used their feathers for flight, but they probably did use them for insulation.

By current estimates, more than 99 percent of all species that ever lived are now **extinct**, which means they no longer have living members. In addition to continuing extinctions of individual species, the fossil record indicates that there have been more than twenty mass extinctions, which are simultaneous losses of many lineages. These include five catastrophic events in which the majority of species on Earth disappeared.

Figure 12.18 The coelacanth: a living fossil.
Photos on the left compare a 320-million-year-old coelacanth fossil found in Montana with a live coelacanth. The diagram on the right shows a few of the coelacanth's unusual ancestral features that have been lost in almost all other fish lineages over evolutionary time.

Top left, Courtesy of The Virtual Fossil Museum, www.fossilmuseum.net; bottom left, AllessandroZocc/Shutterstock.com; right, Raul Martin Domingo/National Geographic Creative.

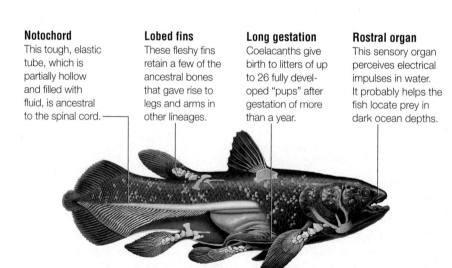

Notochord
This tough, elastic tube, which is partially hollow and filled with fluid, is ancestral to the spinal cord.

Lobed fins
These fleshy fins retain a few of the ancestral bones that gave rise to legs and arms in other lineages.

Long gestation
Coelacanths give birth to litters of up to 26 fully developed "pups" after gestation of more than a year.

Rostral organ
This sensory organ perceives electrical impulses in water. It probably helps the fish locate prey in dark ocean depths.

Akepa (*Loxops coccineus*) Akikiki (*Oreomystis bairdi*) Nihoa finch (*Telespiza ultima*) Akekee (*Loxops caeruleirostris*)

Iiwi (*Vestiaria coccinea*) Akohekohe (*Palmeria dolei*) Hawaii Amakihi (*Hemignathus virens*) Akiapolaau (*Hemignathus munroi*)

Maui parrotbill (*Pseudonestor xanthophrys*) Maui Alauahio (*Paroreomyza montana*) Kauai Amakihi (*Hemignathus kauaiensis*) Apapane (*Himatione sanguinea*)

Figure 12.19 A few Hawaiian honeycreepers.

The bills of these birds are adapted to feed on insects, seeds, fruits, nectar in floral cups, and other foods. All Hawaiian honeycreepers descended from a Eurasian rosefinch that probably resembled modern rosefinches (left).

(Akepa, Nihoa finch, Maui Alauahio) © Jack Jeffrey Photography; (Akekee, Akohekohe, Apapane, Akiapolaau, Maui parrotbill, Kauai Amakihi) © Eric VanderWerf/Pacific Rim Photos; (Iiwi) Michael Ord/Science Source; (Hawaii Amakihi) James A. Hancock/Science Source; (Akikiki) U.S. Geological Survey/photo by Carter Atkinson; (rosefinch) Andrzej Sliwinski/Shutterstock.

With **adaptive radiation**, one lineage rapidly diversifies into several new species. Adaptive radiation can occur after a population colonizes a new environment that has a variety of different habitats and few competitors. Speciation occurs along with adaptation to the different habitats. The Lake Victoria cichlids that you learned about in the previous section arose this way, as did the Hawaiian honeycreepers (Figure 12.19). Some finch species migrate far outside of their normal range when food becomes scarce in a preferred overwintering spot, traveling in flocks of thousands or even tens of thousands of individuals. About 5.8 million years ago in southern Asia, one of these migratory flocks was caught up in the winds of a huge storm. The birds—rosefinches—were blown at least 7,000 miles (11,000 kilometers) across the open ocean to the islands of the Hawaiian archipelago (right). Enough individuals survived the journey to found a new population. The birds' arrival had been preceded by insects and plants, but no predators, and their descendants spread into habitats along the coasts, through dry lowland forests, and into highland rain forests. Isolation from gene flow with mainland finch populations allowed the island colonizers to diverge. Over many generations, populations living in the different habitats became hundreds of separate species—the honeycreepers—as unique forms and behaviors evolved in them. These unique traits helped the birds exploit special opportunities presented by their particular island habitats.

Hawaiian archipelago

Figure 12.20 Example of coevolved species.

(A) © Brian Raine, www.flickr.com/people/25801055@N00. (B) © Jeremy Thomas/Natural Visions.

A. A *Maculinea arion* butterfly emerges from a pupa to mate and lay eggs on wild thyme flowers. Larvae that emerge from the eggs will survive only if a colony of *Myrmica sabuleti* ants adopts them.

B. A *Maculinea arion* caterpillar interacting with a *Myrmica sabuleti* ant. This beguiled ant is preparing to carry the honey-exuding, hunched-up caterpillar back to its nest, where the caterpillar will feed on ant larvae for the next 10 months until it becomes a pupa.

Adaptive radiation may also occur after a key innovation evolves. A **key innovation** is a trait that allows its bearer to exploit a habitat more efficiently or in a novel way. The evolution of lungs offers an example, because lungs were a key innovation that opened the way for an adaptive radiation of vertebrates on land. A geologic or climatic event that eliminates some species from a habitat can spur adaptive radiation; species that survive the event then have access to resources from which they had previously been excluded. This is the way mammals were able to undergo an adaptive radiation after the dinosaurs disappeared 66 million years ago.

Two species that have close ecological interactions may evolve jointly, a pattern called **coevolution**. One species acts as an agent of selection on the other, and each adapts to changes in the other. Over evolutionary time, the two species may become so interdependent that they can no longer survive without one another. Relationships between coevolved species can be quite intricate. Consider the large blue butterfly (*Maculinea arion*), a parasite of ants. After hatching, the butterfly larvae (caterpillars) feed on wild thyme flowers (Figure 12.20A) and then drop to the ground. An ant that finds a caterpillar strokes it, which makes the caterpillar exude honey. The ant eats the honey and continues to stroke the caterpillar, which secretes more honey. This interaction continues for hours, until the caterpillar suddenly hunches itself up (Figure 12.20B). The ant then picks up the caterpillar and carries it back to its nest, where, in most cases, other ants kill it—except if the ants are of the species *Myrmica sabuleti*. The caterpillar secretes the same chemicals as *Myrmica sabuleti* larvae, and makes the same sounds as their queen—behaviors that deceive the ants into adopting the caterpillar and treating it better than their own larvae. The adopted caterpillar feeds on ant larvae for about 10 months, then undergoes metamorphosis, changing into a butterfly that emerges from the ground to mate. Eggs are deposited on wild thyme near another *M. sabuleti* nest, and the cycle starts anew. This relationship between ant and butterfly is typical of coevolved relationships in that it is extremely specific. Any increase in the ants' ability to identify a caterpillar in their nest selects for caterpillars that better deceive the ants, which in turn select for ants that can better identify the caterpillars. Each species exerts directional selection on the other.

Evolutionary Theory Biologists have no doubt that macroevolution occurs, but many disagree about how it occurs. However we choose to categorize evolutionary processes, the very same genetic change may be at the root of all evolution—fast or slow, large-scale or small-scale. Dramatic jumps in form, if they are not artifacts of gaps in the fossil record, may be the result of mutations in homeotic or other regulatory genes. Macroevolution may include more processes than microevolution, or it may not. It may be an accumulation of many microevolutionary events, or it may be an entirely different process. Evolutionary biologists may disagree about these and other hypotheses, but all of them are trying to explain the same thing: how all species are related by descent from common ancestors.

Take-Home Message 12.7

What is macroevolution?

- Large-scale patterns of evolutionary change such as adaptive radiation, the origin of major groups, and mass extinctions are called macroevolution.

12.8 Phylogeny

REMEMBER: A "species" is a convenient but artificial construct of the human mind; with taxonomy, we systematically name species and rank them into higher taxa based on shared traits (Section 1.4). A homeotic gene is a type of master gene, and its expression directs the formation of a specific body part during embryonic development (7.7). Humans reconstruct history by studying physical evidence of past events (11.1). Body parts that appear alike may have evolved independently in lineages that have faced similar environmental pressures (11.6). Similarities in the nucleotide sequence of a shared gene (or in the amino acid sequence of a shared protein) are often used as evidence of an evolutionary relationship (11.7).

Classifying life's tremendous diversity into a series of taxonomic ranks is a useful endeavor, in the same way that it is useful to organize a telephone book or contact list in alphabetical order: The result is convenient. Traditional (Linnaean) classification schemes rank species into higher taxa based on shared traits—birds have feathers, cacti have spines, and so on—but these rankings do not necessarily reflect evolutionary relationships.

Today's biologists work from the premise that every living thing is related if you just look back far enough in time. Grouping species according to evolutionary relationships is a way to fill in the details of this bigger picture of evolution. Thus, reconstructing **phylogeny**, the evolutionary history of a species or a group of species, is a priority. Phylogeny is a kind of genealogy that follows evolutionary relationships through time. Instead of ranking species by shared traits, evolutionary biologists spend their time pinpointing what makes the species share the traits in the first place: a common ancestor. They determine common ancestry by looking for a derived trait—one that is present in a group under consideration, but not in any of the group's ancestors.

A group whose members share one or more defining derived traits is called a **clade**. By this definition, each species is a clade. Many higher taxonomic rankings are also equivalent to clades—flowering plants, for example, are both a phylum and a clade—but some are not. For example, the traditional Linnaean class Reptilia ("reptiles") includes crocodiles, alligators, tuataras, snakes, lizards, turtles, and tortoises. While it is convenient to classify these animals together, they would not constitute a clade unless birds are also included, as you will see in Chapter 15.

All species are interconnected in the big picture of evolution; an evolutionary biologist's job is to figure out where the connections are. All species bear traces of their evolutionary history in their traits. For example, humans and bacteria use some of the same proteins to repair DNA. Even so, humans and bacteria are not close relatives. It is the relative newness of a shared trait that defines a clade. Consider how alligators look a lot more like lizards than birds. In this case, the similarity in appearance does indicate shared ancestry, but it is a more distant relationship than alligators have with birds. A unique set of traits that include a gizzard and a four-chambered heart evolved in the lineage that gave rise to alligators and birds, but not in the lineage that gave rise to lizards.

Making hypotheses about evolutionary relationships among clades is called **cladistics**. A **cladogram** is an evolutionary tree diagram that visually summarizes a hypothesis about how a group of clades are related (Figure 12.21). Data from an outgroup (a species not closely related to any member of the group under study) may be included in order to "root" the tree. Each line is a lineage, which may branch into two lineages. The branch point represents a common ancestor of two lineages.

A. Evolutionary connections are represented as lines on a cladogram. Each line is a lineage, and each branch point represents a common ancestor.

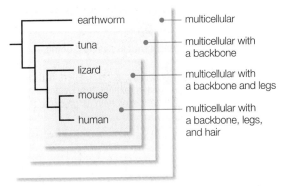

B. A cladogram can be viewed as "sets within sets" of derived traits. Each set (an ancestor together with all of its descendants) is a clade.

Figure 12.21 An example of a cladogram.

© National Geographic/SuperStock.

---

clade A group whose members share one or more defining derived traits.

cladistics Making hypotheses about evolutionary relationships among clades.

cladogram Evolutionary tree diagram that summarizes hypothesized relationships among a group of clades.

coevolution The joint evolution of two closely interacting species; each species is a selective agent for traits of the other.

key innovation An evolutionary adaptation that gives its bearer the opportunity to exploit a particular environment more efficiently or in a new way.

phylogeny Evolutionary history of a species or group of species.

A. The palila has an adaptation that allows it to feed on seeds of a native Hawaiian plant that are toxic to most other birds. The one remaining palila population is declining because these plants are being trampled by cows and eaten by goats and sheep. Only about 1,200 palila remained in 2010.

B. The lower bill of the akekee points to one side, allowing this bird to pry open buds that harbor insects. Avian malaria is wiping out the last population of this species. Between 2000 and 2007, the number of akekee dropped from 7,839 birds to 3,536.

C. This poouli—rare, old, and missing an eye—died in 2004 from avian malaria. There were two other poouli alive at the time, but neither has been seen since then.

Figure 12.22 Three honeycreeper species: going, going, and gone.
The genetic diversity of Hawaiian honeycreepers is dwindling along with their continued extinctions. Deciphering their evolutionary connections may help us preserve the remaining species.

(A) © Eric VanderWerf/Pacific Rim Photos; (B) Courtesy of © Lucas Behnke; (C) Bill Sparklin/Ashley Dayer.

Evolutionary history does not change because of events in the present: A species' ancestry remains the same no matter how it evolves. However, as with traditional taxonomic rankings, we can make mistakes grouping organisms based on incomplete information. Thus, a clade or cladogram may change when new discoveries are made. As with all hypotheses, the more data in support of an evolutionary grouping, the less likely it is to require revision.

Applications of Phylogeny   Studies of phylogeny reveal how species relate to one another and to species that are now extinct. In doing so, they inform our understanding of how shared ancestry interconnects all species—including our own.

The story of the Hawaiian honeycreepers offers an example of how finding ancestral connections can help species that are still living. The first Polynesians arrived on the Hawaiian islands sometime before 1000 A.D.; Europeans followed in 1778. Hawaii's rich ecosystem was hospitable to the newcomers and their domestic animals and crops. Entire forests were cleared to grow imported crops, and plants that escaped cultivation began to crowd out native plants. Escaped livestock ate and trampled rain forest plants that had provided the honeycreepers with food and shelter. Mosquitoes accidentally introduced in 1826 spread diseases such as avian malaria from imported chickens to native bird species. Stowaway rats ate their way through populations of native birds and their eggs; mongooses deliberately imported to eat the rats preferred to eat birds and bird eggs.

The isolation that had allowed honeycreepers to arise by adaptive radiation also made them vulnerable to extinction. Divergence from the ancestral species had led to the loss of unnecessary traits such as defenses against mainland predators and diseases. Traits that had been adaptive—such as a long, curved beak matching the flower of a particular plant—became hindrances when habitats suddenly changed or disappeared. Thus, at least 43 Hawaiian honeycreeper species that had thrived on the islands before humans arrived were extinct by 1778. Conservation efforts began in the 1960s, but another 43 species have since disappeared.

Today, the few remaining Hawaiian honeycreepers are still being heavily pressured by established populations of nonnative species of plants and animals (Figure 12.22A). Rising global temperatures are also allowing mosquitoes to invade high-altitude habitats that had previously been too cold for the insects, so honeycreeper species remaining in these habitats are now succumbing to mosquito-borne diseases (Figure 12.22B). Of the 18 remaining honeycreeper species, only two are not in danger of extinction.

As more and more honeycreeper species become extinct, the group's reservoir of genetic diversity dwindles. The lowered diversity means the group as a whole is less resilient to change, and more likely to suffer catastrophic losses. Deciphering their phylogeny can tell us which honeycreeper species are most different from the others—and those are the ones most valuable in terms of preserving the group's genetic diversity. Such research allows us to concentrate our resources and conservation efforts on those species whose extinction would mean a greater loss to biodiversity. For example, we now know the poouli (Figure 12.22C) to be the most distant relative in the Hawaiian honeycreeper family. Unfortunately, the knowledge came too late; the poouli is probably extinct now. Its extinction means the loss of a large part of evolutionary history of the group: One of the longest branches of the honeycreeper family tree is gone forever.

Cladistics analyses are also used to correlate past evolutionary divergences with behavior and dispersal patterns of existing populations. Such studies are useful in conservation efforts. For example, a decline in antelope populations in African savannas is at least partly due to competition with domestic cattle. A cladistic analysis of mitochondrial DNA sequences suggested that current populations of blue wildebeest (Figure 12.23) are genetically less similar than they should be, based on other antelope groups of similar age. Combined with behavioral and geographic data, the analysis helped conservation biologists realize that a patchy distribution of preferred food plants is preventing gene flow among blue wildebeest populations. The absence of gene flow can lead to a catastrophic loss of genetic diversity in populations under pressure. Restoring appropriate grasses in intervening, unoccupied areas of savanna would allow isolated wildebeest populations to reconnect.

Researchers often study the evolution of viruses and other infectious agents by grouping them into clades based on biochemical traits. Even though viruses are not alive, they can mutate every time they infect a host, so their genetic material changes quickly. Consider the H5N1 strain of influenza (flu) virus, which infects birds and other animals. H5N1 has a very high mortality rate in humans, but human-to-human transmission has been rare to date. The virus replicates in pigs without causing symptoms. Pigs transmit the virus to other pigs—and apparently to humans too. A phylogenetic analysis of H5N1 isolated from pigs showed that the virus "jumped" from birds to pigs at least three times since 2005, and that one of the isolates had acquired the potential to be transmitted among humans. Our increased understanding of the evolutionary history of this virus is helping us develop strategies to prevent it from spreading to humans again.

Figure 12.23 A blue wildebeest in Africa.
Conservation biologists discovered that a patchy availability of preferred food was hampering gene flow among wildebeest populations. The biologists recommended restoring grasses in some areas that had been cleared, to reestablish gene flow among isolated wildebeests.

Alan Lucas/Shutterstock.

Every living thing is related if you just look back far enough in time.

Take-Home Message 12.8

Why do we study evolutionary history?

- Evolutionary biologists study phylogeny in order to understand how all species are connected by shared ancestry.
- Among other applications, phylogeny research can help us to prioritize efforts to preserve endangered species, and to understand the spread of infectious diseases.

Summary

Section 12.1 Populations tend to change along with the selection pressures that operate on them. Our overuse of antibiotics exerts directional selection favoring resistant bacterial populations, which are now common in the environment. We are running out of effective antibiotics to use as human drugs.

Section 12.2 Alleles, which arise by mutation, are the basis of differences in the forms of traits shared by a species. All alleles of all genes in a population constitute a **gene pool**. **Microevolution**, which is change in **allele frequency** in a gene pool, occurs constantly in natural poplations because processes that drive it are always operating in nature.

Section 12.3 Natural selection can occur in patterns. In **directional selection**, forms of a trait at one end of a range of variation are most adaptive. An intermediate form of a trait is most adaptive in **stabilizing selection**. In **disruptive selection**, extreme forms of a trait are adaptive and midrange forms are selected against.

Section 12.4 **Sexual selection** is a mode of natural selection in which the adaptive traits are those that make their bearers better at securing mates. Any mode of natural selection can maintain multiple alleles at relatively high frequency in a population.

Section 12.5 Allele frequency can change due to chance alone. This **genetic drift**, which is most pronounced in small populations, can lead to the loss of genetic diversity and cause alleles to become **fixed**. Genetic diversity may be reduced in populations that are **inbred**, and also in those that undergo an evolutionary **bottleneck** or have been founded by a small group of individuals (the **founder effect**). **Gene flow** can stabilize or change allele frequency.

Section 12.6 The details of **speciation** differ every time it occurs, but **reproductive isolation**, the end of gene flow between populations, is always a part of the process. With **allopatric speciation**, a geographic barrier arises and interrupts gene flow between populations. After gene flow ends, genetic divergences that occur independently in the separated populations result in separate species. Speciation can also occur in the absence of a barrier to gene flow. **Sympatric speciation** occurs by genetic divergence within a population.

Section 12.7 **Macroevolution** refers to large-scale patterns of evolution. A lineage may change very little over evolutionary time. In some cases, a body structure used for a particular purpose in a lineage served a different purpose when it first evolved in an ancestor. A **key innovation** can result in an **adaptive radiation**, or rapid diversification of a lineage into several new species. **Coevolution** occurs when two species act as agents of selection upon one another. A lineage with no more living members is **extinct**.

Section 12.8 Evolutionary biologists reconstruct evolutionary history (**phylogeny**) by looking for derived traits. A **clade** consists of an ancestor in which a derived trait evolved, together with all of its descendants. Making hypotheses about the evolutionary history of a clade is called **cladistics**. These hypotheses are often represented as **cladograms**, which are diagrams of evolutionary connections among a group of clades. Each line in a cladogram represents a lineage, and a point where one lineage branches into two represents a shared ancestor.

Reconstructing phylogeny, which is based on the premise that all organisms are connected by shared ancestry, helps us preserve endangered species. It is also useful for studying the spread of viruses and other agents of infectious diseases.

Self-Quiz

Answers in Appendix I

1. _____ is the original source of new alleles.
   - a. Mutation
   - b. Natural selection
   - c. Genetic drift
   - d. Gene flow
   - e. All are original sources of new alleles

2. Which is required for evolution to occur in a population?
   - a. genetic diversity
   - b. selection pressure
   - c. gene flow
   - d. none of the above

3. Match the modes of natural selection with their best descriptions.
   - _____ stabilizing
   - _____ disruptive
   - a. eliminates extreme forms of a trait
   - b. eliminates midrange forms of a trait

4. Sexual selection frequently influences aspects of body form and can lead to _____ .
   - a. a sexual dimorphism
   - b. male aggression
   - c. exaggerated traits
   - d. all of the above

5. The persistence of the sickle allele at high frequency in a population is a case of _____ .
   a. bottlenecking
   b. balanced polymorphism
   c. the founder effect
   d. inbreeding

6. _____ among populations can keep them similar to one another.
   a. Genetic drift
   b. Gene flow
   c. Mutation
   d. Natural selection

7. The theory of natural selection does not explain _____ .
   a. genetic drift
   b. the founder effect
   c. gene flow
   d. how mutations arise
   e. inheritance
   f. any of the above

8. Which of the following is *not* part of how we define a species?
   a. Its individuals appear different from other species.
   b. It is reproductively isolated from other species.
   c. Its populations can interbreed.
   d. Fertile offspring are produced.

9. Which of the following statements is correct?
   a. Genetic drift occurs only in small populations.
   b. Inbreeding increases genetic diversity.
   c. Gene flow can introduce new alleles into a population.

10. After fire devastates all of the trees in a wide swath of forest, populations of a species of tree-dwelling frog on either side of the burned area diverge to become separate species. This is an example of _____ .
    a. allopatric speciation
    b. adaptive radiation
    c. sympatric speciation
    d. an evolutionary bottleneck

11. Sex in many birds is typically preceded by an elaborate courtship dance. If a male's movements are unrecognized by the female, she will not mate with him. This is an example of _____ .
    a. reproductive isolation
    b. natural selection
    c. sexual selection
    d. all of the above

12. _____ is a way of reconstructing evolutionary history based on derived traits.
    a. Natural selection
    b. Phylogeny
    c. Gene flow
    d. Cladistics

13. The evolution of wings helped the insect clade to be very successful. In this example, wings are a(n) _____ .
    a. derived trait
    b. adaptive trait
    c. key innovation
    d. all of the above

14. In evolutionary trees, each line represents a(n) _____ .
    a. lineage
    b. extinction
    c. point of divergence
    d. adaptive radiation

15. Match the evolution concepts.
    _____ gene flow
    _____ sexual selection
    _____ derived trait
    _____ extinct
    _____ genetic drift
    _____ natural selection
    _____ cladogram
    _____ adaptive radiation
    _____ phylogeny
    _____ coevolution

    a. can lead to interdependent species
    b. changes in a population's allele frequencies due to chance alone
    c. alleles enter or leave a population
    d. evolutionary history
    e. adaptive traits make their bearers better at securing mates
    f. burst of divergences from one lineage into many
    g. no more living members
    h. diagram of sets within sets
    i. present in a group, but not in any of the group's ancestors
    j. operates on variations in shared traits

Critical Thinking

1. Species have traditionally been characterized as "primitive" and "advanced." For example, mosses were considered to be primitive, and flowering plants advanced; crocodiles were primitive and mammals were advanced. Why do most biologists of today think it is incorrect to refer to any modern species as primitive?

2. Rama the cama, a llama–camel hybrid, was born in 1997. The idea was to breed an animal that has the camel's strength and endurance, and the llama's gentle disposition. However, instead of being large, strong, and sweet, Rama is smaller than expected and has a camel's short temper. The breeders plan to mate him with Kamilah, a female cama. What potential problems with this mating should the breeders anticipate?

3. Two species of antelope, one from Africa, the other from Asia, are put into the same enclosure in a zoo. To the zookeeper's surprise, individuals of the different species begin to mate and produce healthy, hybrid baby antelopes. Explain why a biologist might not view these offspring as evidence that the two species of antelope are in fact one.

4. Some human traits may have arisen by sexual selection. Over thousands of years, women attracted to charming, witty men perhaps prompted the development of human intellect beyond what was necessary for mere survival. Men attracted to women with juvenile features may have shifted the species as a whole to be less hairy and softer featured than any of our simian relatives. Can you think of a way to test these hypotheses?

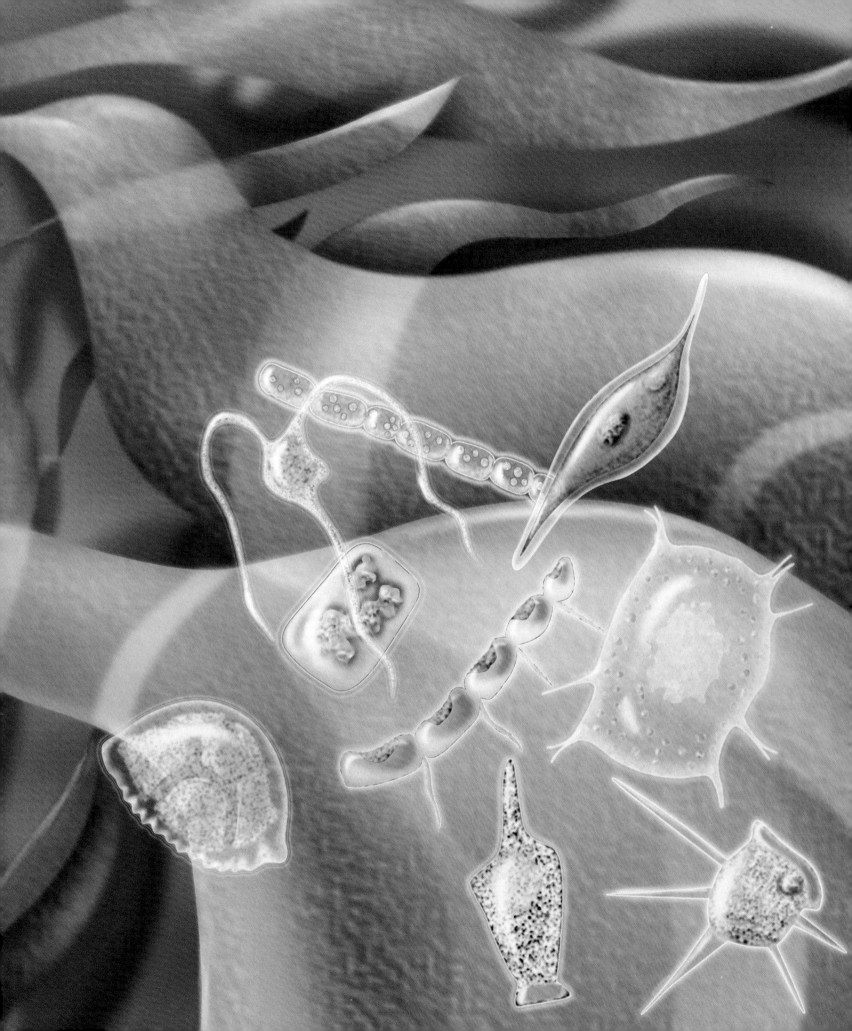

Application ⟩ 13.1 The Human Microbiome

REMEMBER: Biologists divide all life into three domains: Bacteria, Archaea, and Eukarya (Section 1.4).

Figure 13.1 Microbes that can live in the human gut.

(A) Dr. Fred Hossler/Visuals Unlimited, Inc.; (B) CDC/Dr. Stan Erlandsen.

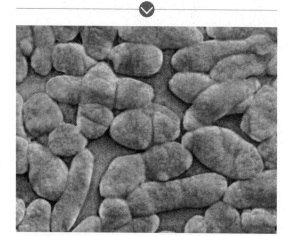

A. Bacteria (*Bilophila wadsworthia*). This species is most abundant in people who eat a lot of animal fat, and it may promote irritable bowel disease.

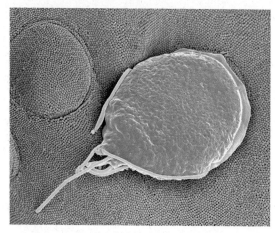

B. The flagellated protist *Giardia lamblia*. It helps fend off parasitic worms, but causes diarrhea and abdominal pain in some infected people.

microbiome Collection of microorganisms that inhabits a specific habitat, such as a human body.

pathogen Disease-causing agent.

The first forms of life lived as single cells and the overwhelming majority of modern organisms do the same. Consider that the single-celled organisms living in and on your body outnumber your cells by about ten to one. Even when healthy, a person is host to bacteria, archaea, protists, fungi, and viruses. The Human Microbiome Project, a collaborative endeavor that began in 2007, aims to identify the types of microorganisms we support—our **microbiome**—and how their presence affects our health and well-being. The project has already turned up some interesting findings.

Each person has a unique microbiome and its composition changes over time. The womb is sterile, so we acquire our first microorganism during birth. During normal childbirth, an infant becomes coated with bacteria from its mother's vagina. Some of these bacteria are swallowed and colonize the infant's digestive tract, where they help the infant digest milk sugars. Picking up maternal bacteria during birth may also kick-start the infant's immune system. Surgical delivery (delivery by a cesarean section or C-section) increases the risk of allergies and other immune disorders later in life. By one hypothesis, this heightened susceptibility to immune problems results from the failure of a surgical delivery to provide the infant with the types of bacteria that normally encourage immune development.

The current mix of species in your gut depends in part on your diet, and that species assortment may in turn affect your health. For example, some diets promote the growth of bacterial **pathogens**, or agents of disease. Consider that a diet rich in grain products selects for gut bacteria that can break down complex carbohydrates. An abundance of one such species, *Prevotella copri*, is associated with an increased risk for rheumatoid arthritis, a disorder in which the immune system attacks the joints. On the other hand, eating an excessive amount of animal fat selects for bacteria that can metabolize bile, an acidic substance essential to fat digestion. One bile-tolerant species, *Bilophila wadsworthia* (Figure 13.1A), increases the risk of irritable bowel disease (IBD). With IBD, chronic inflammation of the large intestine causes cramping, diarrhea, and weight loss. An estimated 14 million Americans have IBD and the prevalence of the disorder is on the rise worldwide. Increasing consumption of fat-rich, processed foods that encourage the growth of *B. wadsworthia* may be a factor in this increase.

The overwhelming majority of organisms that can live in the human gut are bacteria, but some protists also call the human gut home. The flagellated protozoan *Giardia lamblia* (Figure 13.1B) is one example. *G. lamblia* was first described in the late 1600s by Antoni van Leeuwenhoek, a pioneer in the use of microscopy. Van Leeuwenhoek noticed the random jerky movements of *G. lamblia* while examining a sample of his own feces. Today, *G. lamblia* infects an estimated 2 to 5 percent of adults in the United States. In nations where sanitation is poor, the adult infection rate can be as high as 30 percent. Unfiltered drinking water from streams or lakes and unpasteurized dairy products are common sources of infection.

G. lamblia infection sometimes cause IBD-like symptoms, but many infected people have no ill effects. Some even benefit from the protist's presence. Having *G. lamblia* in the gut reduces one's risk of infection by parasitic worms. Whether people infected by *G. lamblia* become ill depends both on their general health and the particular strain of *G. lamblia* that infects them. (In microbiology, the term "strain" refers to a genetically distinct subtype of a particular microorganism.)

13.2 On the Road to Life

REMEMBER: All organisms consist of complex organic compounds that they assembled from organic monomers (Section 2.6).

Conditions on the Early Earth Scientists estimate that Earth formed by about 4.6 billion years ago through the aggregation of dust and rock bits that were orbiting our sun. The composition of Earth's early atmosphere remains a matter of debate, but geologic evidence suggests our planet started out with little or no free oxygen (O_2). Had O_2 been present early on, we would see evidence of iron oxidation (rust formation) in Earth's most ancient rocks. However, these rocks show no sign of such oxidation. The apparent lack of O_2 interests scientists because it would have facilitated some proposed steps on the path to life. Had O_2 been present, oxidation reactions would have broken apart small organic compounds as quickly as they formed.

Liquid water is essential to life as we know it because molecules that carry out metabolic reactions have to be dissolved in water. At first, Earth's surface was molten rock, so all water was in the form of vapor. However, examination of crystals in ancient rocks indicates that by 4.3 billion years ago, Earth had cooled enough for water to pool on its surface.

Origin of the Building Blocks of Life Until the early 1800s, chemists thought that organic molecules possessed a special "vital force" and could only be made by living organisms. Then in 1825, a German chemist synthesized urea, a molecule abundant in urine. Later, another chemist made alanine, an amino acid. These synthetic reactions showed that nonliving mechanisms could yield organic molecules.

Today, there are three main hypotheses concerning the source of the organic building blocks for Earth's first life.

1. *Lightning fueled atmospheric reactions.* In 1950s, Stanley Miller and his colleagues tested the hypothesis that lightning-fueled atmospheric reactions could have produced simple organic compounds. They filled a reaction chamber with a mix of gases designed to simulate Earth's early atmosphere, then circulated the mixture while zapping it with sparks from electrodes (Figure 13.2). Within a week, this process produced simple organic compounds, including some amino acids present in living organisms.

2. *Delivery from space via meteorites.* The presence of amino acids, sugars, and nucleotide bases in meteorites that fell to Earth suggests an alternative origin for life's building blocks. Organic monomers that formed in interstellar clouds of ice, dust, and gases could have been delivered to Earth by meteorites. Keep in mind that during Earth's early years, meteorites fell to Earth thousands of times more frequently than they do today.

3. *Reactions at deep-sea hydrothermal vents.* Life's building blocks may also have formed in the sea, fueled by heat from hydrothermal vents. A **hydrothermal vent** is like an underwater geyser, a place where mineral-rich water heated by geothermal energy streams out through a rocky opening in the seafloor (Figure 13.3). Amino acids form spontaneously in a simulated vent environment.

Note that the three possible sources of organic monomers discussed above are not mutually exclusive. Most likely all three contributed to an accumulation of simple organic compounds in Earth's early seas.

Figure 13.2 Stanley Miller's experimental apparatus. It was used to test whether lightning-fueled reactions could have formed organic monomers in Earth's early atmosphere. Water vapor, hydrogen gas (H_2), methane (CH_4), and ammonia (NH_3) circulated in a glass chamber to simulate the atmosphere. Sparks provided by an electrode simulated lightning.

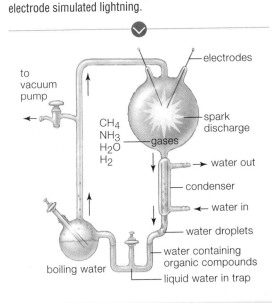

Figure It Out: Why didn't Miller include oxygen (O_2) in the mix of gases?

Answer: Earth's early atmosphere lacked oxygen.

Figure 13.3 A hydrothermal vent on the seafloor. At such vents, mineral-rich water heated by geothermal energy streams out into cold ocean water.

Courtesy of the University of Washington.

hydrothermal vent Underwater opening from which mineral-rich water heated by geothermal energy streams out.

Origin of Metabolism Modern cells take up small organic molecules, concentrate them, and assemble them into larger organic polymers. Before there were cells, a nonbiological process that concentrated organic subunits would have increased the chance of polymer formation.

By one hypothesis, organic polymers began to form on clay-rich tidal flats. Clay particles have a slight negative charge, so positively charged molecules in seawater stick to them. At low tide, evaporation would have concentrated the subunits even more, and energy from sunlight might have induced the formation of polymers. Amino acids do form short chains under simulated tidal flat conditions.

The **iron–sulfur world hypothesis** proposes that early metabolic reactions took place in rocks around hydrothermal vents. Such rocks are porous, with many tiny chambers about the size of cells. Metabolism may have begun when iron sulfide in the rocks donated electrons to dissolved carbon monoxide (CO), setting in motion reactions that led to formation of larger organic compounds. In simulations of vent conditions, organic compounds such as pyruvate do form and accumulate. In addition, iron–sulfur clusters serve as cofactors in all modern organisms. The clusters function as electron donors in essential metabolic reactions. A universal requirement for iron–sulfur cofactors may be a legacy of life's rocky beginnings.

Origin of Genetic Material DNA is the genetic material in all modern cells. Cells pass copies of their DNA to descendant cells, which use instructions encoded in DNA to build proteins. Some of these proteins aid synthesis of new DNA, which is passed along to descendant cells, and so on. Protein synthesis depends on DNA, which is built by proteins. How did this cycle begin?

In the 1960s, Francis Crick and Leslie Orgel addressed this dilemma by proposing the **RNA world hypothesis**: Early on, RNA served a dual role, functioning both

> Simulations and experiments cannot prove how life or cells began, but they can show us what is plausible.

Figure 13.4 Protocells.
Scientists test hypotheses about protocell formation by carrying out laboratory simulations and field experiments.

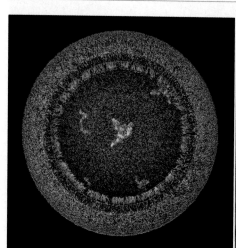

A. Illustration of a laboratory-produced protocell with a bilayer membrane of fatty acids and strands of RNA inside.

© Janet Iwasa.

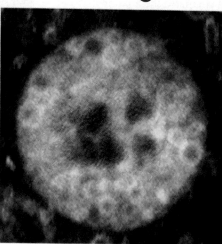

B. Micrograph of a laboratory-formed protocell with RNA-coated clay (red) surrounded by fatty acids and alcohols.

From Hanczyc, Fujikawa, and Szostak, "Experimental Models of Primitive Cellular Compartments: Encapsulation, Growth, and Division"; www.sciencemag.org, *Science* 24 October 2003; 302;529, Fig. 2, p. 619. Reprinted with permission of the authors and AAAS.

C. Field-testing a hypothesis about protocell formation. David Deamer pours a mix of small organic molecules and phosphates into a pool heated by volcanic activity.

Photo by Tony Hoffman, courtesy of David Deamer.

as a genome and as a catalyst. Evidence that RNA can both store genetic information and function like an enzyme in protein synthesis supports this hypothesis. RNAs that function as enzymes (called ribozymes) are common in living cells. For example, some ribozymes cut noncoding bits (introns) out of newly formed RNAs (Section 7.3), and the rRNA in ribosomes speeds formation of peptide bonds during protein synthesis (Section 7.5).

If the earliest self-replicating genetic systems were RNA-based, why do all cells now have a genome of DNA? What was the selective advantage of DNA-based systems? The difference in stability of the two nucleic acids was probably a factor. Compared to a double-stranded DNA molecule, a single-stranded RNA breaks more easily and is more prone to replication errors. Thus, a switch from RNA to DNA would have made larger, more stable genomes possible.

Origin of Cell Membranes   Self-replicating molecules and products of other early synthetic reactions would have floated away from one another unless something enclosed them. In modern cells, a plasma membrane serves this function. If the first reactions took place in tiny rock chambers, rock would have acted as a boundary. Over time, lipids produced by reactions inside such a chamber could have accumulated and lined the chamber wall, forming a protocell. A **protocell** is a membrane-enclosed collection of interacting molecules that can take up material and replicate. Scientists hypothesize that protocells were the ancestors of cellular life.

Researchers have combined organic molecules in the laboratory to yield synthetic protocells that have some lifelike properties. For example, some experiments have produced vesicle-like spheres in which a bilayer of fatty acids surrounds molecules of RNA (Figure 13.4A, B). These spheres "grow" by taking up and incorporating fatty acids and nucleotides from their surroundings. Mechanical force causes the spheres to divide into smaller spheres that have the same composition.

Biochemist David Deamer thinks that the conditions in hot, acidic pools near ancient volcanoes would have encouraged formation of protocells. To test this hypothesis, he carries out experiments both in the laboratory and in the field near currently active volcanoes (Figure 13.4C). His results show that conditions in these pools do favor formation of fatty acids that can self-assemble as vesicles.

Simulations and experiments cannot prove how life or cells began, but they can show us what is plausible. A variety of investigations by many researchers tell us this: Chemical and physical processes that operate today can produce simple organic compounds, concentrate them, and assemble them into protocells (Figure 13.5). Billions of years ago, the same processes may have led to the first life.

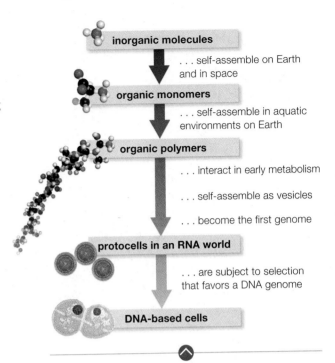

Figure 13.5 Proposed sequence for the evolution of cells. Scientists carry out experiments and simulations that test the feasibility of each step.

inorganic molecules
. . . self-assemble on Earth and in space

organic monomers
. . . self-assemble in aquatic environments on Earth

organic polymers
. . . interact in early metabolism
. . . self-assemble as vesicles
. . . become the first genome

protocells in an RNA world
. . . are subject to selection that favors a DNA genome

DNA-based cells

Take-Home Message 13.2

What do scientific studies reveal about the origin of life?

- Small organic subunits could have formed on the early Earth, or formed in space and fallen to Earth on meteorites.
- Complex organic molecules could have self-assembled from simpler ones.
- The first genetic material may have been RNA rather than DNA.
- Protocells—chemical-filled membranous sacs that grow and divide—may have been the ancestors of the first cells.

iron–sulfur world hypothesis   Hypothesis that life began in rocks rich in iron sulfide near deep-sea hydrothermal vents.

protocell   Membranous sac that contains interacting organic molecules; hypothesized to have formed prior to the earliest cells.

RNA world hypothesis   Hypothesis that RNA served as the first material of inheritance.

Figure 13.6 Stromatolites.
The filament shown in the inset may be a chain of
3.5-billion-year-old fossil bacteria from an ancient
stromatolite. The underlying photo shows modern
stromatolites in Australia's Shark Bay. Each consists
of living photosynthetic bacteria atop the remains of
countless earlier generations of cells and the sediment
that they trapped.

Background, Michael Aw/Lonely Planet Images/Getty Images; inset, Courtesy of John Fuerst,
University of Queensland. Originally published in *Archives of Microbiology* vol. 175, p. 413–29,
Lindsay MR, Webb RI, Strous M, Jetten MS, Butler MK, Forde RJ, Fuerst JA. Cell compart-
mentalisation in planctomycetes: Novel types of structural organization for the bacterial cell.
Arch. Microbiol. 2001 Jun, 175(6):413–29.

13.3 Origin of the Three Domains

REMEMBER: Photosynthesis releases oxygen (Section 5.3), and aerobic respiration
requires it (5.5). UV radiation can cause mutations (6.4).

Reign of the Prokaryotes The processes described in the previous section may
have produced cellular life more than once. If so, all but one of those early lineages
have become extinct. Studies of modern genomes tell us that all modern species
descended from a common single-celled ancestor, a cell that lived perhaps as early
as 4 billion years ago. Given what scientists know about relationships among modern species, most assume that this ancestor was prokaryotic, meaning it did not have
a nucleus. Oxygen was scarce on the early Earth, so the ancestral cell must also have
been anaerobic (capable of living without oxygen).

The two domains of prokaryotic cells, Bacteria and Archaea, diverged very early
in the history of life. Shortly after this divergence, some bacteria began to capture
and use light energy in photosynthetic pathways that did not produce oxygen. The
photosynthetic bacteria grew in the sea as dense mats that trapped sediments. Over
many years, cell growth and sediment deposition formed dome-shaped, layered
structures called **stromatolites**, some of which were preserved as the earliest known
fossils (Figure 13.6).

By 2.7 billion years ago, one lineage of bacteria began to carry out photosynthesis by the oxygen-releasing pathway. As a result of their activity, oxygen
began to accumulate in the air and water. The rise in oxygen had two important
consequences. First, oxygen created a new selective pressure, putting organisms
that thrived in higher-oxygen conditions at an advantage. The pathway of aerobic
respiration evolved and became widespread. This pathway requires oxygen, and it is
far more efficient at releasing energy from organic molecules than other pathways.
Second, ozone gas (O_3) formed and accumulated as the **ozone layer** in the upper
atmosphere. The ozone layer prevents much of the sun's ultraviolet (UV) radiation
from reaching Earth's surface. Such radiation can damage DNA and other biological

molecules. Water screens out some UV radiation, but without the ozone layer to protect it, life could not have moved onto land.

Origin of Eukaryotes Eukaryotes first appear in the fossil record about 1.8 million years ago. All eukaryotes have a nucleus and an associated endomembrane system. These structures probably evolved from infoldings of the plasma membrane in a prokaryotic ancestor (Figure 13.7 ❶, ❷). No prokaryotes have a nuclear envelope, but some bacteria do have infoldings of their plasma membrane. Such folds provide multiple advantages. They increase the surface area that can hold membrane-associated enzymes and divide the cytoplasm into compartments inside which specific reactions can take place. If the membrane infoldings enclose a cell's DNA, they also protect the genetic material.

Mitochondria and chloroplasts resemble bacteria in their size and shape, and they replicate independently of the cell that holds them. Like bacteria, they have their own DNA in the form of a single circular chromosome. They also have at least two outer membranes, with the innermost membrane structurally similar to a bacterial plasma membrane. The **endosymbiont hypothesis** explains these similarities by proposing that mitochondria and chloroplasts are descended from bacteria that entered and lived inside a host cell. (*Endo–* means within; *symbiosis* means living together.) Endosymbionts that live inside a cell can be passed to the cell's descendants when the cell divides.

Evolution of mitochondria began when aerobic bacteria were taken up by or invaded an archaeal cell, then lived and replicated inside it ❸. When the host cell divided, it passed some endosymbionts along to its offspring. As the two species lived together over many generations, genes carried by both partners were free to mutate and to move between the host and its guests. Eventually, the host and endosymbionts could not live independently—the endosymbionts had become mitochondria ❹.

The endosymbiont hypothesis explains why gene sequence comparisons indicate that eukaryotes have both archaeal and bacterial ancestors. Eukaryotic genes that govern basic genetic processes (DNA replication, transcription, and translation) were passed down from an archaeal ancestor. By contrast, genes governing some metabolic processes came from the bacterial ancestors of mitochondria.

Although all eukaryotic lineages have mitochondria or organelles derived from them, only some have chloroplasts. Thus, biologists think that the two types of organelles were acquired independently. The first chloroplasts evolved from oxygen-producing bacteria that were engulfed by and lived in an early eukaryote ❺.

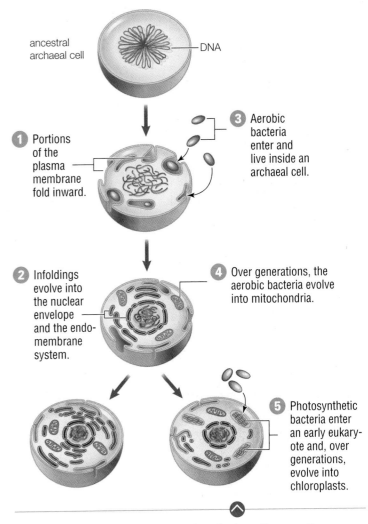

Figure 13.7 Evolution of eukaryotic organelles.
The nuclear envelope and endomembrane components are derived from infoldings of the plasma membrane. Mitochondria and chloroplasts evolved from bacteria that lived inside a host cell.

1 Portions of the plasma membrane fold inward.

2 Infoldings evolve into the nuclear envelope and the endo-membrane system.

3 Aerobic bacteria enter and live inside an archaeal cell.

4 Over generations, the aerobic bacteria evolve into mitochondria.

5 Photosynthetic bacteria enter an early eukary-ote and, over generations, evolve into chloroplasts.

ancestral archaeal cell — DNA

Take-Home Message 13.3

How did the three domains arise?

- The first cells may have arisen as early as 4 billion years ago; they were anaerobic and prokaryotic.
- An early divergence separated ancestors of modern bacteria and archaea.
- Eukaryotes have a mixed ancestry. Their basic genetic apparatus is derived from archaea, but mitochondria and chloroplasts are descendants of bacteria.

endosymbiont hypothesis Hypothesis that mitochondria and chloroplasts evolved from free-living bacteria that entered and lived inside another cell.

ozone layer Atmospheric layer with a high concentration of ozone that prevents much UV radiation from reaching Earth's surface.

stromatolites Dome-shaped structures composed of layers of prokaryotic cells and sediments; form in shallow seas.

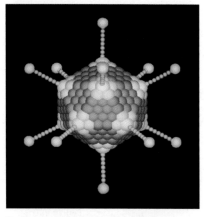

A. Tobacco mosaic virus, a helical virus that infects tobacco and related plants.

After Stephen L. Wolfe.

B. An adenovirus, a polyhedral virus that infects animals. The 20-sided coat encloses double-stranded DNA.

© Dr. Richard Feldmann/National Cancer Institute.

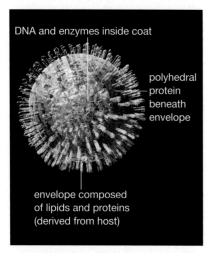

C. A herpesvirus, an enveloped virus that infects animals. The envelope is derived from a host cell.

© Russell Knightly/Science Source.

Figure 13.8 Examples of virus structure.

13.4 Viruses

REMEMBER: Viruses that infect bacteria played a role in the discovery that DNA is the molecule of inheritance (Section 6.2).

In the late 1800s, biologists studying diseased tobacco plants discovered a previously unknown pathogen. It was smaller than even the smallest cells, and could not be seen with a light microscope. The scientists called this unseen infectious agent a virus, a term that means "poison" in Latin. Today, we define a **virus** as a noncellular infectious particle that replicates only inside a living cell.

We do not know how viruses are related to cellular life. The fact that they can only replicate inside cells suggests that they may have evolved from cells. Alternatively, viruses may be remnants of a time before cells.

Viral Structure and Replication A free viral particle (one that is not inside a cell) always includes a viral genome enclosed within a protein coat. The viral genome may be RNA or DNA, and it may be single-stranded or double-stranded. The viral coat consists of many protein subunits that bond together in a repeating pattern, producing a helical rod (Figure 13.8A) or many-sided (polyhedral) structure (Figure 13.8B). The coat protects the viral genetic material and plays a role in infection. In all viruses, components of the viral coat bind to proteins at the surface of a host cell. The coat may also enclose some viral enzymes that will act within the host. In many animal-infecting viruses, the protein coat is enclosed within a **viral envelope** (Figure 13.8C). The viral envelope is layer of cell membrane derived from the host cell in which the viral particle formed.

Viral replication cycles vary in their details, but nearly all include the following steps. The virus first attaches to an appropriate host cell by binding to a specific protein or proteins in the host's plasma membrane. Once a virus comes into contact with and attaches to an the host cell, the viral genome, and in some cases other viral components, enter into that cell.

A viral infection is like a cellular hijacking. Viral genes take over a host's cellular machinery. They direct the cell to replicate viral DNA or RNA and to build viral proteins. These viral components self-assemble to form new viral particles. The particles may be released when the infected host cell bursts (lyses) or they may bud from the host cell, taking some of its plasma membrane with them.

Bacteriophages **Bacteriophages**, sometimes called phages, are nonenveloped viruses that infect bacteria. You learned earlier how Hershey and Chase used one type of bacteriophage to identify DNA as the genetic material of all organisms (Section 6.2). This bacteriophage, called lambda, has a complex structure. A headlike protein coat encloses the viral DNA. Other protein components allow the virus to bind to a bacterium, pierce it, and inject viral DNA into it.

Bacteriophages replicate in bacteria by two pathways. Both pathways begin when a bacteriophage attaches to a bacterial cell and injects its DNA (Figure 13.9). In the lytic pathway, viral genes are expressed immediately ➊. The infected host first produces viral components that self-assemble as virus particles. Then a viral-encoded enzyme breaks down the host's cell wall. Breakdown of the cell wall kills the cell and releases viral particles into the environment.

In the lysogenic pathway, viral DNA becomes integrated into the host cell's genome and viral genes are not immediately expressed, so the cell remains healthy ➋. When the cell reproduces, viral DNA is copied and passed to the cell's

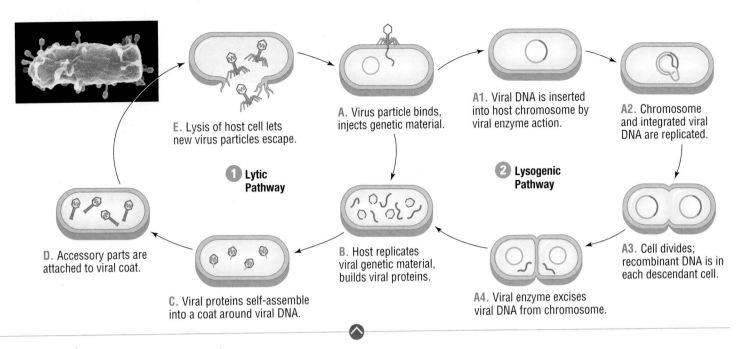

A1. Viral DNA is inserted into host chromosome by viral enzyme action.

A2. Chromosome and integrated viral DNA are replicated.

E. Lysis of host cell lets new virus particles escape.

A. Virus particle binds, injects genetic material.

1 Lytic Pathway

2 Lysogenic Pathway

A3. Cell divides; recombinant DNA is in each descendant cell.

D. Accessory parts are attached to viral coat.

B. Host replicates viral genetic material, builds viral proteins.

C. Viral proteins self-assemble into a coat around viral DNA.

A4. Viral enzyme excises viral DNA from chromosome.

Figure 13.9 The two bacteriophage replication pathways.

Left photo, Science Photo Library/Science Source.

descendants along with the host's genome. Like miniature time bombs, the viral DNA inside the new cells awaits a signal to enter the lytic pathway.

Some bacteriophages can only replicate by the lytic pathway. They always kill their host cell quickly and are not passed from one bacterial generation to the next. Others embark upon either the lytic or lysogenic pathway, depending on conditions in the host cell.

Plant Viruses Plant viruses are typically nonenveloped, with a helical structure and a genome of single-stranded RNA. The tobacco mosaic virus, illustrated in Figure 13.8A, is an example. Plant cells have a thick wall, so plants usually become infected only after insects, pruning, or some other mechanical injury creates a wound that allows the virus into a cell. Sucking insects such as aphids and whiteflies are the vector for many viral plant diseases. A **disease vector** is an organism that transmits a pathogen from one host to the next. Once a plant has become infected by a virus, little can be done to treat it. Thus, protecting crop plants from viruses depends mainly upon controlling insect vectors of viral diseases and breeding plants that are virus resistant.

Viruses and Human Health Some viruses have a beneficial effect on human health. For example, certain bacteriophages in the mucus that coats our airways and our gut help keep bacterial pathogens from infecting us.

Other viruses are themselves human pathogens. Most of these viruses produce mild symptoms and trouble us only briefly. For example, some rhinoviruses infect membranes of our upper respiratory system and cause common colds. Such a cold ends when the immune system eliminates all virus-infected cells. A minority of viral diseases are more persistent. Herpesviruses cause cold sores, genital herpes, mononucleosis, or chicken pox. Typically the initial infection causes symptoms for only a short time. However, the virus remains in the body in a latent state, and can reawaken later on. The herpes simplex virus 1 (HSV-1) can remain latent in nerve

A viral infection is like a cellular hijacking. Viral genes take over a host's cellular machinery.

bacteriophage Virus that infects bacteria.

disease vector Organism that carries a pathogen from one host to the next.

viral envelope A layer of cell membrane derived from the host cell in which an enveloped virus was produced.

virus A noncellular infectious particle with a protein coat and a genome of RNA or DNA; replicates only in living cells.

viral glycoprotein
(binds to host proteins)

viral coat
proteins

one of two
strands of
viral RNA

lipid envelope
with proteins

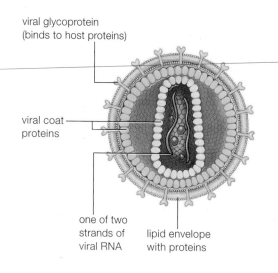

A. Structure of the virus.

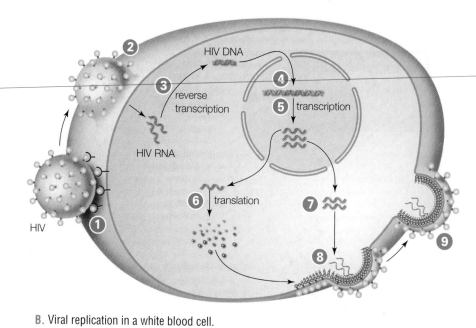

B. Viral replication in a white blood cell.

Figure 13.10 HIV, an enveloped RNA virus.

1️⃣ Viral protein binds to proteins at the surface of a white blood cell.

2️⃣ Viral RNA and enzymes enter the cell.

3️⃣ Viral reverse transcriptase uses viral RNA to make double-stranded viral DNA.

4️⃣ Viral DNA enters the nucleus and becomes integrated into the host genome.

5️⃣ Transcription produces viral RNA.

6️⃣ Some viral RNA is translated to produce viral proteins.

7️⃣ Other viral RNA forms the new viral genome.

8️⃣ Viral proteins and viral RNA self-assemble at the host plasma membrane.

9️⃣ New virus buds from the host cell, with an envelope of host plasma membrane.

Figure It Out: What is the product of reverse transcription of HIV RNA?

Answer: Double-stranded DNA

cells for years. When activated, the virus replicates and causes painful "cold sores" on the edge of the lips. Another type of herpesvirus causes genital herpes.

Some viruses can cause cancer. A few strains of human papillomavirus (HPV) can cause cancers of the cervix, penis, anus, or mouth. Infection by some hepatitis viruses increases the risk of liver cancer.

HIV—The AIDS Virus HIV (**human immunodeficiency virus**) is an enveloped RNA virus that replicates inside human white blood cells (Figure 13.10). It attaches to a cell via a glycoprotein that extends out beyond the viral envelope 1️⃣. After attachment, the viral envelope fuses with the blood cell's plasma membrane, releasing viral enzymes and RNA into the cell 2️⃣. A viral enzyme called reverse transcriptase uses viral RNA as a template to synthesize a double-stranded DNA 3️⃣. This DNA enters the nucleus together with another viral enzyme that inserts the DNA into one of the host's chromosomes 4️⃣. Once integrated, the viral DNA is replicated and transcribed along with the host genome 5️⃣. Some of the resulting viral RNA is translated into viral proteins 6️⃣ and some becomes the genetic material of new HIV particles 7️⃣. The particles self-assemble at the plasma membrane 8️⃣. As the virus buds from the host cell, some of the host's plasma membrane becomes the viral envelope 9️⃣. Each new virus can then infect another white blood cell. New HIV-infected cells are also produced when an infected cell replicates. The disease AIDS (acquired immune deficiency syndrome) arises as a result of HIV's detrimental effects on the immune system. We consider these effects in detail in Chapter 22.

The most common strain of HIV (HIV-1) evolved in west central Africa from a virus that infects nonhuman primates. In the mid-1960s, HIV-1 was introduced to Haiti, where it diversified and acquired distinctive mutations. By 1969, HIV-1 with Haiti-specific mutations reached the United States. It spread quietly until AIDS was identified as a threat in 1981. Today, more than 20 million people worldwide have died from AIDS. About 30 million are currently infected with HIV.

Drugs that fight HIV take aim at steps in viral replication. Some interfere with the way HIV binds to a host cell. Others impair reverse transcription or assembly of new virus particles. These antiviral drugs lower the number of HIV particles, so a person stays healthier. Lowering the concentration of HIV in body fluids also reduces the risk of passing the virus to others.

Ebola   Like HIV, the Ebola virus is an enveloped RNA virus that emerged in Africa. It was identified in 1976. The virus infects fruit bats and nonhuman primates. New outbreaks in the human population arise when the virus gets into a person who has close contact with an infected animal, as by butchering it for food. The virus kills more than half of those of those it infects. Within three weeks of infection, a person develops flulike symptoms, followed by a rash, vomiting, diarrhea, and bleeding from the eyes, nose, mouth and other body openings. The virus is transmitted among people by direct contact with body fluids, so those who care for the sick must wear protective gear (Figure 13.11).

Until recently, all Ebola outbreaks had been confined to limited regions within Africa and had affected fewer than 500 people. However, an outbreak that began in Guinea in December of 2013 killed thousands and raised fears of a widespread epidemic. As of late 2014, there was no vaccine against Ebola and a limited supply of the experimental drugs that could be used to fight the disease.

New Flus   Flus are caused by enveloped RNA viruses called influenza viruses. To keep up with ongoing mutations in influenza viruses, scientists create a new flu shot every year. The flu shot is a vaccine designed to protect against the influenza strains that scientists predict are most likely to pose a threat during the upcoming flu season. Unfortunately, determining which flu strains will be circulating in the future is not an exact science. Even after a flu shot, a person remains susceptible to a virus that differs from the strains targeted by the vaccine.

New influenza strains arise both through mutation and by **viral reassortment**, the swapping of genes between related viruses that infect a host at the same time (Figure 13.12). Consider what could happen if two influenza strains currently circulating underwent such a reassortment. The H5N1 strain is a bird flu that occasionally infects people who have direct contact with birds. When the virus does infect people, the death rate is high, about 60 percent. Fortunately, person-to-person transmission of the H5N1 virus is exceedingly rare. By contrast H1N1, commonly referred to as "swine flu," is easily passed between people, but only rarely deadly. The coexistence of these viruses raises the possibility of a potentially disastrous gene exchange. If H1N1 and H5N1 infected the same host simultaneously, the result could be a flu virus that is easily transmissible and deadly.

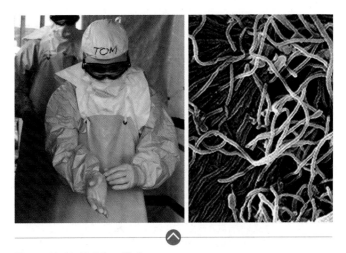

Figure 13.11 Fighting Ebola
Dr. Tom Frieden, head of U.S. Center for Disease Control, at an Ebola treatment center in Liberia during the 2014 outbreak. The micrograph on the right shows the virus, which has a threadlike structure.

Left, CDC/Sally Ezra; right, CDC/NIAID.

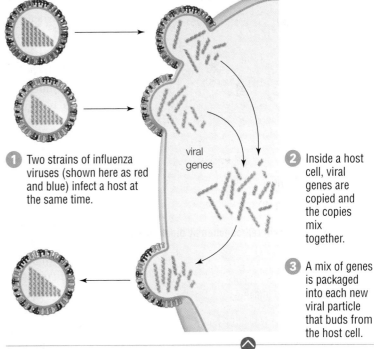

1 Two strains of influenza viruses (shown here as red and blue) infect a host at the same time.

viral genes

2 Inside a host cell, viral genes are copied and the copies mix together.

3 A mix of genes is packaged into each new viral particle that buds from the host cell.

Figure 13.12 Viral reassortment.
When a host cell is infected by two viruses of the same type, such as two influenza viruses, viral genes recombine to form viruses with new gene combinations.

Take-Home Message 13.4

What are viruses and how do they affect us?

- Viruses are noncellular particles that consist of genetic material wrapped in a protein coat. They replicate only inside living cells, and each type of virus infects and replicates inside a specific type of host.
- A virus harms and eventually kills a host cell. Viral genes direct the host cell's metabolic machinery to produce new viral particles.
- Viral genomes can be altered by mutation. Viruses with new combinations of genes also arise as a result of viral reassortment.

HIV (human immunodeficiency virus) Enveloped RNA virus that causes AIDS.

viral reassortment Two viruses of the same type infect an individual at the same time and swap genes.

13.5 Bacteria and Archaea

REMEMBER: All organisms are producers or consumers, and nutrients cycle between the two groups (Section 1.3).

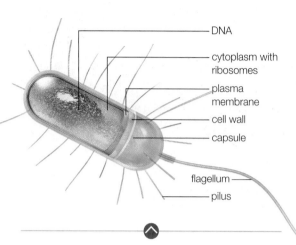

Figure 13.13 Generalized prokaryotic body plan.

Biologists have historically divided all life into two groups. Cells without a nucleus were prokaryotes and those with a nucleus were eukaryotes. More recently we learned that "prokaryotes" actually constitute two distinct lineages, now referred to as the domains Bacteria and Archaea. **Bacteria** are the more well-known and widespread group of cells that do not have a nucleus. **Archaea** are more closely related to eukaryotes than to bacteria, and many live in extreme habitats.

Structure and Function Bacteria and archaea are small and, with rare exceptions, cannot be seen without a light microscope. Figure 13.13 shows a typical bacterial cell. It has no nucleus or membrane-enclosed organelles like those of eukaryotes. The prokaryotic chromosome (a ring of DNA) lies in the cytoplasm, as do the ribosomes. Nearly all bacteria and archaea have a porous cell wall around their plasma membrane. The wall gives the cell its shape, which may be spherical, spiral, or rod-shaped. A spherical cell is a coccus, a spiral-shaped one is a spirillum, and a rod-shaped one a bacillus. The cell depicted in Figure 13.13 is a bacillus. Like many bacteria, it has a capsule of secreted material around its cell wall.

Most bacteria can move from place to place. Some have one or more bacterial flagella that rotate like a propeller. Other bacteria glide along surfaces by using thin protein filaments called pili (singular, pilus) as grappling hooks. The pilus is extended out to a surface, sticks to it, then shortens, drawing the cell forward. Another type of pilus is used to draw cells together for gene transfers.

Reproduction and Gene Transfers Bacteria and archaea have staggering reproductive potential. Division most commonly occurs by **binary fission**, a mechanism of asexual reproduction that yields two equal-sized, genetically identical descendant cells (Figure 13.14). The process begins when the cell replicates its single chromosome, which is attached to the inside of the plasma membrane ❶. The DNA replica attaches to the plasma membrane adjacent to the parent molecule. Addition of new membrane and wall material elongates the cell and moves the two DNA molecules apart ❷. Then, membrane and cell wall material is deposited across the cell's midsection ❸, yielding two identical descendant cells ❹.

Bacteria and archaea do not reproduce sexually, but they can transfer genetic material among existing individuals. Three mechanisms permit such exchanges. With **transformation** (Figure 13.15A), a prokaryote takes up free DNA, such as that from a dead cell, from its environment. With **transduction** (Figure 13.15B), a virus picks up DNA from one host, then passes that DNA along to its next host. With **conjugation** (Figure 13.15C), one cell donates a small circle of DNA called a plasmid to another. A **plasmid** is a circle of double-stranded DNA with a few genes. Conjugation begins when a cell with a particular plasmid cell uses a special sex pilus to draw a cell without that plasmid close. The donor cell passes one strand of plasmid DNA to the recipient cell, then each cell makes the missing strand of DNA.

The ability of prokaryotic cells to acquire genetic information from other cells has important health implications. Suppose a gene for antibiotic resistance arises in one bacterial cell. Not only can this gene be passed on to that cell's descendants, but it can also be transferred to other existing cells. Such transfers speed the rate at which a gene spreads through a population.

Figure 13.14 Asexual reproduction by binary fission.

❶ A bacterium has one circular chromosome that attaches to the inside of the plasma membrane.

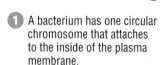

❷ The cell duplicates its chromosome, attaches the copy beside the original, and adds membrane and wall material between them.

❸ When the cell has just about doubled in size, a new membrane and wall are deposited across its midsection.

❹ Two genetically identical cells result.

Figure 13.15 Mechanisms of gene exchange between prokaryotic cells.

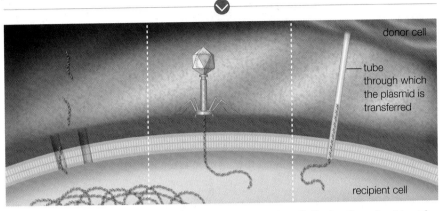

A. Transformation: taking up DNA from the environment.

B. Transduction: transfer of DNA by means of a virus.

C. Conjugation: direct transfer of a plasmid between cells.

Figure It Out: Which of these processes doubles the number of cells?

Answer: None of them. All are means of gene exchange, not reproduction.

archaea Lineage of prokaryotes most closely related to eukaryotes; many live in extreme environments.

autotroph Organism that uses carbon dioxide as its carbon source; obtains energy from light or breakdown of minerals.

bacteria Most diverse and well-known lineage of prokaryotes.

binary fission Method of asexual reproduction in which a prokaryote divides into two identical descendant cells.

conjugation Mechanism of gene transfer. One prokaryotic cell directly transfers a plasmid to another.

decomposer Organism that breaks down organic material into its inorganic subunits.

heterotroph Organism that obtains both carbon and energy by breaking down organic compounds.

plasmid Of many prokaryotes, a small ring of nonchromosomal DNA.

transduction Mechanism of gene transfer. A virus moves genes from one host cell to another.

transformation Mechanism of gene transfer. A prokaryotic cell takes up and uses DNA from its environment.

Metabolic Diversity A high degree of metabolic diversity contributes to the wide distribution of the prokaryotic lineages. Organisms obtain energy and nutrients from the environment in four different ways (Figure 13.16). All four nutritional modes occur among bacteria or archaea or in both.

Autotrophs are producers that build their own food using carbon dioxide (CO_2) as their carbon source. There are two subgroups: photoautotrophs and chemoautotrophs. Photoautotrophs are photosynthetic. They use the energy of light to assemble organic compounds from CO_2 and water. Many bacteria are photoautotrophs, as are plants and photosynthetic protists. Chemoautotrophs obtain energy by oxidizing (removing electrons from) inorganic molecules such as hydrogen sulfide or methane and use it to build organic compounds from CO_2. Chemoautotrophic bacteria and archaea are the main producers in dark environments such as the seafloor. So far, no eukaryotic chemoautotroph is known.

Heterotrophs cannot use inorganic sources of carbon. Instead, they obtain carbon by taking up organic molecules from their environment. As with autotrophs, there are two types. Photoheterotrophs harvest energy from light, and carbon from alcohols, fatty acids, or other small organic molecules. Heliobacteria that live in the soils of rice paddies are an example. Chemoheterotrophs obtain both energy and carbon by breaking down carbohydrates, lipids, and proteins. Most bacteria and some archaea are chemoheterotrophs, as are animals, fungi, and nonphotosynthetic protists. All pathogenic bacteria are chemoheterotrophs that extract the organic compounds they need to live from their host. Other prokaryotic chemoheterotrophs serve as **decomposers**, meaning they break down organic molecules into inorganic ones. By their actions, decomposers make nutrients that were tied up in wastes and remains accessible to producers.

Most eukaryotic organisms are aerobic, meaning they rely on aerobic respiration (Section 5.5) and thus need oxygen. By contrast, many bacteria and most archaea are anaerobes, which means they can tolerate an oxygen-free environment. Some are obligate anaerobes, meaning oxygen either slows their growth or kills them outright. Anaerobes are harmed by oxygen because oxidation reactions damage their biological molecules and, unlike aerobic cells, they do not have enzymes

Figure 13.16 Nutritional classification of organisms.

CARBON SOURCE	ENERGY SOURCE	
	Light	Chemicals
Inorganic source such as CO_2	Photoautotrophs bacteria, archaea, photosynthetic protists, plants	Chemoautotrophs bacteria, archaea
Organic source such as glucose	Photoheterotrophs bacteria, archaea	Chemoheterotrophs bacteria, archaea, fungi, animals, nonphotosynthetic protists

A. Thermally heated waters. Pigmented archaea color rocks in waters of this Nevada hot spring.

B. Highly salty waters. Pigmented extreme archaea color brine in this California lake.

C. The gut of many animals. Cows belch to expel methane produced by archaea in their digestive system.

Figure 13.17 Examples of archaeal habitats.

(A) © Savannah River Ecology Laboratory; (B) Courtesy of Benjamin Brunner; (C) Dr. John Brackenbury/Science Source.

that can repair that damage. We find obligate anaerobes in aquatic sediments and the animal gut. They can also infect deep wounds.

Many bacteria and some archaea can respond to adverse conditions by shutting down their metabolism and forming a dormant resting structure. Depending on the group and how the structure forms, it may be called a spore or a cyst. For example, some bacteria, including those that cause the diseases tetanus and anthrax, produce a resilient resting structure consisting of a stripped-down bacterial cell with a thick protective covering. This structure, called an endospore, can withstand heating, freezing, drying out, and exposure to ultraviolet radiation. Scientists have extracted bacterial endospores from the gut of a bee that had been encased in amber (fossilized tree sap) for at least 20 million years. When given nutrients and moisture, the endospores germinated and the cells within them became active once again.

Domain Archaea Archaea were discovered in the 1970s. Many thrive in seemingly hostile habitats. The **extreme thermophiles** live in very hot places. For example, archaea have been found in scalding hot water near deep-sea hydrothermal vents and in thermal springs (Figure 13.17A). The archaea that are **extreme halophiles** live in highly salty environments (Figure 13.17B).

Many archaea, including some extreme halophiles and thermophiles, are **methanogens**, chemoautotrophs that produce methane, an odorless flammable gas, as a by-product of their metabolic reactions. Methane-producing archaea abound in sewage, marsh sediments, and the animal gut (Figure 13.17C). About a third of the human population has significant numbers of methanogens in their intestine, so their flatulence (farts) contains methane.

As biologists continue to explore archaeal diversity, they are finding that these organisms are not restricted to extreme environments. They live alongside bacteria nearly everywhere. So far, scientists have not found any archaea that pose a major threat to human health. However, some that live in the mouth may encourage gum disease, and some that live in the gut may encourage weight gain.

Domain Bacteria Many bacteria play important roles in nutrient cycles. Photosynthesis evolved in many bacterial lineages, but only cyanobacteria (Figure 13.18A) use a pathway that produces oxygen as a by-product. Biologists infer that ancient cyanobacteria were the ancestors of modern chloroplasts. Thus, we have cyanobacteria and their chloroplast relatives to thank for nearly all the oxygen we breathe.

Some cyanobacteria also carry out **nitrogen fixation**, meaning they incorporate nitrogen from the air into ammonia (NH_3). Nitrogen fixation is an important ecological service provided only by bacteria. Photosynthetic eukaryotes need nitrogen, but they cannot use the gaseous form ($N \equiv N$) because they do not have an enzyme that can break the molecule's triple bond. They can, however, take up ammonia released by nitrogen-fixing bacteria.

Nitrogen-fixing bacteria of the genus *Rhizobium* live inside the roots of legumes, a group of plants that includes peas, alfalfa, and clover. The plants benefit from the presence of the bacteria, which provide them with ammonia. The bacteria benefit by living in the shelter of the roots and receiving sugar from the plant.

Bacteria also help cycle nutrients by acting as decomposers. Together with fungal decomposers, bacteria ensure that nutrients in wastes and remains of organisms return to the soil in a form that plants can use. Lactate-fermenting bacteria are among the decomposers. Sometimes these bacteria get into our food and spoil it, as when they cause milk to go sour. On the other hand, we use some lactate fermenters to make sauerkraut, pickles, cheese, and yogurt (Figure 13.18B). Lactate fermenters

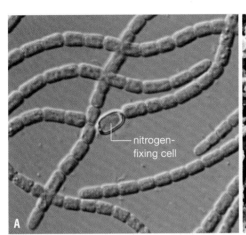
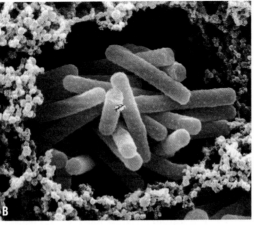

nitrogen-
fixing cell

A

B

Figure 13.18 Ecologically important bacteria.
A. Aquatic cyanobacteria. Cyanobacteria release oxygen as a by-product of photosynthesis. This species grows as long chains of cells connected by a secreted mucous sheath. Some specialized cells in the chain fix nitrogen.

B. Lactate-fermenting bacteria used to produce yogurt. Other lactate-fermenting bacteria live in the human gut or serve as decomposers in the soil.

(A) Michael Abbey/Visuals Unlimited, Inc.; (B) SciMAT/Science Source.

are also present in a healthy human gut and vagina. The acidity of the lactate these bacteria produce helps keep disease-causing organisms from taking hold. Other intestinal bacteria benefit us by producing essential vitamins or by breaking down materials we could not otherwise digest. For example, most of the vitamin K you need is produced by *Escherichia coli* bacteria in your large intestine.

Escherichia coli is the best-studied species of bacteria. Researchers often investigate genetic and metabolic processes in this species because it is easily grown in laboratories. *E. coli* is also used in industrial biotechnology. Recombinant *E. coli* now make hormones and other proteins for medical use.

When biotechnologists want to alter a plant's genome, they may turn to *Agrobacterium*. These soil bacteria have a plasmid that gives them the capacity to infect plants and cause a tumor. Scientists produce recombinant plants by inserting genes into the tumor-inducing plasmid, then infecting a plant with recombinant bacteria.

Bacteria cause many common diseases (Table 13.1). Some, such as whooping cough (pertussis) and tuberculosis, spread when a person with an active infection coughs or sneezes, distributing bacteria-laden droplets into the environment.

Impetigo, a skin disease, is caused by *Streptococcus* and *Staphylococcus* bacteria that infect outer skin layers. *Streptococcus* can also cause strep throat. Gonorrhea, syphilis, and chlamydia are bacterial diseases transmitted by sexual contact. Bacteria also enter our body in tainted food or water. Cholera, which kills about 100,000 people per year, spreads when bacteria-tainted feces contaminate drinking water. Lyme disease is a vector-borne bacterial disease. Ticks carry the bacteria that cause the disease between vertebrate hosts. Lyme disease may initially cause a bull's-eye-shaped rash at the site of the tick bite. Later, flulike symptoms occur.

Table 13.1 Examples of Bacterial Diseases

Disease	Description
Whooping cough	Childhood respiratory disease
Tuberculosis	Respiratory
Impetigo, boils	Blisters, sores on skin
Strep throat	Sore throat, can damage heart
Cholera	Diarrheal illness
Syphilis	Sexually transmitted disease
Gonorrhea	Sexually transmitted disease
Chlamydia	Sexually transmitted disease
Lyme disease	Rash, flulike symptoms, spread by ticks
Botulism, tetanus	Muscle paralysis by bacterial toxin

Take-Home Message 13.5

What are prokaryotes?

- Prokaryotes are cells that do not have a nucleus. They reproduce mainly by binary fission and they swap genes by conjugation and other processes.
- Archaea are the most recently discovered prokaryotic domain. Many archaea live in extremely hot or salty habitats.
- Bacteria benefit other organisms by releasing oxygen, fixing nitrogen, and serving as decomposers. We use them to produce foods and in biotechnology. Some bacteria cause human disease.

extreme halophile Organism that lives where the salt concentration is high.

extreme thermophile Organism that lives where the temperature is very high.

methanogen Organism that produces methane gas as a metabolic by-product.

nitrogen fixation Process of combining nitrogen gas with hydrogen to form ammonia.

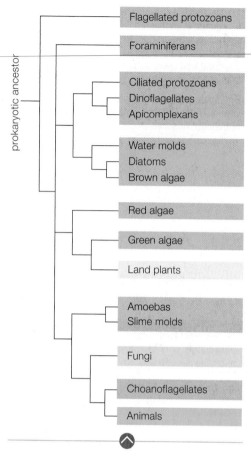

Figure 13.19 Evolutionary tree for the eukaryotes.
Orange boxes indicate the protist lineages.

Figure 13.20 Euglena, a freshwater flagellated protozoan.
The species depicted here has chloroplasts that evolved from a green alga.

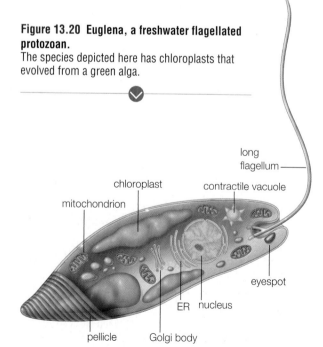

13.6 Protists

REMEMBER: Eukaryotic cells have a nucleus and other membrane-bound organelles and they use cilia and flagella to propel themselves (Section 3.5). Water moves across a plasma membrane by osmosis (4.5).

A diverse array of eukaryotic lineages are collectively referred to as **protists**. Protists were historically lumped together in a kingdom between prokaryotes and the "higher" forms of life (fungi, plants, and animals). Now that scientists have improved methods of comparing genomes, protists are being reclassified in ways that reflect their evolutionary relationships. Figure 13.19 shows where the protist lineages that we cover in this book fit in the eukaryote family tree. There are many additional protist lineages, but learning about a representative few will give you a good idea of the diversity of protist forms and of their importance in health and ecology. Notice that some of the protists are actually more closely related to plants, animals, or fungi than they are to other protists.

Being eukaryotes, all protist cells have a nucleus and a cytoskeleton with micro-tubules. Most also have mitochondria, endoplasmic reticulum, and Golgi bodies. All protists have multiple chromosomes, each consisting of DNA with proteins attached. Protists reproduce asexually by mitosis, sexually by meiosis, or both.

Most protist lineages include only single-celled species. However, colonial protists exist, and multicellularity evolved independently in several lineages. Cells of a **colonial organism** live together and behave in an integrated fashion, but still remain self-sufficient. Each retains the traits required to survive and reproduce on its own. By contrast, the cells of a **multicellular organism** have a division of labor and rely on one another for survival.

Flagellated Protozoans "Protozoans" is the general term for heterotrophic protists that live as single cells. **Flagellated protozoans** are single, unwalled cells that have one or more flagella. In protists, movement of the flagellum pulls a cell forward, rather than pushing the cell along as the flagellum of an animal sperm does. A pellicle, a layer of elastic proteins just beneath the plasma membrane, helps flagellated protozoans retain their shape.

Euglenoids have a single long flagella and most live in ponds and lakes (Figure 13.20). The interior of the cell has a higher solute concentration than the fresh water the cell lives in, so water tends to enter euglenoids by osmosis. Excess water collects in **contractile vacuoles**, organelles that can also contract and expel the water to the outside. Many euglenoids are heterotrophs, but some such as the one depicted in Figure 13.20 have chloroplasts that evolved from a green algae. Photosynthetic euglenoids are able to detect light using a an eyespot, an organelle near the base of their long flagellum.

Some flagellated protozoans live in the bodies of other organisms. *Giardia*, the intestinal parasite described in Section 13.1, is one example. *Trichomonas* (Figure 13.21A) is another. Its multiple flagella propel it through the human reproductive tract with jerky movements. *Trichomonas* causes trichomoniasis, a sexually trans-mitted disease commonly referred to as "trich." Trypanosomes are long, tapered cells with a single mitochondrion and a flagellum that is attached to the cell body by a membrane (Figure 13.21B). All trypanosomes are parasites of either plants or animals. Trypanosomes that cause human diseases such as sleeping sickness are transmitted by insect bites. Once inside a human host, trypanosomes live in the blood and other body fluids.

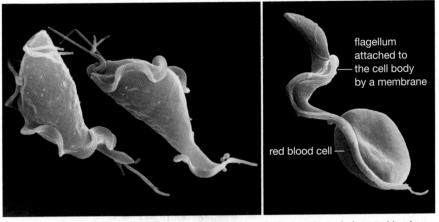

A. *Trichomonas*, a sexually transmitted pathogen.

B. *Trypanosoma* in human blood.

flagellum attached to the cell body by a membrane

red blood cell

Figure 13.21 Flagellated protozoans that parasitize humans.

(A) David M. Phillips/The Population Council/Science Source; (B) Oliver Meckes/Science Source.

ciliate Unwalled, single-celled protist with many cilia.

colonial organism Organism composed of many integrated cells, each capable of surviving and reproducing on its own.

contractile vacuole In freshwater protists, an organelle that collects and expels excess water.

flagellated protozoan Unwalled, single-celled protist that has one or more flagella.

foraminiferan Heterotrophic single-celled protist that secretes a calcium carbonate shell.

multicellular organism Organism composed of a variety of specialized cells, each unable to survive and reproduce on its own.

plankton Community of mostly microscopic drifting or swimming organisms.

protist General term for eukaryote that is not a fungus, plant, or animal.

Foraminifera **Foraminifera**, or forams, are single-celled predators that secrete a shell containing calcium carbonate. Including its shell, an individual foraminiferal cell can be as big as a grain of sand. Threadlike cytoplasmic extensions protrude through openings in the shell. Most forams live on the seafloor, where they probe the water and sediments for prey. Others are part of the marine **plankton**, a collection of tiny organisms that drift or swim in the open sea. Planktonic forams often have photosynthetic protists that live in their cytoplasm (Figure 13.22).

Foraminifera have lived and died in the oceans for more than 500 million years, so remains of countless cells have fallen to the seafloor. Over time, geologic processes transformed some accumulations of foraminiferal shells into chalk and limestone, two types of calcium carbonate–rich sedimentary rock. The giant blocks of limestone that were used to build the great pyramids of Egypt consist largely of the shells of ancient foraminifera.

Modern foraminifera play an important role in the global carbon cycle. By taking up carbon dioxide from seawater and incorporating it into their shells, foraminifera lower the ocean's carbon dioxide concentration and allow it to absorb more carbon dioxide from the air. Removing carbon dioxide from the air is important because the increasing concentration of atmospheric carbon dioxide is currently causing a global climate change.

200 µm

Figure 13.22 A planktonic foraminiferan.
The yellow dots are algae that live in its cytoplasm.

Courtesy of Allen W. H. Bé and David A. Caron.

Ciliates Ciliated protozoans, or **ciliates**, are unwalled cells with many cilia. Most ciliates are predators in seawater or fresh water. They feed on bacteria, algae, and one another. *Paramecium* is a freshwater ciliate commonly found in ponds (Figure 13.23). The cilia that cover its entire surface function in feeding and locomotion. They sweep water laden with bacteria, algae, and other food particles into an oral groove at the cell surface, and then to a gullet. Enzyme-filled vesicles digest food in the gullet.

Other ciliates live in the gut of mammalian grazers such as cattle and sheep. Like some bacteria, these ciliates help their host digest plant material. Only one species of ciliate (*Balantidium coli*) is a known human pathogen. It also infects pigs, and people become infected when pig feces containing a resting form of the ciliate taint drinking water. Infection causes nausea and diarrhea.

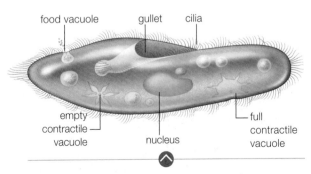

food vacuole — gullet — cilia

empty contractile vacuole — nucleus — full contractile vacuole

Figure 13.23 The freshwater ciliate *Paramecium*.

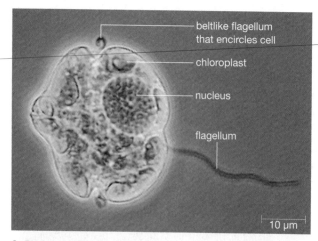

beltlike flagellum that encircles cell

chloroplast

nucleus

flagellum

10 μm

A. Photosynthetic dinoflagellate.

B. Bioluminescent dinoflagellates agitated by wave action give this shoreline an eerie glow.

Figure 13.24 Dinoflagellates.

(A) © Bob Andersen and D. J. Patterson; (B) Travelart/Alamy.

algal bloom Population explosion of single-celled aquatic organisms such as dinoflagellates.

apicomplexan Parasitic protist that enters and lives inside the cells of its host.

bioluminescence Light produced by a living organism.

dinoflagellates Single-celled, aquatic protist typically with cellulose plates and two flagella; may be heterotrophic or photosynthetic.

Dinoflagellates The name **dinoflagellate** means "whirling flagellate." These single-celled protists typically have two flagella, one at the cell's tip and the other running in a groove around the middle of the cell like a belt (Figure 13.24A). Combined action of the two flagella causes the cell to rotate as it moves forward. Most dinoflagellates deposit cellulose just beneath their plasma membrane, and these deposits form thick protective plates.

Some dinoflagellates live in fresh water and others in the oceans. Some prey on bacteria, others are parasites of animals, and still others have chloroplasts that evolved from red algae. A few species of photosynthetic dinoflagellates live inside the cells of reef-building corals. They supply their coral host, which is an invertebrate animal, with essential sugars. In exchange, the coral provides the dinoflagellates with nutrients, shelter, and the carbon dioxide necessary for photosynthesis. The coral cannot live without its protist helpers. If it loses them, it will starve.

In tropical seas, dinoflagellates are a common source of **bioluminescence**, which is light produced by a living organism (Figure 13.24B). Emitting light may protect a cell by startling a predator that was about to eat it. By another hypothesis, the flash of light acts like a car alarm. It attracts the attention of other organisms, including predators that pursue would-be eaters of dinoflagellate.

In nutrient-enriched water, free-living photosynthetic dinoflagellates or other aquatic protists sometimes undergo great increases in population size, a phenomenon known as an **algal bloom**. Algal blooms can harm other organisms. When the cells die, aerobic bacteria that feed on their remains use up all the oxygen in the water, so aquatic animals suffocate. In additions, some dinoflagellates produce toxins that can kill aquatic organisms directly and sicken people.

Apicomplexans **Apicomplexans** are parasitic protists that spend part of their life inside cells of their hosts. Their name refers to a complex of microtubules at their apical (top) end that allows them to enter a host cell. Apicomplexans infect a variety of animals, from worms and insects to humans. In most species, the life cycle is complicated, with multiple hosts and several forms. Consider *Plasmodium*, the apicomplexan that causes malaria (Figure 13.25). A female mosquito transmits the infectious form of *Plasmodium* (called a sporozoite) to a human when she bites ❶. The sporozoite travels through blood vessels to the liver, where it reproduces asexually ❷. Some of the resulting offspring, called merozoites, enter red blood cells, where they reproduce asexually to produce more merozoites ❸. Other merozoites enter into red blood cells and develop into immature gametes, or gametocytes ❹.

When a mosquito bites an infected person, it takes up gametocytes along with blood. The gametocytes mature in the mosquito's gut, then fuse to form zygotes ❺. Zygotes develop into new sporozoites that migrate to the insect's salivary glands, where they await transfer to a new vertebrate host ❻.

Malaria symptoms usually start a week or two after a mosquito bite, when infected liver cells rupture and release *Plasmodium* cells and cellular debris into the blood. Shaking, chills, a burning fever, and sweats result. After the first episode, symptoms may subside for weeks or even months. However, an ongoing infection damages the liver, spleen, kidneys, and brain. If untreated, malaria nearly always results in death. Malaria kills about half a million people each year, mainly in Africa.

The apicomplexan *Toxoplasma* also commonly infects humans. Most people never realize they are infected, but an infection can be deadly in someone with an impaired immune system, and an infection that begins during pregnancy can cause birth defects. Domestic cats that spend time ouside catching birds or rodents can serve as carriers of *Toxoplasma*; their feces may contain infectious cysts.

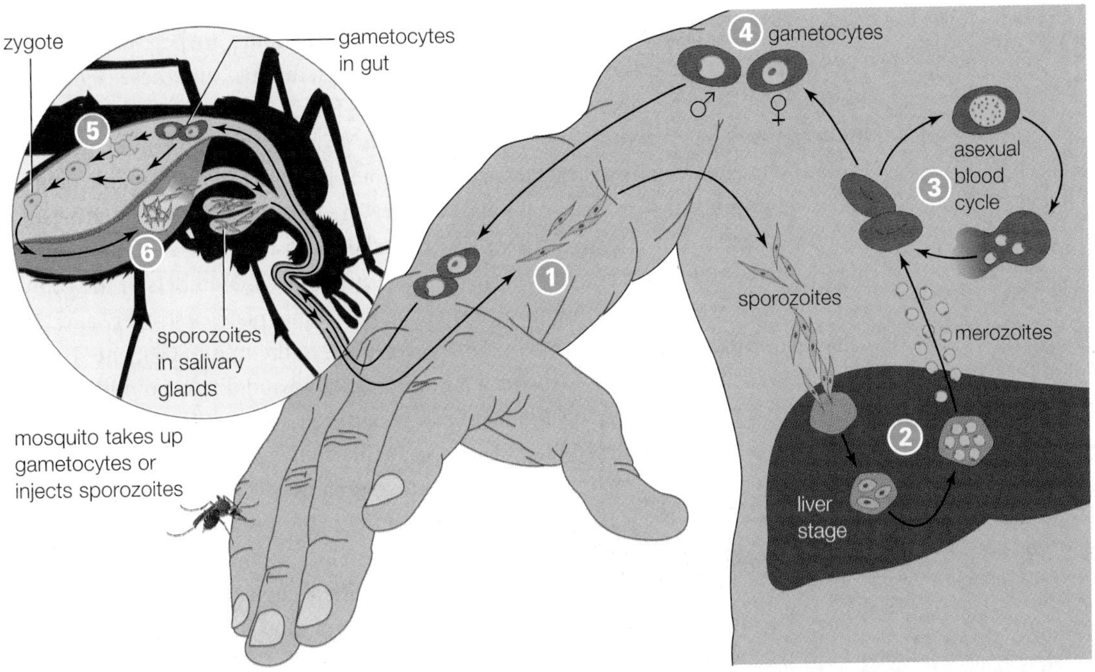

Figure 13.25 Life cycle of *Plasmodium*, the protist that causes malaria.

1 Infected mosquito bites a human. Sporozoites enter the blood, which carries them to the liver.

2 Sporozoites reproduce asexually in liver cells, then mature into merozoites. Merozoites leave the liver and enter the bloodstream, where they infect red blood cells.

3 Inside some red blood cells, merozoites reproduce asexually. These cells burst and release more merozoites into the bloodstream.

4 Inside other red blood cells, merozoites develop into male and female gametocytes.

5 A female mosquito bites and sucks blood from the infected person. Gametocytes in red blood cells enter her gut and mature into gametes, which fuse to form zygotes.

6 Zygotes develop into sporozoites that migrate to the mosquito's salivary glands.

zygote

gametocytes in gut

sporozoites in salivary glands

mosquito takes up gametocytes or injects sporozoites

gametocytes ♂ ♀

asexual blood cycle

sporozoites

merozoites

liver stage

Based on Fig. 1 from "Genetic linkage and association analyses for trait mapping in Plasmodium falciparum," by Xinzhuan Su, Karen Hayton & Thomas E. Wellems, *Nature Reviews Genetics 8*, 497–506 (July 2007).

Digging Into Data

How *Plasmodium* Summons Mosquitoes

Parasites sometimes alter their host's behavior in a way that increases their chances of transmission to another host. *Plasmodium* (the agent of malaria) would benefit by making its human host more attractive to hungry mosquitoes when immature gametes (gametocytes) are present in the host's blood. Such immature gametes are taken up by the mosquito along with blood, and they mature into gametes inside the mosquito's gut. Dr. Jacob Koella and his associates performed an experiment to see whether infection by *Plasmodium* makes a person more attractive to mosquitoes. The researchers recorded the response of mosquitoes to the odor of *Plasmodium*-infected children and uninfected children over the course of 12 trials on 12 separate days. They also recorded which stage of *Plasmodium* life cycle the infected children were carrying at the time. Figure 13.26 shows their results.

1. On average, which group of children was the most attractive to mosquitoes?
2. Did carrying noninfectious, asexual stage *Plasmodium* make children more attractive to mosquitoes than uninfected children?
3. Did the data support the hypothesis that the presence of infectious *Plasmodium* cells (gametocytes) makes an individual more attractive to mosquitoes?
4. Why would it be it beneficial for *Plasmodium* gametocytes to make a host attractive to mosquitoes?

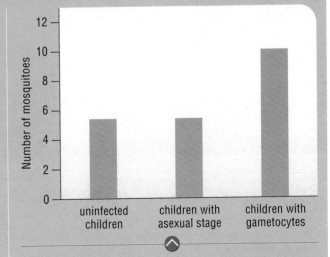

Figure 13.26 Number of mosquitoes (out of 100) attracted to uninfected children, children with asexual stages of *Plasmodium* (sporozoites, merozoites), and children with gametocytes. The bars show the average number of mosquitoes attracted to that category of child over the course of 12 separate trials.

Water Molds, Diatoms, and Brown Algae

Water molds are heterotrophs known to scientists as oomyetes. The term means "egg fungus," and these organisms were once mistakenly grouped with the fungi. Like fungi, the water molds form a mesh of nutrient-absorbing filaments, but the two groups differ in many structural traits and are genetically distinct. Most water molds help decompose organic debris and dead organisms in aquatic habitats, but a few are parasites that have significant economic effects. Some grow as fuzzy white patches on fish in fish farms and aquariums. Others infect land plants, destroying crops and forests. Members of the genus *Phytophthora* are especially notorious. Their name means "plant destroyer" and they cause an estimated $5 billion in crop losses each year. In the mid-1800s, one species destroyed Irish potato crops, causing a famine that killed and displaced millions of people. Today, another *Phytophthora* species is causing an epidemic of sudden oak death in Oregon, Washington, and California. Millions of oaks have already died.

The closest relatives of water molds are two photosynthetic groups: diatoms and brown algae (Figure 13.27). Both have chloroplasts that include a brownish accessory pigment (fucoxanthin) that tints them olive green, golden, or dark brown.

Diatoms have a two-part silica shell, with upper and lower parts that fit together like a box with an overlapping lid. Some cells live individually, and others form chains ❶. Most diatoms float near the surface of seas or lakes, but some live in moist soil or in water droplets that cling to mosses in damp environments.

Diatom cells contain a large amount of oil. Oil is less dense than water, and its presence helps these photosynthetic cells stay afloat in sunlit waters. The oil also serves as a store of energy.

Like foraminifera, marine diatoms have lived and died in the oceans for many millions of years, and their remains form vast deposits on the seafloor. In some places, deposits of ancient diatom oil have been transformed into petroleum, which we extract to produce gasoline. In other places, diatom remains have been transformed into a silica-rich powder called diatomaceous earth. This material is quarried for use in filters, abrasive cleaners, and as an insecticide that is not harmful to vertebrates.

Brown algae are multicelled inhabitants of temperate or cool seas. In size, they range from microscopic filaments to giant kelps that stand 30 meters (100 feet) tall. Giant kelps form forestlike stands in coastal waters of the Pacific Northwest ❷. Like trees in a forest, kelps shelter a wide variety of other organisms. The Sargasso Sea in the North Atlantic Ocean is named for its abundance of *Sargassum*. This kelp forms vast, floating mats that can be up to 9 meters (30 feet) thick. The mats provide food and shelter to fish, sea turtles, and invertebrates.

Sargassum and other brown algae have commercial uses. Alginic acid from the cell walls of brown algae is used to produce algins, which serve as thickeners, emulsifiers, and suspension agents. Algins are used to manufacture ice cream, pudding, jelly beans, toothpaste, cosmetics, and other products.

Figure 13.27 Two related protist groups common in California's coastal waters.

❶ Microscopic diatoms have a silica shell.

❷ Kelp are large, multicelled brown alga.

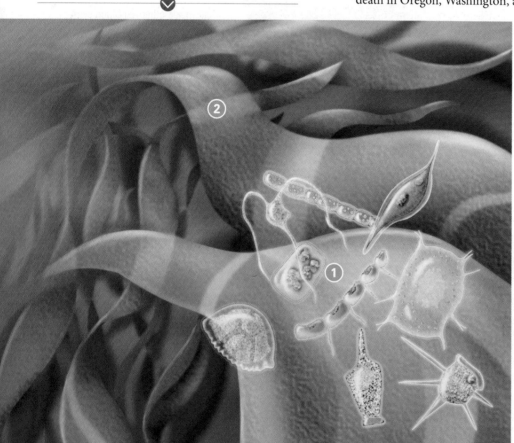

brown alga Multicelled, photosynthetic protist with brown accessory pigments.

diatom Single-celled photosynthetic protist with brown accessory pigments and a two-part silica shell.

green alga Single-celled, colonial, or multicelled photosynthetic protist belonging to the group most closely related to land plants.

red alga Single-celled or multicelled photosynthetic protist with red accessory pigment.

water mold Heterotrophic protist that forms a mesh of nutrient-absorbing filaments.

Although some large brown algae have a plantlike form, this similarity is an example of morphological convergence, rather than evidence of shared ancestry. Brown algae evolved from a different single-celled ancestor than the lineage that includes red algae, green algae, and land plants.

Red Algae Some **red algae** are single cells, but most are multicelled forms that live in tropical seas. Most commonly they have a branching structure, but some form thin sheets (Figure 13.28). Coralline algae (red algae with cell walls hardened by calcium carbonate) are a component of tropical coral reefs. Red algae are tinted red to black by accessory pigments called phycobilins. These pigments absorb the blue-green light that penetrates deep into water. Phycobilins allow red algae to carry out photosynthesis at greater depths than other algae.

Red algae have many commercial uses. Nori, the sheets of seaweed used to wrap some sushi, is a red alga that is grown commercially. Agar and carrageenan are valuable products extracted from the cell walls of other red algae. Agar keeps baked goods and cosmetics moist, helps jellies set, and is used to make capsules that hold medicines. Carrageenan is added to soy milk, dairy foods, and the fluid that is sprayed on airplanes to prevent ice formation.

Green Algae **Green algae** include single-celled, colonial, and multicelled species (Figure 13.29). Most live in fresh water, but some are marine, and some grow on soil, trees, or other damp surfaces. A few single-celled species partner with a fungus to form a lichen.

The single-celled alga *Chlorella* is cultivated in ponds, dried, and sold in powdered or pill form as a nutritional supplement. *Chlorella* is also a promising candidate for biofuel production because it has a high oil content. Many ciliates, including some *Paramecium*, already get an energy boost from *Chlorella*. They feed on sugars produced by *Chlorella* cells that live inside them.

Figure 13.28 Branching and sheetlike red algae growing 75 meters (225 ft) beneath the sea surface in the Gulf of Mexico.

Image courtesy of FGB-NMS/UNCW-NURC.

Figure 13.29 Green algae.

(A) © Wim van Egmond/Visuals Unlimited; (B) Charles Kreb/Science Faction/SuperStock; (C) © Lawson Wood/Corbis.

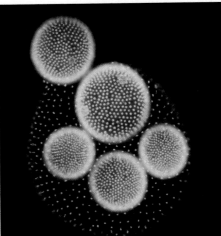

A. Two desmids, a type of single-celled green alga that lives mainly in fresh water.

B. *Volvox*, a colonial, freshwater green alga. Each sphere is a colony made up of flagellated cells linked by thin cytoplasmic strands.

C. Sheets of sea lettuce (*Ulva*), a multicelled marine green alga. Although the sheets can be longer than your arm, they are thinner than a human hair.

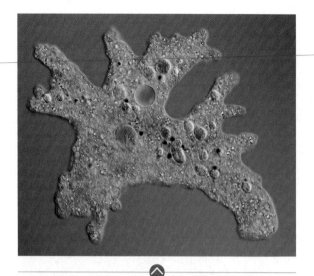

Figure 13.30 *Amoeba proteus*, **a freshwater amoeba.**
Amoebas feed or shift position by extending lobes of
cytoplasm (pseudopods). This amoeba's food vacuoles
contain green algae engulfed by its pseudopods.

iStockphoto.com/micro_photo.

Red algae, green algae, and land plants share a variety of unique traits, including chloroplasts containing a particular type of chlorophyll and a cell wall of cellulose. These similarities are taken as evidence that these three groups share a common ancestor. Both red algae and green algae evolved from an ancestral protist that had chloroplasts descended from cyanobacteria. After the two algal lineages diverged, land plants evolved from one lineage of green algae.

Amoebas and Slime Molds The free-living amoebas and the slime molds are grouped together as amoebozoans. Members of this group do not have a cell wall, shell, or pellicle, so they continually change shape. A compact blob of a cell can extend lobes of cytoplasm called pseudopods (Section 3.5) to move about and to capture food.

Amoebas such as *Amoeba proteus* (Figure 13.30) always live and feed as solitary cells. Most amoebas are predators in freshwater habitats. Others live inside animals, and some cause human disease. Each year, about 50 million people suffer from amebic dysentery after drinking water contaminated by *Entamoeba histolytica* cysts. Inadequate sterilization of contact lenses or swimming with lenses can result in an eye infection by *Acanthamoeba*, an amoeba common in soil, standing water, and even tap water.

Slime molds are sometimes described as "social amoebas." There are two types, plasmodial slime molds and cellular slime molds.

Plasmodial slime molds spend most of their life cycle as a multinucleated mass called a plasmodium. The plasmodium forms when a diploid amoeba-like cell

Figure 13.31 Plasmodial slime mold on a log.
This multinucleated mass (the plasmodium) streams along
at a rate of about a millimeter an hour, engulfing any food it
encounters. As the plasmodium travels, it lays down a trail of
slime. If it later happens across its own trail, it will move off
in a different direction. In this way, the slime mold "remem-
bers" where it has been and avoids revisiting areas where it
has already depleted its food supply.

Edward S. Ross.

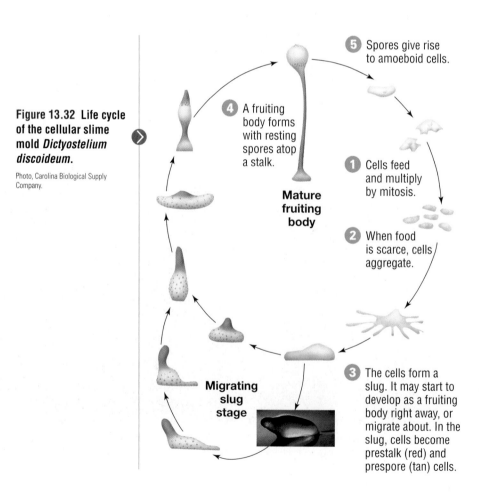

**Figure 13.32 Life cycle
of the cellular slime
mold *Dictyostelium
discoideum*.**

Photo, Carolina Biological Supply
Company.

5 Spores give rise
to amoeboid cells.

4 A fruiting
body forms
with resting
spores atop
a stalk.

**Mature
fruiting
body**

1 Cells feed
and multiply
by mitosis.

2 When food
is scarce, cells
aggregate.

3 The cells form a
slug. It may start to
develop as a fruiting
body right away, or
migrate about. In the
slug, cells become
prestalk (red) and
prespore (tan) cells.

**Migrating
slug
stage**

divides its nucleus repeatedly by mitosis, but does not undergo cytoplasmic division. The resulting mass, which can be as big as a dinner plate, streams along the forest floor engulfing microbes and organic matter (Figure 13.31). When food supplies dwindle, a plasmodium develops into many spore-bearing fruiting bodies.

Cellular slime molds spend the bulk of their existence as individual haploid amoeboid (amoeba-like) cells. *Dictyostelium discoideum* is an example (Figure 13.32). Each cell eats bacteria and reproduces by mitosis ❶. When food runs out, thousands of cells aggregate to form a multicelled mass ❷. Environmental gradients in light and moisture induce the mass to crawl along as a cohesive unit often referred to as a "slug" ❸. When the slug reaches a suitable spot, its component cells differentiate to form a fruiting body. Some cells become a stalk, and others become spores atop it ❹. When a spore germinates, it releases a cell that starts the life cycle anew ❺.

Choanoflagellates The **choanoflagellates** are the protist group with genes most similar to those of animals. Choanoflagellates also look a lot like the feeding cells of sponges, which are among the simplest animals. As a result, choanoflagellates are thought to be the closest living relatives of animals. Note that they are not considered ancestors of animals, but rather a group that shared a common single-celled ancestor with animals long ago.

Choanoflagellate means "collared flagellate." Each cell has a long flagellum surrounded by a ring (or collar) of tiny filaments reinforced with the protein actin (Figure 13.33A). Movement of the flagellum creates a current that draws water through the filaments. After tiny bits of food become entrapped, the cell extends pseudopods to capture them. The food is then digested within the cell body.

Most choanoflagellates live as single cells, but some form colonies (Figure 13.33B). The colonies arise when cells divide and the descendant cells stick together with the help of adhesion proteins. Choanoflagellate adhesion proteins are similar to those found in animals, and researchers have discovered that even solitary choanoflagellates have such proteins. By one hypothesis, the common ancestor of animals and choanoflagellates was a single-celled protist with adhesion proteins that helped it capture prey. Later, these proteins were put to use in a new context, helping cells stick together to form multicelled colonies. Later still, the proteins allowed animal cells to adhere to one another in multicelled bodies. This modification in the use of adhesion proteins is an example of how a trait that evolved to serve one function can later be modified and take on a different function.

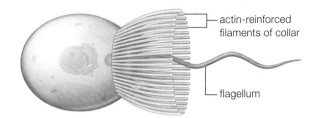

A. Structure of a solitary choanoflagellate.

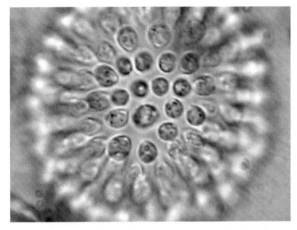

B. A colonial choanoflagellate.

Figure 13.33 Choanoflagellates, the modern protist group most closely related to animals.

(B) Courtesy of Damian Zanette.

Take-Home Message 13.6

What are protists?

- The protists are a diverse collection of eukaryotic lineages, some of which are only distantly related to one another.
- Most protists live as single cells, but there are colonial and multicelled species.
- Protists include photosynthesizers, predators, and decomposers in lakes, seas, and damp places on land. Protists also live inside other eukaryotes, including humans. Some of these protists are helpful, but others are parasites and pathogens.
- Green algae are the closest protist relatives of land plants, and choanoflagellates are the closest protist relatives of animals.

amoeba Solitary heterotrophic protist that feeds and moves by extending pseudopods.

cellular slime mold Heterotrophic protist that usually lives as a single-celled, amoeba-like predator. When conditions are unfavorable, cells aggregate into a cohesive group that can form a fruiting body.

choanoflagellates Heterotrophic protists with a collared flagellum; protist group most closely related to animals.

plasmodial slime mold Heterotrophic protist that moves and feeds as a multinucleated mass; forms a fruiting body when conditions are unfavorable.

Summary

Section 13.1 The human **microbiome**, the collection of cells that call our bodies home, includes a diverse variety of organisms. Diet and other factors affect the gut microbiome. Some organisms in the gut are helpful, but some can be **pathogens** (cause disease).

Section 13.2 Laboratory simulations provide indirect evidence that organic subunits can self-assemble under certain conditions. They also show how complex organic compounds and **protocells** may have formed on the early Earth. The **iron–sulfur world hypothesis** holds that these events took place near **hydrothermal vents**. The **RNA world hypothesis** holds that the first genome was RNA-based.

Section 13.3 Fossil **stromatolites** are evidence of early bacterial life. An early branching separated the bacteria from the archaea. Production of oxygen by some photosynthetic bacteria altered Earth's atmosphere and allowed formation of a protective **ozone layer**. Eukaryotes have a composite ancestry with both bacterial and archaeal genes. According to the **endosymbiont hypothesis**, mitochondria and chloroplasts evolved from bacteria.

Section 13.4 **Viruses** consist of RNA or DNA inside a protein coat. Some also have a **viral envelope**. Viruses replicate only in living cells, as when **bacteriophages** multiply in bacteria. Mutation and **viral reassortment** produce new types of viruses. Insects serve as **disease vectors** that spread some viral diseases. **HIV** (**human immunodeficiency virus**) is an enveloped RNA virus that infects human cells and causes AIDS.

Section 13.5 **Archaea** and **bacteria** are prokaryotic cells, meaning they lack a nucleus. They reproduce asexually by **binary fission**, and exchange genes through **transformation**, **transduction**, and **conjugation** (direct transfer of a **plasmid**).

Bacteria and archaea may be aerobic or anaerobic, and they show great nutritional diversity. **Autotrophs** use carbon dioxide as their carbon source. They include photoautotrophs such as cyanobacteria and chemoautotrophs such as the archaea that live near hydrothermal vents. By contrast, **heterotrophs** obtain carbon from organic compounds. Chemoheterotrophs that serve decomposers and pathogens obtain energy by breaking down organic compounds.

Archaea include heat-loving **extreme thermophiles** or salt-loving **extreme halophiles**. Others live in less extreme environments, such as the human gut. Some people have archaea that produce methane (**methanogens**) in their gut.

Bacteria play important ecological roles by serving as **decomposers**, releasing oxygen into the air, and carrying out **nitrogen fixation**. We use some bacterial species in biological research, biotechnology, and food production. Some bacteria benefit our health, as by living in our gut and producing vitamins. Others are pathogens.

Section 13.6 **Protists** are a diverse collection of lineages. Most lineages are single-celled, but some include **colonial organisms** or **multicellular organisms**. Nearly all protists live in water or in moist habitats, including host tissues.

Flagellated protozoans are single cells with a pellicle that helps hold their shape. Most are heterotrophs, but some euglenoids have chloroplasts. A **contractile vacuole** allows freshwater protozoans to expel excess water. Trypanosomes, *Giardia*, and *Trichomoniasis* are flagellated human pathogens.

Foraminifera are single-celled heterotrophs with calcium carbonate shells. They live on the seafloor or drift as marine **plankton**. Shells of foraminifera contribute to limestone and chalk.

Ciliates are single-celled heterotrophs that use cilia to move and feed. **Dinoflagellates** are single cells that move with a whirling motion and can be heterotrophs or photosynthetic. Some are **bioluminescent** and others cause **algal blooms**. **Apicomplexans**, such as the species that cause malaria, are parasites that spend part of their life in cells of their host.

Water molds are heterotrophs that grow as filaments. Some are pathogens of fish or plants. **Diatoms** are single-celled, silica-shelled, aquatic producers. **Brown algae** include the giant kelps, which are the largest protists.

Red algae can live at greater depths than other algae. Red algae share a common ancestor with **green algae**. Land plants evolved from a green alga.

Amoebas are shape-shifting cells that extend pseudopods to feed and move. They live in aquatic habitats and animal bodies. The related **cellular slime molds** spend part of their life as a single cell and part in a cohesive group that can migrate and differentiate to form a spore-bearing body. **Plasmodial slime molds** feed as a giant multinucleated mass, then form spores when food runs out. The **choanoflagellates** are the protists most closely related to animals.

Self-Quiz

Answers in Appendix I

1. A rise in atmospheric _____ allowed the formation of the ozone layer that screens out UV radiation from the sun.
   a. hydrogen   b. water   c. oxygen   d. ammonia

2. Stanley Miller's experiment demonstrated _____ .
   a. the great age of Earth
   b. that amino acids can assemble under some conditions
   c. that oxygen is necessary for life
   d. all of the above

3. The universal need for iron–sulfur cofactors is taken as evidence that metabolism may have begun _____ .
   a. on a meteorite
   b. on a mudflat
   c. on a rock near a hydrothermal vent

4. Mitochondria are most likely descendants of _____ .
   a. methanogenic archaea
   b. aerobic bacteria
   c. cyanobacteria
   d. green algae

5. The genetic material of HIV is _____ .
   a. protein   b. DNA   c. RNA   d. ATP

6. Viral transfer of genes between bacteria is called _____ .
   a. conjugation
   b. viral reassortment
   c. transduction
   d. transformation

7. All viruses have _____ .
   a. an envelope   b. ribosomes   c. DNA   d. a protein coat

8. Choanoflagellates are most closely related to _____ .
   a. bacteria   b. land plants   c. ciliates   d. animals

9. _____ take up carbon dioxide from seawater and use it to make a chalky shell.
   a. Ciliates   b. Diatoms   c. Foraminifera   d. Euglenoids

10. All _____ are parasitic eukaryotes that live in other cells.
    a. viruses   c. euglenoids   e. both a and b
    b. apicomplexans   d. slime molds   f. all are correct

11. Oil-rich remains of ancient _____ are the main source of the petroleum that we use to make gasoline.
    a. diatoms   b. ciliates   c. foraminifera   d. red algae

12. Some _____ live in corals and supply them with sugars.
    a. ciliates   b. viruses   c. kelps   d. dinoflagellates

13. Only certain _____ can carry out nitrogen fixation.
    a. green algae and land plants   c. bacteria
    b. diatoms   d. archaea

14. Genetic material of a _____ can be either DNA or RNA.
    a. bacteria   b. dinoflagellate   c. ciliate   d. virus

15. Match these terms with the appropriate definition.
    _____ green algae   a. protist population explosion
    _____ virus   b. social amoeba
    _____ bacteria   c. most diverse prokaryotes
    _____ brown algae   d. noncellular infectious agent
    _____ bioluminescence   e. include the largest protists
    _____ euglenoid   f. flagellate with chloroplasts
    _____ algal bloom   g. closest relative of plants
    _____ dinoflagellate   h. layered prokaryotes and sediment
    _____ slime mold   i. biologically produced light
    _____ stromatolite   j. whirling cell

Critical Thinking

1. Researchers looking for fossils of the earliest life forms face many hurdles. For example, few sedimentary rocks date back more than 3 billion years. Review what you learned about plate tectonics (Section 11.5). Explain why so few remaining samples of the earliest rocks remain.

2. The antibiotic penicillin acts by interfering with the production of new bacterial cell walls. Bacteria exposed to penicillin do not die immediately, but they cannot reproduce. Explain why.

3. Viruses that do not have a lipid envelope tend to remain infectious outside the body longer than enveloped viruses. "Naked" viruses are also less likely to be rendered harmless by soap and water. Can you explain why?

4. The apicomplexan that causes malaria had a photosynthetic ancestor and contains an organelle that evolved from its ancestral chloroplast. The organelle no longer functions in photosynthesis, but it does carry out some essential metabolic tasks. Why would targeting this organelle yield an antimalarial drug that would be likely to have minimal side effects?

Visual Question

1. The graphic below depicts the first steps in one of the two bacteriophage replication pathways. Which pathway is it and how can you tell?

PLANTS AND FUNGI

Application ❯ 14.1 **Fungal Threats to Crops**

REMEMBER: Plants are producers (Section 1.3) that make food by photosynthesis, whereas fungi are consumers that must obtain ready-made sugars.

Plants are the main producers on land and thus serve as an important food source for both animals and fungi. As a result, we often find ourselves in competition with fungi for plant foods. For example, wheat is both an important agricultural crop and the host for wheat stem rust fungus (*Puccinia graminis*). Wheat stem rust fungus is an obligate plant parasite, meaning it can grow and reproduce only in a living plant.

Fungi disperse by releasing microscopic spores that travel on the wind. A wheat stem rust infection begins when a fungal spore lands on the leaf of a wheat plant. The spore germinates (becomes active), and a fungal filament grows into the plant. As fungal filaments extend through the plant's tissues, the filaments take up photosynthetic sugars that the plant would normally use to meet its own needs. As a result, an affected plant is stunted and produces little or no wheat. About a week after infection, tens of thousands of rust-colored spore sacs appear on the stem of the infected plant (Figure 14.1). Each spore can disperse and infect a new plant.

Outbreaks of wheat stem rust disease routinely destroyed wheat crops worldwide until the 1960s, when a plant breeding program headed by Norman Borlaug produced resistant strains of wheat. In 1970, Borlaug received the Nobel Peace Prize for his role in preventing food shortages that contribute to global instability. Planting rust-resistant wheats developed by Borlaug prevented outbreaks of wheat stem rust for decades. Then, in 1999, scientists discovered a new strain of wheat stem rust in Uganda. This strain, Ug99, has mutations that allow it to infect about 90 percent of the wheat varieties that were previously resistant to wheat stem rust.

Ug99 is now spreading. Fungal spores are microscopic, so they can lodge in crevices on dust particles. When winds lift these particles aloft, the spores go along for the ride. Dustborne fungal spores disperse long distances riding winds that swirl high above Earth's surface. As of 2014, windblown spores of Ug99 had reached Kenya, Ethiopia, and Sudan, crossed the Red Sea to Yemen, and from there crossed the Persian Gulf to Iran. Given the prevailing winds, India, the world's second largest wheat producer, is expected to be affected soon. Most likely, winds will eventually distribute Ug99 worldwide. Fungicides can minimize the damage, but are too expensive for farmers in developing nations.

The threat fungal diseases pose to crop plants is heightened by current agricultural practices. Farmers often plant a single variety of a crop in dense stands that cover an extensive area. The proximity of many genetically identical host plants allows a fungus to spread quickly through a field. In addition, farmers in different parts of the world often buy seeds from the same large companies and plant the same few varieties of a crop.

Protecting our food supply from the threat of plant diseases requires maintaining the genetic diversity of our crop plants and their wild relatives. Having many varieties of a crop increases the likelihood that at least one variety will be immune to any given disease. That variety can be planted or used to create new disease-resistant varieties, either through traditional plant breeding or by using genetic engineering to transfer disease-resistance genes. Maintaining the wild relatives of crop plants provides another potential source of disease-resistance genes. For example, researchers have discovered genes that confer Ug99 resistance in a wild grass related to wheat, and in a primitive wheat. Transferring these genes into widely used bread wheats could help ensure the safety of the world's wheat supply.

Figure 14.1 Wheat stem rust fungus, a current threat to world food supplies.
Wheat stem, with rust-colored fungal sporangia (spore-producing structures) on its surface. The inset micrograph shows spores (red) escaping from the sporangia.

Background, Photo by Yue Jin/USDA; inset, Courtesy of Charles Good, Ohio State University at Lima.

Figure 14.2 Traits of modern plant groups and relationships among them.

Photos, Courtesy of © Christine Evers.

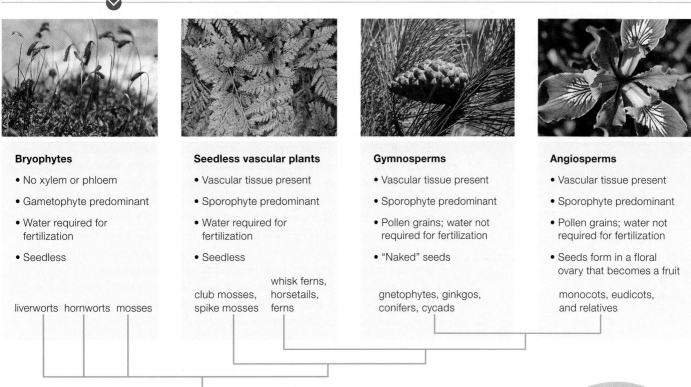

Bryophytes

- No xylem or phloem
- Gametophyte predominant
- Water required for fertilization
- Seedless

liverworts hornworts mosses

Seedless vascular plants

- Vascular tissue present
- Sporophyte predominant
- Water required for fertilization
- Seedless

club mosses,   whisk ferns,
spike mosses   horsetails,
               ferns

Gymnosperms

- Vascular tissue present
- Sporophyte predominant
- Pollen grains; water not required for fertilization
- "Naked" seeds

gnetophytes, ginkgos,
conifers, cycads

Angiosperms

- Vascular tissue present
- Sporophyte predominant
- Pollen grains; water not required for fertilization
- Seeds form in a floral ovary that becomes a fruit

monocots, eudicots,
and relatives

14.2 Plant Traits and Evolution

REMEMBER: Green algae are the closest relatives of land plants (Section 13.6).

Plants are a lineage of land-dwelling, multicelled, photosynthetic eukaryotes. They evolved from freshwater green algae (charophyte algae), and share many traits with this group. The defining trait that sets plants apart is a multicelled embryo that forms, develops within, and is nourished by the parent plant. For this reason, the clade of land plants is referred to as the embryophytes. Figure 14.2 summarizes the relationships among modern plants and their defining traits.

The story of plant evolution is one of adaptation to increasingly drier environments. An aquatic green alga lives surrounded by water, so it can absorb water and dissolved nutrients across its entire body surface. Water also buoys the alga's parts, thus helping it stand upright. Land plants, however, face the constant threat of drying out and must hold themselves upright. Changes in life cycle, structure, and the mechanisms of reproduction and dispersal adapted plants to life on land.

Life Cycle All plants have an **alternation of generations**, meaning their life cycle alternates between haploid and diploid multicelled bodies (Figure 14.3). The diploid generation, called the **sporophyte**, produces spores by meiosis ❶. A plant spore is a single diploid cell that undergoes mitosis and develops into the multicelled, haploid generation. The haploid **gametophyte** ❷ produces gametes by mitosis ❸. Gametes unite at fertilization to form a zygote ❹ that will develop into a diploid sporophyte ❺.

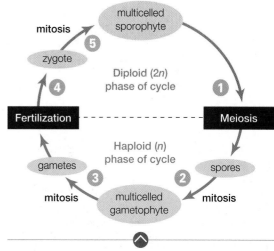

Figure 14.3 Plant life cycle (alternation of generations).

alternation of generations As in plants, a life cycle that alternates between a diploid spore-producing body and a haploid, gamete-producing one.

gametophyte Haploid gamete-forming body that forms in a plant life cycle.

plant Multicelled, typically photosynthetic organism; develops from an embryo that forms on the parent and is nourished by it.

sporophyte Diploid spore-forming body that forms in a plant life cycle.

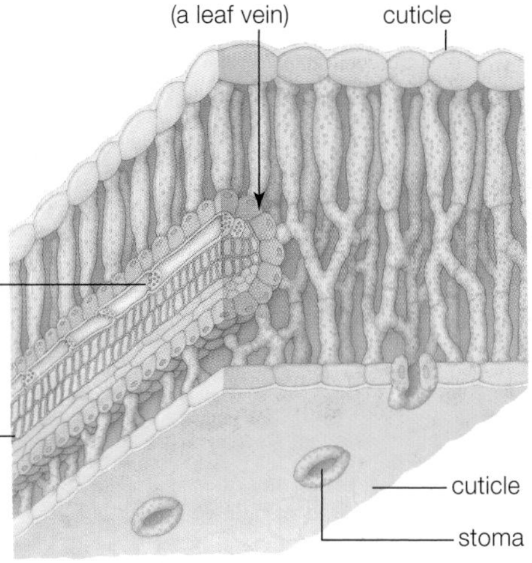

Figure 14.4 Cross-section of a leaf, showing some traits that adapt plants to land.

The relative size, complexity, and longevity of the sporophyte and gametophyte vary among plants lineages. In the oldest surviving plant lineages, collectively referred to as nonvascular plants or **bryophytes**, the haploid gametophyte is larger and longer-lived than the diploid sporophyte. In all other plants, sporophytes dominate the life cycle and gametophytes are tiny and short-lived.

Structural Adaptations to Life on Land The aboveground parts of most land plants are covered by a **cuticle** (Figure 14.4). This layer of waxy secretions slows evaporation across the body surface. Adjustable pores called **stomata** (singular, stoma) extend across the cuticle. Depending on environmental conditions, these pores either open to allow gas exchange or close to conserve water.

Bryophytes have threadlike structures that hold them in place, but they do not have true roots. Such roots not only anchor a plant, they also take up water and essential minerals from the soil. Plants with true roots have a vascular system with two types of tissues that serve as internal pipelines. **Xylem** is the vascular tissue that distributes water and mineral ions. **Phloem** is the vascular tissue that distributes sugars produced by photosynthetic cells. More than 90 percent of modern plants have xylem and phloem and thus are grouped as **vascular plants**.

In addition to distributing materials, vascular tissue provides structural support. Organic compounds called **lignins** stiffen the walls of xylem. Evolution of stems with lignin-stiffened tissue allowed vascular plants to stand taller than nonvascular ones and to branch. Most vascular plants also have leaves. Leaves are flattened, aboveground organs that increase the surface area available for intercepting sunlight and for gas exchange. They contain veins of vascular tissue.

Reproduction and Dispersal All bryophytes and some vascular plants produce sperm that have one or more flagella. To reach eggs, the sperm must swim through a film of water. As a result, these plants can reproduce only when their surroundings are damp. By contrast, seed-bearing vascular plants (seed plants) produce pollen grains. A **pollen grain** is a walled, immature male gametophyte. Winds or animals can carry pollen grains to a female gametophyte, thus fertilization can occur even in dry environments.

Bryophytes and seedless vascular plants disperse by releasing spores, but seed plants disperse by releasing seeds. A **seed** consists of an embryo sporophyte and food to support it, enclosed within a protective coat. There are two kinds of seed plants, gymnosperms and angiosperms. Angiosperms are the only plants that make flowers and disperse their seeds inside of fruits.

bryophyte Member of a plant lineage that does not have vascular tissue; for example, a moss.

cuticle Secreted covering at a body surface. In plants it is waxy and helps conserve water.

lignin Compound that stiffens walls of some cells (including xylem) in vascular plants.

phloem Vascular tissue that distributes dissolved sugars.

pollen grain Immature male gametophyte of a seed plant.

seed Embryo sporophyte of a seed-bearing plant packaged with nutritive tissue inside a protective coat.

stomata Adjustable pores in a plant cuticle.

vascular plant A plant that has xylem and phloem.

xylem Vascular tissue that distributes water and dissolved mineral ions.

Take-Home Message 14.2

What adaptive traits allow plants to live on land and in dry places?

• Plants are multicelled, photosynthetic eukaryotes that protect and nourish their multicelled embryos on their body.

• Adaptations to life on land include a waxy cuticle with stomata, true roots, and vascular tissues that distribute materials and provide structural support.

• The plant life cycle alternates between two multicelled generations: a haploid gametophyte generation and a diploid sporophyte generation.

• Evolution of pollen grains and seeds gave seed plants the capacity to live in drier places than other plants.

14.3 Nonvascular Plants

REMEMBER: Gametes form by meiosis (Section 8.5).

Modern bryophytes (nonvascular plants) include 24,000 species belonging to three lineages: mosses, hornworts, and liverworts. Some mosses have tubes that conduct water and sugar, but none has the lignin-stiffened vascular pipelines that define vascular plants. As a result, nearly all nonvascular plants stand less than 20 centimeters (8 inches) high.

Mosses **Mosses** are the most diverse and familiar nonvascular plants. We will use the life cycle of one moss (*Polytrichum*) to illustrate a bryophyte life cycle (Figure 14.5). Like all bryophytes, this moss has a gametophyte-dominated life cycle. As in other mosses, the gametophyte has leaflike photosynthetic parts that grow from a central stalk ❶. Threadlike structures called **rhizoids** hold the gametophyte in place.

The moss sporophyte is not photosynthetic, so it must obtain nourishment from the gametophyte to which it is attached. The sporophyte consists of a stalk with a spore-producing organ (a sporangium) at its tip ❷. Meiosis of cells inside a spore chamber yields haploid spores ❸.

After dispersal by the wind, a spore germinates (becomes active) and grows into a gametophyte that has multicellular gamete-producing organs (gametangia). The moss we are using as our example has separate sexes, with each gametophyte producing either eggs or sperm ❹. Other bryophytes are bisexual. In either case, rain triggers the release of flagellated sperm that swim through a film of water to eggs ❺. Fertilization occurs inside the egg chamber and produces a zygote ❻ that develops into a new sporophyte ❼. Mosses also reproduce asexually by fragmentation when a bit of gametophyte breaks off and develops into a new plant.

There are about 14,000 named species of moss. Many of them can colonize rocky areas where the lack of soil prevents other plants from becoming established. Over time, decomposition of dead moss helps to create a layer of soil in which vascular plants can take root.

The 350 or so species of moisture-loving peat mosses (*Sphagnum*) are of great ecological and commercial importance. They are the main plants in peat bogs, a type of plant community that covers more than 350 million acres in high-latitude areas of Europe, Asia, and North America. Many peat bogs have existed for thousands of years, and layer upon layer of partially decayed plant remains have become compressed to form deposits of a carbon-rich material called peat. Blocks of peat can be cut, dried, and used as a clean-burning fuel. Freshly harvested peat moss is also an important commercial product. It is dried and added to planting mixes to help soil retain moisture.

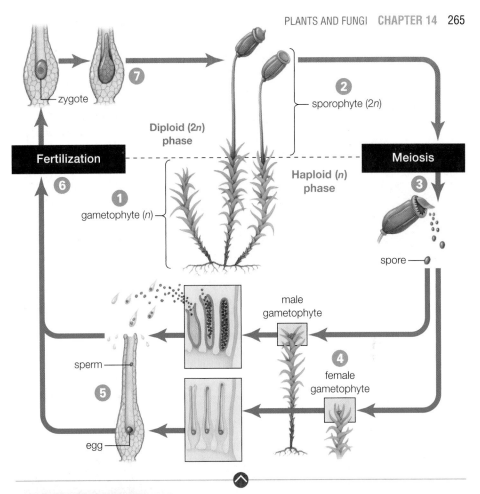

Figure 14.5 Life cycle of the moss *Polytrichum*, shown at left.

❶ The leafy green part of a moss is the haploid gametophyte.

❷ The diploid sporophyte has a stalk and a capsule (sporangium). It is not photosynthetic.

❸ Haploid spores form by meiosis in the capsule, are released, and drift with the winds.

❹ Spores germinate and develop into male or female gametophytes with gametangia that produce eggs or sperm by mitosis.

❺ Sperm swim to eggs.

❻ Fertilization occurs in the egg chamber on the female gametophyte and produces a zygote.

❼ The zygote grows and develops into a sporophyte while remaining attached to and nourished by its female parent.

Bottom photo, Jane Burton/Bruce Coleman Ltd.

mosses Most diverse group of bryophytes (nonvascular plants). Low-growing plants that have flagellated sperm and disperse by producing spores.

rhizoid Threadlike structure that anchors some plants.

Figure 14.6 Liverwort.
Umbrella-like female gametangia of a liverwort (*Marchantia*).
The attached sporophytes on their lower surface have spores
(yellow) at their tips.

Dr. Annkatrin Rose, Appalachian State University.

Figure 14.7 Hornworts.
Elongated photosynthetic sporophytes grow attached to the
flat, ribbonlike gametophytes. Spores will form at the tip of
these "horns."

age fotostock/SuperStock.

Figure It Out: Do spores of liverworts develop into
sporophytes or gametophytes?

Answer: All plant spores develop into gametophytes.

Liverworts and Hornworts Liverworts and hornworts are less familiar nonvascular lineages. They often live alongside mosses in damp places.

Liverworts may be the most ancient of the surviving plant lineages. The oldest known fossils of land plants are spores that resemble those of modern liverworts. In addition, genetic comparisons put liverworts near the base of the plant family tree.

Marchantia is a widespread genus of liverworts that reproduces sexually by producing sperm and eggs in umbrella-like structures that grow from the surface of the flattened gametophyte (Figure 14.6). After fertilization, the sporophyte develops while still attached to the gametophyte. *Marchantia* species also reproduce asexually by producing small clumps of cells in cups on the gametophyte surface. The clumps are dispersed by rain and develop into new plants.

Hornworts have flattened, ribbonlike gametophytes. Fertilization of eggs that form on the gametophyte body produces a zygote that develops into a tall, horn-shaped sporophyte (Figure 14.7). Hornwort sporophytes, unlike those of mosses and liverworts, contain chloroplasts and grow continually. They can be several centimeters tall. These sporophyte traits, together with evidence from gene comparisons, suggest that hornworts are probably the nonvascular plants that are most closely related to vascular plants.

Take-Home Message 14.3

What are bryophytes?

- Bryophytes include three lineages of low-growing plants (mosses, hornworts, and liverworts). All have flagellated sperm and disperse by releasing spores.
- Bryophytes are the only modern plants in which the gametophyte dominates the life cycle and the sporophyte is dependent upon it.
- Liverworts are probably the oldest plant lineage. Hornworts are most likely the closest living relatives of vascular plants.

14.4 Seedless Vascular Plants

Seedless vascular plants include ferns, horsetails, and club mosses. Like bryophytes, these plants have flagellated sperm that require a film of water to swim to eggs. Also like bryophytes, they disperse by releasing spores.

Seedless vascular plants differ from bryophytes in other aspects of their life cycle and structure. The gametophyte is reduced in size and is relatively short-lived. Although the sporophyte develops on the gametophyte body, it survives on its own after the gametophyte dies. Lignin stiffens a sporophyte's body, and a system of vascular tissue distributes water, sugars, and minerals through it. These innovations in support and plumbing allow seedless vascular sporophytes to be large and structurally complex, with roots, stems, and leaves.

Ferns We begin our survey of the seedless vascular plants with the most diverse and familiar lineage, the **ferns**. Figure 14.8 shows the life cycle of a common North American fern. The leafy plant we envision when we think of a fern is a sporophyte ❶. In most ferns, roots and fronds (leaves) sprout from **rhizomes**, which are stems that grow along or just below the ground. Spores form by meiosis in capsules that cluster as **sori** (singular, sorus) on the underside of leaves ❷. When the spore

Figure 14.8 Life cycle of a common North American fern (*Woodwardia*).

1. The familiar leafy form is the diploid sporophyte.

2. Meiosis in cells on the underside of fronds produces haploid spores.

3. After their release, the spores germinate and grow into tiny gametophytes that produce eggs and sperm.

4. Sperm swim to eggs and fertilize them, forming a zygote.

5. The sporophyte begins its development attached to the gametophyte, but it continues to grow and live independently after the gametophyte dies.

Photo, A. & E. Bomford/Ardea, London.

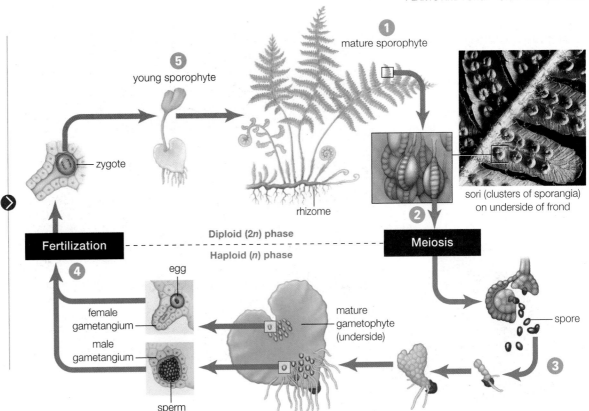

1 mature sporophyte

5 young sporophyte

zygote

rhizome

sori (clusters of sporangia) on underside of frond

Diploid (2*n*) phase

Fertilization

Haploid (*n*) phase

4

egg

female gametangium

male gametangium

sperm

mature gametophyte (underside)

Meiosis

2

spore

3

capsule pops open, wind disperses the spores. A spore develops into a gametophyte a few centimeters wide that forms eggs and sperm in chambers on its underside ❸. In the fern we are discussing, eggs and sperm form on the same gametophyte. In some other ferns, each gametophyte produces either sperm or eggs. In either case, sperm swim to eggs to fertilize them ❹. After fertilization, the resulting zygote develops into a new sporophyte, and its parental gametophyte dies ❺.

In many ferns, asexual reproduction occurs more frequently than sexual reproduction. New shoots and roots develop from a rhizome as it grows through the soil. Eventually the connection to the parent plant breaks, and that segment of rhizome becomes an independent plant.

Fern sporophytes vary enormously in their size and form. Some float on freshwater ponds and have fronds less than 1 centimeter wide. Many tropical ferns are **epiphytes**, plants that live attached to the trunk or branches of another plant but do not withdraw nutrients from it. The largest ferns are tree ferns that can be 25 meters (80 feet) tall.

Horsetails and Club Mosses Horsetails (*Equisetum*) are close relatives of ferns. They thrive along streams and roadsides, and in disturbed areas. A horsetail sporophyte has rhizomes and hollow stems with tiny nonphotosynthetic leaves at the joints (Figure 14.9A). Photosynthesis occurs in stems and leaflike branches. The stems contain silica, a gritty mineral that helps the plant fend off insects and snails. Before the invention of modern abrasive cleansers, silica-rich stems of some *Equisetum* species were used to scrub pots and polish metals. Depending on the species, spore-bearing structures form either at tips of photosynthetic stems or on specialized reproductive stems that do no have chlorophyll.

Club mosses are nonvascular plants common on the floor of temperate forests (Figure 14.9B). Their branching form makes them look like tiny pine trees. Roots

Figure 14.9 Seedless vascular plants.

A. Horsetail   B. Club moss

(A) © William Ferguson; (B) © Martin LaBar, www.flickr.com/photos/martinlabar.

epiphyte Plant that grows on the trunk or branches of another plant but does not harm it.

ferns Most diverse lineage of seedless vascular plants.

rhizome Stem that grows horizontally along or just below the ground.

sorus Cluster of spore-forming chambers on a fern frond.

Figure 14.10 Coal forest.
An artist's depiction of a swamp forest during the Carboniferous period. An understory of ferns ❶ is shaded by tree-sized club mosses ❷ and horsetails ❸.

Over millions of years, remains of such plants were transformed to coal (below).

Noraluca013/Shutterstock.com.

and upright stems with tiny leaves grow from a rhizome. Club moss spores have a waxy coating that makes them ignite easily. They were used to create flashes for early photography and are still used to produce special effects in theaters.

Ferns are currently the most diverse seedless vascular plants, but during the Carboniferous period (359–299 million years ago) relatives of modern club mosses and horsetails were the dominant plants in swamp forests (Figure 14.10). Some stood 40 meters (more than 130 feet) high. After forests of these plants formed, climates changed, and the sea level rose and fell many times. When the waters receded, the forests flourished. After the sea moved back in, submerged trees became buried in sediments that protected them from decomposition. As layers of sediments accumulated one on top of the other, their weight squeezed the water out of the saturated, undecayed remains, and the compaction generated heat. Over time, pressure and heat transformed the compacted organic remains into **coal**.

It took millions of years of photosynthesis, burial, and compaction to form coal. When you hear about annual production of coal or other fossil fuels, keep in mind that we do not really "produce" these materials, we only extract them. This is why fossil fuels are said to be nonrenewable sources of energy.

Take-Home Message 14.4

What are seedless vascular plants?

- Seedless vascular plants include ferns, club mosses, and horsetails.
- These plants disperse by releasing spores, and the life cycle is dominated by a sporophyte that has vascular tissue and lignin. The gametophyte is small and relatively short-lived.
- Like bryophytes, seedless vascular plants have flagellated sperm that must swim through a film of water to reach eggs.

coal Fossil fuel consisting primarily of the carbon-rich remains of seedless nonvascular plants.

14.5 Rise of the Seed Plants

Seed plants evolved from a lineage of seedless vascular plants about 400 million years ago. They survived alongside bryophytes and seedless nonvascular plants until the late Carboniferous, then rose to dominance as the climate became drier. Unique traits of seed plants give them a competitive advantage over seedless plants in places where water is scarce.

Gametophytes of seedless vascular plants are free-living, meaning they develop from spores that were released into the environment. By contrast, gametophytes of seed plants develop within the protection of spore-forming chambers (pollen sacs and ovules) on a sporophyte (Figure 14.11).

All nonvascular plants and most seedless vascular plants produce spores of just one type. By contrast, all seed plants produce two types of spores that differ in size. Meiosis of a cell inside a **pollen sac** produces four microspores ❶. Each **microspore** develops into a pollen grain, which is the sperm-producing gametophyte ❷. In **ovules**, meiosis and unequal cytoplasmic division produce three small cells that disintegrate and one large **megaspore** ❸. The megaspore develops into an egg-producing gametophyte ❹.

A seed plant releases its pollen grains, but holds onto its eggs. Wind or animals deliver pollen from one seed plant to the ovule of another, a process known as **pollination** ❺. Sperm of seed plants do not need to swim through a film of water to reach eggs, so these plants can reproduce in dry times.

After pollination, a pollen tube grows into the ovule and delivers a sperm, which in most seed plants is not flagellated, to the egg ❻. Fertilization produces a zygote within the ovule, which matures into a seed ❼. Releasing seeds puts seed plants at an advantage over plants that release spores. A seed contains a multicelled embryo sporophyte and stored food that the embryo can draw on during early development. By contrast, a plant spore is a single cell that does not have any food reserves. In addition, seeds often also have adaptations that aid in their long-distance dispersal. For example, the seed may have a winglike structure that helps it catch the wind.

Structural traits also gave seed plants an advantage. Some seed plants undergo secondary growth (growth in diameter) and produce wood. **Wood** is lignin-stiffened tissue that strengthens and protects older stems and roots. The giant nonvascular plants that lived in Carboniferous forests did not produce wood, and their trunks were supported by overlapping leaf bases. As a result, their trunks were softer and more flexible than those of woody seed plants, and they could not grow as tall as the tallest modern trees do.

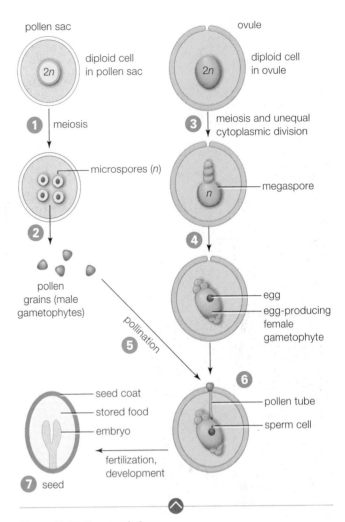

Figure 14.11 How seeds form.
This diagram shows the process in gymnosperms. The process is slightly different in angiosperms, as described in Section 14.7.

Figure It Out: Do pollen sacs form on sporophytes or on gametophytes?

Answer: Pollen sacs form on sporophytes. They produce spores that develop into male gametophytes.

Take-Home Message 14.5

What gave seed plants an adaptive advantage over spore-bearing plants?
- Seed plant spores develop into gametophytes while protected by the sporophyte body.
- Male gametophytes are pollen grains that can be carried to egg-bearing ovules even in dry environments.
- A seed contains a plant embryo and a food supply that it can draw upon during its early development.
- Some seed plants undergo secondary growth (they thicken) and become woody.

megaspore In seed plants, a haploid cell that gives rise to a female gametophyte.

microspore In seed plants, a haploid cell that gives rise to a male gametophyte (pollen grain).

ovule Of seed plants, chamber inside which megaspores form and develop into female gametophytes; after fertilization this chamber becomes a seed.

pollen sac Of seed plants, chamber in which microspores form and develop into male gametophytes (pollen grains).

pollination Delivery of a pollen grain to the egg-bearing part of a seed plant.

wood Lignin-reinforced tissue produced by secondary growth of some seed plants.

14.6 Gymnosperms

Gymnosperms are seed plants that produce seeds on the surface of ovules. Their seeds are said to be "naked," because unlike those of angiosperms, they are not inside a fruit. (*Gymnos* means naked and *sperma* is taken to mean seed.) However, many gymnosperms enclose their seeds in a fleshy or papery covering. There are four modern gymnosperm lineages.

Conifers Conifers (phylum Pinophyta) are the most diverse gymnosperms with more than 600 species. All are woody trees or shrubs with needlelike or scalelike leaves. Most conifers are evergreen, meaning they do not shed their leaves all at once. Evergreen conifers are the main plants in cool Northern Hemisphere forests. Their conical shape helps many conifers shed snow easily. Needlelike leaves with a heavy wax coating help them minimize water loss during a long winter when soil is frozen, or during a dry hot season. Conifers include the tallest trees in the Northern Hemisphere (Figure 14.12) and the longest-lived trees; some bristlecone pines are more than 4,000 years old.

A pine tree's life cycle is typical of conifers. The tree is a sporophyte and its cones are specialized spore-bearing structures. There are two types of cones: small, soft pollen cones and large, woody, ovulate cones (Figure 14.13). Both have scales (modified leaves) arranged around a central axis. Pollen made by a pollen cone is released and drifts on the wind. Ovulate cones produce a sticky substance that traps

ovule, which may become seed

Figure 14.13 Pine cones.
A. Pollen cone. **B.** Cross section of an ovulate cone.

(A) R. J. Erwin/Science Source; (B) © Stan Elems/ Visuals Unlimited.

Figure 14.12 Composite photo of one of the world's tallest trees, a coast redwood (*Sequoia sempervirens*) that stands 116 meters (379 feet) high in a Northern California forest.
Can you spot the three blue-shirted people climbing among its branches?

© James Balog/Aurora Photos.

pollen grains. After pollen is trapped, a tube grows into the ovulate cone scale and delivers a nonflagellated sperm cell to an egg in an ovule inside it. The pollen tube grows very slowly, so fertilization can occur as long as a year after pollination. After fertilization, the ovule develops into a seed.

Conifers are of great economic importance. For example, pines are the main source of lumber for building homes, and we use fir bark as mulch in our gardens. Some pines make a sticky resin that deters insects from boring into them. We use this resin to make turpentine, which is a paint thinner and solvent. We use oils from cedar in cleaning products, and we eat seeds, or "pine nuts," of pinyon pines.

Cycads and Ginkgos   Cycads (phylum Cycadophyta) and ginkgos (phylum Ginkgophyta) are ancient lineages that were at their most diverse when dinosaurs walked the Earth. They are the only modern gymnosperms that have flagellated sperm. In both groups, a male plant produces pollen that is carried by the wind to a female plant. As in conifers, the ovule of the female plant secretes fluid that traps pollen grains. Sperm emerge from pollen grains, then swim through the secreted fluid to eggs in the plant's ovule.

About 130 species of cycads survive and they live mainly in tropical and subtropical regions. Cycads resemble palms or ferns but are not close relatives of either (Figure 14.14A). "Sago palms" commonly used in landscaping and as houseplants are actually cycads.

The only living ginkgo species is *Ginkgo biloba*, the maidenhair tree. It is native to China, but its pretty fan-shaped leaves (Figure 14.14B) and resistance to air pollution make it popular in many cities in the United States. Typically, male trees are planted because female trees make seeds with a fleshy covering that has a strong, unpleasant odor. Ginkgos are **deciduous plants**, meaning they shed all their leaves at the end of their growing season and spend the winter leafless and dormant.

Gnetophytes   Gnetophytes (phylum Gnetophyta) include about 70 species. One of these, *Welwitschia mirabilis*, is an unusual-looking plant that lives in Africa's Namib desert (Figure 14.14C). It has a taproot, woody stem, and two long, straplike leaves that split lengthwise repeatedly, giving the plant a shaggy appearance. The leaves grow continually as a result of cell divisions at their base. *Welwitschia* is very long-lived, with some plants reaching an age of more than a thousand years. Members of the genus *Ephedra* are shrubs with jointed, photosynthetic branches and tiny, scalelike leaves (Figure 14.14D). Eurasian species of *Ephedra* contain large amounts of the amphetamine-like compound ephedrine and are used in traditional medicine. *Ephedra* species native to North America do not make this compound.

A. Cycad with palmlike leaves and fleshy seeds.

B. Ginkgo tree with fleshy seeds and fan-shaped seeds.

C. *Welwitschia* with seed cones and two long leaves.

D. *Ephedra* with photosynthetic stems and pollen cones.

Figure 14.14 Gymnosperm diversity.

(A, C) Fletcher and Baylis/Science Source; (B) left, KPG_Payless/Shutterstock.com; right, picturepartners/Shutterstock.com; (D) Christine Evers.

Take-Home Message 14.6

What are gymnosperms?

- Gymnosperms are one of the two lineages of seed-bearing vascular plants. They form seeds on the surface of cones or other spore-producing structures.
- Conifers, the most diverse gymnosperm group, are evergreen trees with needlelike leaves. They dominate cool, high-latitude forests.
- Cycads and ginkgos are two lineages with pollen grains that release flagellated sperm.
- Gnetophytes include diverse forms.

conifer Woody gymnosperm with needlelike leaves.

deciduous plant Plant that sheds all its leaves in preparation for a seasonal dormancy.

gymnosperm Seed plant that produces "naked" seeds (seeds that are not encased within a fruit).

Figure 14.15 Floral structure.

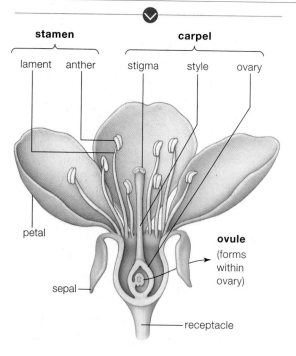

14.7 Angiosperms—Flowering Plants

REMEMBER: Coevolution refers to the joint evolution of two species as a result of their close ecological interaction (Section 12.7).

Floral Structure and Function **Angiosperms** are vascular seed plants, and the only plants that make flowers and fruits. A **flower** is a specialized reproductive shoot (Figure 14.15). Sepals, which usually have a green leaflike appearance, ring the base of a flower and enclose it until it opens. They surround a ring of petals, which are often brightly colored.

Stamens, a flower's pollen-producing parts, surround a **carpel** that captures pollen and produces eggs. Typically a stamen consists of a tall stalk, called the filament, topped by an **anther** that holds two pollen sacs.

A carpel has a sticky **stigma**, a region specialized for receiving pollen, at its tip. The stigma is located atop a stalk, called the **style**. At the base of the style is an **ovary**, a chamber containing one or more egg-producing ovules. After fertilization, an ovule matures into a seed and the ovary becomes the **fruit**. The name angiosperm refers to the fact that seeds form within the protection of the ovary. (*Angio*– means enclosed chamber, and *sperma*, seed.)

Figure 14.16 Angiosperm life cycle.

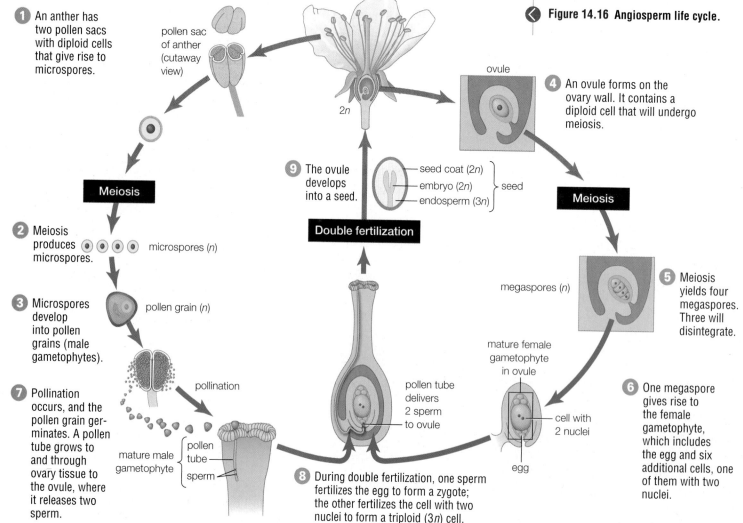

1 An anther has two pollen sacs with diploid cells that give rise to microspores.

pollen sac of anther (cutaway view)

Meiosis

2 Meiosis produces microspores.

microspores (n)

3 Microspores develop into pollen grains (male gametophytes).

pollen grain (n)

7 Pollination occurs, and the pollen grain germinates. A pollen tube grows to and through ovary tissue to the ovule, where it releases two sperm.

pollination

mature male gametophyte

pollen tube

sperm

9 The ovule develops into a seed.

seed coat (2n)
embryo (2n)   seed
endosperm (3n)

Double fertilization

pollen tube delivers 2 sperm to ovule

8 During double fertilization, one sperm fertilizes the egg to form a zygote; the other fertilizes the cell with two nuclei to form a triploid (3n) cell.

ovule

4 An ovule forms on the ovary wall. It contains a diploid cell that will undergo meiosis.

Meiosis

megaspores (n)

5 Meiosis yields four megaspores. Three will disintegrate.

mature female gametophyte in ovule

cell with 2 nuclei

egg

6 One megaspore gives rise to the female gametophyte, which includes the egg and six additional cells, one of them with two nuclei.

A Flowering Plant Life Cycle Figure 14.16 shows a generalized life cycle for a flowering plant. A flower forms on the sporophyte body. Pollen sacs in the anthers hold diploid cells ❶ that produce microspores by meiosis ❷. The microspores develop into pollen grains (immature male gametophytes) ❸. Ovules form on the wall of an ovary at the base of a carpel ❹. Meiosis of cells in ovules yields haploid megaspores ❺. A megaspore develops into a female gametophyte consisting of a haploid egg, a cell with two nuclei, and a few other cells ❻.

Pollination occurs when a pollen grain arrives on a receptive stigma, the uppermost part of the carpel ❼. The pollen grain germinates, and a pollen tube grows through the style (the structure that elevates the stigma) to the ovary at the base of the carpel. Two nonflagellated sperm form inside the pollen tube as it grows.

Double fertilization occurs when a pollen tube delivers the two sperm into the ovule ❽. One sperm fertilizes the egg to create a zygote. The other sperm fuses with the cell that has two nuclei, forming a triploid ($3n$) cell. After double fertilization, the ovule matures into a seed ❾. The zygote develops into an embryo sporophyte and the triploid cell develops into **endosperm**, a nutritious tissue that will serve as a source of food for the developing embryo.

Keys to Angiosperm Diversity Flowering plants constitute 90 percent of all modern plant species. What accounts for angiosperm success? For one thing, they tend to grow faster than gymnosperms. Think of how a plant like a dandelion or a grass can grow from a seed and produce seeds of its own within a few months. In contrast, most gymnosperms take years to mature and produce seeds.

Evolution of flowers gave angiosperms an edge by facilitating animal-assisted pollination. After pollen-producing plants evolved, some insects began feeding on the plants' highly nutritious pollen. Plants gave up some pollen but gained a reproductive edge when insects unknowingly moved pollen, thus facilitating pollination. Many traits of flowering plants are adaptations that attract **pollinators**, animals that move pollen of one plant species onto female reproductive structures of the same species. Insects are the most common pollinators (Figure 14.17), but birds, bats, and other animals also serve in this role. Producing sugary nectar encourages more pollinator visits, which improves pollination rates and enhances seed production.

A variety of fruit structures help angiosperms disperse their seeds. Fruits ride the winds, stick to animal fur, or entice animals to eat them and release seeds in their feces. Gymnosperm seeds have less diverse dispersal mechanisms.

Major Groups The vast majority of flowering plants belong to one of two lineages. The 80,000 or so **monocots** include orchids, palms, lilies, and grasses, such as rye, wheat, corn, rice, sugarcane, and other important crop plants. **Eudicots** include most herbaceous (nonwoody) plants such as tomatoes, cabbages, roses, daisies, most flowering shrubs and trees, and cacti. Monocots and eudicots derive their group names from the number of seed leaves (cotyledons) in the embryo. Monocots have one seed leaf, and eudicots have two. The two lineages also differ in many structural details such as the arrangement of their vascular tissues and the number of flower petals. Although some eudicots undergo secondary growth (growth in width) and become woody, no monocots produce true wood. Section 27.2 describes the differences between eudicots and monocots in more detail.

Figure 14.17 A pollinator.
A bee unknowingly transfers pollen from flower to flower as it gathers pollen and sips nectar.

Courtesy of Christine Evers.

angiosperm Seed plant that produces flowers and fruits.

anther Part of the stamen that contains pollen sacs.

carpel Ovule-containing part of a flower.

double fertilization In flowering plants, one sperm fertilizes the egg, forming the zygote, and another fertilizes a diploid cell, forming what will become endosperm.

endosperm Nutritive tissue in an angiosperm seed.

eudicots Largest lineage of angiosperms; includes herbaceous plants, woody trees, and cacti.

flower Specialized reproductive shoot of a flowering plant.

fruit Mature ovary tissue that encloses a seed or seeds.

monocots Lineage of angiosperms that includes grasses, orchids, and palms.

ovary Of flowering plants, a floral chamber that holds one or more ovules.

pollinator Animal that moves pollen from one plant to another, thus facilitating pollination.

stamen Pollen-producing part of a flower. Consists of an anther that contains pollen sacs, atop a filament.

stigma Pollen-receiving part of a carpel.

style Elongated portion of a carpel that holds the stigma above the ovary.

A. Mechanized harvesting of wheat.
Photo USDA.

B. A field of cotton ready for harvest.
Photo by Scott Bauer, USDA/ARS.

Figure 14.18 Angiosperms as crops.

chytrid Fungus that produces flagellated spores.
club fungus Fungus that produces spores by meiosis in club-shaped cells.
fungus Spore-producing heterotroph that has cell walls of chitin and feeds by extracellular digestion and absorption.
hypha A single filament in a fungal mycelium.
mycelium Mass of threadlike filaments (hyphae) that compose the body of a multicelled fungus.
zygote fungus Fungus that usually grows as a mold; sexual reproduction yields a thick-walled zygospore.

Ecology and Human Uses of Angiosperms It would be nearly impossible to overestimate the importance of angiosperms. As the dominant plants in most land habitats, they provide food and shelter for land animals. They also supply many products that meet human needs (Figure 14.18).

Angiosperms provide nearly all of our food, directly or as feed for livestock. Cereal crops are the most widely planted. In the United States, more acreage is devoted to corn than to any other plant. Worldwide, rice feeds the greatest number of people. Wheat, barley, and sorghum are other widely grown grains. All are grasses. Legumes are the second most important source of human food. They can be paired with grains to provide all the amino acids the human body needs to build proteins. Soybeans, lentils, peas, and peanuts are examples of legumes.

Name a part of a plant, and humans eat it. In addition to the seeds of grains and legumes, we dine on leaves of lettuce and spinach, stems of asparagus, developing flowers of broccoli, modified roots of potatoes, carrots, and beets, and fruits of tomatoes, apples, and blueberries. Stamens of crocus flowers provide the spice saffron, and the bark of a tropical tree provides cinnamon.

Fibers used to make clothing come from two main sources, petroleum and plants. Plant fibers include cotton, flax, ramie, and hemp. We also use fibers from flowering plants to weave rugs, and in many places to thatch roofs. Oak and other hardwoods derived from angiosperms provide flooring and furniture.

We extract medicines and psychoactive drugs from plants. Aspirin is derived from a compound discovered in willows. Digitalis from foxglove strengthens a weak heartbeat. Coffee, tea, and tobacco are widely used plant-derived stimulants. Worldwide, cultivation of opium poppies (the source of heroin) and coca (the source of cocaine) have wide-reaching health, economic, and political effects.

Take-Home Message 14.7

What are angiosperms?

- Angiosperms are plants in which seeds develop inside ovaries that become fruits.
- They are the most diverse plant lineage. Adaptations that contributed to their success include a short life cycle, coevolution with insect pollinators, and a variety of fruit structures that aid in dispersal of seeds.
- There are two major angiosperm lineages: monocots and dicots.
- The overwhelming majority of plants grown as crops are angiosperms.

14.8 Fungal Traits and Diversity

REMEMBER: Glycogen is a complex carbohydrate (Section 2.7).

Yeasts, Molds, Mildews, and Mushrooms With this section, we begin our survey of another major lineage of eukaryotes, the fungi. Like plants, **fungi** have walled cells, spend their lives fixed in place, and produce haploid spores by meiosis. However, fungi are more closely related to animals than to plants. Like animals, fungi are heterotrophs and they store excess sugars as glycogen, rather than starch. Most are decomposers that feed on organic wastes and remains. A lesser number live on or in other living organisms. As a fungus grows in or over organic matter, it secretes digestive enzymes, then absorbs the resulting breakdown products.

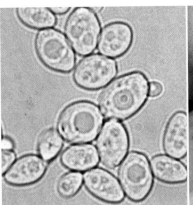
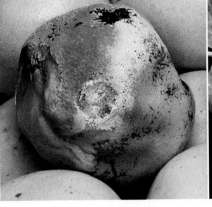

A. Yeast (single-celled fungus). **B.** Mold growing on a grapefruit. **C.** Mildew growing on leaves. **D.** Mushrooms on a forest floor.

Figure 14.19 Variety of fungal forms.

(A) Dr. John D. Cunningham/Visuals Unlimited, Inc.; (B) Photo by Scott Bauer/USDA; (C) Nigel Cattlin/Science Source; (D) Robert C. Simpson/Nature Stock.

Fungi have a variety of growth habits. Yeasts live as single cells (Figure 14.19A). However, most fungi are multicelled. Molds, mildews, and mushrooms are familiar examples of multicelled fungi (Figure 14.19B–D). These fungi grow as a **mycelium** (plural, mycelia), a network of microscopic interwoven filaments. Each filament, or **hypha** (plural, hyphae), is a strand of walled cells arranged end to end (Figure 14.20). Fungal cell walls consist primarily of chitin, a complex carbohydrate also found in the body covering of crabs and insects.

Fungi do not have vascular tissue, but walls between cells in a hypha are porous, so materials flow among cells. As a result, nutrients or water taken up in one part of the mycelium can be shared with cells in other regions.

Some fungal bodies are enormous. The largest organism we know about is a soil fungus in Oregon. Its hyphae extend through the soil over an area of about 2,200 acres, and it is still growing. Given its growth rate and size, researchers estimate that this individual has been alive for 8,000 years.

Lineages and Life Cycles The oldest fungal fossils are about 500 million years old. They resemble **chytrids**, a group of modern fungi that are primarily aquatic and produce flagellated spores. The fossils and the existence of the aquatic, flagellated chytrids are taken as evidence that fungi, like plants and animals, evolved from an aquatic protist. By about 400,000 years ago, fungi had moved onto land.

Scientists estimate there are than a million species of fungus, although they have only named about 70,000. Most fungi are distributed among five major subgroups. In addition to the chytrids (phylum Chytridiomycota), there are zygote fungi (phylum Zygomycota), glomeromycete fungi (Glomeromycota), sac fungi (Ascomycota), and club fungi (Basidiomycota).

Most **zygote fungi** are molds that grow on or through organic matter as a mass of hyphae. As long as food is plentiful, the fungus reproduces asexually by making spores by mitosis at the tips of special hyphae. When food runs low, sexual reproduction occurs. As in all multicelled fungi, sexual reproduction begins with the fusion of haploid hyphae from two individuals of different mating strains. In zygote fungi, this fusion leads to formation of a thick-walled diploid spore called a zygospore (Figure 14.21). Meiosis of cells inside the zygospore yields haploid cells that develop into new mycelia.

Many sac fungi and club fungi produce spores on a multicelled fruiting body. A **club fungus** produces spores in club-shaped cells on a fruiting body such as a

Figure 14.20 Micrograph (top) and diagram (bottom) of club fungus hyphae.

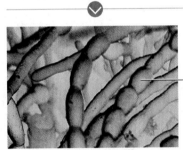

one cell (part of one hypha of a club fungus mycelium)

Garry T. Cole, University of Texas, Austin/BPS.

pore in cross-wall

From Russell/Wolfe/Hertz/Starr, *Biology*, 3e, © Cengage Learning®.

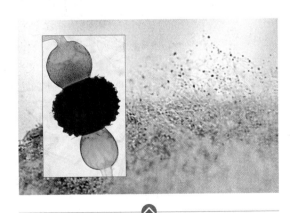

Figure 14.21 Black bread mold, a zygote fungus.
The fungus usually grows as a mass of asexually reproducing hyphae. The inset micrograph shows a zygospore.

Background, © Micrograph J. D. Cunningham/Visuals Unlimited; inset, Ed Reschke.

Figure 14.22 Generalized life cycle for a club fungus. Hyphae of different mating strains often grow through the same patch of soil.

1 Two haploid hyphal cells meet and their cytoplasm fuses, forming a dikaryotic (*n+n*) cell.

2 Mitotic cell divisions form a mycelium that produces a mushroom.

3 Spore-making cells form at the edges of the mushroom's gills.

4 Inside these dikaryotic cells, nuclei fuse, making the cells diploid (2*n*).

5 The diploid cells undergo meiosis, forming haploid (*n*) spores.

6 Spores are released and give rise to a new haploid mycelium.

After T. Rost, et al., Botany, Wiley, 1979.

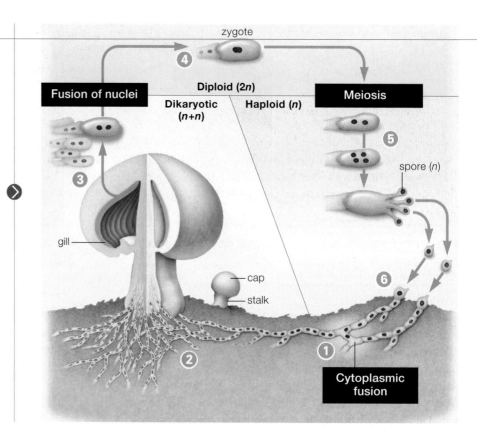

Figure It Out: Are cells that make up the stalk of a mushroom haploid, diploid, or dikaryotic?

Answer: Dikaryotic

Figure 14.23 Morels, edible fruiting bodies of a sac fungus.

© unverdorben jr/Shutterstock.com.

common button mushroom (Figure 14.22). When haploid hyphae of two different club fungus individuals meet in the soil cytoplasmic fusion produces a dikaryotic cell **1**. Dikaryotic means "having two nuclei," being (*n+n*). Mitotic divisions produce a dikaryotic mycelium **2**. Multicelled club fungi spend most of their life cycle growing as a dikaryotic mycelium. When conditions favor reproduction, a sudden burst of hyphal growth produces a mushroom that emerges above the ground. A typical mushroom has a stalk and a cap, with thin sheets of tissue called gills on the cap's underside **3**. Fusion of haploid nuclei inside cells at the edges of the gills forms diploid zygotes **4**. Each zygote undergoes meiosis, producing haploid spores **5**. After a spore germinates, repeated mitotic divisions give rise to a new haploid mycelium **6**.

Some **sac fungi** form fruiting bodies that somewhat resemble those of club fungi (Figure 14.23). However, a sac fungus fruiting body does not have gills and sac fungus spores do not form in club-shaped cells. Instead, sac fungi form spores by meiosis inside a saclike cell. Hence the name "sac" fungus.

Take-Home Message 14.8

What are fungi?

• Fungi are heterotrophs that absorb nutrients from their environment. They live as single cells or as a multicelled mycelium and disperse by producing spores.

• Zygote fungi grow as molds that most often produce spores asexually.

• Mushrooms are spore-producing fruiting bodies of club fungi. Some sac fungi also produce large fruiting bodies that we eat.

sac fungus Fungus that produces spores by meiosis in saclike cells.

14.9 **Ecological Roles of Fungi**

REMEMBER: Fermentation is an anaerobic energy-releasing pathway (Section 5.6).

Decomposers Most fungi provide an important ecological service by breaking down complex compounds in organic wastes and remains. When a fungus secretes digestive enzymes onto these materials, some soluble nutrients escape into nearby soil or water. Plants and other producers can then take up these substances to meet their own nutrient needs. Bacteria also serve as decomposers, but they tend to grow mainly on surfaces. By contrast, fungal hyphae can extend deep into a dead log or another bulky food source and break it down from the inside (Figure 14.24).

Parasites Many sac fungi and club fungi are plant parasites. Powdery mildews (sac fungi) and rusts and smuts (club fungi) grow only in living plants. Wheat stem rust is an example. Hyphae of such fungi extend into cells of stems and leaves, where they suck up photosynthetically produced sugars. The resulting loss of nutrients stunts the plant, prevents it from producing seeds, and may eventually kill it. However, a plant usually does not die before the fungus has produced spores on the surface of its infected parts. Other pathogenic fungi produce toxins that kill plant tissues, then feed on the resulting remains. The club fungus *Armillaria* causes root rot by infecting trees and woody shrubs in forests worldwide. Once an infected tree dies, the fungus decomposes the stumps and logs left behind.

Animals are less vulnerable to fungal infections than plants and, among animals, those with a high body temperature are least susceptible. About 50,000 species of fungus can infect insects, whereas only a few hundred can infect mammals. Like

Figure 14.24 Fungi as decomposers.
Club fungus (*Armillaria*) fruiting bodies on pine logs. Fungal hyphae extend through the logs and digest them.

Chris Moody/Shutterstock.com.

Digging Into Data

Removing Fungus-Infected Stumps to Save Trees

The club fungus *Armillaria ostoyae* infects living trees and acts as a parasite, withdrawing nutrients from them. When the tree dies, the fungus continues to dine on its remains. Fungal hyphae grow out from the roots of infected trees and roots of dead stumps. If these hyphae contact roots of a healthy tree, they can invade and cause a new infection.

Canadian forest pathologists hypothesized that removing stumps after logging could help prevent tree deaths. To test this hypothesis, they carried out an experiment. In half of a forest, they removed stumps after logging. In a control area, they left stumps behind. For more than 20 years, they recorded tree deaths and whether *A. ostoyae* caused them. Figure 14.25 shows the results.

1. Which tree species was most often killed by *A. ostoyae* in control forests? Which was least affected by the fungus?
2. For the species most affected, what percentage of deaths did *A. ostoyae* cause in control and in experimental forests?
3. Do the overall data support the hypothesis that stump removal helps protect living trees from infection by *A. ostoyae*?

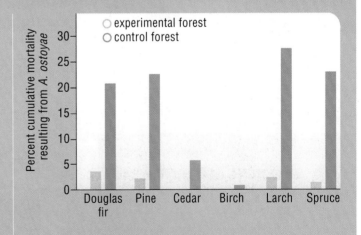

Figure 14.25 Results of a long-term study of how logging practices affect tree deaths caused by the fungus *A. ostoyae*.
In the experimental forest, whole trees—including stumps—were removed (brown bars). The control half of the forest was logged conventionally, with stumps left behind (blue bars).

After graph from www.pfc.forestry.ca.

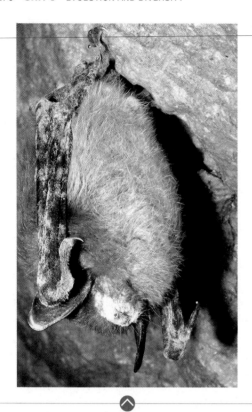

Figure 14.26 White-nose syndrome.
Affected bats have white filaments of a parasitic sac
fungus on their wings, ears, and muzzle.

USFWS/ScienceSource.

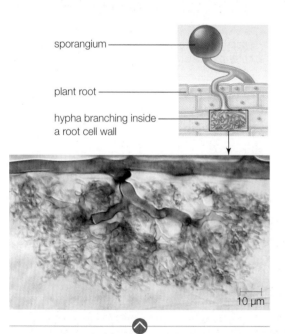

Figure 14.27 Mycorrhiza.
Specialized hyphae of glomeromycete fungus enter and
branch inside the cell wall of a plant root cell.

© Dr. Mark Brundrett, The University of Western Australia.

insects, amphibians do not maintain a high body temperature, and many species are
currently threatened by a pathogenic chytrid inadvertently spread by humans.

In mammals, fungi seldom cause fatal infections. White-nose syndrome, a
fungal disease currently decimating North American bats, is an exception (Figure 14.26). The first signs of this disease occurred in New York in 2006. By early
2014, white-nose syndrome had been discovered in 21 states and four Canadian
provinces, and it had killed millions of North American bats. Scientists think the
sac fungus that causes the disease was recently introduced to North America from
Europe, where the bats have evolved resistance to it and are not sickened by it.

Human fungal infections most frequently involve body surfaces. Typically a
fungus feeds on the outer layers of skin, secreting enzymes that dissolve keratin, the
main skin protein. Infected areas become raised, red, and itchy. For example, several
species of fungus infect skin between the toes and on the sole of the foot, causing
"athlete's foot." Fungi also cause skin infections misleadingly known as "ringworm."
No worm is involved. The ring-shaped lesion is caused by the growth of hyphae
outward from the initial infection. Fungal vaginitis (a vaginal yeast infection) occurs
when single-celled fungi that normally live in the vagina in low numbers undergo a
population explosion. Symptoms of the infection include itching or burning sensations; a thick, odorless, whitish vaginal discharge; and pain during intercourse.

Fungi seldom cause systemwide disease in otherwise healthy people, but infections can be life-threatening in people whose immune system is impaired by AIDS,
chemotherapy, or drugs that must be taken after an organ transplant.

Fungal Partnerships Nearly all plants form mutually beneficial relationships,
or **mutualisms**, with fungi. For example, many soil fungi take part in a **mycorrhiza**
(plural, mycorrhizae), a mutually beneficial relationship with plant root cells (Figure
14.27). All **glomeromycete fungi** take part in a mycorrhiza in which hyphae penetrate the root cell wall and share space with the cell. An estimated 80 percent of vascular plants have a glomeromycete partner. Club fungi and sac fungi also take part
in mycorrhizae. When they do, their hyphae grow into a root and between its cells.
Hyphae of all mycorrhizal fungi functionally increase the absorptive surface area of
their plant partner. Hyphae are thinner than even the smallest roots and can grow
more easily between soil particles. The fungus shares water and nutrients taken up
by its hyphae with root cells. In return, the plant supplies sugar to the fungus.

Fungal partners also enhance the nutrition of some animals. Chytrid fungi that
live in the stomachs of grazing hoofed mammals such as cattle, deer, and moose

Figure 14.28 Structure of a leaflike lichen.
Photo Gary Head.

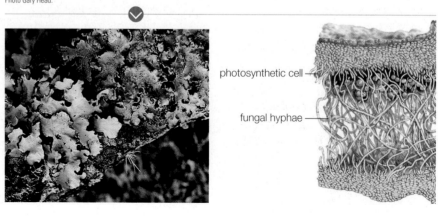

aid their hosts by breaking down otherwise indigestible cellulose. Similarly, fungal partners of leafcutter ants serve as an external digestive system. The ants gather bits of leaf to sustain the fungus that they cultivate in their nest. The ants cannot digest leaves, but they do eat the fungus.

Lichens are composite organisms consisting of a fungus and a single-celled photosynthetic species, either a green alga or a cyanobacterium. The fungus makes up most of the lichen's mass and shelters the photosynthetic species, which shares nutrients with the fungus. Lichens grow on many exposed surfaces (Figure 14.28). They are ecologically important as colonizers in places that are too hostile for other organisms. By releasing acids and retaining water that freezes and thaws, lichens help break down rocks and form soil.

Human Uses of Fungi Some fungi are important as human food crops. Mushroom farms produce many varieties of club fungi by inoculating sterilized compost with spores of the desired crop species. Mycorrhizal fungi, including chantrelles, morels, and truffles, cannot be cultivated, so they are gathered from the wild. Mushrooms taken from the wild should always be identified by an experienced mushroom forager before they are eaten because many edible species have poisonous look-alikes.

In addition to eating fungi, we make use of fungal fermentation reactions to produce food and drinks (Figure 14.29). A package of baker's yeast contains spores of a sac fungus (*Saccharomyces cerevisiae*). Set bread dough out to rise, and yeast cells carry out fermentation reactions that produce carbon dioxide, causing the dough to expand (rise). Another strain of *Saccharomyces* is used in the production of beer and wine.

Geneticists and biotechnologists also make use of the yeast *S. cerevisiae*. Like *E. coli* bacteria, *S. cerevisiae* grows readily in laboratories and it offers the added advantage of being eukaryotic like us. Checkpoint genes that regulate the eukaryotic cell cycle (Section 8.3) were first discovered in *S. cerevisiae*. This discovery was the first step toward our current understanding of how mutations of these genes cause human cancers. Genetically engineered *S. cerevisiae* and other yeasts are also now used to produce proteins that serve as vaccines or other medicines.

Some medicines and drugs are compounds first isolated from fungi. Most famously, the initial source of the antibiotic penicillin was the sac fungal mold *Penicillium*. Fungi have also yielded drugs used to lower blood pressure, reduce cholesterol levels, or to prevent rejection of transplanted organs. The hallucinogen LSD was first isolated from ergot, a club fungus that infects grains. So-called "magic mushrooms" contain compounds called psilocybins that induce a dreamlike state.

Figure 14.29 Products made with the help of fungi.
Sac fungal yeasts carry out fermentation necessary to make bread and wine. The blue material in "blue cheese" is hyphae of a sac fungal mold.

© CameronWhitman/iStockphoto.com.

Take-Home Message 14.9

How do fungi interact with other species?

- Mycorrhizal fungi live in or on plant roots in a mutually beneficial relationship. Fungi also live with single-celled photosynthetic cells as lichens.
- Fungi benefit other organisms when they feed on wastes and remains, releasing nutrients into the soil. They harm other organisms, including humans, by infecting and feeding on their tissues.
- We use fungi as food, as sources of drugs, and as a model for studying genetic processes.

glomeromycete fungus Soil fungus whose hyphae extend inside the cell wall of a plant root cell.

lichen Composite organism consisting of a fungus and a green alga or cyanobacterium.

mutualism Species interaction that benefits both species.

mycorrhiza Fungus–plant root partnership.

Summary

Section 14.1 Plants are food for both people and fungi. Winds spread spores of fungal pathogens, so outbreaks of fungal disease can often spread to crops over a wide area.

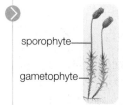

Section 14.2 **Plants** evolved from freshwater green algae. Their life cycle, an **alternation of generations**, includes two multicelled forms, a haploid **gametophyte** and a diploid **sporophyte**. The gametophyte dominates the life cycle of **bryophytes**, but the sporophyte dominates in **vascular plants**.

Key adaptations to dry habitats include a waterproof **cuticle** with **stomata**, and internal pipelines of **xylem** and **phloem**. Xylem reinforced by **lignin** helps vascular plants stand upright. **Seeds** and male gametophytes that can be dispersed without water (**pollen grains**) evolved in seed plants.

Section 14.3 Bryophytes include three lineages of low-growing plants: **mosses**, liverworts, and hornworts. The photosynthetic moss gametophyte is held in place by threadlike **rhizoids**. Flagellated sperm swim to eggs. The sporophyte remains attached to and often dependent upon the gametophyte even when mature.

sporophyte—

gametophyte—

Section 14.4 **Ferns** are seedless vascular plants. Sporophytes dominate their life cycle and produce spores in **sori**. Gametophytes produce flagellated sperm. Ferns grow from **rhizomes** (horizontal stems). Some live on trees as **epiphytes**. Other seedless vascular plants include club mosses and horsetails. **Coal** formed from the remains of ancient nonvascular seed plants.

Section 14.5 Seed-bearing vascular plants make two types of spores. **Microspores** give rise to pollen grains in a **pollen sac**. **Megaspores** form in **ovules**, and give rise to egg-producing female gametophytes. Even in the absence of water, winds or pollinators can move pollen, thus facilitating **pollination**. The seed is a mature ovule, with an embryo sporophyte and some nutritive tissue inside it. Some seed plants undergo secondary growth and produce **wood**.

Section 14.6 **Conifers**, cycads, ginkgos, and gnetophytes are among the **gymnosperms**. They are adapted to dry climates and bear seeds on exposed surfaces of spore-bearing structures. In conifers, these spore-bearing structures are distinctive cones. Conifers tend to be evergreen. Ginkgos are **deciduous plants**.

Section 14.7 **Angiosperms** are the dominant land plants. They alone have **flowers**. Pollen forms in **anthers**, the part of a **stamen** that holds pollen sacs. Many flowering plants coevolved with **pollinators** that deliver pollen to a receptive **stigma**. After pollination, a pollen tube grows through a **style** of the flower's **carpel** to the **ovary** at its base and **double fertilization** occurs. The ovary becomes a **fruit** containing one or more seeds. A flowering plant seed includes an embryo sporophyte and **endosperm**, a nutritious tissue. Most crops are angiosperms. There are two main lineages of flowering plants. **Monocots** include grasses and palm trees. **Eudicots** include most flowering trees and shrubs, as well as most herbaceous plants.

Section 14.8 **Fungi** are single-celled or multicelled heterotrophs. They secrete enzymes onto organic material and absorb the resulting breakdown products. Multicelled fungi grow as a **mycelium** composed of many filaments called **hyphae**. Fungi produce spores both sexually and asexually. A mushroom is a spore-producing body of a **club fungus**. Fungi of some **sac fungi** are also used as food. Bread mold is an example of a **zygote fungus**.

Section 14.9 Most fungi are decomposers. A **mycorrhiza** is a **mutualism**, as between a **glomeromycete fungus** and plant root cells. Fungi also partner with photosynthetic cells to form **lichens**. Some fungi infect plants or animals, causing disease. We eat fungi and use them to produce foods, drinks, and medicines.

Self-Quiz

Answers in Appendix I

1. Which of the following statements is not correct?
   a. Angiosperms produce pollen and seeds.
   b. Mosses are nonvascular plants.
   c. Ferns and angiosperms are vascular plants.
   d. Only gymnosperms produce fruits.

2. Which does *not* apply to all seed plants?
   a. vascular tissues   c. single spore type
   b. diploid dominance   d. all of the above

3. Bryophytes have independent _____ and dependent _____ .
   a. sporophytes; gametophytes   b. gametophytes; sporophytes

4. Ferns are classified as _____ plants.
   a. multicelled aquatic
   c. seedless vascular
   b. nonvascular seed
   d. seed-bearing vascular

5. The _____ produce flagellated sperm.
   a. ferns
   c. monocots
   b. conifers
   d. a and c

6. The _____ produced in the male cones of a conifer develop into pollen grains.
   a. ovules
   c. megaspores
   b. ovaries
   d. microspores

7. A seed is _____ .
   a. a female gametophyte
   c. a mature pollen tube
   b. a mature ovule
   d. an immature spore

8. Match the terms appropriately.
   _____ gymnosperm
   a. gamete-producing body
   _____ sporophyte
   b. help control water loss
   _____ horsetail
   c. "naked" seeds
   _____ bryophyte
   d. spore-producing body
   _____ gametophyte
   e. nonvascular land plant
   _____ stomata
   f. seedless vascular plant
   _____ angiosperm
   g. flowering plant

9. All fungi _____ .
   a. are multicelled
   c. are heterotrophs
   b. form flagellated spores
   d. all of the above

10. Fungal decomposers derive nutrients from _____ .
    a. organic wastes and remains
    c. living animals
    b. living plants
    d. photosynthesis

11. A mushroom is _____ .
    a. the food-absorbing part of a fungus
    b. the only part of the fungal body not made of hyphae
    c. a reproductive structure that releases sexual spores
    d. the longest-lived part of the fungal life cycle

12. Human fungal infections most commonly involve _____ .
    a. the brain
    c. the digestive system
    b. the heart
    d. body surfaces

13. A _____ is a composite organism composed of a fungus and a single-celled photosynthetic species.
    a. mycorrhiza
    c. decomposer
    b. lichen
    d. ringworm

14. Cell walls of fungi are composed of _____ .
    a. cellulose
    c. lignin
    b. keratin
    d. chitin

15. Match the terms appropriately.
    _____ decomposer
    a. filament made of walled cells
    _____ yeast
    b. club fungus fruiting body
    _____ mushroom
    c. fungus with flagellated spores
    _____ chytrid
    d. mesh of fungal filaments
    _____ hypha
    e. partners with a plant root
    _____ mycelium
    f. single-celled fungus
    _____ glomeromycete fungus
    g. breaks down organic matter

Critical Thinking

1. Early botanists admired ferns but found their life cycle perplexing. In the 1700s, they learned to propagate them by sowing what appeared to be tiny dustlike "seeds" from the undersides of fronds. Despite many attempts, the scientists could not find the pollen source, which they assumed must stimulate the "seeds" to develop. Imagine you could write to one of these botanists. Compose a note that would clear up their confusion.
2. Fungi play an important role in the breakdown of cellulose into its component simple sugars. What simple sugar is the final product of these reactions?
3. Fungal skin diseases are persistent, in part because fungi can penetrate deeper layers of skin than can ointments and creams. There are fewer antifungal drugs than antibacterial ones, and antifungals often have more severe side effects. Reflect on the evolutionary relationships among bacteria, fungi, and humans. Why it is harder to fight fungi than bacteria?

Visual Question

Match each term with a structure in the diagram below.
_____ stigma
_____ style
_____ sepal
_____ anther
_____ filament
_____ petal
_____ ovule in ovary
_____ ovary

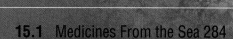

Application ◉ 15.1 Medicines From the Sea

Animal life began in the sea, and Earth's oceans still hold the greatest store of animal diversity. In the sea, as on land, the majority of animals are **invertebrates**, meaning they do not have a backbone. Of all animals, only about 5 percent have a backbone and thus are described as **vertebrates**.

Many marine invertebrates produce compounds that are toxic to other organisms. Such toxins can protect an animal from predators, help it fend off pathogens, or assist in the capture of prey. Some invertebrate toxins also have effects in the human body, and thus can be useful as medicines. Consider the venom that some fish-eating cone snails use to subdue their prey (Figure 15.1). The snail's venom anesthetizes and paralyzes a fish, thus preventing the fish from struggling and possibly harming the snail as it captures and consumes it. Humans evolved from a fish ancestor and our nerves use the same chemical communication signals that fish nerves do. As a result, a toxin that acts on fish nerves also affects the function of the human nervous system. A person accidentally stung by a fish-eating cone snail may become numb at the site of venom injection, suffer from temporary paralysis, or even die.

A synthetic version of a peptide extracted from the venom of one cone snail is now used as a pain reliever. The drug, ziconotide (Prialt), is injected into the spinal cord to suppress pain that cannot be controlled by other means. Additional peptides isolated from cone snail venom are being tested as treatments for epilepsy, diabetes, and cancer.

Compounds derived from other marine invertebrates are also in use. AZT (azidothymidine), the first drug successfully used to treat AIDS, is a synthetic version of a molecule first discovered in a sponge. Other compounds made by sponges can be used to treat infections caused by herpesviruses. Sea whips, which are relatives of sea anemones, produce a variety of anti-inflammatory compounds, one of which is used in face creams.

Figure 15.1 Cone snail with its fish prey.
After administering a toxin that puts the fish into a stupor, the snail engulfs and devours it.

© K.S. Matz.

Finding a compound that might have medicinal value is only the first step in developing a new drug. For a compound to be used in clinical tests, researchers must obtain a sufficient amount of it. This can be difficult because many compounds of interest occur only at very low concentrations in animals. Consider the drug eribulin (Halaven), which is now used to treat some advanced-stage breast cancers. Sponges synthesize the compound on which eribulin is modeled, but only in tiny amounts. Obtaining enough of the sponge compound to test its efficiency as a cancer drug (300 milligrams) required processing more than one metric ton of sponge tissue.

Once the structure of a useful compound has been determined, chemists can usually manufacture that compound or one with similar structure and properties. Use of synthetically produced compounds prevents overharvesting of the species in which the compound was found. The breast cancer drug eribulin is a synthetic molecule with a slightly different structure than the molecule extracted from sponges.

15.2 Origins and Diversification

REMEMBER: Animals are eukaryotes with unwalled cells (Section 3.5). Choanoflagellates are protists closely related to animals (13.6). Homeotic genes regulate animal development (7.7). Tectonic plate movement causes supercontinents to form, rotate, and break up (11.5). Cutting off gene flow between populations can lead to speciation (12.6). With adaptive radiation, a lineage gives rise to many others (12.7).

Animals are multicelled consumers that take food into their body, where they digest it and absorb the released nutrients. An animal develops from an embryo (an early developmental stage) to an adult. Most animals reproduce sexually, some reproduce asexually, and some do both. Nearly all animals are motile (can move from place to place), during part or all of their life cycle.

Animal Origins

The **colonial theory of animal origins** states that animals evolved from a heterotrophic protist that formed colonies. At first, all cells in the colony were identical. Each could survive and reproduce on its own. Later, mutations produced cells that specialized in some tasks and did not carry out others. Perhaps some cells captured food more efficiently, but did not make gametes, whereas other made gametes but did not catch food. The division of labor among interdependent cells made colonies more efficient, allowing them to obtain more food and produce more offspring. Over time, cells became increasingly interdependent and more specialized cell types evolved, producing the first animal.

Studies of modern protists support the colonial theory. Among choanoflagellates, the modern protists most closely related to animals, some species can live either as individual cells or as a colony. Colonies form when a cell undergoes mitosis and the resulting descendant cells stick together. Thus, cells of a choanoflagellate colony are, like the cells of an animal body, genetically identical.

Evidence of Early Animals

We have no fossil evidence of the earliest animals. Most likely they were microscopic and, like the protist from which they evolved, had no hard parts that would fossilize easily. However, a collection of 570-million-year-old fossil organisms provides evidence of an early animal diversification. These fossils are collectively referred to as Ediacarans because they were found in Australia's Ediacara Hills. They include the remains of many small soft-bodied organisms that are widely thought to be marine invertebrates (Figure 15.2). Some of these fossil species may be early representatives of modern groups. Others belong to lineages that have no surviving members.

A great adaptive radiation of animals occurred during the Cambrian (541–485 million years ago). By the end of this period, all major animal lineages were present in the seas. Environmental factors encouraged diversification. During the Cambrian, global climate warmed and the amount of oxygen in the seas increased, making the environment more hospitable to animal life. (With rare exceptions, animals require oxygen.) Also during the Cambrian, the supercontinent Gondwana underwent a dramatic rotation. Movement of this landmass interrupted gene flow between populations, increasing the likelihood that speciation events would occur.

Biological factors also have encouraged animal diversification. After predatory animals arose, evolution of novel prey defenses, and predators able to overcome these, would have been favored. Duplications and divergence of homeotic genes probably facilitated such modifications. Changes in these genes can have dramatic effects on body plans.

Figure 15.2 Fossils of early animals.

(A) © DK Limited/Corbis; (B) Dr. Chip Clark.

A. Fossil of *Spriggina*, an Ediacaran that lived in the sea about 570 million years ago. It was a soft-bodied animal about 3 centimeters (1 inch) long. It is considered a possible ancestor of arthropods.

B. Fossil trilobite, an early arthropod. Trilobites arose during the Cambrian and thrived in the oceans for 270 million years before going extinct.

A great adaptive radiation of animals occurred during the Cambrian (542–488 million years ago).

animal A eukaryotic consumer that is made up of unwalled cells and develops through a series of stages. Most ingest food, reproduce sexually, and can move from place to place.

colonial theory of animal origins Well-accepted hypothesis that animals evolved from a colonial protist.

invertebrate Animal without a backbone.

vertebrate Animal with a backbone.

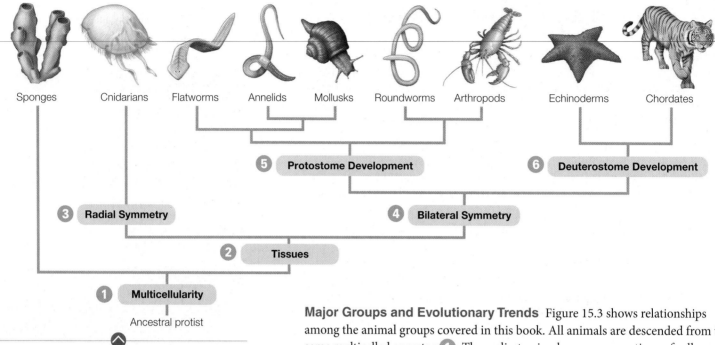

Figure 15.3 Family tree for major animal phyla based on body form and genetic comparisons.
Vertebrates (animals with a backbone) are a subgroup of the chordates.

Figure It Out: What type of body symmetry do the arthropods have?

Answer: Bilateral symmetry

Major Groups and Evolutionary Trends

Figure 15.3 shows relationships among the animal groups covered in this book. All animals are descended from the same multicelled ancestor ❶. The earliest animals were aggregations of cells, and sponges still show this level of organization. However, most animals have tissues ❷. A **tissue** consists of one or more types of cells that are organized in a specific pattern and that carry out a particular task. In the early animal lineages, embryos had two tissue layers: an outer ectoderm and an inner endoderm. In later lineages, cell movements produced a middle embryonic layer called mesoderm. Evolution of a three-layer embryo allowed an increase in structural complexity.

The structurally simplest animals such as sponges are asymmetrical, meaning you cannot divide their body into halves that are mirror images. Sea anemones and other cnidarians have **radial symmetry**: Body parts are repeated around a central axis, like the spokes of a wheel ❸. Radial animals have no front or back end. They attach to an underwater surface or drift along, so their food can arrive from any direction. Most animals have **bilateral symmetry**. They have a right and left half, with body parts repeated on either side of the body ❹. Bilateral animals have a distinctive "head end" that has a concentration of nerve cells.

There are two lineages of animals that have a three-layer embryo, and they differ somewhat in how they develop. In **protostomes**, the first opening that forms on an embryo becomes the mouth ❺. *Proto–* means first and *stoma* means opening. In **deuterostomes**, the second opening becomes the mouth ❻.

All animals take in and digest food, but the digestive process varies among groups. In sponges, digestion is intracellular. Cnidarians and flatworms digest food inside a saclike gut called a **gastrovascular cavity**. Food enters this cavity through the same opening that expels wastes, and the cavity also functions in gas exchange. Most bilateral animals have a tubular gut, or **complete digestive tract**, with an opening at either end. A tubular gut has advantages. Parts of the tube can be specialized for taking in food, digesting food, absorbing nutrients, or compacting waste. Unlike a saclike cavity, a tubular gut carries out all these tasks simultaneously.

A mass of tissues and organs surrounds a flatworm's gut (Figure 15.4A). However, most animals have a "tube within a tube" body plan, with a fluid-filled body cavity around the gut. Typically, this cavity is lined with tissue derived from mesoderm, and is called a **coelom**. Earthworms have this type of body plan (Figure 15.4B). Sheets of tissue (called mesentery) suspend the gut in the center of a coelom.

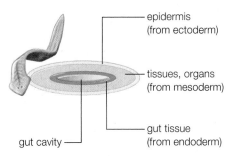

epidermis
(from ectoderm)

tissues, organs
(from mesoderm)

gut cavity — gut tissue
(from endoderm)

A. Flatworm, with no body cavity other than the gut.

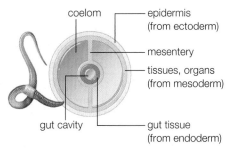

coelom — epidermis
(from ectoderm)

mesentery

tissues, organs
(from mesoderm)

gut cavity — gut tissue
(from endoderm)

B. Annelid, with a mesoderm-lined, fluid-filled coelom. Sheets of mesentery hold organs in place.

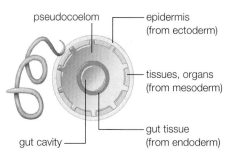

pseudocoelom — epidermis
(from ectoderm)

tissues, organs
(from mesoderm)

gut cavity — gut tissue
(from endoderm)

C. Roundworm, with a pseudocoelom.

Figure 15.4 Variations in body plans among bilateral animals.
Diagrams show a cross-section through the body. Relative width of tissue layers is not to scale.

Figure It Out: Which of the animals shown above develops from an embryo with three tissue layers?

Answer: All three do.

A few invertebrates, such as roundworms, have a pseudocoelom, with only a partial mesodermal lining (Figure 15.4C).

Evolution of a fluid-filled body cavity provided a number of benefits. First, materials can diffuse through coelomic fluid to body cells. Second, muscles can redistribute the fluid to alter the shape of body parts in ways that allow movement. Finally, internal organs were no longer hemmed in by a mass of tissue so, over evolutionary time, they could become larger and move relative to one another.

Gases and nutrients diffuse quickly through the body of a small animal. However, diffusion alone cannot move substances through a large body fast enough to keep its cells alive. In most animals, a circulatory system speeds the distribution of substances through the body. In an **open circulatory system**, a heart pumps fluid out of vessels into internal spaces, from which it is taken up again by a heart. With a **closed circulatory system**, a heart or hearts propel blood through a continuous system of vessels. Materials carried by the blood diffuse out of vessels and into cells, and vice versa. A closed circulatory system allows for faster distribution of materials than an open one.

Segmentation is common in bilateral animals, meaning similar units are repeated along the length of the body. We clearly see body segments in annelids such as earthworms. Segmentation allows evolutionary innovations in body form. When many segments have organs that carry out the same function, some segments can become modified without endangering the animal's survival.

Take-Home Message 15.2

What are animals?

- Animals are multicelled heterotrophs that typically ingest food. Their cells are unwalled.
- Animals reproduce sexually and, in many cases, asexually. They go through a period of embryonic development, and most move about during at least part of the life cycle.
- Most animals have bilateral symmetry and are composed of tissues and organs.
- An animal's digestive cavity may be tubular or saclike, and the circulatory system (if present) may be open or closed.

bilateral symmetry Having right and left halves with similar parts, and a front and back that differ.

closed circulatory system System in which blood never leaves blood vessels and exchanges with cells take place across vessel walls.

coelom A body cavity completely lined by tissue derived from mesoderm.

complete digestive tract Tubular gut.

deuterostomes Animal lineage with a three-layer embryo in which the mouth is the second opening to form; includes echinoderms and chordates.

gastrovascular cavity Saclike gut.

open circulatory system System in which circulatory fluid leaves open-ended vessels and flows among tissues before returning to the heart.

protostomes Animal lineage with a three-layer embryo in which the first opening to form is the mouth; includes most bilateral invertebrates.

radial symmetry Having parts arranged around a central axis, like spokes around a wheel.

tissue One or more types of cells that are organized in a specific pattern and that carry out a specific task.

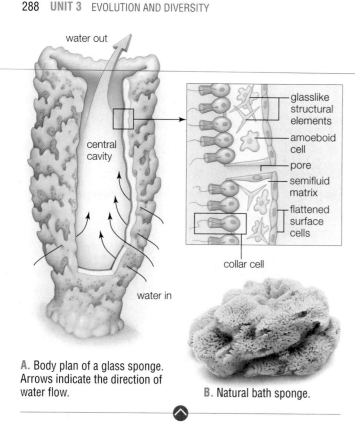

A. Body plan of a glass sponge. Arrows indicate the direction of water flow.

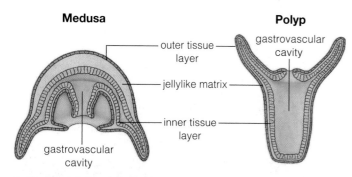

B. Natural bath sponge.

Figure 15.5 Sponges, animals without body symmetry or tissues.

(B) Image © ultimathule/Shutterstock.

Medusa   **Polyp**

- outer tissue layer
- jellylike matrix
- inner tissue layer
- gastrovascular cavity
- gastrovascular cavity

A. Structure of the two body plans.

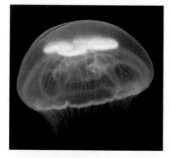

B. Jelly, a medusa.
© Boris Pamikov/Shutterstock.

C. Sea anemone, a polyp.
Ethan Daniels/Shutterstock.

Figure 15.6 Cnidarian body plans.

15.3 Invertebrate Diversity

Sponges **Sponges** (phylum Porifera) are aquatic animals with a hollow, asymmetrical body (Figure 15.5A). Adults are sessile, meaning they do not move about. Flattened cells cover a sponge's outer body surface, and flagellated collar cells line its internal cavities. A jellylike extracellular matrix lies between these cell layers. Movement of the collar cells' flagella causes water to flow in through many pores, then out through one or more larger openings.

In sponges, unlike most animals, digestion is intracellular. Collar cells filter bits of food from the water, engulf them by endocytosis, then break them down in vesicles. Amoeba-like cells in the matrix receive digested food from collar cells and distribute it to other cells. In many species, cells in the matrix also secrete fibrous proteins or glassy spikes (called spicules) that structurally support the body and discourage predators. Some protein-rich sponges are harvested from the sea, dried, cleaned, and bleached, then sold for bathing and cleaning (Figure 15.5B).

A typical sponge is a **hermaphrodite**: an individual that produces both eggs and sperm. Usually sperm are released into the water, and eggs are retained by the parent. After fertilization, the zygote develops into a ciliated larva. A **larva** (plural, larvae) is a free-living, sexually immature form in an animal life cycle. Sponge larvae exit the parental body, swim briefly, then settle and develop into adults.

Cnidarians **Cnidarians** (phylum Cnidaria) are aquatic, radially symmetrical animals with stinging tentacles. The two common body plans, medusa and polyp, both consist of two tissue layers with a secreted jellylike matrix between them (Figure 15.6A). A bell-shaped **medusa** (plural, medusae) swims or drifts about. Jellies (jellyfish) are medusae (Figure 15.6B). A **polyp**, such as a sea anemone, is tubular, and one end usually attaches to a surface (Figure 15.6C). In both polyps and medusae, tentacles surround the entrance to the gastrovascular cavity. This saclike area takes in and digests food, expels wastes, and functions in gas exchange.

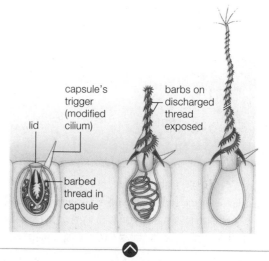

- lid
- capsule's trigger (modified cilium)
- barbs on discharged thread exposed
- barbed thread in capsule

Figure 15.7 Action of a cnidarian's stinging cells (cnidocytes).

The phylum name Cnidaria is derived from *cnidos*, the Greek word for nettle, a kind of stinging plant. The name refers to unique stinging cells (cnidocytes) at the surface of the tentacles. The cells have a capsule-like organelle (a nematocyst) with coiled threadlike structure inside it (Figure 15.7). Touch causes the capsule to pop open, forcing the thread outward. The thread entangles prey or pierces it and delivers venom. Tentacles then push the prey through the mouth into the gastrovascular cavity, where gland cells secrete enzymes that digest it.

Reef-building corals are polyps that capture prey with their tentacles, but also rely on sugars made by photosynthetic protists (dinoflagellates) that live in their tissues. Each polyp secretes a hard, calcium carbonate–rich skeleton around its base. Over time, skeletal remains of many generations of polyps accumulate as a reef, with a layer of living polyps at its surface. Coral reefs are of great ecological importance because they provide food and shelter for many types of animals.

Flatworms **Flatworms** (phylum Platyhelminthes) are flat-bodied worms that do not have a body cavity. They are the simplest animals that develop from a three-layered embryo. Many flatworms live in the sea or in fresh water, but a few live in damp places on land. Still others live as parasites inside animals.

Planarians (Figure 15.8) are free-living flatworms common in ponds. Cilia on their surface allow them to glide along. A muscular tube (a pharynx) sucks food into a highly branched gastrovascular cavity. Nutrients and oxygen diffuse from the fine branches to body cells. Clusters of nerve cells in the head serve as a simple brain. The head also has chemical receptors and light-detecting eyespots.

Tapeworms are hermaphroditic flatworms that infect the vertebrate gut. The tapeworm has a segmented body, and it adds new segments (called proglottids) as it grows. Figure 15.9 shows the life cycle of the beef tapeworm, which can infect people who eat undercooked beef.

cnidarian Radially symmetrical invertebrate with two tissue layers; uses tentacles with stinging cells to capture food.

flatworm Bilaterally symmetrical invertebrate with organs but no body cavity; for example, a planarian or tapeworm.

hermaphrodite Animal that makes both eggs and sperm.

larva Preadult stage in some animal life cycles.

medusa Bell-shaped, free-swimming cnidarian body form.

polyp Tubular, typically sessile, cnidarian body form.

sponge Aquatic invertebrate that has no tissues or organs and filters food from the water.

Figure 15.8 Body plan of a planarian.

muscular tube that sucks up food and expels waste

branching gastro-vascular cavity

Figure 15.9 Life cycle of the beef tapeworm.

Right photo, Andrew Syred/Science Source.

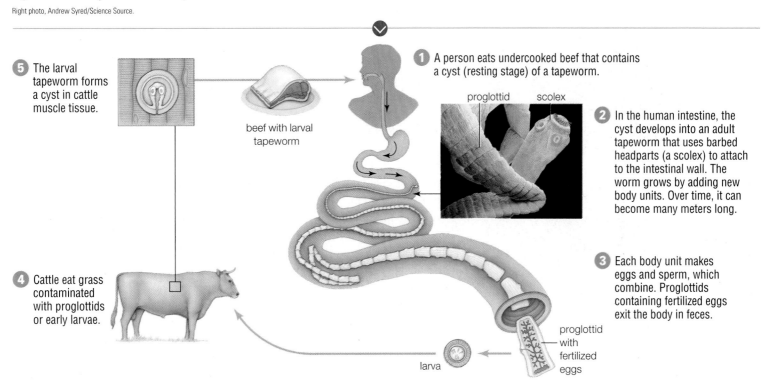

5 The larval tapeworm forms a cyst in cattle muscle tissue.

beef with larval tapeworm

1 A person eats undercooked beef that contains a cyst (resting stage) of a tapeworm.

proglottid   scolex

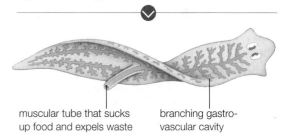

2 In the human intestine, the cyst develops into an adult tapeworm that uses barbed headparts (a scolex) to attach to the intestinal wall. The worm grows by adding new body units. Over time, it can become many meters long.

4 Cattle eat grass contaminated with proglottids or early larvae.

3 Each body unit makes eggs and sperm, which combine. Proglottids containing fertilized eggs exit the body in feces.

proglottid with fertilized eggs

larva

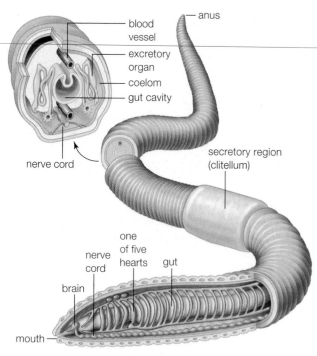

A. Body plan of an earthworm, an oligochaete.

B. Sandworm (*Nereis*), a marine polychaete.

Darlyne A. Murawski, National Geographic Creative.

C. Blood-sucking leech swollen with blood.

J. A. L. Cooke/Oxford Scientific Films.

Figure 15.10 Annelids.

Flukes are parasites that have an unsegmented body with suckers. Some tropical freshwater snails carry blood flukes that infect human intestinal veins and cause schistosomiasis, a potentially fatal disease. People become infected when they wade in water containing larval flukes, which enter the body through a cut.

Annelids **Annelids** (phylum Annelida) are segmented worms with a coelom, a complete digestive system, and a closed circulatory system. The phylum name comes from the Greek word *annulus*, which means ringed. There are three major subgroups: oligochaetes, polychaetes, and leeches.

Earthworms are oligochaetes that live on land. Most have more than one hundred segments (Figure 15.10A). Each earthworm body segment has a fluid-filled coelomic chamber with paired excretory organs that remove waste from the fluid. An earthworm moves when muscles in the body wall change the shape of segments in a coordinated fashion. An earthworm eats its way through the soil, digesting the organic material that it takes in. The digestive tract extends through the coelom. A nerve cord runs the length of the body on the ventral (lower) side and connects to a simple brain. Five hearts pump blood through a closed system of blood vessels. Earthworms are hermaphrodites. Mucus produced by a secretory region called the clitellum glues worms together while they exchange sperm. Later, the clitellum secretes a silky case for the fertilized eggs that the worm deposits in the soil.

Most marine annelids are polychaetes, a lineage of annelids that typically have many bristles per segment. (*Poly–* means many and *chaete* means bristle.) Some polychaetes, including the sandworms often sold as bait, are active predators (Figure 15.10B). Others remain fixed in place and have feathery appendages specialized for filtering food from currents.

Leeches typically live in fresh water, although some are found in damp places on land. Most are scavengers and predators of invertebrates. An infamous few suck blood from vertebrates (Figure 15.10C). A protein in leech saliva keeps blood from clotting while the leech feeds. For this reason, doctors who reattach a severed finger or ear sometimes apply leeches to it. As the leeches feed, their saliva prevents unwanted clots from forming in blood vessels of the reattached body part.

Mollusks **Mollusks** (phylum Mollusca) have a small coelom and a soft, unsegmented body. *Mulluscus* is Latin for soft. The **mantle**, a skirtlike extension of the upper body wall, drapes over a visceral mass that contains the organs (Figure 15.11A). In most mollusks, the mantle secretes a hard, calcium-rich shell. A large, muscular foot functions in locomotion.

Among animals, mollusks are second only to arthropods in diversity. Most live in the seas, but some have adapted to life in fresh water or on land. With 60,000 species of snails and slugs, **gastropods** are the largest subgroup. Gastropod means "belly foot," and members of this group glide about on a broad muscular foot that makes up most of their lower body mass. The gastropod shell, when present, is one piece and usually coiled. Most gastropods scrape up algae with a **radula**, a tongue-like organ hardened with chitin, a tough polysaccharide. Some snails have adapted to a predatory lifestyle. For example, cone snails use a modified harpoonlike radula to inject a paralyzing venom into prey such as small fish.

Most mollusks that end up on dinner plates, including mussels, oysters, clams, and scallops, are **bivalves** (Figure 15.11B). About 15,000 species of bivalves live in fresh water and the seas. Their hinged, two-part shell encloses a body that does not have a distinct head or a radula. Bivalves typically attach to a surface or burrow into sediment. They feed by drawing water into the mantle cavity and filtering out bits of

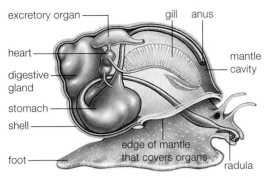

excretory organ — gill anus

heart —

digestive gland —

stomach —

shell —

mantle cavity

edge of mantle that covers organs

foot —

radula

A. Body plan of an aquatic snail (a gastropod).

B. Scallop, a bivalve with a two-part, hinged shell.

C. Octopus, a cephalopod.

Figure 15.11 Three types of mollusks.

(B) Frank Park/ANT Photo Library; (C) NURC/UNCW and NOAA/FGBNMS.

food. Being filter-feeders, bivalves occasionally take in toxins or pathogens that can sicken people who eat them. Risk of illness is greatest when bivalves are eaten raw or only partially cooked.

Cephalopods include cuttlefish, squids, nautiluses, and octopuses (Figure 15.11C). Cephalopod means "head-footed," and their foot has been modified into tentacles and/or arms that extend from the head. All cephalopods are predators and most have beaklike, biting mouthparts in addition to a radula. Cephalopods move by jet propulsion. They draw water into their mantle cavity, then force it out through a funnel-shaped siphon. Of all mollusks, only cephalopods have a closed circulatory system.

Cephalopods include the fastest (squids), largest (giant squid), and smartest (octopuses) invertebrates. Of all invertebrates, octopuses have the largest brain relative to body size, and show the most complex behavior.

Roundworms **Roundworms**, or nematodes (phylum Nematoda), are cylindrical, unsegmented worms with a pseudocoelom (Figure 15.12). They have a complete digestive system, excretory organs, and a nervous system, but no circulatory or respiratory organs. Like insects, roundworms secrete a protective cuticle, a body covering that is periodically molted (shed) and replaced as the worm grows.

Roundworms live in the seas, in fresh water, in damp soil, and inside other animals. Most of the nearly 20,000 species are free-living decomposers less than a millimeter long. Like the fruit fly, the free-living soil roundworm *Caenorhabditis elegans* is often used in scientific studies. It has the same tissue types as more complex organisms, but it is transparent and has fewer than 1,000 body cells. Such traits make it easy for scientists to monitor each cell's fate during development. Several scientists have made Nobel Prize–winning discoveries while studying *C. elegans*.

A few types of parasitic roundworms infect people. In the tropics, mosquitoes transmit parasitic roundworms that enter into human lymph vessels. The worms

Figure 15.12 Body plan of a free-living roundworm.

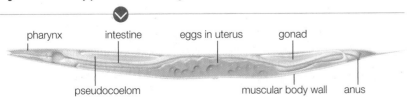

pharynx   intestine   eggs in uterus   gonad

pseudocoelom   muscular body wall   anus

annelid Segmented worm with a coelom, complete digestive system, and closed circulatory system.

bivalve Mollusk with a hinged two-part shell.

cephalopod Predatory mollusk with a closed circulatory system; moves by jet propulsion.

gastropod Mollusk that moves about on its enlarged foot.

mantle Skirtlike extension of tissue in mollusks; covers the mantle cavity and secretes the shell in species that have a shell.

mollusk Invertebrate with a reduced coelom and a mantle.

radula Tonguelike organ of many mollusks.

roundworm Unsegmented worm with a pseudocoelom and a cuticle that is molted as the animal grows.

A. A man with elephantiasis of his left leg. The swelling arises after roundworms damage lymph vessels.

B. Plant-infecting roundworm entering a root.

Figure 15.13 Parasitic roundworms.

(A) Courtesy of © Emily Howard Staub and The Carter Center; (B) William Wergin and Richard Sayre. Colorized by Stephen Ausmus.

Figure 15.14 Grasshopper body plan.
The body has three distinct regions. Compound eyes and a pair of antennae on the head provide sensory information.

From Russell/Wolfe/Hertz/Starr, *Biology*, 1e. © 2008 Cengage Learning®.

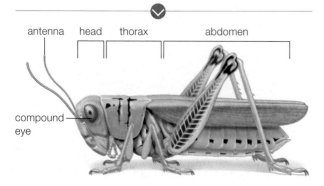

injure valves in these vessels, allowing lymph to pool in the lower limbs. The resulting condition is called elephantiasis, in reference to the fluid-filled, "elephant-like" legs (Figure 15.13A). The swelling is permanent because lymph vessels remain damaged even after the worms have been eliminated.

Pinworms (*Enterobius vermicularis*) are infectious roundworms common in the United States. Most infections occur in children. The worms, about the size of a staple, live in the rectum. At night, females crawl out and lay eggs on skin around the anus. Their movement produces an itching sensation, and scratching puts eggs onto fingertips and under fingernails. If swallowed, the eggs start a new infection.

Roundworms also affect our livestock, pets, and crops. A roundworm that infects pigs can also infect humans who eat undercooked pork, causing trichinosis. Dogs are susceptible to heartworms transmitted by mosquitoes. Cats usually become infected after eating an infected rodent that carries an infectious form of the worm. Roundworms can be important agricultural pests. Some roundworms suck nutrients from plant roots and others actually enter the plant (Figure 15.13B). Either way, a roundworm infection stunts the plant's growth and lowers crop yields.

Arthropods   Arthropods (phylum Arthropoda) are invertebrates with jointed legs. They have a complete digestive system and an open circulatory system. Researchers have identified more than a million living species.

Arthropods secrete a cuticle hardened with chitin, the same material that hardens the mollusk radula. Arthropod cuticle is an external skeleton—an **exoskeleton**. It helps fend off predators and serves as a point of attachment for muscles. In land arthropods, the exoskeleton also helps conserve water and supports the animal's weight. A hard exoskeleton does not restrict growth, because, like roundworms, arthropods molt their cuticle after each growth spurt. A new cuticle forms under the old one, which is shed.

If an arthropod's cuticle were uniformly hard like a plaster cast, it would prevent movement. However, the cuticle thins at joints, where two hard body parts meet. "Arthropod" means jointed leg. Body parts move when the muscles that attach to the exoskeleton on either side of a joint contract.

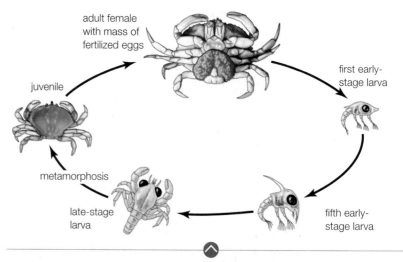

Figure 15.15 Development of a Dungeness crab (a crustacean arthropod).
Fertilized eggs develop into planktonic larvae that grow and molt. Metamorphosis (change in body form) occurs during the molt from a late-stage larva to a juvenile with the adult form. Additional growth and molting produces a sexually mature adult.

In early arthropods, body segments were distinct and all appendages were alike. In many groups, segments later fused into structural units such as a head, thorax, and abdomen (Figure 15.14). Specialized appendages such as wings developed on some segments.

Most arthropods have paired eyes. Insects and crustaceans have **compound eyes** that consist of many units, each with a lens. Such eyes excel at detecting movement. Many arthropods also have one or two pairs of **antennae**, sensory structures on the head that detect touch, odor, and vibrations.

The body plan of many arthropods changes during the life cycle. Individuals undergo **metamorphosis**: Tissues are remodeled as larvae develop into adults. For example, crab larvae swim near the ocean surface and filter food from the water, but adults are bottom-feeders (Figure 15.15). Each stage is specialized for a different lifestyle. Having such different body forms helps prevent adults and juveniles from competing with one another for resources.

Horseshoe crabs (Figure 15.16) are the oldest surviving arthropod lineage. They have lived in Earth's oceans for more than 400 million years. Their common name refers to the fused head and thorax segments that form a horseshoe-shaped region called the cephalothorax. A long spike (a telson) extends from the last abdominal segment and helps the animal right itself if it is turned over by the waves. Horseshoe crabs are bottom-feeders that live in shallow, nearshore waters. In the spring, they venture onto a sandy beach to mate and lay eggs in the damp sand. The billions of eggs they leave behind are an essential food source for migratory shorebirds.

antenna Of some arthropods, sensory structure on the head that detects touch and odors.

arthropod Invertebrate with jointed legs and a hardened exoskeleton that is periodically molted.

compound eye Eye that consists of many individual units, each with its own lens.

exoskeleton External skeleton.

metamorphosis Dramatic remodeling of body form during the transition from larva to adult.

Figure 15.16 Atlantic horseshoe crab (*Limulus*).

Ethan Daniels/Shutterstock.com.

Digging Into Data

Sustainable Use of Horseshoe Crabs

The Atlantic horseshoe crab, *Limulus polyphemus*, is an economically important species. The horseshoe crab's blue blood (right) is used to test for the presence of potentially deadly bacteria in injectable drugs and on medical implants. A liter of horseshoe crab blood sells for about $16,000. To keep horseshoe crab populations stable, blood is extracted from captured animals, which are then returned to the wild. Concerns about the survival of animals after bleeding led researchers to do an experiment. They compared survival of animals captured and maintained in a tank with that of animals captured, bled, and kept in a similar tank. Figure 15.17 shows the results.

Mark Thiessen/National Geographic Creative

1. In which trial did the most control crabs die? In which did the most bled crabs die?
2. Looking at the overall results, how did the mortality of the two groups differ?
3. Based on these results, would you conclude that bleeding harms horseshoe crabs more than capture alone does?

Trial	Control Animals		Bled Animals	
	Number of Crabs	Number That Died	Number of Crabs	Number That Died
1	10	0	10	0
2	10	0	10	3
3	30	0	30	0
4	30	0	30	0
5	30	1	30	6
6	30	0	30	0
7	30	0	30	2
8	30	0	30	5
Total	200	1	200	16

Figure 15.17 Mortality of young male horseshoe crabs kept in tanks for the two weeks after their capture.
Blood was drawn from half of the animals on the day of their capture. Control animals were handled, but not bled. The experiment was repeated eight times with eight different sets of horseshoe crabs.

Data: Walls, E., Berkson, J., *Fish. Bull.* 101:457–459 (2003).

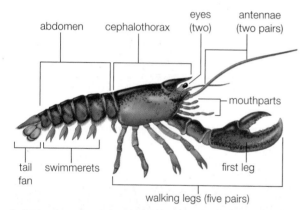

abdomen   cephalothorax

palps

A. Spider
(tarantula).
It kills prey
with a venomous bite.

stinger

palps

B. Scorpion. It delivers venom through a stinger.

C. Tick. It sucks blood from vertebrates.

Figure 15.18 Arachnids. All have four pairs of legs.

(A) © Eric Isselée/Shutterstock.com; (B) © Frans Lemmens/The Image Bank/ GettyImages; (C) James Gathany/CDC.

abdomen   cephalothorax   eyes
(two)   antennae
(two pairs)

mouthparts

tail
fan   swimmerets   first leg

walking legs (five pairs)

A. Body plan of an American lobster.

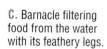

B. Krill swimming in Antarctic waters.

C. Barnacle filtering
food from the water
with its feathery legs.

**Figure 15.19
Marine crustaceans.**

(B) © David Tipling/Photographer's Choice/
Getty Images; (C) © Peter Parks/Image-
quest Marine.

Despite their common name, horseshoe crabs are not true crabs, but rather close relatives of the arachnids. **Arachnids** have four pairs of walking legs, a pair of touch-sensitive appendages called palps, and no antennae. They include spiders, scorpions, ticks, and mites. Spiders and scorpions are venomous predators. Spiders deliver venom through fanglike mouthparts. They have a two-part body, with the cephalothorax (fused head and thorax) separated from the abdomen by a narrow "waist" (Figure 15.18A). The abdomen contains silk-making glands. Scorpions catch their prey with clawlike palps and subdue it with a venom-producing stinger on the final abdominal segment (Figure 15.18B). They eat insects, spiders, and even small lizards. Ticks suck blood from vertebrates (Figure 15.18C). Some serve as vectors for bacterial diseases, such as Lyme disease. Mites, the smallest arachnids, are usually less than a millimeter long. Many are scavengers, but some are parasites. Mites that burrow beneath the skin cause scabies in humans and mange in dogs. Some larval mites, commonly called chiggers, crawl into a hair follicle, secrete an enzyme that breaks down proteins, then suck up liquefied tissues.

Crustaceans are mostly marine arthropods that have two pairs of antennae. Decapod crustaceans are bottom-feeding scavengers with five pairs of walking legs. Many, such as shrimps, crabs, and lobsters (Figure 15.19A), are harvested as human food. Other animals also depend on crustaceans for food. Shrimplike crustaceans called krill abound in cool ocean waters (Figure 15.19B). Each is only a few centimeters long, but krill are so abundant and nutritious that a 100-ton blue whale can subsist almost entirely on the krill that it filters from seawater.

Barnacles are marine crustaceans that secrete a calcium-rich external shell (Figure 15.19C). Barnacle larvae swim, then settle and develop into adults that attach to a surface. Adults filter food from seawater with their feathery legs. Some barnacles are notable for the length of their penis, which can be eight times that of their body.

Isopods are a mostly marine group of crustaceans, but some live in damp land habitats. The species commonly known as pill bugs defend themselves from threats by rolling into a ball (right).

Centipedes and millipedes constitute another arthropod lineage. These nocturnal ground dwellers have an elongated body with many similar segments (Figure 15.20). The head has paired antennae and two simple eyes. Centipedes are venomous predators with a flat, low-slung body and one pair of legs per segment. Most millipedes eat plant material. Their cylindrical body has two legs per segment and their cuticle is hardened with calcium carbonate.

© Chris Howey/
Shutterstock.

Insects, which have three pairs of legs and one pair of antennae, are the most diverse arthropod group. Early insects were wingless, ground-dwelling scavengers that did not undergo metamorphosis. Modern bristletails and silverfish retain this type of body form and development. However, most modern insects have wings and undergo metamorphosis. With *incomplete* metamorphosis, an egg hatches into a nymph that differs somewhat in form from the adult. The nymph changes into the adult form over the course of several molts. Cockroaches, grasshoppers, and dragonflies develop in this way. With *complete* metamorphosis, a larva grows and molts without altering its form, then undergoes pupation. A pupa is a nonfeeding body in which larval tissues are remodeled into the adult form. Members of the most diverse insect orders such as flies, beetles, and the butterflies and moths all have two pairs of wings on the thorax and undergo complete metamorphosis (Figure 15.21).

Insects have important ecological roles. They serve as pollinators to many flowering plants (Figure 15.22A). They also serve as food for a variety of wildlife. Larval moths and butterflies (caterpillars) feed songbird nestlings. Aquatic larvae of dragonflies and mayflies serve as food for fish. Most amphibians and reptiles feed mainly on insects. In addition, insects dispose of wastes and remains. Flies and beetles are quick to find an animal corpse or a pile of feces (Figure 15.22B). They lay eggs in or on this organic material, and the larvae that hatch devour it. By their actions, these insects help distribute nutrients through the ecosystem. On the other hand, insects are our main competitors for plant foods. For example, Mediterranean fruit flies, or medflies, damage citrus crops (Figure 15.22C). Insects also transmit dangerous diseases. Mosquitoes spread malaria and carry parasitic roundworms between human hosts. Fleas can transmit plague and lice can transmit typhus. Bedbugs (Figure 15.22D) do not transmit disease, but their bites itch and infestations can have severe psychological and economic effects.

A. Centipede, a speedy predator.

B. Millipede, a scavenger of decaying plant material.

Figure 15.20 Centipede and millipede.

© Eric Isselée/Shutterstock.

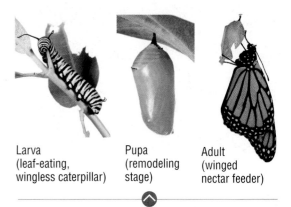

Larva (leaf-eating, wingless caterpillar)   Pupa (remodeling stage)   Adult (winged nectar feeder)

Figure 15.21 Complete metamorphosis in a butterfly.

Left and middle, © Jacob Hamblin/Shutterstock.com right, © Laurie Barr/Shutterstock.com.

A. Bee serving as a pollinator.

B. Dung beetle gathering feces.

C. Medfly, a threat to citrus.

D. Bedbug, a human parasite.

Figure 15.22 Ecological roles of insects.

(A) Photo by Jack Dykinga, USDA, ARS; (B) Gregory G. Dimijian, M.D./Science Source; (C) Photo by Scott Bauer/USDA; (D) CDC/Piotr Naskrecki.

arachnids Land-dwelling arthropods with four pairs of walking legs and no antennae; for example, a spider, scorpion, or tick.

crustaceans Mostly marine arthropods with two pairs of antennae; for example, a shrimp, crab, lobster, or barnacle.

insects Land-dwelling arthropods with a pair of antennae, three pairs of legs, and—in the most diverse groups—wings.

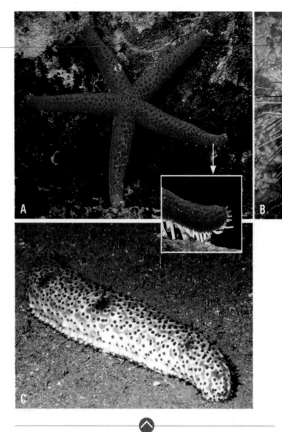

Figure 15.23 Echinoderms.

A. Sea star, and a close-up of its arm with little tube feet.

B. Sea urchin, which moves about on spines and a few tube feet. Sea urchins typically graze on algae.

C. Sea cucumber, with rows of tube feet along its elongated body. The hard parts have been reduced to microscopic plates embedded in a soft body.

(A) Herve Chaumeton/Agence Nature; (B) © Derek Holzapfel/photos.com; (C) Andrew David, NOAA/NMFS/SEFSC Panama City, Lance Horn, UNCW/NURC-Phantom II ROV operator.

Echinoderms **Echinoderms** (phylum Echinodermata) include about 6,000 marine invertebrates such as sea stars, sea urchins, and sea cucumbers (Figure 15.23). Their phylum name means spiny-skinned and refers to interlocking spines and plates of calcium carbonate embedded in the skin. The plates form an internal skeleton called an **endoskeleton**. Echinoderm adults are radially symmetrical.

Sea stars (sometimes called starfish) are the most familiar echinoderms. They do not have a brain, but they do have a decentralized nervous system. Eyespots at the tips of their arms detect light and movement. A typical sea star moves about on tiny, fluid-filled tube feet (Figure 15.23A). Tube feet are part of a **water–vascular system**, a system of fluid-filled tubes unique to echinoderms. The system includes a central ring and fluid-filled canals that extend into each arm. Side canals deliver coelomic fluid into muscular bulbs that function like the bulb on a medicine dropper. Contraction of a bulb forces fluid into the attached tube foot, extending the foot. A sea star glides along as coordinated contraction and relaxation of the bulbs redistribute fluid among hundreds of tube feet.

Most sea stars prey on bivalve mollusks. To feed, the sea star slides its stomach out through its mouth and into a bivalve's shell. The stomach secretes acid and enzymes that kill the mollusk and begin to digest it. Partially digested food enters the stomach and digestion is completed with the aid of digestive glands in the arms. A sea star's reproductive organs are also in its arms. Sexes are separate and eggs or sperm are released into the water. Fertilization produces an embryo that develops into a ciliated, bilaterally symmetrical larva. The larva swims about briefly, then develops into an adult. The bilateral symmetry of echinoderm larva, along with evidence from genetic studies, indicates that the ancestor of echinoderms was a bilateral animal. Developmental studies show that echinoderms have a deuterostome developmental pattern, and thus belong to the same lineage as chordates.

Sea stars and other echinoderms have a remarkable ability to regenerate lost body parts. If a sea star is cut into pieces, any portion with some of the central disk can regrow the missing body parts.

chordates Animal phylum characterized by a notochord, dorsal nerve cord, pharyngeal gill slits, and a tail that extends beyond the anus. Includes invertebrate and vertebrate groups.

echinoderms Invertebrates with a water–vascular system and an endoskeleton made of hardened plates and spines.

endoskeleton Internal skeleton.

lancelets Invertebrate chordates that have a fishlike shape and retain their defining chordate traits into adulthood.

notochord Stiff rod of connective tissue that runs the length of the body in chordate larvae or embryos.

tunicates Invertebrate chordates that lose their defining chordate traits during the transition to adulthood.

water–vascular system Of echinoderms, a system of fluid-filled tubes and tube feet that function in locomotion.

Take-Home Message 15.3

What are the traits of the main invertebrate groups?

- Sponges are filter-feeders that have specialized cells but no tissues. Cnidarians have two tissue layers and a radial body plan. Most other invertebrates have a bilateral body plan with three tissue layers that form organ systems.
- Three groups of "worms" differ in their traits and are not closely related. Flatworms do not have a coelom. Annelids have a coelom and a segmented body. Roundworms have a pseudocoelom and are the only worms that molt.
- Mollusks and arthropods are the two most diverse invertebrate phyla. Mollusks are soft-bodied, although many make a shell. Arthropods have a hardened exoskeleton and jointed legs. Insects, an arthropod subgroup, are the most diverse invertebrates and the only winged ones.
- Echinoderms are radial animals with bilateral larvae. They are on the same branch of the animal family tree as the chordates.

15.4 Introducing the Chordates

REMEMBER: Studies of similarities between genes of different organisms can be used to determine relationships among those organisms (Section 11.7).

Chordate Traits Chordates (phylum Chordata) are defined by four embryonic traits: (1) A **notochord**, a rod of stiff but flexible connective tissue, extends the length of the body and provides support. (2) A dorsal, hollow nerve cord parallels the notochord. (3) Gill slits open across the wall of the pharynx (throat region). (4) A muscular tail extends beyond the anus. Depending on the chordate group, some, all, or none of these traits persist in the adult.

Invertebrate Chordates There are two groups of invertebrate chordates, tunicates and lancelets. Both are marine and both filter food from currents of water that pass through gill slits in their pharynx.

Lancelets (subphylum Cephalochordata) have a fishlike shape and are 3 to 7 centimeters long (Figure 15.24). Lancelets retain all characteristic chordate traits as adults. The dorsal nerve cord extends into the head, where a single eyespot at its tip detects light. However, the head does not have a brain or any paired sensory organs similar to those of fishes.

Tunicates (subphylum Urochordata) are named for the secreted carbohydrate-rich covering or "tunic" that encloses the adult body (Figure 15.25A,B). Larval tunicates have all the typical chordate traits (Figure 15.25C). They swim about briefly, then undergo metamorphosis to the adult form. Of the four typical chordate traits, the adult retains only the pharynx with gill slits. Most adult tunicates attach to an undersea surface and filter food from the water. As water flows in an oral opening and past gill slits, bits of food stick to mucus on the gills and are then sent to a gut. Water leaves through another body opening.

Which invertebrate chordate is most closely related to vertebrates? An adult lancelet looks more like a fish than an adult tunicate does, but such superficial similarities are sometimes deceiving. Studies of developmental processes and gene sequences indicate that tunicates are the closest invertebrate relatives of vertebrates. Keep in mind that neither tunicates nor lancelets are ancestors of vertebrates. These groups share a recent common relative, but each has unique traits that put it on a separate branch of the animal family tree.

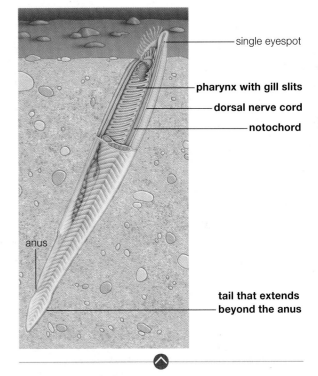

single eyespot

pharynx with gill slits

dorsal nerve cord

notochord

anus

**tail that extends
beyond the anus**

Figure 15.24 Lancelet.
These marine invertebrates retain all chordate traits (labeled with bold text) into adulthood. Lancelets live in sandy sediments, and filter food from the water with their pharynx.

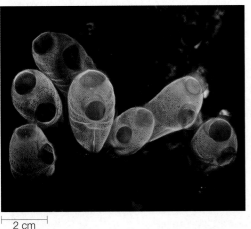

2 cm

A. Adult tunicates.

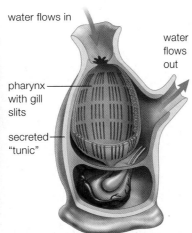

water flows in

water flows out

pharynx with gill slits

secreted "tunic"

B. Body plan of an adult tunicate.

Figure 15.25 Tunicates.
Larval tunicates are free-swimming and have all the characteristic chordate features. After metamorphosis, the adult retains only the pharynx with gill slits.

(A) Ethan Daniels/Shutterstock; (B,C) From Russell/Worlfe/Hertz/Starr, *Biology*, 1e © 2008 Cengage Learning®.

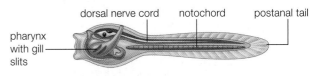

dorsal nerve cord   notochord   postanal tail

pharynx with gill slits

C. Body plan of a tunicate larva.

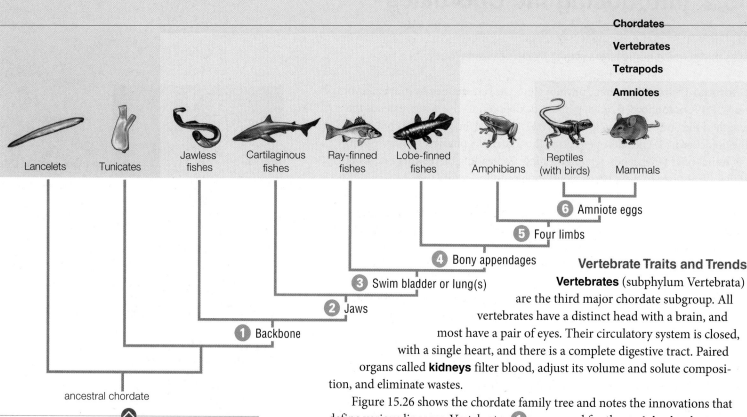

Figure 15.26 The chordate family tree.
The vast majority of the chordates are vertebrates.

Figure It Out: Do the cartilaginous fishes have a backbone?

Answer: Yes

bony fish Jawed fish with a skeleton composed mainly of bone.

cartilaginous fish Fish that has jaws, paired fins, and a skeleton made of cartilage; for example, a shark.

jawless fish Fish that has a skeleton of cartilage, but no jaws or paired fins; for example, a lamprey.

kidney Organ of the vertebrate urinary system that filters blood and adjusts its composition.

lung Saclike organ inside which blood exchanges gases with the air.

scales Hard, flattened elements that cover the skin of reptiles and some fishes.

tetrapod Vertebrate with four limbs.

vertebral column Backbone.

vertebrate Animal with a backbone; a fish, amphibian, reptile, bird, or mammal.

Vertebrate Traits and Trends

Vertebrates (subphylum Vertebrata) are the third major chordate subgroup. All vertebrates have a distinct head with a brain, and most have a pair of eyes. Their circulatory system is closed, with a single heart, and there is a complete digestive tract. Paired organs called **kidneys** filter blood, adjust its volume and solute composition, and eliminate wastes.

Figure 15.26 shows the chordate family tree and notes the innovations that define various lineages. Vertebrates ❶ are named for the **vertebral column** (backbone) that replaces the notochord as a vertebrate embryo develops. This flexible but sturdy structure encloses the spinal cord that develops from the nerve cord. The vertebral column is part of the vertebrate endoskeleton.

The next major innovation was jaws, which are hinged skeletal elements used in feeding ❷. Jaws evolved by modification of bony parts that supported the gill slits of early jawless fishes. Evolution of jaws opened up new feeding opportunities, allowing an adaptive radiation. The vast majority of modern fishes have jaws.

Some jawed fishes evolved internal air sacs ❸. In some species, such a sac functioned as a simple **lung**, a respiratory organ in which blood exchanges gases with the air. One lineage of fish with lungs also had bony paired fins ❹. Their descendants evolved bony paired limbs, becoming the first **tetrapods**, or four-legged walkers ❺. A special type of egg (the amniote egg) allowed one tetrapod group to lay eggs on land ❻. Members of that lineage, the amniotes (reptiles, birds, and mammals), are the most successful tetrapods on land.

Take-Home Message 15.4

What trends shaped chordate evolution?

• The earliest chordates were invertebrates, and two groups of invertebrate chordates (tunicates and lancelets) still exist. They are aquatic filter-feeders.

• Vertebrates evolved from an invertebrate chordate ancestor. All have a vertebral column, or backbone.

• Jaws, lungs, limbs, and waterproof eggs were key innovations that led to the adaptive radiation of vertebrates, first in the seas and then on the land.

15.5 Fishes and Amphibians

REMEMBER: Homologous structures are structures that are similar in different lineages because they evolved in a common ancestor. (Section 11.6).

We begin our survey of vertebrate diversity with fishes, the first vertebrate lineages to evolve and the most fully aquatic.

Jawless Fishes Modern **jawless fishes** have a smooth, elongated body without paired fins. Their skeletal elements are made of cartilage, the same type of connective tissue that supports your nose and ears. Figure 15.27 shows the distinctive mouth of one jawless fish, a lamprey. As adults, most lampreys feed on other fish. Lacking jaws, a lamprey cannot bite. Instead, it attaches to a fish using an oral disk ringed by horny teeth made of the protein keratin. Once attached, the lamprey secretes enzymes and scrapes up bits of its host's flesh with a tooth-covered tongue. The host fish often dies from blood loss or a resulting infection.

Jawed Fishes Jaws evolved from gill arches, which are skeletal elements that support a fish's gills (Figure 15. 28). Most modern jawed fishes have paired fins and **scales**: hard, flattened structures that grow from and often cover the skin. Scales and an internal skeleton make a fish denser than water and prone to sinking. Highly active swimmers have fins with a shape that helps lift them, something like the way that wings help lift up an airplane. Friction slows movement through water, so speedy swimmers typically have a streamlined body that reduces friction.

There are two groups of jawed fishes: cartilaginous fishes and bony fishes (Figure 15.29). As their name implies, **cartilaginous fishes** are jawed fishes with a skeleton made of cartilage. They include 850 species, with sharks being the best known. Some sharks are predators that swim in upper ocean waters. Others strain plankton from the water or suck up food from the seafloor. Human surfers and swimmers resemble typical prey of some predatory sharks, and rare attacks by a few species give the group as a whole an undeserved bad reputation. Worldwide, shark attacks kill about 25 people a year. For comparison, dogs kill about 30 people each year in the United States alone.

In **bony fishes**, an embryonic skeleton of cartilage becomes transformed to an adult skeleton consisting mainly of bone. Both cartilaginous fishes and bony fishes have paired fins, but only bony fishes can move those fins. Bony fishes also differ from other fishes in having their gill slits hidden beneath a gill cover. In jawless and cartilaginous fish, gill slits are visible at the body surface.

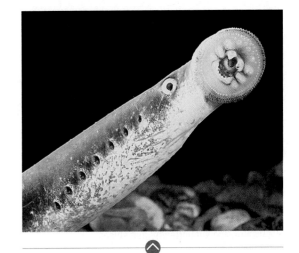

Figure 15.27 Lamprey, a jawless fish.
A lamprey has no paired fins and its gill slits are visible at the body surface. It attaches to other fishes with its oral disk and scrapes at their flesh.

Heather Angel/Natural Visions.

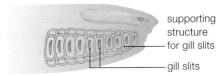

supporting structure for gill slits

gill slits

Jawless fish with skeletal elements supporting gills.

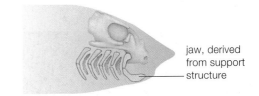

jaw, derived from support structure

Early jawed fish.

Figure 15.28 Evolution of jaws from gill supports.

Figure 15.29 Jawed fishes.

A. Cartilaginous fish with jaws and paired fins. This species of shark is a swift predator.

© Luiz A. Rocha/Shutterstock.

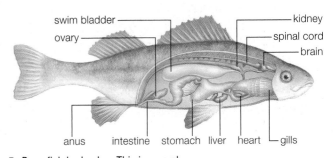

swim bladder
ovary
kidney
spinal cord
brain
anus   intestine   stomach   liver   heart   gills

B. Bony fish body plan. This is a perch.

After E. Solomon, L. Berg, and D.W. Martin, *Biology*, Seventh Edition, © Cengage Learning®.

There are two lineages of bony fishes. **Ray-finned fishes** have flexible fins supported by thin rays derived from skin. A gas-filled swim bladder helps them adjust their buoyancy. With about 30,000 species, ray-finned fishes constitute nearly half of the vertebrates. They include most familiar freshwater fish (Figure 15.30A) as well as marine species such as tuna, halibut, and cod.

Modern **lobe-finned fishes** include coelacanths (Section 12.7) and lungfishes (Figure 15.30B). These fish have thick, fleshy pelvic and pectoral fins with internal bony supports. As their name suggests, lungfishes have one or two lungs in addition to their gills. The fish inflates its lungs by gulping air, then air in the lungs exchanges gases with blood. Having lungs in addition to gills allows lungfishes to survive in low-oxygen waters.

Early Tetrapods Genome comparisons indicate that, of the two modern lobe-finned groups, lungfishes are closest to tetrapods. Fossils reveal how the skeleton became modified as fishes adapted to swimming evolved into four-legged walkers (Figure 15.31). The bones inside a lobe-finned fish's pelvic and pectoral fins are homologous (Section 11.6) with amphibian limb bones. However, the transition to land was not only a matter of skeletal changes. Fishes have a two-chamber heart: one chamber receives blood, the other pumps it out. In amphibians, the heart became divided into three chambers: one to receive blood, one to pump blood to the lungs, and one to pump blood to the body. This change increased the rate of blood flow to the body and the efficiency of gas exchange. Changes in the inner ear improved the ability to detect airborne sounds, and eyelids prevented the eyes from drying out.

What was the advantage of living on land? An ability to survive out of water is useful in seasonally dry places. Individuals on land were also safe from aquatic predators and had access to a new food—insects—which had recently evolved.

A. Goldfish (carp), a ray-finned bony fish. Its fins consist of a web of skin supported by thin spines.

pelvic fin   pectoral fin

B. Lungfish, a lobe-finned bony fish. Its thick, fleshy fins have sturdy bony supports inside them.

Figure 15.30 Two types of fins in bony fishes.

(A) © iStockphoto.com/GlobalP; (B) © Wernher Krutein/photovault.com.

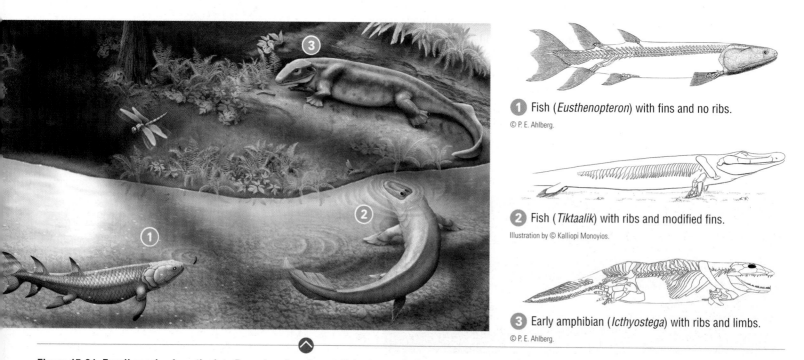

1 Fish (*Eusthenopteron*) with fins and no ribs.
© P. E. Ahlberg.

2 Fish (*Tiktaalik*) with ribs and modified fins.
Illustration by © Kalliopi Monoyios.

3 Early amphibian (*Icthyostega*) with ribs and limbs.
© P. E. Ahlberg.

Figure 15.31 Fossil species from the late Devonian show how vertebrates made the transition from water to land.

**Figure 15.32
Amphibians.**
All have a
scaleless body.

(A) Photo by James Bet-
taso, US Fish and Wild-
life Service; (B) Stephen
Dalton/Science Source;
(C) © iStockphoto.com/
Tommounsey.

A. Salamander, with equal-sized front and back limbs.

B. Frog, with long, muscular legs and short forelimbs.

C. Tadpole, a swimming frog larva with gills and a tail.

Modern Amphibians **Amphibians** are scaleless tetrapods that spend time on
land, but require water to breed. Fertilization is typically external. Eggs and sperm
are released into water through a **cloaca**, a body opening that also serves as the
exit for urinary and digestive wastes. (Sharks, reptiles, birds, and egg-laying mammals also have a cloaca.) Amphibian larvae are aquatic and have gills. Most species
lose their gills and develop lungs during the transition to adulthood. The adults are
active predators with a three-chambered heart.

Of all amphibians, the 530 or so species of salamanders and newts most closely
resemble early tetrapods in body form. Their forelimbs and back limbs are similarly
sized, and they have a long tail (Figure 15.32A). Frogs and toads belong to the most
diverse amphibian lineage, with more than 5,000 species. Long, muscular hindlimbs
allow the tailless adults to swim, hop, and make spectacular leaps (Figure 15.32B).
The much smaller forelimbs help absorb the impact of landings.

Larvae of salamanders look like small adults, except for the presence of gills. In
contrast, frog and toad larvae differ markedly from adults. The larvae, commonly
called tadpoles, have gills and a tail but no limbs (Figure 15.32C).

Many amphibian populations are in decline. Shrinking or deteriorating habitats
are part of the problem. People commonly fill low-lying areas where pooling of seasonal rains could provide breeding grounds for amphibians. Climate change causes
additional harm by fostering the spread of pathogens and parasites, and increasing
the amount of UV radiation that reaches Earth's surface. Amphibians are also highly
sensitive to water pollution. Their thin skin, unprotected by scales, allows waste
carbon dioxide to diffuse out of their body. Unfortunately, it also allows chemical
pollutants to enter.

Take-Home Message 15.5

What are the traits of fishes and amphibians?

- Lampreys are modern fishes that do not have jaws or paired fins.
- Cartilaginous fishes such as sharks have jaws and immobile paired fins. Like
  lampreys, they have a skeleton made of cartilage.
- Bony fishes are the most diverse fishes. They have paired movable fins.
- The lobe-finned fishes, a subgroup of the bony fishes, are the closest living relatives
  of the tetrapods.
- Amphibians are tetrapods with a three-chambered heart and a scaleless body.
  Fertilization typically takes place in water. Gilled larvae typically develop into adults
  that have lungs.

amphibian Tetrapod with a three-chambered heart
and scaleless skin that typically develops in water,
then lives on land as a carnivore with lungs.

cloaca Of some vertebrates, body opening that
releases urinary and digestive waste, and also
functions in reproduction.

lobe-finned fish Bony fish that has bony supports
inside its fins.

ray-finned fish Bony fish with fins supported by thin
rays derived from skin.

A. Hognose snakes hatching from leathery amniote eggs.

B. Komodo dragon, the largest lizard.

C. Turtle in a defensive posture.

D. Crocodile with its fish prey.

Figure 15.33 Diversity of nonbird reptiles.

(A) Z. Leszczynski/Animals Animals; (B) Soren Egeberg Photography/Shutterstock.com;
(C) Joel Sartore/National Geographic Creative; (D) Johan Swanepoel/Shutterstock.com.

15.6 Escape From Water—Amniotes

REMEMBER: A clade includes all descendants of the ancestor in which the unique trait that defines the clade arose (Section 12.8). Dinosaurs perished in a mass extinction about 66 million years ago (11.1).

Amniote Innovations **Amniotes**, a vertebrate group adapted to life on dry land, branched off from an amphibian ancestor during the Carboniferous. Amniotes have lungs throughout their life, and their skin is rich in keratin, a protein that makes it waterproof. A pair of well-developed kidneys help conserve water, and fertilization usually takes place inside the female's body. Sexes are typically separate and fixed for life. Amniotes produce eggs in which an embryo develops bathed in fluid, a trait that allows them to develop on dry land (Figure 15.33A).

An early branching of the amniote lineage separated ancestors of mammals from the common ancestor of all modern reptiles. You probably do not think of birds as reptiles, but from an evolutionary perspective, they are. To a biologist, the **reptiles** (clade Reptilia) include modern lizards, snakes, turtles, crocodilians, and birds, as well as the extinct dinosaurs.

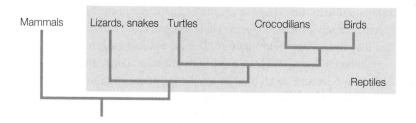

The ability to regulate internal body temperature evolved in some early amniotes, including the ancestors of birds and of mammals. Amphibians, turtles, lizards, and snakes are **ectotherms**, which means "heated from outside." Ectotherms adjust their internal temperature by altering their behavior. They bask on a warm rock to heat up, or retreat into a burrow to cool off. In contrast, birds and mammals are **endotherms** that maintain their body temperature by varying their production of metabolic heat. Endotherms use energy staying warm, so they require more food than ectotherms. A bird or mammal requires far more calories than a lizard or snake of the same weight. However, because endotherms warm themselves, they can remain active at lower temperatures than ectotherms.

Nonbird Reptiles Lizards are the most diverse reptiles. Most are predators, although iguanas eat plants. The largest lizard, the Komodo dragon (Figure 15.33B), grows up to 3 meters (10 feet) long. It has a venomous bite, and trails its prey for hours or days until the poisoned animal collapses. Lizards typically lay eggs that develop outside the body, but some species give birth to live young. In live-bearers, eggs develop inside the mother, but they are not nourished by her tissues.

Snakes evolved from lizards during the Cretaceous and some snakes such as pythons retain bony remnants of ancestral hind limbs. All snakes are predators and all have teeth, but only some have fangs. Rattlesnakes and other fanged species bite and subdue prey with venom made in modified salivary (saliva-making) glands. Other snakes are constrictors that suffocate a prey animal by wrapping around it so tightly that it cannot expand its chest to inhale. Like lizards, most snakes lay eggs, but some hold eggs inside the body and give birth to live young.

Turtles have a bony, protective shell attached to their backbone (Figure 15.33C). Modern turtles do not have teeth. Instead, a thick layer of the protein keratin covers their jaws and forms a horny beak. Most turtles live either in the sea or in freshwater. Those that live entirely on land are commonly called tortoises.

Crocodilians live in or near water and include crocodiles, alligators, and caimans. They are predators with powerful jaws, a long snout, and sharp teeth (Figure 15.33D). Like birds, all crocodilians have a highly efficient four-chambered heart. They are the closest living relatives of birds.

The Jurassic and Cretaceous periods (201–66 million years ago) are sometimes referred to as the "Age of Reptiles." During this time the dinosaurs underwent a great adaptive radiation and became the dominant animals on land. They became extinct at the end of the Cretaceous, most likely as a result of an asteroid impact. Birds, which had earlier evolved from feathered dinosaurs by about 160 million years ago, survived this extinction event.

Birds **Birds** are the only modern animals with feathers, which are modified scales. The bird body is adapted to flight. Bird wings are homologous to our arms. Contraction of one set of muscles produces a powerful downstroke that lifts the bird (Figure 15.34). A less powerful set of muscles contracts to raise the wing. Birds have the most efficient respiratory system of any vertebrate. This system provides the oxygen necessary to produce a steady supply of ATP that can power flight. A four-chambered heart pumps oxygen-poor blood to the lungs and oxygen-rich blood to the rest of the body in separate circuits. Flying also requires good eyesight and a great deal of coordination. Compared to a lizard of a similar body mass, a bird has much larger eyes and a larger brain. Most birds are surprisingly lightweight. Air cavities inside a bird's bones keep its body weight low, as does the lack of a bladder (an organ that stores urinary waste in many other vertebrates). Rather than heavy, bony teeth, bird jaws are covered with a lightweight beak made of keratin. The structure of the beak adapts a bird to feed on a particular type of food.

In birds, as in other reptiles, fertilization is internal. Unlike most male reptiles, male birds do not typically have a penis. Thus, to inseminate a female, a male bird must press his cloaca against hers, in a maneuver poetically described as a cloacal kiss. A female bird lays an egg that has four membranes characteristic of **amniote eggs** (Figure 15.35A). Nutrients from the egg's yolk and water from the egg white (albumen) sustain the developing embryo. Like some turtles and all crocodilians, birds encase their eggs within a rigid shell hardened by calcium carbonate. Birds are endotherms, so their egg must be kept warm in order for the embryo to develop. In nearly all bird species, one or both parents incubate the eggs until they are ready to hatch. Unlike other reptiles, many birds hatch in a relatively undeveloped state and require extensive parental care before they can live on their own (Figure 15.35B).

Many birds make a seasonal migration. Migration typically involves spring travel to a breeding site where insects are abundant in the summer, then an autumn flight to a site where the bird spends the winter. Such flights can require remarkable endurance. One shorebird monitored by researchers flew from Alaska to New Zealand, a distance of 11,500 kilometers (7,145 miles), without stopping to feed or rest.

Mammals **Mammals** are the only amniotes in which females nourish their offspring with milk secreted from mammary glands. The group name is derived from the Latin *mamma*, meaning breast. Mammals are also the only animals that have hair or fur. Both are modifications of scales. Like birds, mammals are endotherms. A coat of fur or head of hair helps them maintain their core temperature. Mammals

Figure 15.34 Bird in flight.
Birds flap their wings to fly. The downstroke provides lift.
Eric Isselée/Shutterstock.com.

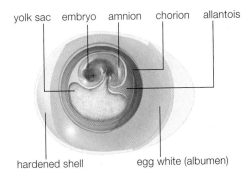
yolk sac   embryo   amnion   chorion   allantois

hardened shell   egg white (albumen)

A. Developing embryo inside an egg.

B. Newly hatched parrots.
Jane Burton/naturepl.com.

Figure 15.35 Bird development.

amniote Vertebrate that produces amniote eggs; a reptile, bird, or mammal.

amniote egg Egg with four membranes that allows an embryo to develop away from water.

bird Modern amniote with feathers.

ectotherm Animal that gains heat from the environment; commonly called "cold-blooded."

endotherm Animal that produces its own heat; commonly called "warm-blooded."

mammal Vertebrate that nourishes its young with milk from mammary glands.

reptile Amniote subgroup that includes lizards, snakes, turtles, crocodilians, and birds.

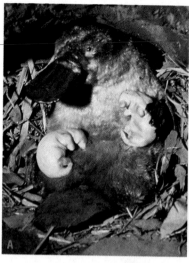

Figure 15.36 Mammal mothers and young.

A. Egg-laying mammal, the platypus. Her young hatched from eggs laid outside her body. They lick milk that oozes from her skin.

B. Marsupial, a kangaroo. Her young developed to an early stage in her body, then climbed into a pouch on her belly to nurse and complete development.

C. Placental mammal, a bear. Her young developed to a late stage inside her body. After birth, they suck milk from nipples on her chest and belly.

(A) Jean Phillipe Varin/Jacana/Science Source; (B) Craig Dingle/iStockphoto.com; (C) Sergey Gorshkov/Nature Picture Library.

have a single lower jawbone and most have more than one type of tooth. By contrast, reptiles have a hinged two-part jaw, and all teeth are similarly shaped. Having a variety of different kinds of teeth allows mammals to eat more types of foods than most other vertebrates.

Mammals evolved early in the Jurassic, and early mouselike species coexisted with dinosaurs. By 130 million years ago, three lineages had evolved: **monotremes** (egg-laying mammals), **marsupials** (the pouched mammals), and **placental mammals** (mammals in which an organ called the placenta provides nutrients to developing offspring). Figure 15.36 shows examples of each group.

Only three species of monotremes survive, and most pouched mammals live in Australia or New Zealand. In contrast, placental mammals occur worldwide. What gives placental mammals their competitive edge? They have a higher metabolic rate, better body temperature control, and a more efficient way to nourish embryos. Compared to other mammals, placental mammals develop to a far more advanced stage inside their mother's body.

Rats and bats are the most diverse mammals. About half the 4,000 species of placental mammals are rodents and, of those, about half are rats. The next most diverse group is the bats, with about 375 species. Bats are the only flying mammals. Although some may look like flying mice, bats are more closely related to carnivores such as wolves and foxes than to rodents.

marsupial Mammal in which offspring complete development in a pouch on the mother's body.

monotreme Egg-laying mammal.

placental mammal Mammal in which developing offspring are nourished within the mother's body by way of a placenta.

primate Mammalian group with grasping hands; includes lemurs, tarsiers, monkeys, apes, and humans.

Take-Home Message 15.6

What are amniotes?

- Amniotes are vertebrates that are adapted to life on land by their unique eggs, waterproof skin, and highly efficient kidneys.
- One amniote lineage includes snakes, lizards, turtles, and crocodiles as well as the birds. Birds are the only amniotes with feathers. Like mammals, birds are endotherms; metabolic heat maintains their internal temperature.
- Mammals are amniotes that nourish their young with milk. Most have hair or fur. The three lineages are egg-laying monotremes, pouched marsupials, and placental mammals. Placental mammals are the most widespread and diverse.

15.7 Human Evolution

REMEMBER: Radiometric dating can reveal the age of fossils (Section 11.4). Mutation (12.2) and founder effects cause allele differences among populations (12.5).

Primate Traits **Primates** are an order of placental mammals that includes humans, apes, monkeys, and their close relatives. Primates first evolved in tropical forests, and many of the group's characteristic traits arose as adaptations to life among the branches. Primate shoulders have an extensive range of motion that facilitates climbing (Figure 15.37). Unlike most mammals, a primate can extend its arms out to its sides, reach above its head, and rotate its forearm at the elbow. Both hands and feet are capable of grasping. Many types of mammals have claws, whereas the tips of primate fingers and toes typically have touch-sensitive pads protected by flattened nails.

Compared to other mammals, primates have a large brain for their body size. The regions of the brain devoted to vision and to information processing are expanded, and the area devoted to smell is reduced. Most mammals have widely spaced eyes set toward the side of the skull, but primate eyes tend to be at the front of the head. As a result, both eyes view the same area, each from a slightly different vantage point. The brain integrates the differing signals it receives from the two eyes to produce a three-dimensional image. A primate's excellent depth perception adapts it to a life spent leaping or swinging from limb to limb.

Most primates spend their life in a social group that includes adults of both sexes. Females usually give birth to only one or two young at a time and provide care for an extended period after birth.

Primate Origins and Diversification Figure 15.38 shows the relationships among modern primate subgroups. Primates most likely arose before the demise of

Figure 15.37 Adapted to climbing.
A female orangutan (an Asian ape) demonstrates her wide range of shoulder motion and the grasping ability of her hands and feet. Eyes situated at the front of her face provide excellent depth perception.

© Thomas Marent/ ardea.com.

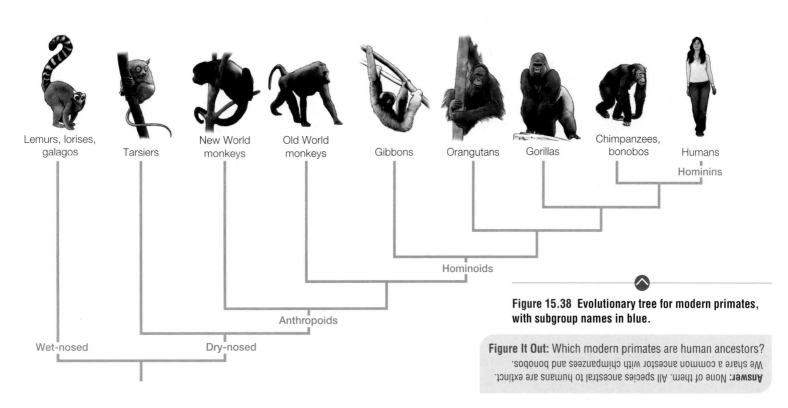

Lemurs, lorises, galagos — Tarsiers — New World monkeys — Old World monkeys — Gibbons — Orangutans — Gorillas — Chimpanzees, bonobos — Humans

Hominins

Hominoids

Anthropoids

Wet-nosed

Dry-nosed

Figure 15.38 Evolutionary tree for modern primates, with subgroup names in blue.

Figure It Out: Which modern primates are human ancestors?

Answer: None of them. All species ancestral to humans are extinct. We share a common ancestor with chimpanzees and bonobos.

A. Lemur, with a wet nose and cleft upper lip.

B. Tarsier, with a dry nose and uncleft upper lip.

C. Squirrel monkey (New World monkey), with a flat face and prehensile tail.

D. Baboon (Old World monkey), with a long nose and short tail.

E. Gorilla, the largest living great ape.

F. Chimpanzee, one of our two closest living relatives.

Figure 15.39 Diversity of modern, nonhuman primates.

(A) toos/iStockphoto.com; (B) roc8jas/iStockphoto.com; (C) primates.com; (D) © iStockphoto.com/
JasonRWarren; (E) © Dallas Zoo, Robert Cabello; (F) Kenneth Garrett/National Geographic Image
Collection.

the dinosaurs, sometime between 85 and 66 million years ago. Their closest living relatives are colugos, a group of nocturnal mammals that live in the forests of Southeast Asia and glide from tree to tree like flying squirrels.

Lemurs are members of the oldest existing primate lineage. Like dogs and most other mammals, lemurs are "wet-nosed"; they have a moist nose and an upper lip that attaches tightly to the underlying gum (Figure 15.39A). A dry nose and a movable, noncleft upper lip evolved in the common ancestor of tarsiers (Figure 15.39B) and all other modern primates. This innovation gave dry-nosed primates a wider range of facial expressions and allowed more complex vocalizations.

Anthropoids include monkeys, apes, and humans; "anthropoid" means humanlike. Nearly all anthropoids are diurnal (active during the day) and have good eyesight, including color vision.

New World monkeys (Figure 15.39C) climb through forests of Central and South America in search of fruits. They have a flat face and a nose with widely separated nostrils. A long tail helps them maintain balance. In many species, the tail is prehensile, meaning it grasps things.

Old World monkeys live in Africa, the Middle East, and Asia. They tend to be larger than New World monkeys, with a longer nose and closely set nostrils. Some are tree-climbing forest dwellers. Others, such as baboons (Figure 15.39D), spend most of their time on the ground in grasslands and deserts. Not all Old World monkeys have a tail, but in those that do the tail is typically short and never prehensile.

Tailless nonhuman primates are commonly called apes. About 15 species of small apes called gibbons inhabit Southeast Asian forests. Gibbons are sometimes referred to as "lesser apes," in comparison with the larger apes, or "great apes." The forest-dwelling orangutan of Sumatra and Borneo is the only surviving Asian great ape. All other great apes (gorillas, chimpanzees, and bonobos) are native to central Africa and spend most of their time on the ground. When walking, African apes lean forward and support their weight on their knuckles (Figure 15.39E). Gorillas, the largest living primates, live in forests and feed mainly on leaves. Chimpanzees (Figure 15.39F) and the bonobos are our closest living relatives. The chimpanzee/bonobo lineage and the lineage leading to humans diverged 6 to 13 million years ago.

Australopiths Humans and their closest extinct relatives are now grouped together as **hominins** (previously called hominids). The defining trait of hominins is **bipedalism**—habitual upright walking. Thus, researchers interested in human origins look for fossil evidence that a species walked upright. Some fossils of primates that may have been bipedal date back as long ago as 7 million years ago. However, the fossil record of these species is patchy and little is known about them.

The best-known early hominins are the **australopiths**, who lived in Africa from about 4 million to 1.2 million years ago. One australopith genus, *Australopithecus*, includes several species considered likely human ancestors. The fossil history of australopiths shows trends toward smaller teeth and improvements in the ability to walk upright, but little increase in brain size.

Fossil footprints in Tanzania document the passage of a bipedal species 3.6 million years ago (Figure 15.40A). The prints reveal that the walkers' feet had a pronounced arch and a big toe in line with the other toes—both adaptations to upright walking. The prints were probably made by *Australopithecus afarensis*, an australopith that lived in Tanzania and other parts of eastern Africa from 3.9 to 3 million years ago. An *A. afarensis* skeleton known as Lucy (Figure 15.40B) is the best-known representative of this species. *A. afarensis* is considered a possible human ancestor.

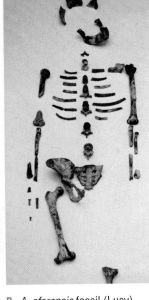

A. 3.6-million-year-old foot-prints of a bipedal species.

B. *A. afarensis* fossil (Lucy) from 3.5 million years ago.

Figure 15.40 Evidence of early hominins.

(A) Louise M. Robbins; (B) Dr. John D. Cunningham/Visuals Unlimited, Inc.

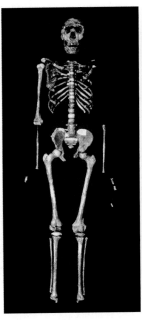

A. Fossil of a male who lived in Kenya 1.5 million years ago.

B. Reconstruction based on a 700,000-year-old skull discovered in China.

Figure 15.41 *Homo erectus*, the first hominin species to leave Africa.

(A) Science VU/NMK/Visuals Unlimited, Inc.; (B) Philippe Plailly & Atelier Daynes/Science Source.

Early Humans **Humans** are members of the genus *Homo*. Fossils of the oldest species in this genus, **Homo habilis**, date from 2.3 million years to 1.4 million years before the present. The species was named based on a fossil discovered in Kenya in 1964. The name *Homo habilis* means "handy man"; it refers to the stone tools found near the fossil. At the time of the fossil's discovery, the ability to make stone tools was considered a distinctly human trait. We now know that hominins used sharp stone edges to scrape meat from bones as early as 3.4 million years ago. Given the age of the bones, these early tool users were most likely australopiths. Classification of *H. habilis* continues to inspire debate. The species is australopith-like in its body proportions and brain size, so some scientists think it should be reclassified as a species of *Australopithecus*. Other scientists argue for the species' inclusion in the genus *Homo*, pointing out that the hands and arms are similar to those of modern humans.

Homo erectus, the first hominin with body proportions like our own, arose in Africa by about 2 million years ago. The species name means "upright man," and like us, *H. erectus* stood on legs that were longer than its arms. The most complete *H. erectus* fossil found thus far is a skeleton of a young male who lived about 1.5 million years ago in Kenya (Figure 15.41A). This individual, informally known as Turkana boy, was under age 14 when he died. However, he was already was 5 feet 2 inches (1.60 meters) tall and his brain was twice the size of a chimpanzee's. Foot-prints from the same region and time suggest that *H. erectus* had a gait like that of modern humans. As far as we know, *H. erectus* was the first hominin to venture out of Africa. By 1.75 million years ago, a population had become established in what is now the Eurasian country of Georgia. *H. erectus* went on to colonize Indonesia by 1.6 million years ago, and China by 1.15 million years ago (Figure 15.41B). At the same time, African populations continued to thrive.

anthropoids Monkeys, apes, and humans.
australopiths Informal name for a lineage of chimpanzee-sized hominins that lived in Africa between 4 million and 1.2 million years ago.
bipedalism Habitually walking upright.
hominin Humans and extinct humanlike species.
Homo erectus Human species that dispersed out of Africa.
Homo habilis Earliest named human species.
humans Members of the genus *Homo*.

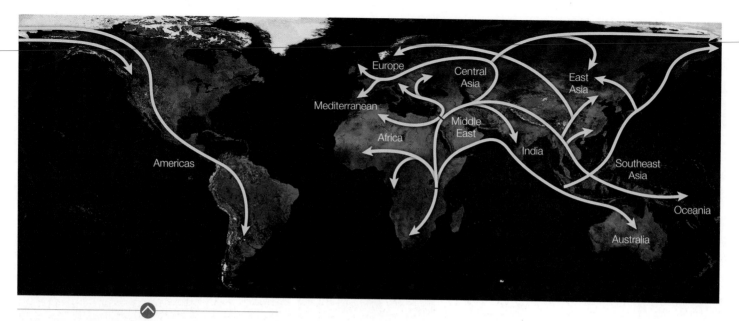

Figure 15.42 Human dispersal routes.
Exposure of a temporary land bridge from Siberia to North America allowed entry into the Americas.

Photo, NASA; data, National Geographic.

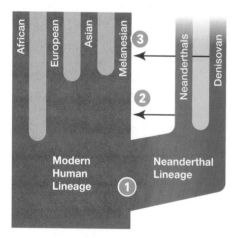

1. Perhaps as long as 700,000 years ago, a divergence separated the ancestors of modern humans from the lineage leading to Neanderthals and Denisovans.

2. About 60,000 years ago, Neanderthals living in the Middle East mated with modern humans who were venturing out of Africa.

3. About 40,000 years ago, ancestors of the modern humans who would later populate New Guinea and Australia mated with Denisovans.

Figure 15.43 One current model of human evolution.
The model is based on observed genetic similarities between some populations of modern humans and two fossil members of the genus *Homo*: Neanderthals and Denisovans.

Homo erectus is considered the most likely ancestor of our own species, and of Neanderthals, and some other more recently discovered hominins.

Homo Sapiens **Homo sapiens** is the species of anatomically modern humans. Compared to *H. erectus*, our species has a higher, rounder skull, a larger brain, and a flatter face with smaller jaw and teeth. We are also the only hominins with a chin, a protruding area of thickened bone in the middle of the lower jawbone.

To date, the oldest *H. sapiens* fossils discovered are two partial male skulls from Ethiopia, in East Africa. Known as Omo I and Omo II, the fossil individuals date to 195,000 years ago. Fossil remains of two adult males and a child who lived 160,000 years ago were also unearthed in this region. Other fossils show that *H. sapiens* reached South Africa by 115,000 years ago.

Genetic studies also support an African origin for our species. Modern Africans are more genetically diverse than people of any other region. This diversity indicates that African populations have existed for a very long time—long enough to accumulate a very large number of random mutations compared with other populations. Furthermore, most genetic variations seen in people native to regions outside of Africa are subsets of the variation found in Africa. This pattern is evidence that founder effects occurred after some people left their homeland to colonize the rest of the world. Over many generations, pioneers traveled along the coasts of Africa, then Eurasia and Australia. We know from fossilized feces deposited in an Oregon cave that they reached North America by about 14,000 years ago. Today, the distribution of specific mutations among different ethnic groups provides evidence of routes taken by the ancient travelers. By mapping the frequency of maternal and paternal genetic markers in modern peoples, geneticists have created a picture of how *Homo sapiens* dispersed across the world (Figure 15.42).

Neanderthals and Denisovans As early modern humans left Africa and expanded their range to Eurasia, they met and interbred with at least two other groups of hominins who had already become established there (Figure 15.43). These hominins had shared an ancestor with *H. sapiens* between 500,000 and 700,000 years ago 1. About 60,000 years ago, as some modern humans ventured into the Middle East, they met up with and interbred with Neanderthals (**Homo**

neanderthalensis) **2**. As a result of this interbreeding, modern human populations native to regions outside of Africa have distinctive Neanderthal alleles that are rare among populations native to Africa. A second interbreeding event took place about 40,000 years ago somewhere in Asia. Here, humans interbred with Denisovans, a recently discovered, genetically distinct group of hominins **3**. As a result, we find unique Denisovan alleles in some modern Asians. Melanesians, such as people of Papua New Guinea, have an especially high percentage of Denisovan DNA.

Neanderthals have long been considered our closest extinct relatives. They first appeared about 230,000 years ago, and they left an extensive fossil record in the Middle East, Europe, and central Asia. The common conception of Neanderthals as brutish cavemen with poor posture arose from an early reconstruction based on a fossil of an individual deformed by arthritis. More recent reconstructions reveal Neanderthals were shorter than modern humans, but stood upright. They lived in regions where winters are cold, and a short, stocky body minimized the surface area available for heat loss. Modern Arctic peoples have a similar body shape. A recent reconstruction of a Neanderthal male, based on material from multiple fossils, stands about 164 centimeters (5′4″) tall (Figure 15.44). The Neanderthal braincase was longer and lower than that of modern humans, but their brain was as big as or bigger than ours. Their face had pronounced brow ridges, a large nose with widely spaced nostrils, and no chin.

Fossils of Neanderthal individuals who survived despite disabilities such as the loss of a limb testify to a compassionate social structure. Some simple burials suggest possible symbolic thought. Several lines of evidence suggest Neanderthals were able to speak. The most recent evidence of Neanderthals dates to about 40,000 years ago. Neanderthals may have been outcompeted by newly arrived *H. sapiens* or killed by diseases these migrants brought with them. Climate changes and volcanic eruptions may have contributed to their demise by altering the abundance of the animals they hunted. Meat made up the bulk of the Neanderthal diet.

Denisovans were first identified when researchers sequenced DNA extracted from a fossil pinky (small finger) bone discovered in Denisova Cave in Siberia. The researchers expected to find typical Neanderthal DNA. Instead, the finger's DNA contained unique sequences never seen before. As a result, the fossil was assigned to a group—the Denisovans—named for the site of its discovery. At this writing, Denisovans are not considered a distinct species, but rather a subgroup of Neanderthals.

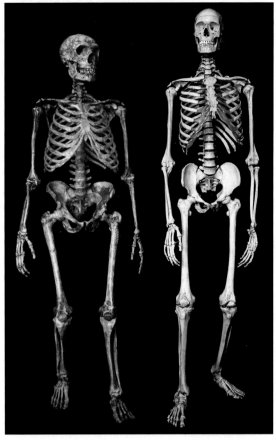

Homo neanderthalensis   *Homo sapiens*

Figure 15.44 Skeletal comparison of Neanderthal and modern human males.
The Neanderthal skeleton is a reconstruction based on multiple fossils. Each color denotes a different fossil.

Courtesy of @ Blaine Maley, Washington University, St. Louis.

Take-Home Message 15.7

How did humans evolve?

- Humans, like other primates, have grasping hands and shoulders adapted to climbing. Unlike most primates, they walk upright.
- Australopiths are a group of upright-walking species that probably include human ancestors. They evolved in Africa by about 4 million years ago.
- The first humans, or members of the genus *Homo*, also arose in Africa. *Homo erectus* migrated out of Africa and into Europe and Asia.
- Modern humans (*Homo sapiens*) and the extinct Neanderthals are descendants of *Homo erectus*. By the currently favored model, modern humans evolved in Africa, then migrated worldwide.

Homo neanderthalensis Neanderthals. Closest extinct relatives of modern humans; had large brain, stocky body.

Homo sapiens Modern humans; evolved in Africa, then expanded their range worldwide.

Summary

Section 15.1 **Invertebrates** (animals without a backbone) are the most diverse animal group. They far outnumber **vertebrates**. Some compounds that invertebrates produce can be used as medicines.

Section 15.2 **Animals** are multicelled heterotrophs that ingest food and move about. The **colonial theory of animal origins** states that they evolved from a colonial protist. Animals with **tissues** have **radial symmetry** or **bilateral symmetry**. Most animals are either **deuterostomes** or **protostomes**, two groups that differ in their development. The animal gut is a saclike **gastrovascular cavity** or a tubelike **complete digestive tract**. Most bilateral animals have a **coelom** around the gut. Some have an **open circulatory system** or a more efficient **closed circulatory system**.

Section 15.3 **Sponges** are sessile filter-feeders with no body symmetry or tissues. Each individual is a **hermaphrodite**; it makes eggs and sperm. The sponge **larva** is free-swimming.

The radially symmetrical **cnidarians** have two body types: **medusa** (as in jellies) and **polyp** (as in sea anemones). Stinging cells help cnidarians capture prey.

Flatworms have simple organ systems, a gastrovascular cavity, and no coelom. Planarians are free-living flatworms, whereas the tapeworms and flukes are parasites.

Annelids are segmented worms that have a coelom, a complete digestive system, and a closed circulatory system. Earthworms are oligochaete annelids. The annelid lineage also includes leeches and polychaetes.

Mollusks have a **mantle**, a skirtlike extension of the body. In many groups, the mantle secretes a shell. **Gastropods** (such as snails) and **cephalopods** (such as octopuses) have a head with a **radula** used in feeding. **Bivalves** such as clams are filter-feeders.

Roundworms (nematodes) are unsegmented worms with a complete gut and an incompletely lined coelom. They may be free-living or parasitic.

Arthropods have a jointed **exoskeleton**. Horseshoe crabs are marine bottom-feeders. **Arachnids** include spiders, mites, ticks, and scorpions. Most **crustaceans**, such as lobsters, krill, and barnacles, are aquatic. Centipedes and millipedes have an elongated body and live on land. **Insects** have paired **antennae** and **compound eyes**, and most undergo **metamorphosis**. They are the most diverse animals and the only winged invertebrates.

Echinoderms such as sea stars have an **endoskeleton** of spines, spicules, or plates of calcium carbonate. A **water–vascular system** with tube feet allows adults to move about. Adult echinoderms are radial, but larva are bilateral, indicating the group has a bilateral ancestor.

Section 15.4 Four embryonic traits define the **chordates**: a **notochord**, a dorsal hollow nerve cord (which becomes a brain and spinal cord), a pharynx with gill slits, and a tail extending past the anus. Depending on the group, some or all of the features persist in adults.

Lancelets and **tunicates** are invertebrate chordates. Most chordates are vertebrates; they have a **vertebral column** of cartilage or bone. Jaws, **lungs**, limbs, and waterproof eggs are innovations that made the adaptive radiation of **vertebrates** possible. **Tetrapods** have four limbs. All vertebrates have a complete digestive system, closed circulatory system, and **kidneys**.

Section 15.5 The earliest fishes were **jawless fishes**. **Cartilaginous fishes** and **bony fishes** have jaws, **scales**, and paired fins. Jaws evolved from gill supports. The two bony fish lineages are the highly diverse **ray-finned fishes**, which includes most familiar fishes, and the **lobe-finned fishes**. **Amphibians** evolved from a lobe-finned fish and require water to reproduce. Existing amphibian groups include salamanders, frogs, and toads. Fertilization is external. Eggs and sperm exit the body through a **cloaca** that also expels digestive and urinary wastes.

Section 15.6 **Amniotes** were the first vertebrates that did not need external water for reproduction. Their skin and kidneys conserve water, and they produce **eggs** that have distinctive membranes. Mammals are one amniote lineage. **Reptiles**, including **birds**, are another. Most reptiles are **ectotherms**, but birds and mammals are **endotherms**; they maintain their temperature by metabolic production of heat). There are three lineages of **mammals**: egg-laying **monotremes**, pouched **marsupials**, and placental mammals. **Placental mammals** are the most diverse group.

Section 15.7 **Primates** are a lineage adapted to climbing and all have hands capable of grasping. Lemurs have a moist, doglike snout, but tarsiers and **anthropoids** (monkeys, apes, humans) have a dry nose and movable upper lip. Our closest living relatives are the chimpanzees and bonobos. **Bipedalism** defines the **hominins**, which include **humans** and their extinct close relatives. **Australopiths** were early hominins and some are considered likely human ancestors. The first named members of our genus, *Homo habilis*, resembled australopiths. *Homo erectus* had a larger brain and some populations became established outside of Africa. Modern humans (*Homo sapiens*) arose in Africa. As modern humans dispersed, some interbred with typical Neanderthals (*Homo neanderthalensis*) or with Denisovans. As a result, some human genomes contain alleles that can be traced to these groups.

Self-Quiz

1. True or false? Animal cells do not have walls.

2. The colonial theory of animal origins states that _____ .
   a. animals are more closely related to plants than to fungi
   b. animals evolved from a colonial protist
   c. most animals live in colonies
   d. all animals have a backbone

3. A body cavity that is fully lined with tissue derived from mesoderm is a _____ .
   a. pseudocoelom   c. coelom
   b. kidney   d. gastrovascular cavity

4. Most animal bodies have _____ symmetry.
   a. radial   b. bilateral   c. no

5. Earthworms are most closely related to _____ .
   a. insects   c. leeches
   b. tapeworms   d. roundworms

6. The _____ include the only winged invertebrates.
   a. flatworms   c. arthropods
   b. annelids   d. cnidarians

7. List the four distinguishing chordate traits. Which of these traits are retained by an adult tunicate?

8. Jaws evolved from _____ of jawless fishes.
   a. fins   b. gill supports   c. the backbone

9. All vertebrates are _____ but only some are _____ .
   a. tetrapods; mammals   c. amniotes; hominins
   b. chordates; amniotes   d. bipedal; australopiths

10. Amniote adaptations to land include _____ .
    a. waterproof skin   d. specialized eggs
    b. internal fertilization   e. a and c
    c. highly efficient kidneys f. all of the above

11. Birds and placental mammals _____ .
    a. are endotherms   c. have mammary glands
    b. lay eggs   d. have an open circulatory system

12. True or false? Bipedalism is the defining trait of primates.

13. *Homo erectus* is a likely ancestor of *H. sapiens* and _____ .
    a. *Homo habilis*   c. the great apes
    b. australopiths   d. Neanderthals

14. Match the organisms with the appropriate description.
    _____ sponges   a. most diverse vertebrates
    _____ cnidarians   b. no true tissues, no organs
    _____ flatworms   c. jointed exoskeleton
    _____ roundworms   d. mantle over body mass
    _____ annelids   e. segmented worms
    _____ arthropods   f. tube feet, spiny skin
    _____ mollusks   g. have specialized stinging cells
    _____ echinoderms   h. lay amniote eggs
    _____ amphibians   i. feed young secreted milk
    _____ fishes   j. unsegmented, molting worms
    _____ birds   k. first terrestrial tetrapods
    _____ mammals   l. tailless primates
    _____ apes   m. saclike gut, no coelom

15. Arrange the events in order, from most ancient to most recent.
    _____ 1   a. Cambrian explosion of diversity
    _____ 2   b. Origin of animals
    _____ 3   c. Tetrapods move onto the land
    _____ 4   d. Extinction of dinosaurs
    _____ 5   e. *Homo erectus* leaves Africa
    _____ 6   f. First jawed vertebrates evolve

Critical Thinking

1. In the summer of 2000, only 10 percent of the lobster population in Long Island Sound survived after a massive die-off. Many lobstermen in New York and Connecticut lost small businesses that their families had owned for generations. Some believe the die-off was the result of increased spraying of pesticides to control mosquitoes that carry West Nile virus. Fisherman saw no similar decrease in their catch. Explain why a chemical substance that targets mosquitoes might also harm lobsters but not fish.

2. Why is it more difficult to determine the sex of a newly hatched canary than a newborn puppy?

3. As human activities put more and more carbon dioxide (CO_2) into the air, the ocean takes up more CO_2 and becomes increasingly acidic. Increased ocean acidity makes it more difficult for marine animals to produce their calcium-rich hard parts. List three types of invertebrates that are likely to be adversely affected by an increase in ocean acidity.

4. One subgroup of ray-finned fishes, the teleosts, is exceptionally diverse. It includes about half of all vertebrate species. Early in their evolution, teleosts underwent a duplication of their entire genome. By one hypothesis, this event opened the way to their diversification. Explain why such a duplication might make the evolution of new traits more likely.

16 POPULATION ECOLOGY

Application ❯ 16.1 A Honkin' Mess

Visit a grassy park or a golf course and you may need to watch where you step. Wide expanses of grass with a nearby body of water attract large numbers of Canada geese, or *Branta canadensis* (Figure 16.1). These plant-eating birds produce slimy, green feces that can soil shoes, stain clothes, and discourage picnics. Goose feces also wash into ponds and waterways. The nutrients they add encourage bacteria and algae to grow, clouding the water and making swimming unappealing and possibly dangerous. Goose feces sometimes contain microbes that can sicken humans.

Geese can also pose a risk to air traffic. In January of 2009, both engines of a US Airways flight failed shortly after the plane took off from a New York City airport. Fortunately, the pilot was able to land the plane in the nearby Hudson River (Figure 16.1B), where boats unloaded all 155 people aboard. Afterwards, investigators from the Federal Aviation Agency found bits of feather, bone, and muscle in the plane's wing flaps and engines. Unique sequences in the DNA identified the tissue in both engines as Canada goose.

The number of Canada geese in the United States has increased dramatically. For example, Michigan had about 9,000 in 1970, and has 300,000 today. Controlling their number is challenging because several different Canada goose populations spend time in the United States. A **population** is a group of organisms of the same species who live in a specific location and breed with one another more often than they breed with members of other populations. In the past, nearly all Canada geese seen in the United States were migratory. They nested in northern Canada, flew to the United States to spend the winter, then returned to Canada. Most Canada geese still migrate, but some populations are permanent residents of the United States. The geese breed where they grew up, and the nonmigratory birds are generally descendants of geese deliberately introduced to a park or hunting preserve. During the winter, migratory birds often mingle with nonmigratory ones. For example, a bird that breeds in Canada and flies to Virginia for the winter finds itself alongside geese that have never left Virginia.

Life is more difficult for migratory geese than for nonmigratory ones. Flying to and from a northern breeding area takes lots of energy and is dangerous. A bird that does not migrate can devote more energy to producing young than a migratory one can. If the nonmigrant lives in a suburban or urban area, it also benefits from an unnatural abundance of food (grass) and an equally unnatural lack of predators. Not surprisingly, the greatest increases in Canada geese have been among nonmigratory birds living where humans are plentiful.

Migratory birds are protected under federal law and by international treaties. However, in 2006, increasing complaints about Canada geese led the U.S. Fish and Wildlife Service to exempt this species from some protections. The agency encouraged wildlife managers to look for ways to reduce nonmigratory Canada goose populations, without unduly harming migratory birds. To do so, these biologists need to know about the traits that characterize different goose populations, as well as how these populations interact with one another, and with other species.

These sorts of questions are the focus of the science of **ecology**, the study of interactions among organisms, and between organisms and their physical environment. Ecology is not the same as environmentalism, which is advocacy for protection of the environment. However, environmentalists often cite the results of ecological studies when drawing attention to environmental concerns.

A. Geese overrun a California park.

B. An airliner downed by a collision with geese.

Figure 16.1 Problems caused by Canada geese.

(A) Courtesy of Joel Peter; (B) AP Images/Steven Day.

A. Clumped distribution of hippopotamuses.　　**B.** Near-uniform distribution of nesting seabirds.　　**C.** Random distribution of dandelions.

Figure 16.2 Population distributions.

(A) Michael Poliza/National Geographic Creative; (B) © Eric and David Hosking/Corbis; (C) Elizabeth A. Sellers/life.nbii.gov.

16.2 Characteristics of Populations

REMEMBER: Sampling error can lead to misleading results (Section 1.6).

When studying a population, ecologists collect information about its gene pool, reproductive traits, and behavior of its component individuals. They also look at **demographics**—vital statistics that describe the population.

Demographic Traits **Population size** refers to the number of individuals of a species in a population. **Population density** is the number of individuals in some specified area or volume of a habitat, such as the number of frogs per acre of rain forest or the number of amoebas per liter of pond water.

　Population distribution describes where individuals are relative to one another. Most populations have a clumped distribution, meaning members of the population are closer to one another than would be predicted by chance alone. Often, a patchy distribution of an essential resource draws individuals together, as when hippopotamuses gather in muddy river shallows (Figure 16.2A). Similarly, a cool, damp, north-facing slope may be covered with ferns, whereas an adjacent drier, south-facing slope has none. Limited dispersal ability increases the likelihood of a clumped distribution: The nut does not fall far from the tree. Asexual reproduction also results in clumping. It produces colonies of coral and vast stands of some trees. In addition, some animals are social; they benefit by living in a group.

　Competition for resources can produce a near-uniform distribution, with individuals more evenly spaced than would be expected by chance. Creosote bushes in deserts of the American Southwest grow in this pattern. Competition for water among the plants' root systems prevents them from growing near one another. Seabirds in breeding colonies often show a near-uniform distribution too. Each bird repels others that get within reach of its beak as it sits on its nest (Figure 16.2B).

　A random population distribution is rare in nature. Random distribution arises when resources are uniformly available, and proximity to others neither benefits nor harms individuals. For example, when wind-dispersed dandelion seeds land on the uniform environment of a suburban lawn, dandelions grow in a random pattern (Figure 16.2C).

　The scale of a study area and timing of the study can influence the observed pattern of distribution. For example, although seabirds may be spaced almost

demographics Statistics that describe a population.

ecology The study of interactions among organisms, and between organisms and their environment.

population A group of organisms of the same species who live in a specific location and breed with one another more often than they breed with members of other populations.

population density Number of members of a population in a given area.

population distribution The way in which members of a population are dispersed in their environment.

population size Number of individuals in a population.

age structure Of a population, distribution of individuals among various age groups.

exponential model of population growth Model for population growth when resources are unlimited. The per capita growth rate remains constant as population size increases.

per capita growth rate The number of individuals added during some interval divided by the initial population size.

uniformly at a nesting site, nesting sites are clustered along a shoreline. Also, these birds crowd together in the breeding season, but disperse at other times.

Age structure refers to the distribution of individuals among various age categories. A population's age structure affects its capacity for growth. A population with a large proportion of young individuals who have not begun to breed has a greater potential for growth than one composed mainly of older individuals.

Collecting Demographic Data Scientists often cannot directly count all the members of a population. Instead, they sample a population, then use data from that sample to estimate characteristics of the population as a whole. Plot sampling estimates the total number of individuals in an area on the basis of direct counts in part of the area. For example, to determine the number of daisies in a prairie or clams in a mudflat, ecologists begin by measuring the number of individuals in several 1-meter-by-1-meter-square plots. To calculate the total population size, researchers multiply the average number of individuals in the sample plots by the number of plots in the area the population inhabits. Size estimates from plot sampling are most accurate for organisms that do not move about and live in an area where conditions are uniform. Mark–recapture sampling is used to study mobile animals. Researchers capture animals, mark them, then release them. Sometime later, animals are captured again. The proportion of marked animals in the second sample is then taken to be representative of the proportion marked in the whole population:

$$\frac{\text{marked individuals in sampling at time 2}}{\text{total captured in sampling 2}} = \frac{\text{marked individuals in sampling at time 1}}{\text{total population size}}$$

Digging Into Data

Monitoring Iguana Populations

In 1989, Martin Wikelski started a long-term study of marine iguana populations in the Galápagos Islands. He marked the iguanas on two of the islands—Genovesa and Santa Fe—and collected data on how their body size, survival, and reproductive rates varied over time. The iguanas eat algae and have no predators, so deaths are usually the result of food shortages, disease, or old age. His studies showed that the iguana populations decline during El Niño events, when water surrounding the islands heats up.

In January 2001, an oil tanker ran aground and leaked a small amount of oil into the waters near Santa Fe. Figure 16.3 shows the number of marked iguanas that Wikelski and his team counted in their census of study populations just before the spill and about a year later.

1. Which island had more marked iguanas at the time of the first census?
2. How much did the population size on each island change between the first and second census?
3. Wikelski concluded that changes on Santa Fe were the result of the oil spill, rather than sea temperature or other climate factors common to both islands. How would the census numbers be different from those he observed if an adverse event had affected both islands?

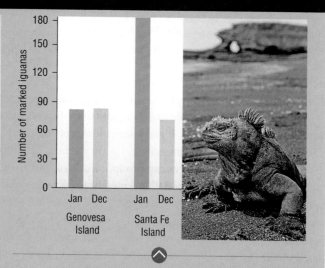

Figure 16.3 Shifting numbers of marked marine iguanas on two Galápagos islands.
An oil spill occurred near Santa Fe just before the January 2001 census (blue bars). A second census was carried out in December 2001 (green bars).

Photo, Bruce Coleman/Photoshot.

Suppose, for example, that scientists capture, mark, and release 100 deer. Later, they return and once again capture 100 deer. Of these, 50 are marked. The proportion of marked deer in the second sample (50 percent) indicates that half of the deer in the population have been marked. Thus the 100 deer initially marked must have be members of a population of 200 deer.

Information about the traits of individuals in a sample plot or capture group can be used to infer properties of the population as a whole. For example, if half of the deer recaptured in a mark–recapture study are of reproductive age, half of the population is assumed to share this trait.

Any study that draws conclusions based on only a sample of a population is susceptible to sampling error (Section 1.6). The larger the sample, the more likely that conclusions drawn from that sample will be accurate.

Neil Burton/Shutterstock.com.

Starting Size of Population		Net Monthly Increase	New Size of Population
2,000	× r =	800	2,800
2,800	× r =	1,120	3,920
3,920	× r =	1,568	5,488
5,488	× r =	2,195	7,683
7,683	× r =	3,073	10,756
10,756	× r =	4,302	15,058
15,058	× r =	6,023	21,081
21,081	× r =	8,432	29,513
29,513	× r =	11,805	41,318
41,318	× r =	16,527	57,845
57,845	× r =	23,138	80,983
80,983	× r =	32,393	113,376
113,376	× r =	45,350	158,726
158,726	× r =	63,490	222,216
222,216	× r =	88,887	311,103
311,103	× r =	124,441	435,544
435,544	× r =	174,218	609,762
609,762	× r =	243,905	853,667
853,667	× r =	341,467	1,195,134

A. Increases in size over time. Note that the net increase becomes larger with each generation.

Take-Home Message 16.2

How do scientists describe populations?

- Members of a population live in the same area and breed with one another.
- Populations can be described in terms of their size, density, and how their members are distributed through their environment.
- Population studies typically utilize sampling methods. Such methods run the risk of sampling error, which can be minimized by using a large sample size.

16.3 Population Growth

Exponential Growth A population grows when its birth rate exceeds its death rate. Ecologists measure births and deaths per individual, or per capita. For example, if the birth rate in a population of 2,000 mice is 1,000 young per month, then the birth rate is 1,000/2,000, or 0.5 per mouse per month. Subtract the per capita death rate from the per capita birth rate and you have the **per capita growth rate**. If the death rate in the population of 2,000 mice is 200 per month (0.1 per mouse per month), then the per capita growth rate is $0.5 - 0.1 = 0.4$ per mouse per month.

The **exponential model of population growth** describes how a population's size changes over time if its per capita growth rate is constant and its resources are unlimited. Under these theoretical conditions, the population growth in any interval (G) can be calculated as follows:

$$N \quad × \quad r \quad = \quad G$$

N	r	G
(number of individuals)	(per capita growth rate)	(population growth per unit time)

Suppose we apply this to our population of 2,000 mice with their per capita growth rate of 0.4 per month. In the first month, the population grows by 2,000 mice × 0.4. This brings the size to 2,800. In the next month, 2,800 × 0.4, or 1,120 mice, are added, and so on (Figure 16.4A). At this growth rate, the number of mice would rise from 2,000 to more than 1 million in under two years! Plotting the size of this population against time produces a J-shaped, or "hockey stick," curve, characteristic of exponential growth (Figure 16.4B).

Exponential population growth is analogous to compounding of interest on a bank account that pays a fixed rate of return. Although the interest rate does not

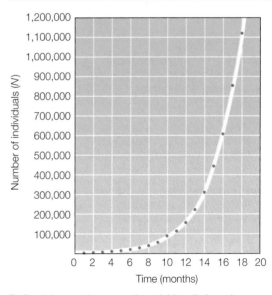

B. Graphing numbers over time yields a J-shaped curve.

Figure 16.4 Exponential growth in population of mice with a per capita growth rate (r) of 0.4 per mouse per month and an initial population size of 2,000.

No population can grow forever. Limiting factors put the brakes on population growth.

change, the amount of interest paid continually increases. Each year, the annual interest paid into the account adds to the size of the balance, and the following year's interest payment is based on that higher balance.

The exponential model of population growth assumes that resources are unlimited, so it cannot accurately predict the long-term growth of real populations. However, it does provide insight into the expected short-term growth of a population with plentiful resources. For example, when a few individuals of a species colonize a new habitat, the resulting population often grows exponentially for a period.

Carrying Capacity and Logistic Growth In the real world, the resources that organisms need to survive and reproduce are always limited. The **logistic model of population growth** addresses this limitation. With this model, the rate of population growth does not remain constant, but rather declines as population density increases. When there are few individuals relative to the amount of resources, the population grows exponentially (Figure 16.5 ❶).

As the number of individuals rises, **density-dependent limiting factors** such as competition begin to put the brakes on growth. As individuals become more and more crowded, they must compete with one another for food, hiding places, nesting sites, and other essential resources. Parasitism and disease also increase with population density, and both hinder population growth. As a result of density-dependent limiting factors, population growth begins to slow ❷.

Population growth continues until it eventually levels off at the environment's carrying capacity ❸. **Carrying capacity** is the maximum number of individuals of a species that a particular environment can sustain indefinitely. The carrying capacity for a species is not constant; rather, it depends on physical and biological factors that can change over time. For example, a prolonged drought can lower the carrying capacity for a plant species and for any animals that depend on that plant. The population size of one species can also affect the carrying capacity of another species with similar resource needs. For example, a grassland can support only so many grazers, so the presence of more than one grass-eating species lowers the carrying

Figure 16.5 One example of logistic growth: what happens when a few deer are introduced to a new habitat with finite resources.

❶ When the population is small, individuals have access to all the resources they require and the population grows exponentially.

❷ As population size grows, the growth rate begins to slow as density-dependent limiting factors begin to have an effect.

❸ Eventually, population size levels off. Population size plotted against time produces an S-shaped curve.

capacity for all of them. Human activities also affect carrying capacity. For example, human harvest of horseshoe crabs has decreased the carrying capacity for red knot sandpipers, a type of migratory bird. Horseshoe crab eggs are the sandpipers' main food during their long-distance migration.

Density-Independent Factors   The logistic model of population growth describes what happens when density-dependent limiting factors influence population growth. However, other factors can affect population size, including harsh weather such as hurricanes, and natural disasters such as tsunamis or landslides. Such **density-independent limiting factors** increase the death rate in crowded and uncrowded populations alike; they do not arise as an effect of crowding.

In nature, density-dependent and density-independent factors often interact to determine the fate of a population. Consider what happened after the 1944 introduction of 29 reindeer to St. Matthew Island, an uninhabited island off the coast of Alaska. When biologist David Klein first visited the island in 1957, he found 1,350 well-fed reindeer munching on lichens (Figure 16.6). When Klein returned in 1963, he counted 6,000 reindeer. The population had soared far above the island's carrying capacity. Although a population can temporarily exceed the carrying capacity of its environment, the high density cannot be sustained. Klein observed that density-dependent negative effects were already apparent. For example, the average body size of the reindeer had decreased.

When Klein returned in 1966, only 42 reindeer survived. The single male had abnormal antlers, and was thus unlikely to breed. There were no fawns. Klein figured out that thousands of reindeer had starved to death during the winter of 1963–1964. That winter was unusually harsh, with low temperatures, high winds, and 140 inches of snow. Most reindeer, already in poor condition as a result of increased competition, starved when deep snow covered their food. A population decline had been expected—a population that exceeds its carrying capacity usually shrinks and falls below that capacity—but bad weather magnified the extent of the crash. By the 1980s, there were no reindeer on the island.

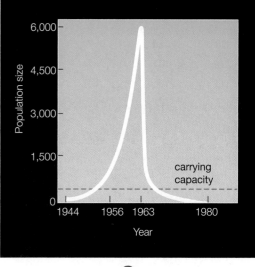

Figure 16.6 Changes in the size of a reindeer population. The reindeer were introduced to an island off the coast of Alaska in 1944.

Top, © Jacques Langevin/Corbis Sygma.

Take-Home Message 16.3

What factors affect population growth?

- The exponential model of population growth describes the growth of a population with unlimited resources. In this idealized circumstance, per capita growth rate remains constant, but the population grows faster and faster.
- The logistic model of population growth describes the growth of a population affected by density-dependent limiting factors. Such a population grows exponentially at first, then its growth rate declines as a result of competition for resources, infectious disease, and other negative effects of crowding.
- A population undergoing logistic growth levels off at the environment's carrying capacity for that species.
- Density-independent factors such as harsh weather are not addressed by models for population growth, but they too affect natural populations.

carrying capacity Maximum number of individuals of a species that a specific environment can sustain.

density-dependent limiting factor Factor whose negative effect on growth is felt most in dense populations; for example, infectious disease or competition for food.

density-independent limiting factor Factor that limits growth in populations regardless of their density; for example a natural disaster or harsh weather.

logistic model of population growth Model for growth of a population limited by density-dependent factors; numbers increase exponentially at first, then the growth rate slows and population size levels off at carrying capacity.

Table 16.1 Annual Plant* Life Table

Age Interval (days)	Survivorship (number surviving at start of interval)	Number Dying During Interval	Death Rate (number dying/number surviving)
0–63	996	328	0.329
63–124	668	373	0.558
124–184	295	105	0.356
184–215	190	14	0.074
215–264	176	4	0.023
264–278	172	5	0.029
278–292	167	8	0.048
292–306	159	5	0.031
306–320	154	7	0.045
320–334	147	42	0.286
334–348	105	83	0.790
348–362	22	22	1.000
362–	0	0	0
		996	

* Phlox drummondii; data from W. J. Leverich and D. A. Levin, 1979.

16.4 Life History Patterns

REMEMBER: Natural selection shapes the traits of a population (Section 12.3).

Biotic Potential The exponential growth rate for a population under ideal conditions is its **biotic potential**. This is the theoretical value that would hold if shelter, food, and other essential resources were unlimited and there were no predators or pathogens. Populations seldom reach their biotic potential because of limiting factors. Biotic potential is determined by **life history traits**, which are a set of heritable traits such as rate of development, age at first reproduction, number of breeding events, and life span. In this section, we look at how such life history traits vary and the evolutionary basis for this variation.

Describing Life Histories Each species has a characteristic life span, but only a few individuals survive to the maximum age possible. Death is more likely at some ages than others, as is reproduction. To gather information about age-specific risk of death, researchers focus on a **cohort**, a group of individuals that are all born at about the same time. Members of the cohort are tracked from birth until the last member of the cohort dies. Mortality data from a cohort study can be summarized in a life table. Table 16.1 provides an example of a life table.

A **survivorship curve** plots how many members of a cohort remain alive over time, providing information about age-specific death rates. Ecologists describe three generalized types of survivorship curves. A type I curve is convex (bulges outward), indicating survivorship is high until late in life (Figure 16.7A). This pattern is characteristic of humans and other large mammals that produce one or two young and care for them. A diagonal type II curve indicates that the death rate of the population does not vary much with age (Figure 16.7B). A type II curve is characteristic of lizards, small mammals, and large birds. In these groups, old individuals are as likely to die of disease or predation as young ones. A type III curve is concave (bulges inward), indicating that the death rate for a population is highest early in life (Figure 16.7C). Marine animals that release eggs into water have this type of curve, as do plants that release enormous numbers of tiny seeds.

Figure 16.7 Types of survivorship curves.

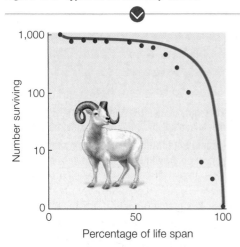

A. Type I curve. Mortality is highest very late in life. Data are for Dall sheep (*Ovis dalli*).

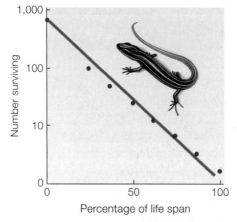

B. Type II curve. Mortality does not vary with age. Data are for a small lizard (*Eumeces fasciatus*).

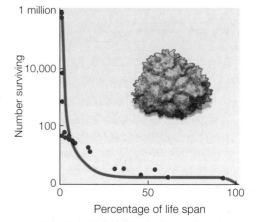

C. Type III curve. Mortality is highest early in life. Data are for a desert shrub (*Cleome droserifolia*).

Figure It Out: Which type of curve best fits the plant mortality data shown in Table 16.1?

Answer: Type III. Mortality is highest early in life.

Opportunistic life history	Equilibrial life history
shorter development	longer development
early reproduction	later reproduction
fewer breeding episodes, many young per episode	more breeding episodes, few young per episode
less parental investment per young	more parental investment per young
higher early mortality, shorter life span	low early mortality, longer life span

A, B

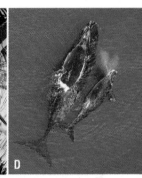

C, D

Figure 16.8 Opportunistic versus equilibrial life histories.
A, B. Dandelions and flies are opportunists.
C, D. Whales and coconut palms are equilibrial species.

Evolution of Life Histories When producing offspring, an individual uses resources that it could otherwise use to grow and maintain itself. Species differ in the way in which they distribute their parental investment among offspring and over the course of a lifetime. Some invest little in each of many offspring, others invest a lot in only a few offspring. Some reproduce once, others many times. In studying this variation, ecologists have come to recognize two general life history strategies (Figure 16.8). Both strategies maximize the number of offspring that will be produced and survive, but each does so under different environmental conditions.

When a species lives where conditions vary in an unpredictable manner, its populations seldom reach their carrying capacity. As a result, there is usually little competition for resources among members of the same species. Such conditions favor an **opportunistic life history**, in which individuals produce as many offspring as possible, as quickly as possible. Opportunistic species (also called *r*-selected species) tend to have a short generation time and small body size. Because parental investment is spread across many offspring, each offspring receives a relatively small share. Opportunistic species usually have a type III survivorship curve, with mortality heaviest early in life. Annual plants such as dandelions (Figure 16.8A) have an opportunistic life history. They mature within weeks, produce many tiny seeds, then die. Flies are opportunistic animals. A female fly can lay hundreds of small eggs (Figure 16.8B) in a temporary food source such as a rotting tomato or a pile of feces.

When a species lives in a stable environment, its populations are often near their carrying capacity for that environment. Under these circumstances, competition for resources can be fierce and an **equilibrial life history**, in which parents produce a few, high-quality offspring, is adaptive. Equilibrial species (also called *K*-selected species) tend to have a large body and a long generation time. Consider a coconut palm, which grows for years before beginning to produce a few coconuts at a time (Figure 16.8C). Large mammals such as whales have this sort of life history too. They take years to reach adult body size and begin reproducing. When mature, a female whale produces only one large calf at a time (Figure 16.8C), and she continues to invest in the calf by nursing it after its birth. In both coconut palms and whales, a mature individual produces young for many years.

Some species such as century plants (a type of agave) and bamboo are large and long-lived, but reproduce only once. Atlantic eels and Pacific salmon are unusual among vertebrates in also having a one-shot reproductive strategy. Such a strategy can evolve when opportunities for reproduction are unlikely to be repeated. In the century plant and bamboo, climate conditions that favor reproduction occur only rarely. In the eels and salmon, physiological changes related to migration between fresh water and salt water make a repeat journey impossible.

biotic potential Maximum possible population growth under optimal conditions.

cohort Group of individuals born during the same interval.

equilibrial life history Life history favored in stable environments; individuals grow large, then invest a lot in each of a few offspring.

life history traits Set of traits related to growth, survival, and reproduction such as life span, age-specific mortality, age at first reproduction, and number of breeding events.

opportunistic life history Life history favored in unpredictable environments; individuals reproduce while young and invest little in each of many offspring.

survivorship curve Graph showing the decline in numbers of a cohort over time.

A. Biologist David Reznick at his study site, a freshwater stream in Trinidad that is home to guppies and their predators.

B. Killifish. It preys on small guppies, thus selecting for individuals that grow quickly to large size before reproducing.

C. Pike cichlid. It preys on big guppies, thus selecting for individuals that reproduce early, while still young and small.

Figure 16.9 Effects of predation on guppy life history.

(A) Helen Rodd, inset David Reznick/University of California–Riverside, computer enhanced by Lisa Starr; (B,C) Hippocampus Bildarchiv.

Figure 16.10 A fisherman with a large Atlantic codfish. A human preference for big codfish selected for fish that matured while still young and small.

© Bruce Bornstein, www.captbluefin.com.

Predation and Life History Evolution Often, different populations within a species live in slightly different environments, and their life history traits reflect these differences. Consider one long-term study in which biologists David Reznick and John Endler documented the evolutionary effects of predation on the life history traits of guppies, a type of small freshwater fish. The study involved field-work in the mountains of Trinidad, an island in the southern Caribbean Sea. Here, guppies live in shallow freshwater streams (Figure 16.9A). Waterfalls in the streams function as barriers that keep guppies from moving from one part of the stream to another. As a result, guppy populations are genetically isolated. Waterfalls also restrict the movement of predatory fish, so different predators prey upon different guppy populations. Two kinds of guppy predators live in the streams. Killifish (Figure 16.9B) are relatively small and they prey on small, juvenile guppies but ignore full-grown adults. Pike cichlids (Figure 16.9C) are larger. They pursue large, mature guppies, while ignoring juveniles.

Reznick and Endler suspected that predation shapes guppy life history patterns through natural selection, and they devised an experiment to test this hypothesis. They found a pool that held guppies and adult guppy–eating pike cichlids, but no juvenile guppy–eating killifish. They left some of these guppies in place as a control group, and moved others to a pool that contained only killifish. They predicted that exposure to this previously unfamiliar predator would cause life history traits of the experimental population to evolve.

The results supported their hypothesis. Eleven years later, the guppy population at the experimental site had evolved. Compared to the control population, guppies in the experimental population grew faster, reproduced when larger, and produced bigger offspring. Selective pressure exerted by predation on the smallest fishes favored individuals that put their energy into growth rather than reproduction, until they reached the size at which they were too big to be eaten.

Evolution of life history traits in response to predation is not merely of theoretical interest. It has economic importance. Just as guppies evolved in response to predators, a population of Atlantic codfish (*Gadus morhua*) evolved in response to human fishing pressure. From the mid-1980s to early 1990s, the number of boats fishing for cod in the North Atlantic increased. As it did, the proportion of fishes that reproduced while young and small increased. Such individuals were at an advantage because both commercial fisherman and sports fishermen preferentially caught and kept larger fish (Figure 16.10). Fishing pressure continued to rise until 1992, when declining cod numbers caused the Canadian government to ban cod fishing in some areas. That ban, and later restrictions, came too late to stop the Atlantic cod population from crashing. In some areas, the population declined by 97 percent and has not recovered.

Take-Home Message 16.4

What factors shape life history patterns?

- Life history traits such as the age at which an organism first reproduces, the number of offspring produced at a time, and potential life span have a heritable basis.
- Rapid production of many offspring is adaptive when environmental factors keep population density low. Producing offspring that are fewer in number but better able to compete is adaptive if populations are often near carrying capacity.
- Predator preferences can influence the life history traits of their prey.

16.5 **Human Populations**

REMEMBER: Our species arose about 200,000 years ago in Africa (Section 15.7).

Population Size and Growth Rate For most of its history, the human population grew very slowly (Figure 16.11). The growth rate began to pick up about 10,000 years ago, and during the past two centuries, it soared. Three trends promoted the large increases. First, humans migrated into new habitats and expanded into new climate zones. Second, they developed technologies that increased the carrying capacity of existing habitats. Third, they sidestepped some limiting factors that typically restrain population growth.

Modern humans evolved in Africa approximately 200,000 years ago, and about 60,000 years ago they began to spread out across the globe. A large brain and the capacity to master a variety of skills gave humans an unmatched ability to live in a broad range of habitats. Humans learned how to start fires, build shelters, make clothing, manufacture tools, and cooperate in hunts. With the advent of language, knowledge of such skills did not die with the individual.

The invention of agriculture about 11,000 years ago provided a more dependable food supply than traditional hunting and gathering. A pivotal factor was the domestication of wild grasses, including species ancestral to wheat and rice. In the middle of the eighteenth century, people learned to harness energy in fossil fuels to operate machinery. This innovation opened the way to high-yielding mechanized agriculture and to improved food distribution systems. Food production was further enhanced in the early 1900s, when chemists discovered a way to convert gaseous nitrogen to ammonia. Previously, this process had been carried out primarily by nitrogen-fixing bacteria. Use of synthetic nitrogen fertilizers dramatically increased

Figure 16.11 Growth curve (red) for the world human population.
The gray box lists how long it took for our number to increase from 5 million to 7 billion.

Photo, NASA.

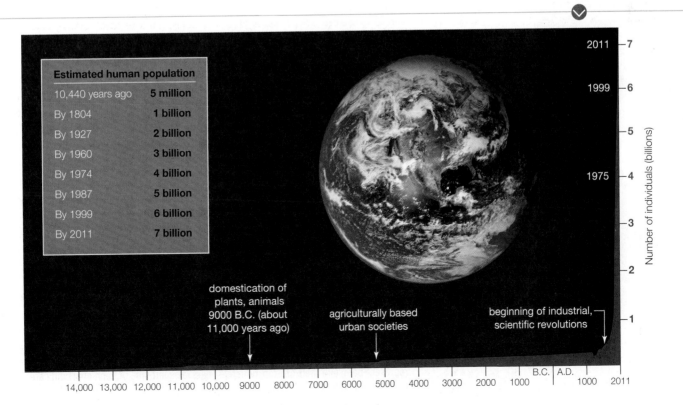

Estimated human population	
10,440 years ago	5 million
By 1804	1 billion
By 1927	2 billion
By 1960	3 billion
By 1974	4 billion
By 1987	5 billion
By 1999	6 billion
By 2011	7 billion

domestication of plants, animals 9000 B.C. (about 11,000 years ago)

agriculturally based urban societies

beginning of industrial, scientific revolutions

14,000 13,000 12,000 11,000 10,000 9000 8000 7000 6000 5000 4000 3000 2000 1000 B.C. | A.D. 1000 2011

Number of individuals (billions)

crop yields. Invention of synthetic pesticides in the mid-1900s also contributed to increased food production.

Disease has historically dampened human population growth. For example, during the mid-1300s, one-third of Europe's population was lost to a pandemic known as the Black Death. Beginning in the mid-1800s, an increased understanding of the link between microorganisms and illness led to improvements in food safety, sanitation, and medicine. People began to pasteurize foods and drinks, heating them to reduce the numbers of harmful bacteria. They also began to protect their drinking water. The first modern sewer system was constructed in London, England, in the late 1800s. By diverting wastewater downstream of the city's water source, the system lowered the incidence of waterborne diseases such as cholera and typhoid fever. Beginning in the early 1900s, new methods of sterilizing drinking water contributed to a further decline in these diseases in the most industrialized nations.

Advances in sanitation also lowered the death rate associated with medical treatment. In the mid-1800s, a German physician began urging doctors to wash their hands between patients. His advice was largely ignored until after his death, when Louis Pasteur popularized the idea that unseen organisms cause disease. Acceptance of this idea also revolutionized surgery, which had previously been carried out with little regard for cleanliness.

Vaccines and antibiotics also helped lower death rates. Vaccinations became widespread in developed countries during the 1800s. Antibiotics are a more recent development. Large-scale production of penicillin, the first antibiotic to be widely used, did not begin until the 1940s.

A worldwide decline in death rates without an equivalent drop in birth rates is responsible for the ongoing explosion in human population size. It took more than 100,000 years for the human population to reach 1 billion in number. Since then, the rate of increase has risen steadily. The population is now about 7 billion and it is expected to reach 9 billion by 2050.

Fertility Rates and Future Growth A population's **total fertility rate** is the average number of children born to the females of a population during their reproductive years. In 1950, the worldwide total fertility rate averaged 6.5 children per woman. By 2011, it had declined to 2.5 but it remains above the replacement level, which is the number of children a woman must bear to ensure that two children grow to maturity and replace her and her partner. At present, this replacement level is 2.1 for developed countries and as high as 2.5 in some developing countries. (It is higher in developing countries because more female children die before reaching the age when they can reproduce.)

At present, China (with 1.3 billion people) and India (with 1.2 billion) hold 38 percent of the world population. The United States is the next largest, with 318 million. Compare the age structure, or distribution of individuals among age groups, for these three countries (Figure 16.12). Notice especially the size of the age groups that will be reproducing during the next fifteen years. The broader the base of an age structure diagram, the greater the proportion of young people, and the greater the expected growth. Government policies that favor couples who have only one child have narrowed China's pre-reproductive base.

Figure It Out: Which country has the largest number of people aged 45 to 49?

Answer: China

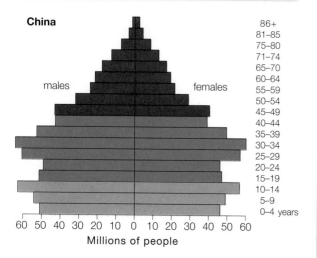

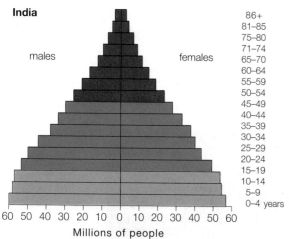

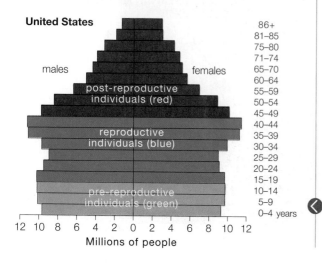

Figure 16.12 Age structure diagrams for the world's three most populous countries. The width of each bar represents the number of individuals in a 5-year age group. Green bars represent people in their pre-reproductive years. The left side of each chart indicates males; the right side, females.

Even if every couple now alive decides to bear no more than two children, world population growth will not slow for many years, because more than one-third of the world population is now in the broad pre-reproductive base. About 1.9 billion people are about to enter the reproductive age bracket.

Effects of Industrial Development   Industrial development affects the demographics of human populations. The most highly developed countries have the lowest fertility rate, the lowest infant mortality, and the highest life expectancy. The **demographic transition model** describes how birth and death rates change over the course of four stages of development. Living conditions are harshest during the preindustrial stage, before technological and medical advances become widespread. Birth and death rates are both high, so the growth rate is low. As industrialization begins and food production and health care improve, the death rate drops fast, but the birth rate declines more slowly. As a result, the population growth rate increases rapidly. India is in this stage. Once industrialization is in full swing, the birth rate moves closer to the death rate, and the population grows less rapidly. Mexico is currently in this stage. In the postindustrial stage, a population's growth rate becomes negative. The birth rate falls below the death rate, and population size slowly decreases. In some developed countries, the decreasing total fertility rate and increasing life expectancy have resulted in a high proportion of older adults.

Resource consumption rises with economic and industrial development. Ecological footprint analysis is one way to compare resource use. An **ecological footprint** is the amount of Earth's surface required to support a particular level of development and consumption in a sustainable fashion. Figure 16.13 shows per capita global footprint data for a few nations. Note that the average person in the United States has an ecological footprint nearly three times that of an average world citizen, and about nine times that of an average person living in India.

The average person in an industrialized nation uses far more nonrenewable resources than one in a less developed country. For example, the United States accounts for about 4.6 percent of the world's population, yet it uses about 25 percent of the world's minerals and energy supply. Billions of people living in India, China, and other less developed nations would like to own the same kinds of goods that people in developed countries enjoy. However, given current technology, Earth may not have the resources to make that possible. The World Resources Institute estimates that for everyone now alive to have an average American lifestyle would require the resources of four Earths. Finding ways to meet the wants and needs of expanding populations with limited resources will be a challenge.

For everyone now alive to live like an average American would require the resources of four Earths.

Country	Hectares per Capita
United States	8.0
Canada	7.0
France	5.0
United Kingdom	4.9
Japan	4.7
Russian Federation	4.4
Mexico	3.0
Brazil	2.9
China	2.2
India	0.9
World Average	2.7

* Data from www.footprintnetwork.org

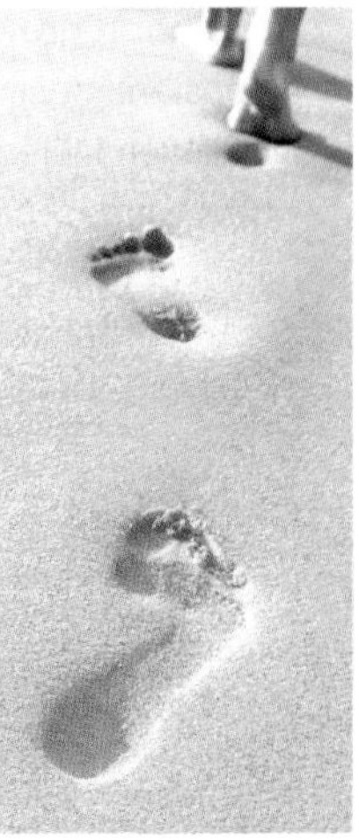

Figure 16.13 Ecological footprints.
A nation's ecological footprint estimates the amount of Earth's surface that a nation uses at its current level of development and consumption. It includes the area needed to secure food and manufacture products, as well as the natural area needed to take up the excess CO_2 produced by human activities.

Background photo, KonstantinChristian/Shutterstock.com.

Take-Home Message 16.5

What factors affect human population growth?

- Through expansion into new regions, invention of agriculture, and technological innovation, the human population has sidestepped environmental resistance to growth. Its size has been skyrocketing since the industrial revolution.
- Historically, death rates, and later birth rates, have fallen as nations have become more industrialized.
- Earth's resources are limited, so the current exponential growth of the human population is unsustainable.

demographic transition model   Model describing the changes in human birth and death rates that occur as a region becomes industrialized.

ecological footprint   Area of Earth's surface required to sustainably support a particular level of development and consumption.

total fertility rate   Average number of children born to females of a population over the course of their lifetimes.

Summary

Section 16.1 A **population** is a group of individuals that live in a given area and tend to mate with one another. Human actions have influenced the growth of some Canada goose populations. The study of populations is one aspect of the field of biology known as **ecology**.

Section 16.2 Populations vary in their **demographics**, which are characteristics such as **population size**, **population density**, **population distribution**, and **age structure**. In most populations, individuals have a clumped distribution because of limited dispersal, a need for resources that are clumped, and/or the benefits of living in a social group.

Section 16.3 Birth and death rates determine how fast a population grows. The **exponential model of population growth** describes what happens with a constant, positive **per capita growth rate**. The population increases by a fixed percentage of the whole with each successive intervals, so a plot of numbers over time produces a J-shaped curve.

Number / *Time*

An essential resource that is in short supply is a **limiting factor** for growth. The **logistic model of population growth** model describes how population growth is affected by **density-dependent limiting factors**, such as disease or competition for resources. The population slowly increases in size, goes through a rapid growth phase, then stabilizes once carrying capacity is reached. **Carrying capacity** is the maximum number of individuals that can be sustained indefinitely by the resources available in the environment. Adverse weather and other **density-independent limiting factors** can affect any population regardless of its size.

Section 16.4 The maximum theoretical rate of population growth is the **biotic potential**. Limiting factors prevent populations from attaining this potential. Biotic potential is affected by aspects of an organism's **life history** such as age at maturity, number of reproductive events, offspring number per event, and life span. Life histories are often studied by following a **cohort**, a group of individuals that were born in the same time interval.

Three types of **survivorship curves** are common: a high death rate late in life, a constant rate at all ages, or a high rate early in life. Life histories have a genetic basis and are subject to natural selection. Depending on the environment, a population may be more successful if its individuals reproduce once, or many times. At low population density, an **opportunistic life history**, in which individuals make many offspring fast, is selectively advantageous. At a higher population density, an **equilibrial life history**, in which individuals invest more in fewer, higher-quality offspring) is most favored. Predation can affect life history patterns because predators (including humans) act as selective agents that affect the traits of prey populations.

Section 16.5 The human population has now surpassed 7 billion. Expansion into new habitats and the invention of agriculture allowed early increases. Later, improved sanitation and technological innovations such as the invention of synthetic fertilizer raised the carrying capacity and minimized limiting factors that adversely affect other species.

A population's **total fertility rate** is the average number of children born to women during their reproductive years. The global total fertility rate is declining but it remains above the replacement rate. A population's age structure influences its growth. The pre-reproductive base of the human population is so large that the population is expected to grow for many years.

The **demographic transition model** describes how population growth rates have historically responded to industrialization. Developed nations have a lower growth rate, but larger **ecological footprint** than developing nations. Given current technology, Earth does not have enough resources to support the current population in the style of the most developed nations.

Self-Quiz

Answers in Appendix I

1. Most commonly, individuals of a population have a _____ distribution.
   - a. clumped
   - b. random
   - c. nearly uniform
   - d. none of the above

2. All members of a population _____ .
   - a. are the same age
   - b. reproduce
   - c. belong to the same species
   - d. all of the above

3. The exponential model of population growth assumes _____ .
   - a. the death rate declines as population density increases
   - b. per capita growth rate does not change
   - c. industrialization causes a fall in birth rates
   - d. resources are limited

4. Competition for resources and disease are _____ controls on population growth rates.
   - a. density-independent
   - b. density-dependent

5. For a given species, the maximum rate of population increase under ideal conditions is the _____ .
   a. biotic potential
   b. carrying capacity
   c. opportunistic strategy
   d. density control

6. An increase in the population of a prey species would most likely _____ the carrying capacity for that species' predators.
   a. increase
   b. decrease
   c. not affect
   d. stabilize

7. Members of a species with an _____ life history have many offspring and invest little in each one.
   a. equilibrial
   b. opportunistic

8. The logistic model of population growth takes into account _____ , but not _____ .
   a. density-dependent factors; density-independent factors
   b. density-independent factors; density-dependent factors

9. The human population is now about _____ .
   a. 7 billion   b. 7 million   c. 70 billion   d. 70 million

10. Compared to the less developed countries, the highly developed ones have a higher _____ .
    a. death rate
    b. birth rate
    c. total fertility rate
    d. ecological footprint

11. Carrying capacity _____ .
    a. is the same for all species that share a habitat
    b. is constant for a given species regardless of its habitat
    c. varies among both species and habitats
    d. is constant over time

12. The total fertility rate of a population is _____ .
    a. always higher than the replacement fertility rate
    b. the average number of offspring a female has in her lifetime
    c. the maximum number of children a woman could have if her resources were unlimited

13. If an exponentially growing population of 1,000 mice has a per capita growth rate (r) of 0.3 mice per month, how many mice will there be one month from now?
    a. 3,000
    b. 3,300
    c. 1,300
    d. 300

14. Human population growth was encouraged by _____ .
    a. invention of agriculture
    b. medical advances
    c. improved sanitation
    d. all of the above

15. Match each term with its most suitable description.
    _____ carrying capacity
    _____ logistic growth
    _____ exponential growth
    _____ demographic transition
    _____ limiting factor
    _____ cohort

    a. change in birth and death rates with industrialization
    b. group of individuals born during the same period of time
    c. population growth plots out as an S-shaped curve
    d. largest number of individuals sustainable by the resources in a given environment
    e. population growth plots out as a J-shaped curve
    f. essential resource that restricts population growth when scarce

Critical Thinking

1. Each summer, the giant saguaro cacti in deserts of the American Southwest produce tens of thousands of tiny black seeds apiece. Most die, but a few land in a sheltered spot and sprout the following spring. The saguaro is a CAM plant (Section 5.4) and it grows very slowly. After 15 years, a saguaro may be only knee high, and it will not flower until it is about age 30. The cactus may survive to age 200. Saguaros share their desert habitat with annuals such as poppies, which sprout just after the seasonal rains, produce seeds, and die in just a few weeks. How would you describe these two life history patterns? How could such different life histories both be adaptive in this environment?

2. A biologist catches 50 butterflies, marks them, then releases them. Later, the biologist returns and again catches 50 butterflies, 10 of them marked. What is the size of the population?

Visual Question

1. Consider the shape of the two age structure diagrams to the right. Which population will grow the fastest? Which includes the greater proportion of older people?

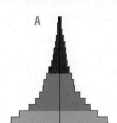

17 COMMUNITIES AND ECOSYSTEMS

Application ◉ 17.1 Fighting Foreign Fire Ants

A. One red imported fire ant worker. A colony can contain thousands of such workers, each with a stinger.

B. Red imported fire ants attack and kill eggs and hatchlings of ground-nesting birds such as quail.

C. A phorid fly attempting to lay its eggs in a red imported fire ant.

Figure 17.1 Red imported fire ant (*Solenopsis invicta*), an introduced threat to native species.

Red imported fire ants, or RIFAs (Figure 17.1A), are native to South America but, with a bit of human assistance, they have dramatically expanded their range. The ants first arrived in the southeastern United States in the 1930s, probably as stowaways on a cargo ship. They flourished in their new environment and eleven states now have well-established populations. In addition, the ants have used the United States as a springboard for dispersal to the Caribbean, Australia, New Zealand, and several Asian countries.

Red imported fire ants are considered pests because of their negative health and economic effects. The ants live in the ground and form moundlike nests. Accidentally step on one of these nests and large numbers of ants will rush out to bite and sting you. The venom injected by a RIFA's stinger causes an extremely painful burning sensation, followed by formation of a raised itchy bump at the site of the sting. As a result, a field or lawn colonized by red imported fire ants is inhospitable to humans, pets, and livestock. The ants' attraction to electricity also causes problems. For unknown reasons, large numbers of RIFAs sometimes congregate inside electrical motors, appliances, or switches, causing these devices to malfunction. Efforts to prevent the spread of red imported fire ants center on quarantines—prohibitions against moving soil that could contain the ants from affected areas to unaffected ones. Thus, arrival of RIFAs can spell disaster for commercial plant or sod growers.

Fire ants also have a negative impact on natural communities. A biological **community** includes all the species in a region. Competition with red imported fire ants typically causes a region's native ant populations to decline, and the resulting change in species composition can harm ant-eating animals. For example, the Texas horned lizard feeds mainly on native harvester ants, but it does not eat the red imported fire ants that have largely replaced its natural prey. The spread of RIFAs is also contributing to population declines in some birds, including bobwhite quails and vireos (a type of songbird). The ants decrease the abundance of insects that the birds would normally feed to their young. They also feed on birds' eggs and on nestlings. Ground-nesting birds are at special risk (Figure 17.1B). The presence of RIFAs can even affect native plants. The ants interfere with pollination by displacing or preying on native pollinators such as ground-nesting bees. The ants also interfere with the dispersal of native plants whose seeds are normally spread by native ant species that RIFAs have replaced.

Given all the difficulties RIFAs cause in the United States, you might wonder what happens in their native South America. The ants are not considered much of a problem there, in part because they are far less common. In South America, parasites and predators keep RIFA numbers far lower than those in affected parts of the United States. When some RIFAS left South America, they benefited by leaving their many natural enemies behind.

Biologists have now turned to some of those natural enemies to help them fight the RIFAs. Phorid flies are one such enemy. These flies lay their eggs in a RIFA's body (Figure 17.1C). After a fly egg hatches, it develops into a larva that grows inside the ant and feeds on the ant's soft tissues. When the fly larva is ready to undergo metamorphosis, it enters the ant's head and causes the head to fall off. The larva then develops into an adult fly inside remains of the head.

Phorid flies are not expected to kill off all *S. invicta* in affected areas. Rather, the hope is that the flies will reduce the density of the invader's colonies. When the flies are present, RIFAs try to avoid them and so spend less time foraging. The phorid flies did not evolve with our native ants, so they do not harm them.

17.2 Community Structure

Communities vary in size and often nest one inside another. For example, we find a community of microbial organisms inside the gut of a termite. That termite is part of a larger community of organisms that live on a fallen log. This community is in turn part of a still larger forest community.

Even communities that are similar in scale differ in their species diversity. There are two components to **species diversity**: The first, species richness, refers to the number of species that are present; the second is species evenness, or the relative abundance of each species. A pond that contains similar numbers of five species of fish has a higher species evenness, and thus a higher species diversity, than a pond with one abundant fish species and four rare ones.

Community structure is dynamic, which means that the array of species and their relative abundances change over time. Communities change over a long time span as they form and then age. Some change suddenly as a result of natural or human-induced disturbances.

Nonbiological Factors Factors related to geography and climate affect community structure. These factors include soil quality, sunlight intensity, rainfall, and temperature, which vary with latitude, elevation, and—for aquatic habitats—depth. Tropical regions receive the most sunlight energy and have the most even temperature. For most plants and animal groups, the number of species is greatest in the tropical regions near the equator, and declines as you move toward the poles. Tropical rain-forest communities are highly diverse (Figure 17.2), and forest communities in temperate regions are less so. Similarly, tropical reef communities are more diverse than comparable marine communities farther from the equator.

Biological Factors The evolutionary history and adaptations of the species in a community also influence community structure. Each species evolved in and is adapted to a specific **habitat**, the type of place where it typically occurs. All species of a community share the same habitat, but each has a unique ecological role that sets it apart. This role is the species' **niche**, which we describe in terms of the conditions, resources, and interactions necessary for survival and reproduction. Aspects of an animal's niche include temperatures it can tolerate, the kinds of foods it can eat, and the types of places where it can breed or hide. A description of a plant's niche would include details of its requirements for soil, water, light, and pollinators.

Species interactions also affect community structure. In some cases, the effect is indirect. For example, when songbirds eat caterpillars, the birds indirectly benefit the trees that the caterpillars feed on. Other interactions are direct; the actions of one species helps or harms another. Such direct interactions are the subject of the next section.

Figure 17.2 Tropical diversity.
Two of the 12 or so fruit-eating pigeon species that live in New Guinea's tropical rain forests (left). Top, a Victoria crowned pigeon, which is the size of a turkey. Below, the smaller superb crowned fruit pigeon.

Left, Donna Hutchins; right, top, © Martin Harvey, Gallo Images/Corbis; bottom, © Len Robinson, Frank Lane Picture Agency/Corbis.

Take-Home Message 17.2

What factors shape a community?

- Communities vary in their species diversity as a result of nonbiological factors such as differences in incoming sunlight, temperature, and soil quality.
- Biological factors such as species requirements for survival and reproduction, as well as interactions with other species, also influence community structure.

community All populations of all species in some area.

habitat The type of place in which a species lives.

niche The role of a species in its community.

species diversity The number of species and their relative abundance within a community.

Table 17.1 Interspecific Interactions

Type of Interaction	Direct Effect on Species 1	Direct Effect on Species 2
Commensalism	Benefits	None
Mutualism	Benefits	Benefits
Competition	Harmed	Harmed
Predation	Benefits	Harmed
Parasitism	Benefits	Harmed

Figure 17.3 Commensalism.
This tree provides orchids with an elevated perch from which they can capture sunlight. The presence of the orchids has no effect on the tree.

joloei/Shutterstock.

17.3 Direct Species Interactions

REMEMBER: Species interactions influence population size (Section 16.3) and can result in directional selection (12.3) and coevolution (12.7).

There are five types of direct interactions among species in a community: commensalism, mutualism, competition, predation, and parasitism (Table 17.1). Three of these—parasitism, commensalism, and mutualism—can be a **symbiosis**. Symbiosis means "living together." Symbiotic species, also known as symbionts, spend most or all of their life cycle in close association with each other. An endosymbiont is a species that lives inside its partner species.

Regardless of whether one species helps or hurts another, two species that interact closely may coevolve over generations. Recall that with coevolution, each species is a selective agent that shifts the range of variation in the other.

Commensalism and Mutualism **Commensalism** benefits one species and does not affect the other. For example, some orchids that live attached to a tree trunk or branch (Figure 17.3) benefit by having a perch in the sun, while the tree that provides this support is unaffected. As another example, commensal bacteria live in the gut of many animals. They benefit by having a warm, nutrient-rich place to live, and their presence neither helps nor harms their host.

Other gut bacteria assist their host by aiding in digestion or synthesizing vitamins. An interaction that benefits both species is a **mutualism**. Flowering plants and their pollinators are a familiar example. In some cases, coevolution of two species results in a mutual dependence. For example, there are several species of yucca plant, each pollinated by a single species of yucca moth, whose larvae develop on that plant species alone (Figure 17.4A). More often, mutualistic relationships are less exclusive. Most flowering plants have more than one pollinator, and most pollinators provide their service to more than one species of plant.

Photosynthetic organisms often supply sugars to their nonphotosynthetic partners, as when plants lure pollinators with nectar. In addition, many plants

Figure 17.4 Mutualism.

(A) Bob and Miriam Francis/Tom Stack & Associates; (B) Sergey Uryadnikov/Shutterstock.com; (C) © Thomas W. Doeppner.

A. Yucca moth on a yucca flower. The yucca plant benefits by being pollinated. The moth benefits by laying eggs that develop within the yucca's fruit.

B. Lichen. It consists of a fungus and a green alga. The fungus supports and shelters the alga, which shares the sugar it makes with the fungus.

C. Anemonefish nestles among the tentacles of a sea anemone. In this mutually beneficial partnership, each species protects the other.

Figure 17.5 Active competition among scavengers.
After facing off over a carcass (left), an eagle attacked a fox with its talons (right). The fox then retreated, leaving the eagle to exploit the carcass.

© Pekka Komi.

Figure 17.6 Scramble competition between members of different kingdoms.
Sundew plants (top) and wolf spiders (bottom) compete for food. Both feed on flies and other insects.

Top, scaners3d/Shutterstock.com; bottom, Cathy Keifer/Shutterstock.com.

make sugary fruits that attract seed-dispersing animals. Plants also provide sugars to mycorrhizal fungi and nitrogen-fixing bacteria. The plants' fungal or bacterial symbionts return the favor by supplying their host with other essential nutrients. Similarly, photosynthetic dinoflagellates provide sugars to corals, and photosynthetic bacteria or algae feed their fungal partner in a lichen (Figure 17.4B).

Mutualisms can also involve defense. A pink anemonefish will be eaten by a predator unless it has a sea anemone to hide in (Figure 17.4C). The anemone's tentacles are covered by stinging cells that do not affect the anemonefish, but help fend off predators that would eat the fish or its eggs. The anemone can survive on its own, but benefits by having a partner that chases away the few fish species that are able to feed on anemone tentacles.

From an evolutionary standpoint, mutualism is best considered as a case of reciprocal exploitation. Each participant increases its own fitness by extracting a resource, such as protection or food, from its partner. If taking part in the mutualism has a cost, selection will favor individuals who minimize that cost. Consider nectar production, which is energetically costly for a flower, but serves as a necessary payoff to pollinators. Natural selection will favor a flower that produces the minimum amount of nectar necessary to attract pollinators over one that expends additional energy to produce a more generous serving of nectar.

Interspecific Competition **Interspecific competition** is competition between members of different species. Members of one species sometimes actively prevent members of another species from using a resource. For example, scavengers such as eagles and foxes sometimes fight over a carcass (Figure 17.5). Some plants also actively interfere with their competitors. For example, a sagebrush plant secretes chemicals into the soil, thus preventing potential competitors from growing nearby.

In other cases, competing species do not directly interfere with one another. Instead, all scramble for a share of a limited resource. For example, wolf spiders and carnivorous plants called sundews both capture and feed on insects (Figure 17.6). Although these organisms do not fight over food, they do compete. By catching and eating insects, each reduces the amount of food available to the other.

commensalism Species interaction that benefits one species and has no effect on the other.

interspecific competition Two species compete for a limited resource and both are harmed by the interaction.

mutualism Species interaction that benefits both species.

symbiosis One species lives on or inside another in a commensal, mutualistic, or parasitic relationship.

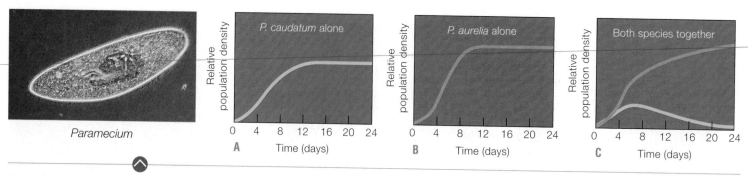

A P. caudatum alone · Relative population density · Time (days) · 0 4 8 12 16 20 24

B P. aurelia alone · Relative population density · Time (days) · 0 4 8 12 16 20 24

C Both species together · Relative population density · Time (days) · 0 4 8 12 16 20 24

Paramecium

Figure 17.7 Competitive exclusion.
Two species of the ciliated protozoan *Paramecium*,
P. caudatum and *P. aurelia*, both feed on bacteria. When
grown in separate test tubes, each does well **A**, **B**.
When grown together **C**, one species drives the other to
extinction.

Left, Michael Abbey/Science Source.

Figure It Out: Which species of *Paramecium* was the
superior competitor?

Answer: *P. aurelia*

Species compete most intensely when the supply of a shared resource is the
main limiting factor for both. In the early 1930s, G. F. Gause carried out a series of
experiments that led him to describe **competitive exclusion**: Whenever two species
require the same limited resource to survive or reproduce, the better competitor
will drive the less competitive species to extinction in that habitat. Gause studied
interactions between two species of ciliated protists (*Paramecium*) that eat the same
bacteria. He grew the species separately and together (Figure 17.7). In the mixed
cultures, population growth of one species always outpaced the other, which eventually died out.

Competitors whose resource needs are similar but not exactly the same can
coexist. However, competition among them suppresses their population growth. In
each species, competition with the other encourages directional selection. Those
individuals who differ most from competing species in terms of resource needs are
at a selective advantage. As a result of this selective pressure, the traits of competing species may evolve so there is less competition between them, an evolutionary
process referred to as **resource partitioning**.

Predator–Prey Interactions In **predation**, one free-living species captures, kills,
and eats another (Figure 17.8). Predators exerts selective pressure on prey, favoring
those with the best anti-predator defenses. Prey defenses in turn favor those predators best able to overcome them. As a result, predators and prey may engage in an
evolutionary arms race that continues over many generations.

You have already learned about some defensive adaptations. Many prey species
have hard or sharp parts that make them difficult to eat. Think of a snail's shell or
a porcupine's quills. Others have chemicals that taste bad or sicken predators. Most
defensive toxins in animals come from the plants they eat. For example, a monarch
butterfly caterpillar feeds on and takes up chemicals from the milkweed plant. If a
bird eats a monarch caterpillar or butterfly, plant-derived chemicals will sicken it.

Some well-defended prey have **warning coloration**, a conspicuous pattern or
color that predators learn to avoid. For example, stinging wasps and bees typically
have black and yellow stripes (Figure 17.9A). Their similar appearance is a type of
mimicry, an evolutionary pattern in which one species comes to resemble another.
In one type of mimicry, well-defended species benefit by looking alike. In another
type of mimicry, prey masquerade as a species that has a defense that they lack.
For example, some flies that cannot sting resemble stinging bees or wasps (Figure
17.9B). Such a fly benefits when predators avoid it after an encounter with the
better-defended look-alike species.

Some prey can startle an attacking predator. Section 1.5 described how eyespots
and a hissing sound protect some butterflies. A lizard's tail may detach from the
body and wiggle a bit as a distraction, while the lizard runs off. Skunks and some
beetles squirt foul-smelling, irritating repellents at potential predators.

Figure 17.8 Predation.
Predators such as this lynx catch, kill, and devour their
prey, in this case a snowshoe hare.

Ed Cesar/Science Source.

A. The coloration of this yellow jacket wasp warns predators that it can sting.

B. This fly, which cannot sting, benefits by mimicking the color pattern of wasps.

Figure 17.9 Warning coloration and mimicry.

(A) © Kletr/Shutterstock.com; (B) Marco Uliana/Shutterstock.com.

Camouflage is a body shape, color pattern, or behavior that allows an individual to blend into its surroundings and avoid detection. Prey benefit when camouflage hides them from predators, and predators benefit when it hides them from prey (Figure 17.10).

In response to prey defenses, predators have evolved sharp teeth and claws that can pierce protective hard parts. Speedy prey select for faster predators. For example, the cheetah, the fastest land animal, can run 114 kilometers per hour (70 mph). Its preferred prey, Thomson's gazelles, run 80 kilometers per hour (50 mph).

Plants and Herbivores In **herbivory**, an animal feeds on a plant, which may or may not die as a result. Some plants can withstand loss of their parts and quickly grow replacements. For example, grasses are seldom killed by herbivores. They have a fast growth rate and store enough resources in their roots to replace the shoots lost to grazers. Other plants have traits that fend off herbivores. Physical deterrents include spines, thorns, and fibrous, difficult-to-chew leaves. Plants can also produce compounds that taste bad to herbivores or sicken them. Capsaicin, a compound that makes some peppers "hot," defends seeds against seed-eating mammals. Rodents find capsaicin-rich pepper fruits unpalatable, leaving them to be eaten by birds, which cannot taste capsaicin. A pepper benefits by deterring rodent seed eaters because rodents chew up and kill seeds, whereas birds excrete seeds alive and intact.

Parasites and Parasitoids **Parasites** withdraw nutrients from a living host. Some bacteria, protists, and fungi are parasites. Tapeworms and flukes are parasitic annelid worms. Some roundworms, insects, and crustaceans are parasites, as are all ticks (Figure 17.11A). Even a few plants are parasitic (Figure 17.11B).

Figure 17.11 Parasites.

(A) © Bill Hilton, Jr., Hilton Pond Center; (B) © The Samuel Roberts Noble Foundation, Inc.

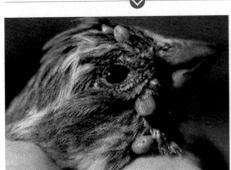

A. Ticks sucking blood from a finch.

B. Dodder (*Cuscuta*). Roots that extend from the leafless golden stems withdraw water and nutrients from another plant.

Figure 17.10 Camouflage.
Frilly pink body parts of a flower mantis hide the mantis from its insect prey, which are attracted to the real flowers.

© Bob Jensen Photography.

camouflage Body shape, pattern, or behavior that helps a plant or animal blend into its surroundings.

competitive exclusion When two species compete for the same resource, the better competitor drives the weaker one to extinction in that habitat.

herbivory An animal feeds on a plant, which may or may not die as a result.

mimicry Two or more species come to resemble one another.

parasite A species that withdraws nutrients from another species (its host), usually without killing it.

predation One species (the predator) captures, kills, and feeds on another (its prey).

resource partitioning Evolutionary process whereby traits of competing species come to differ as a result of the selective pressure imposed by the competition.

warning coloration Distinctive color or pattern that makes a well-defended prey species easy to recognize.

A. Brood parasite. A cowbird chick with its foster parent. A female cowbird minimizes her cost of parental care by laying her eggs in the nests of other bird species.

B. Parasitoid. A commercially raised parasitoid wasp about to deposit a fertilized egg in an aphid. The wasp larva will devour the aphid from the inside.

Figure 17.12 Brood parasites and parasitoids.

(A) E. R. Degginger/Science Source; (B) © Peter J. Bryant/ Biological Photo Service.

Even when a parasite does not cause obvious symptoms, infection can weaken the host, making it more vulnerable to predation or less attractive to potential mates. Some parasitic infections cause sterility.

Adaptations to a parasitic lifestyle include traits that allow the parasite to locate hosts and to feed undetected. For example, ticks that feed on mammals or birds move toward a source of heat and carbon dioxide, which are likely signs of a potential host. A chemical in tick saliva acts as a local anesthetic, preventing the host from noticing the feeding tick. Parasites that live inside other organisms often have adaptations that help them evade a host's immune defenses.

Hosts' defenses against parasites include immune responses (a topic we consider in detail in Chapter 22) and behavioral responses. For example, many primates take turns removing ticks and other parasites from one another. Birds kill external parasites by preening their feathers, and their bill shape reflects this function, as well as its role in feeding. In pigeons, even a slight experimental modification of bill shape that has no effect on feeding can result in an increase in parasite numbers.

Brood parasites do not feed on their hosts, but rather steal parental care. A brood parasite tricks another animal into raising its young. For example, European cuckoos and North American cowbirds (Figure 17.12A) lay their eggs in the nests of other birds. When these eggs hatch, the foster parents care for the young as if they were their own. Freed from the constraints imposed by parental care, a female cowbird can lay as many as thirty eggs in a single season. Some other birds, fish, and insects are also brood parasites.

Parasitoids take the concept of forcing another to care for one's young to an even higher level. These insects lay their eggs inside other insects. The eggs hatch into larvae that devour the host from the inside, eventually killing it. The phorid flies that have been imported to attack red imported fire ants are parasitoids.

Parasites and parasitoids are often used in biological pest control. Such pest control provides an alternative to chemical pesticides, which typically kill or harm a wide variety of nontarget species. Species chosen for use as biological pest control agents target only a specific type of host. For example, the parasitoid wasp in Figure 17.12B is a biological control agent. It lays eggs only in aphids, which are sucking insects that damage many crop plants.

brood parasite An animal that tricks another species into raising its young, for example a cowbird.

parasitoid An insect that lays eggs in another insect, and whose young devour their host from the inside.

Take-Home Message 17.3

How do species interactions affect a community?

- In commensalism, one species benefits and the other is unaffected.
- In mutualism, two species exploit one another in a way that benefits both.
- Competition for resources has a negative effect on both competitors. If both depend on the same limited resource, the stronger competitor may drive the weaker one to local extinction, a process called competitive exclusion.
- Interspecific competition favors individuals of both species whose resource needs are most unlike those of the competing species. Over time, competition alters traits related to resource use and leads to resource partitioning.
- Predators benefit at the expense of their prey, and parasites benefit at the expense of their hosts. As a result, predators and parasites select for defensive traits in prey and hosts. These defenses in turn select for traits that help the predators and parasites overcome them.

17.4 How Communities Change

Ecological Succession Community structure—the kinds of species and their relative abundance in a habitat—changes constantly. In a process called **ecological succession**, the array of species gradually shifts over time as organisms alter their own habitat. One array of species is replaced by another, which is in turn replaced by another, and so on.

Primary succession occurs in habitats that lack soil and have few or no existing species. For example, a rocky area exposed by retreat of a glacier undergoes primary succession (Figure 17.13). At first, no multicellular organisms are present ❶. The community begins to change as pioneer species gain a foothold ❷. **Pioneer species** colonize new or vacated habitats. They often include lichens, mosses, and hardy annual plants with wind-dispersed seeds. As generations of pioneers live and die, they help build and improve the soil. In doing so, they set the stage for their own replacement. Seeds of shrubby species take root in the mats of pioneers ❸. Over time, organic wastes and remains build up and, by adding volume and nutrients to soil, this material allows tall trees to take hold ❹.

Secondary succession occurs after a natural or human disturbance removes the natural array of species, but not the soil. We observe this kind of succession after a fire destroys a forest or a plowed field is abandoned and wild species move in and take over.

The 1980 eruption of Mount Saint Helens, a volcano in Washington State, gave scientists an opportunity to observe succession in action (Figure 17.14). The eruption showered the area around the volcano with volcanic rock and ash, wiping out the existing plant life and covering the mature soil. Since then, plant life has colonized the area and succession is under way. Such field studies of succession have led to a change in the way the process is viewed. The concept of ecological succession was first developed in the late 1800s. At that time, it was viewed as a predictable and directional process that culminated in a "climax community," an array of species that persists over time and is reconstituted in the event of a disturbance. Physical factors such as climate, altitude, and soil type were thought to determine which species appeared in what order. Modern ecologists recognize that three types of factors affect succession: (1) physical factors such as climate, (2) chance events such as the order in which

ecological succession A gradual change in a community in which one array of species replaces another.

pioneer species Species that can colonize a new habitat.

primary succession Ecological succession occurs in an area where there was previously no soil.

secondary succession Ecological succession occurs in an area where a community previously existed and soil remains.

Figure 17.13 Artist's depiction of how primary succession in a previously glaciated area can lead to establishment of a forest community.

Figure 17.14 Ecological succession after a volcanic eruption.

(A) R. Barrick/ USGS; (B) USGS; (C) P. Frenzen, USDA Forest Service.

A. Mount Saint Helens erupted in 1980. Volcanic ash completely buried the community that had previously existed at the base of this volcano.

B. In less than a decade, numerous pioneer species had become established.

C. Twelve years after eruption, Douglas fir seedlings were taking hold in soils enriched by volcanic ash.

**Figure 17.15
A keystone species.**
The sea star *Pisaster* lives along rocky shores. Remove it and the species diversity of this community declines.

lauraslens/Shutterstock.

Figure 17.16 Exotic species that have become pests in the United States.
To learn about others, visit the National Invasive Species Information Center online at www.invasivespeciesinfo.gov.

(A) Angelina Lax/Science Source; (B) © Greg Lasley Nature Photography, www.greglasley.net.

A. Kudzu native to Asia is overgrowing trees across the southeastern United States.

B. Nutrias native to South America are now abundant in freshwater marshes and riversides of the Gulf States, the Chesapeake Bay region, and Oregon.

pioneer species arrive, and (3) the frequency and extent of disturbances. Because the sequence of species arrivals and the frequency and extent of disturbances vary in unpredictable ways, it is difficult to predict exactly how the composition of any particular community will change in the future.

Adapted to Disturbance   In communities that are repeatedly subjected to a particular type of physical disturbance, individuals who withstand or benefit from that disturbance have a selective advantage. For example, some plants in areas subject to periodic fires produce seeds that germinate only after a fire has cleared away potential competitors. Other plants have an ability to resprout quickly after a fire. Because fire affects different species in different ways, the frequency of fire influences competitive interactions. For example, when naturally occurring fires are suppressed, plants adapted to periodic burning lose their competitive edge. The fire-adapted plants can be overgrown by plants that devote all of their energy to growing and reproducing, rather than investing in fire-related adaptations.

Species Losses or Additions   A **keystone species** has a disproportionately large effect on a community relative to its abundance. Robert Paine coined the term to describe the results of his studies of a sea star (*Pisaster*) common along rocky coastlines (Figure 17.15). When Paine experimentally removed the sea star from some plots, species richness within those plots declined. The sea star feeds mainly on mussels, and its presence prevents mussels from overgrowing other species.

Keystone species need not be predators. For example, the large rodents called beavers can be a keystone species. A beaver cuts down trees by gnawing through their trunk, then uses felled trees to build a dam. Construction of a beaver dam creates a deep pool where a shallow stream would otherwise exist. By altering the physical conditions in a section of the stream, the beaver affects the types of fish and aquatic invertebrates that can live there.

Arrival of an **exotic species**—a species that was introduced to a new habitat and became established there—can also alter community structure. In its new home, an exotic species often has no coevolved parasites, pathogens, or predators to keep it in check, so its numbers can soar. More than 4,500 exotic species have become established in the United States. The red imported fire ants are one example. Kudzu is another (Figure 17.16A). Native to Asia, this vine was introduced to the American Southeast as a food for grazers and to control erosion, but it has become an invasive weed. Nutrias, large semiaquatic rodents from South America, are still another (Figure 17.16B). Descendants of nutrias imported for their fur in the 1940s now live wild in marshes where they threaten native plants and contribute to marsh erosion.

Take-Home Message 17.4

What causes changes in community structure?

- In ecological succession, one array of species changes the habitat in a way that allows another array of species to take hold.
- Large and small disturbances shift community structure on an ongoing basis.
- A change in the presence or abundance of a keystone species has a great effect on other species in a habitat.
- An exotic species that leaves behind the predators, parasites, and competitors it evolved with can dramatically affect the structure of its adopted community.

exotic species   A species that has been introduced to a new habitat and become established there.

keystone species   A species that has a disproportionately large effect on community structure.

17.5 The Nature of Ecosystems

REMEMBER: All life requires energy and nutrients (Section 1.3).

Overview of the Participants Organisms of a community interact with their environment as an **ecosystem**. In all ecosystems, there is a one-way flow of energy and a cycling of essential materials (Figure 17.17). An ecosystem's **producers** capture energy and use it to make their own food from nonbiological materials in the environment. Usually the producer's energy source is sunlight and the producers are plants and photosynthetic bacteria and protists. An ecosystem's **consumers** obtain energy and carbon by feeding on tissues, wastes, and remains of producers and one another. Herbivores, predators, and parasites are consumers that feed on living organisms. **Detritivores** such as crabs and earthworms eat tiny bits of organic matter, or detritus. Finally, wastes and remains of organisms are broken down into inorganic building blocks by bacterial, protist, and fungal **decomposers**.

Light energy captured by producers is converted to bond energy in organic molecules, which is then released by metabolic reactions that give off heat as a by-product. This is a one-way process because organisms cannot convert heat back into chemical bond energy.

Unlike energy, with its one-way flow, nutrients cycle within an ecosystem. The cycle begins when producers take up hydrogen, oxygen, and carbon from inorganic sources such as the air and water. They also take up dissolved nitrogen, phosphorus, and other necessary minerals. Nutrients move from producers into the consumers who eat them. Decomposition returns nutrients to the environment, from which producers take them up again.

Food Chains and Webs All organisms of an ecosystem take part in a hierarchy of feeding relationships referred to as **trophic levels** ("troph" means nourishment). When one organism eats another, energy (in the form of chemical bonds) and nutrients are transferred from the eaten to the eater. All organisms at the same trophic level are the same number of transfers away from the energy input into that system.

A **food chain** is one sequence of steps by which energy captured by primary producers moves to higher trophic levels. Consider one food chain in a tallgrass prairie (Figure 17.18). The main producers in this ecosystem—grasses and other plants—are at the first trophic level. Energy flows from the plants to grasshoppers, to sparrows, and finally to hawks. Grasshoppers are primary consumers and are at the second trophic level. Sparrows that eat grasshoppers are second-level consumers and at the third trophic level. Hawks that eat sparrows are third-level consumers and at the fourth trophic level.

consumer Organisms that obtains energy and nutrients from organisms or their remains.

decomposer Consumer that breaks organic remains into their inorganic building blocks.

detritivore Consumer that feeds on small bits of organic material (detritus).

ecosystem A community and its environment.

food chain Sequence of steps by which energy moves from one trophic level to the next.

producer Organism that captures light or chemical energy and makes its food from inorganic materials.

trophic level Position in a food chain.

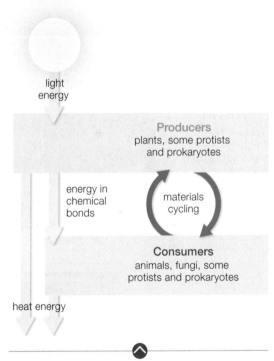

Figure 17.17 Generalized model for the one-way flow of energy (yellow arrows) and the cycling of materials (blue arrows) in an ecosystem.

Figure 17.18 One food chain in a tallgrass prairie.
Species at the first trophic level capture sunlight energy. Arrows represent the transfer of nutrients and energy from one trophic level to the next.

From left, © Van Vives; © D. A. Rintoul; © D. A. Rintoul; © Lloyd Spitalnik/lloydspitalnikphotos.com.

First Trophic Level	Second Trophic Level	Third Trophic Level	Fourth Trophic Level
Producer	Primary Consumer	Second-Level Consumer	Third-Level Consumer
Grass	Grasshopper	Sparrow	Hawk

human (Inuk)

arctic wolf

arctic fox

Higher Trophic Levels

A sampling of carnivores that feed on herbivores and one another

gyrfalcon

snowy owl

ermine

Second Trophic Level

A sampling of primary consumers (herbivores) that eat plants

vole

arctic hare

lemming

mosquito

flea

Parasitic consumers feed at more than one trophic level.

First Trophic Level

Examples of primary producers (plants)

grasses, sedges

purple saxifrage

arctic willow

Detritivores and decomposers (nematodes, annelids, saprobic insects, protists, fungi, bacteria)

Figure 17.19 Arctic food web.
Arrows point from eaten to eater.

From left, top row, © Bryan & Cherry Alexander/Science Source; © Dave Mech; © Tom & Pat Leeson, Ardea London Ltd.; 2nd row, © Tom Wakefield/Bruce Coleman, Inc.; © Paul J. Fusco/Science Source; © E. R. Degginger/Science Source; 3rd row, © Hugo Wilcox/Minden Pictures; © Dave Mech; © Tom McHugh/Science Source; mosquito, Photo by James Gathany, Centers for Disease Control; flea, © Edward S. Ross; 4th row, © Jim Steinborn; © Jim Riley; © Matt Skalitzky; earthworm, © Peter Firus, flagstaffotos.com.au.

Food chains cross-connect with one another as a **food web**. Figure 17.19 shows some participants in an arctic food web. Nearly all food webs include two types of food chains. In grazing food chains, energy stored by producers flows next to herbivores, which tend to be relatively large animals. In a detrital food chain, energy in producers flows to detritivores, which tend to be smaller animals, and to decomposers. In most land ecosystems, detrital food chains predominate. For example, in an arctic ecosystem, grazers such as voles, lemmings, and hares eat some plant parts. However, far more plant matter becomes detritus that sustains soil-dwelling insects and decomposers such as bacteria and fungi. Detrital food chains and grazing food chains interconnect as the ecosystem's overall food web.

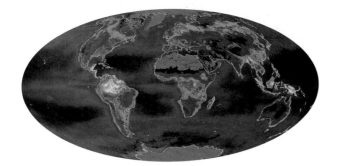

Figure 17.20 Satellite data showing net primary production for land and oceans.
Productivity is coded as red (highest) down through orange, yellow, green, blue, and purple (lowest).
NASA.

Primary Production and Inefficient Energy Transfers The flow of energy through an ecosystem begins with **primary production**: the capture and storage of energy by producers. Primary production, which is measured in terms of the amount of carbon taken up per unit area, varies seasonally and among habitats. On average, primary production is higher on land than it is in the oceans (Figure 17.20). However, because the oceans cover about 70 percent of Earth's surface, they contribute about half of Earth's total primary production.

An **energy pyramid** is a graphic representation of the proportion of the energy captured by producers that reaches higher trophic levels. Energy pyramids always have a large energy base, representing producers, and taper up. Figure 17.21 shows an energy pyramid for a freshwater ecosystem in Florida.

Only about 10 percent of the energy in tissues of organisms at one trophic level ends up in tissues of those at the next trophic level. Several factors limit the efficiency of transfers. All organisms lose energy as metabolic heat and this energy is not available to organisms at the next trophic level. Also, some energy gets stored in molecules that most consumers cannot break down. For example, most carnivores cannot access the energy tied up in bones, scales, hair, feathers, or fur. The inefficiency of energy transfers limits the possible length of food chains.

When people promote a vegetarian diet by touting the benefits of "eating lower on the food chain," they are referring to the inefficiency of energy transfers. A person eating a plant food involves only a single energy transfer. When a plant food is used to grow livestock, the animal uses some of the energy it obtains from plant food to sustain itself, loses some energy as heat, and invests some energy building inedible parts such as bones and hooves. Only a small percentage of the energy in the original plant food ends up as meat that a person can eat. Thus, feeding a population of meat-eaters requires far greater crop production than sustaining a population of vegetarians.

> Oceans contribute about half of Earth's primary production.

Figure 17.21 Energy pyramid for a freshwater ecosystem.
Numbers are energy in kilocalories per square meter per year.

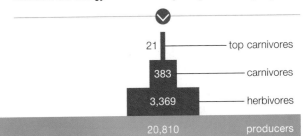

21 — top carnivores
383 — carnivores
3,369 — herbivores
20,810 — producers

Figure It Out: Approximately what percentage of the sunlight energy captured by producers was transferred to the herbivores that ate them?

Answer: 3,389/20,810 × 100 = 16 percent

Take-Home Message 17.5

How do organisms and their environment interact in ecosystems?

- Materials cycle between organisms and their environment, but energy flow is one-way because energy in chemical bonds is converted to heat.
- Energy and raw materials are taken up by producers, then flow to consumers.
- Food chains and food webs describe routes by which energy and nutrients move from one trophic level to another.
- Energy transfers between trophic levels are inefficient because energy is lost as metabolic heat, and some energy becomes tied up in materials that are not easily digested by consumers.

energy pyramid Diagram that illustrates the energy flow in an ecosystem.
food web System of cross-connecting food chains.
primary production The energy captured by an ecosystem's producers.

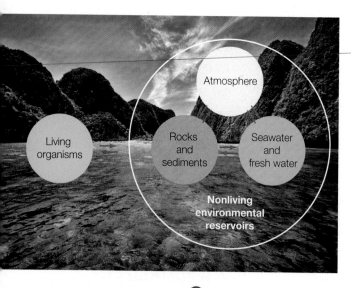

Figure 17.22 Generalized biogeochemical cycle.

Photo, Phaitoon Sutunyawatchai/Shutterstock.

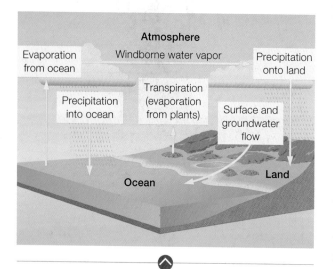

Figure 17.23 The water cycle.
Water moves from the ocean into the atmosphere, onto land, and then back. The graphic below shows the volume of water in the various environmental reservoirs.

© Triff/Shutterstock.com.

Reservoir	Volume (10^3 km³)
Ocean	1,370,000
Polar ice, glaciers	29,000
Groundwater	4,000
Lakes, rivers	230
Atmosphere (water vapor)	14

17.6 Biogeochemical Cycles

REMEMBER: A triple bond holds the two nitrogen atoms in nitrogen gas together (Section 2.3). Movement of tectonic plates uplifts land (11.5). The ozone layer reduces the amount of UV radiation that reaches Earth's surface (13.3). A population explosion of single-cell producers in nutrient-enriched water is an algal bloom (13.6). Coal formed from ancient plants (14.4), and petroleum from diatoms (13.6).

In a **biogeochemical cycle**, an essential element moves from one or more environmental reservoirs, through the living components of an ecosystem, and then back to the reservoirs (Figure 17.22). Depending on the element, environmental reservoirs may include Earth's rocks and sediments, waters, and atmosphere.

Chemical and geologic processes move elements to, from, and among environmental reservoirs. Elements locked in rocks become part of the atmosphere as a result of volcanic activity. Movement of Earth's tectonic plates (Section 11.5) can uplift rocks, so an area that was once seafloor becomes part of a landmass. On land, erosion breaks down rocks, allowing the elements in them to enter rivers and flow to seas. Compared to the movement of elements within a community, movement of elements among nonbiological reservoirs is far slower. Processes such as erosion and uplifting operate over thousands or millions of years.

We focus on four biogeochemical cycles that move important elements: the water cycle, phosphorus cycle, nitrogen cycle, and carbon cycle.

The Water Cycle The **water cycle** moves water from oceans to the atmosphere, onto land and into freshwater ecosystems, and back to the oceans (Figure 17.23). Solar energy drives evaporation of water from the oceans and from freshwater reservoirs. Water that enters the lower atmosphere spends some time aloft as vapor, clouds, and ice crystals. By the process of precipitation, water falls from the atmosphere mainly as rain and snow. Oceans cover about 70 percent of Earth's surface, so most water evaporates from oceans and most precipitation falls on oceans. Most precipitation that falls on land runs into streams or seeps into the ground. Plant roots take up soil water, the water between soil particles. **Transpiration**, the evaporation of water from the aboveground parts of plants, returns most of this water to the atmosphere.

Our planet has a lot of water, but 97 percent of it is salt water. Of the 3 percent that is fresh water, most is frozen in glaciers. **Groundwater**, another freshwater reservoir, includes water in the soil, and water stored in porous rock layers called **aquifers**. About half of the population of the United States relies on aquifers for drinking water. Surface water (water in streams, rivers, lakes, and freshwater marshes) constitutes less than 1 percent of Earth's fresh water.

Movement of water results in the movement of soluble nutrients. Carbon, nitrogen, and phosphorus all have soluble forms that can be moved from place to place by flowing water. As water trickles through soil, it carries nutrient particles from topsoil into deeper soil layers. As a stream flows over limestone, water slowly dissolves the rock and carries carbonates back to the seas where the limestone formed. Flowing water can transport pollutants too; runoff from heavily fertilized lawns and agricultural fields carries dissolved phosphates and nitrates into streams and lakes.

The Phosphorus Cycle Atoms of phosphorus are highly reactive, so phosphorus does not occur naturally in its elemental form. Most of Earth's phosphorus is bonded to oxygen as phosphate (PO_4^{3-}), an ion that abounds in rocks and

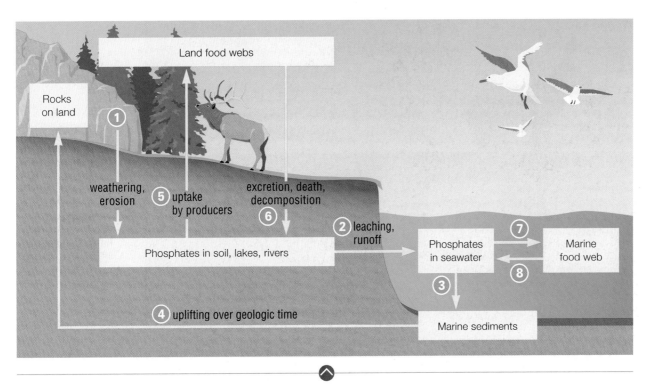

Figure 17.24 The phosphorus cycle.
In this cycle, most of the phosphorus moves in the form of phosphate ions. Earth's main phosphorus reservoir is rocks and sediments.

sediments. There is no commonly occurring gaseous form of phosphorus, so the atmosphere is not one of its reservoirs. In the **phosphorus cycle**, phosphate moves among Earth's rocks, soil, and water, and into and out of food webs (Figure 17.24). In the environmental portion of the cycle, weathering and erosion move phosphate ions from rocks into soil, lakes, and rivers ❶. Leaching and runoff deliver phosphate ions to the ocean ❷, where most of the phosphorus comes out of solution and settles as deposits along the edges of continents ❸. Over millions of years, movements of Earth's crust can uplift parts of the seafloor onto land ❹, where weather releases phosphates from the rocks. Dissolved phosphates enter streams and rivers, and the environmental portion of the phosphorus cycle starts over again.

All organisms require phosphorus to build ATP, nucleic acids, and phospholipids. The biological portion of the phosphorus cycle begins when producers take up phosphate. Roots of land plants take up dissolved phosphate from the soil water ❺. Land animals get phosphate by eating plants or one another. Phosphorus returns to the soil in wastes and remains ❻. In the seas, phosphorus enters food webs when producers take up dissolved phosphate from seawater ❼. As on land, wastes and remains replenish the supply ❽.

Lack of phosphates often limits plant growth, so most fertilizers include phosphate. Phosphorus-rich droppings from seabird or bat colonies can be harvested as a natural fertilizer, but most commercial fertilizer contains phosphorus derived from rock that has been mined and then chemically treated.

Access to phosphorus limits growth of algae and cyanobacteria too, so adding phosphorus to an aquatic habitat allows a population explosion of these organisms. The result is an algal bloom that clouds water and threatens other aquatic species. Humans encourage algal blooms by allowing phosphate-containing detergents, sewage, fertilizer runoff, and waste from livestock to pollute aquatic environments.

aquifer Porous rock layer that holds some groundwater.

biogeochemical cycle A nutrient moves among environmental reservoirs and into and out of food webs.

groundwater Water between soil particles and in aquifers.

phosphorus cycle Movement of phosphorus among rocks, water, soil, and living organisms.

transpiration Evaporation of water from a plant's aboveground parts.

water cycle Water moves from its main reservoir—the ocean—into the air, falls as rain and snow, and flows back to the ocean.

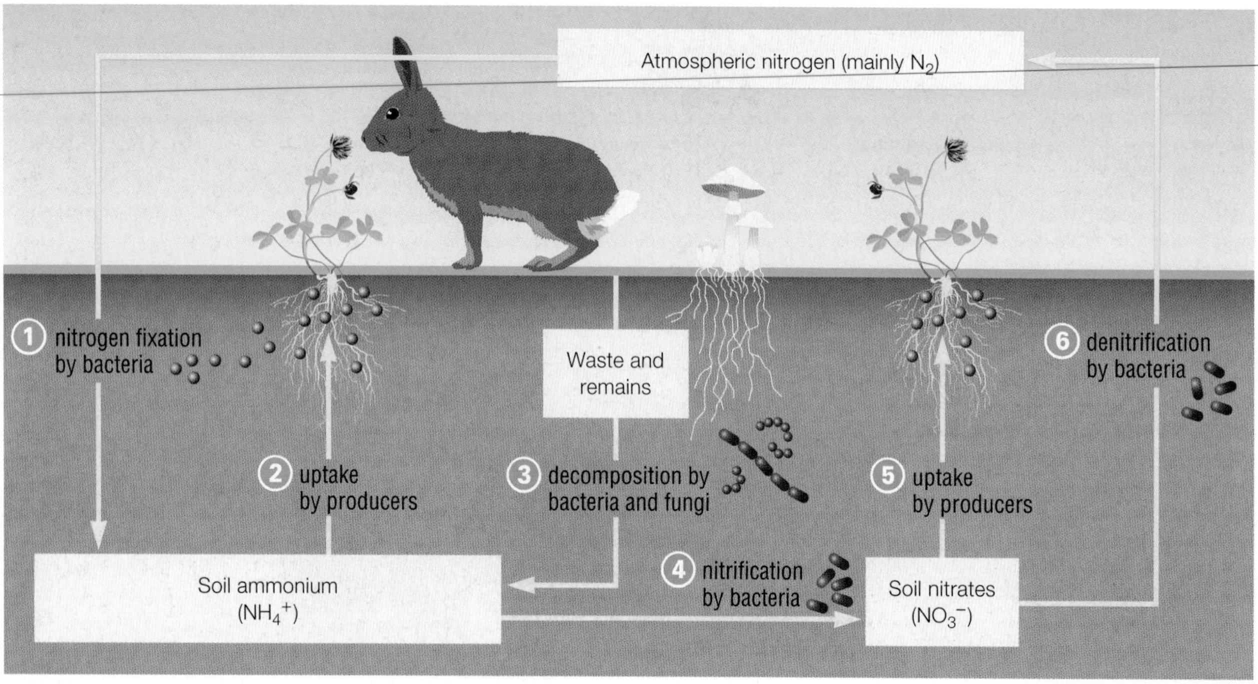

Figure 17.25 The nitrogen cycle. The main reservoir for nitrogen is the atmosphere. Activity of nitrogen-fixing bacteria converts gaseous nitrogen to forms that producers can use.

Figure It Out: What are the two forms of nitrogen that can be taken up by plants?

Answer: Ammonium and nitrate

carbon cycle Movement of carbon among rocks, water, the atmosphere, and living organisms.

nitrogen cycle Movement of nitrogen among the atmosphere, soil, and water, and into and out of food webs.

nitrogen fixation Conversion of nitrogen gas to ammonia.

The Nitrogen Cycle Earth's atmosphere, which is about 80 percent gaseous nitrogen (N_2), is the largest nitrogen reservoir. In the **nitrogen cycle**, nitrogen moves among the atmosphere, through reservoirs in soil and water, and into and out of food webs (Figure 17.25).

Plants cannot use gaseous nitrogen because they do not have an enzyme that can break the triple covalent bond between its two nitrogen atoms. Some bacteria do have such an enzyme, and these organisms carry out **nitrogen fixation**. They break the bonds in N_2, then use the nitrogen atoms to form ammonia, which dissolves and forms ammonium ions (NH_4^+) **1**. Plant roots take up ammonium from the soil **2** and use it in metabolic reactions. Consumers get nitrogen by eating plants or one another.

Additional ammonium is added to the soil when bacterial and fungal decomposers break down the nitrogen-rich wastes and remains of organisms **3**. Other soil bacteria obtain energy by converting ammonium to nitrate (NO_3^-), a process called nitrification **4**. Like ammonium, nitrate can be taken up from the soil and used by producers **5**. Ecosystems lose nitrogen as some types of bacteria convert nitrate to gaseous forms that escape into the atmosphere **6**.

In the early 1900s, German scientists discovered a method of fixing atmospheric nitrogen and producing ammonium on an industrial scale. This process allowed the manufacture of synthetic nitrogen fertilizers that have boosted crop yields and help to feed the rapidly increasing human population. However, use of fertilizers, along with other human activities, has added large amounts of nitrogen-containing compounds to our water and air. Nitrates commonly pollute drinking water in agricultural areas, raising the risk of thyroid cancer. Nitrous oxide, a gas released by bacteria in overfertilized soil and by the burning of fossil fuel, is a greenhouse gas and contributes to destruction of the ozone layer (the layer that filters out dangerous ultraviolet radiation from the sun).

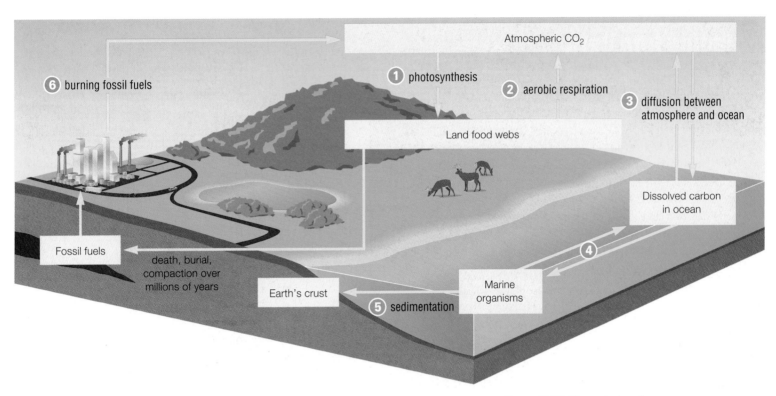

Figure 17.26 The carbon cycle.
Earth's crust is the largest carbon reservoir.

1. Carbon enters land food webs when plants take up carbon dioxide from the air and carry out photosynthesis.

2. Carbon returns to the atmosphere as carbon dioxide when plants and other land organisms carry out aerobic respiration.

3. Carbon diffuses between the atmosphere and the ocean. Carbon dioxide becomes bicarbonate when it dissolves in ocean water.

4. Marine producers take up bicarbonate for use in photosynthesis, and marine organisms release carbon dioxide produced by aerobic respiration.

5. Many marine organisms incorporate carbon into their shells. After they die, these shells become part of the sediments. Over time, these sediments become carbon-rich rocks such as limestone and chalk in Earth's crust.

6. Burning of fossil fuels derived from the ancient remains of plants adds additional carbon dioxide into the atmosphere.

The Carbon Cycle Carbon occurs abundantly in the atmosphere, combined with oxygen as carbon dioxide (CO_2). After water, carbon is the most abundant substance in living organisms. All molecules of life (carbohydrates, fats, lipids, and proteins) have a carbon backbone. In the **carbon cycle**, carbon moves among rocks, water, and the atmosphere, and into and out of food webs (Figure 17.26).

On land, plants take up and use carbon dioxide in photosynthesis ①. Plants and most other land organisms release carbon dioxide back to the atmosphere when they carry out aerobic respiration ②. Bicarbonate ions (HCO_3^-) form when carbon dioxide dissolves in water ③. Aquatic producers take up bicarbonate and convert it to carbon dioxide for use in photosynthesis ④. As on land, most aquatic organisms carry out aerobic respiration and release carbon dioxide.

Soil contains more than twice as much as carbon as the atmosphere. Soil carbon consists of organic wastes and remains along with living soil organisms. Over time, bacteria and fungi in the soil decompose organic material and release carbon dioxide into the atmosphere. The rate of decomposition and the carbon content of the soil varies with the regional climate. In the tropics, decomposition happens fast, so most carbon is stored in living plants, rather than in soil. By contrast, in temperate zone forests and grasslands, soil holds more carbon than the living plants. Soils hold the most carbon in the arctic, where low temperature hampers decomposition, and in peatbogs, where acidic, anaerobic conditions do the same.

Sedimentary rocks such as limestone constitute Earth's largest carbon reservoir. These rocks formed over millions of years by the compaction of the carbon-rich shells of marine organisms ⑤. Plants do not take up dissolved carbon from the soil, so carbon in these rocks is not readily accessible to organisms in land ecosystems.

Deposits of fossil fuels formed over hundreds of millions of years from carbon-rich remains ⑥. High pressure and temperature transformed the remains of ancient land plants to coal. A similar process transformed the remains of marine

Digging Into Data

Changes in Atmospheric Carbon Dioxide

To assess the impact of human activity on the carbon dioxide level in Earth's atmosphere, it helps to take a long view. One useful data set comes from deep core samples of Antarctic ice. The oldest ice core that has been fully analyzed dates back a bit more than 400,000 years. Air bubbles trapped in the ice provide information about the gas content in Earth's atmosphere at the time the ice formed. Combining ice core data with more recent direct measurements of atmospheric carbon dioxide—as in Figure 17.27—can help scientists put current changes in the atmospheric carbon dioxide into historical perspective.

1. What was the highest CO_2 level between 400,000 B.C. and 0 A.D.?
2. During this period, how many times did the CO_2 level exceed the CO_2 level observed in 1980?
3. The industrial revolution began around 1800. How did the CO_2 level change in the 800 years prior to this event? What about in the 175 years after it?
4. Which was greater, the rise in CO_2 level between 1800 and 1975 or the rise between 1980 and 2013?

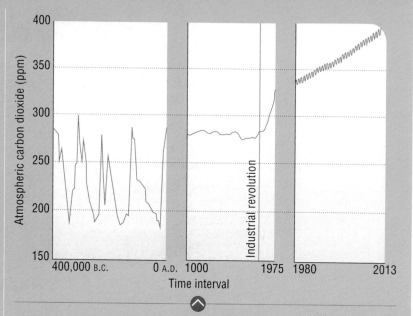

Figure 17.27 **Changes in atmospheric carbon dioxide (in parts per million).** Data from 1980 on are direct measurements. Earlier data are based on ice cores.

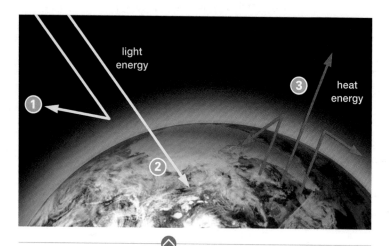

Figure 17.28 The greenhouse effect.

❶ Light energy from the sun is reflected by Earth's atmosphere or surface.

❷ More light energy reaches Earth's surface, and warms the surface.

❸ Earth's warmed surface emits heat energy. Some of this energy escapes through the atmosphere into space. But some is absorbed and then emitted in all directions by greenhouse gases. The emitted heat warms Earth's surface and lower atmosphere.

diatoms to deposits of oil and natural gas. Until recently, the carbon in fossil fuels, like the carbon in rocks, had little impact on ecosystems. At the present time, our burning of this fuel adds billions of tons of CO_2 to the atmosphere every year.

The Greenhouse Effect and Global Climate Change By burning fossil fuels and wood, humans have increased the amount of carbon dioxide and nitrogen oxides that enter the atmosphere. At the same time, we are cutting forests, thus decreasing the global uptake of carbon dioxide by plants. The result of these activities is a well-documented change in the composition of the atmosphere. For example, analysis of air samples taken at Mauna Loa Observatory in Hawaii show that between 1960 and the present the concentration of carbon dioxide rose from about 315 parts per million to nearly 400 parts per million. Analysis of air trapped in arctic ice and the composition of the shells of ancient marine organisms provide indirect evidence that the carbon dioxide content of Earth's atmosphere has not been this high for at least 15 million years.

These atmospheric changes can affect Earth's climate because carbon dioxide and nitrogen oxides are among the greenhouse gases. **Greenhouse gases** are atmospheric gases that slow the movement of heat from Earth to space (Figure 17.28). The process by which heat emitted by Earth's atmosphere warms its surface is called the **greenhouse effect**.

Figure 17.29 Global warming.
Variations in the annual average global temperature. The y-axis (vertical axis) is the deviation in degrees centigrade from the average temperature between 1901 and 2000. Blue bars indicate years that were cooler than this average, and red bars years that were warmer than this average.

Data from http://www.climate.gov/news-features/understanding-climate/climate-change-global-temperature.

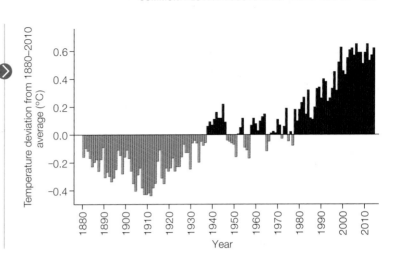

Given the greenhouse effect, we would predict that increases in atmospheric carbon dioxide and other greenhouse gases would raise the temperature of Earth's surface. Evidence from a variety of sources indicates that such global warming is under way (Figure 17.29). In the past 100 years, Earth's average temperature has risen by about 0.74°C (1.3°F), and the rate of warming is accelerating.

A rise of a degree or two in average temperature may not seem like a big deal, but it is enough to increase the rate of glacial melting, raise sea level, alter wind patterns, shift the distribution of rainfall and snowfall, and increase the frequency and severity of hurricanes. Scientists refer to the many climate-related effects of the rise in greenhouse gases as **global climate change**.

Earth's climate has varied greatly over its long history. During ice ages, much of the planet was covered by glaciers. Other periods were warmer than the present, and tropical plants and coral reefs thrived at what are now cool latitudes. Scientists can correlate some historical large-scale temperature changes with shifts in Earth's orbit, which varies in a regular fashion over 100,000 years, and Earth's tilt, which varies over 40,000 years. Changes in solar output and volcanic eruptions also influence Earth's temperature. However, most scientists see no evidence that any of these factors have a role in the current temperature rise.

In 2013, the Intergovernmental Panel on Climate Change reviewed the results of many scientific studies related to climate change. The panel included hundreds of scientists from all over the world. After reviewing all the data, the panel concluded that human activities, rather than any natural process, are the cause of the ongoing climate changes.

Human activities are the cause of ongoing climate change.

Take-Home Message 17.6

How do nutrients cycle between organisms and their environment?

- Water moves on a global scale from the ocean (its main reservoir), through the atmosphere, onto land, then back to the ocean.
- Phosphorus cycles between its main reservoir—rocks and sediments—and soils and water. Phosphorus enters food webs when producers take up dissolved phosphates. The atmosphere does not play a significant role in this cycle.
- Nitrogen moves from its main reservoir—the atmosphere—into soils and water, and into and out of food webs. Bacteria play a pivotal role in the nitrogen cycle by producing forms of nitrogen that producers can take up and use.
- Carbon's main reservoir is rocks, but most carbon enters food webs when producers take up dissolved carbon from water or carbon dioxide from the air.
- Human activities cause nutrient imbalances in land and aquatic habitats. Our activities also add excess greenhouse gases such as carbon dioxide and nitrogen oxides to the air. The rise in greenhouse gases is affecting global climate.

global climate change Wide-ranging changes in rainfall patterns, average temperature, and other climate factors that result from rising concentrations of greenhouse gases.

greenhouse effect Warming of Earth's lower atmosphere and surface as a result of heat trapped by greenhouse gases.

greenhouse gas Atmospheric gas that helps keep heat from escaping into space and thus warms the Earth.

Summary

Section 17.1 All species in a defined area are a **community**. Interactions among members of a community help keep populations in check. A species introduced to a new community without any natural enemies can increase in number and become a pest.

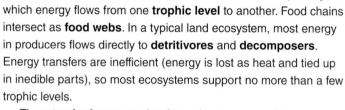

Section 17.2 Each species occupies a certain **habitat** and has a unique **niche**—the conditions and resources it requires, and the interactions it takes part in. The **species diversity** of a community is determined by nonbiological factors such as climate, as well as biological ones such as how species interact.

Section 17.3 Species interactions can result in coevolution. An association in which species live together is a **symbiosis**. In a **commensalism**, one species benefits and it neither helps nor harms the other. In a **mutualism**, two species exploit one another to their mutual benefit.

Interspecific competition harms both participants. **Competitive exclusion** occurs when species with identical resource needs share a habitat. **Resource partitioning** allows similar species to coexist.

Predation occurs when a free-living predator kills and eats its prey. In one type of **mimicry**, well-defended prey species have similar **warning coloration**. Less well-defended species also mimic well-defended ones. **Camouflage** can hide both predators and prey.

Herbivory may or may not kill a plant. **Parasites** withdraw nutrients from a host, usually without killing it. **Brood parasites** lay eggs in another's nest. **Parasitoids** are insects whose larvae develop inside and feed on a host, which they eventually kill.

Section 17.4 **Ecological succession** is the sequential replacement of arrays of species within a community. **Primary succession** occurs in habitats without soil. **Secondary succession** takes place in disturbed habitats. The first species in a community are **pioneer species**. Their presence may help other potential colonists. Community structure is not easily predicted. It is affected by physical factors, but also by random events and disturbances such as fires.

The presence of a **keystone species** has a large effect on community structure. Arrival of an **exotic species** can drastically alter a community.

Section 17.5 **Producers** in most **ecosystems** convert sunlight energy to chemical bond energy and take up nutrients that they and

the ecosystem's **consumers** require. A **food chain** is a path by which energy flows from one **trophic level** to another. Food chains intersect as **food webs**. In a typical land ecosystem, most energy in producers flows directly to **detritivores** and **decomposers**. Energy transfers are inefficient (energy is lost as heat and tied up in inedible parts), so most ecosystems support no more than a few trophic levels.

The rate of **primary production**—the capture and storage of energy by producers—varies with climate, season, and other factors. **Energy pyramids** show how available energy decreases as it is transferred from one trophic level to the next.

Section 17.6 In a **biogeochemical cycle**, water or a nutrient moves through the environment, then through organisms, then back to an environmental reservoir.

In the **water cycle**, water moves from the ocean into the atmosphere, falls on land, and flows back to the ocean, its main reservoir. **Transpiration** from plants releases water into the atmosphere. **Aquifers** and soil store **groundwater**, but most of Earth's fresh water is in the form of glacial ice.

In the **phosphorus cycle**, living things take up dissolved forms of phosphorus released by Earth's rocks and sediments. No gaseous form of phosphorus plays a role in this cycle.

The atmosphere is the main reservoir in the **nitrogen cycle**. Bacteria and the fertilizer industry carry out **nitrogen fixation** (convert atmospheric nitrogen to ammonium that plants can take up). Bacteria and fungi that act as decomposers also release ammonium. Nitrogen fertilizers commonly pollute water in agricultural regions.

The global **carbon cycle** moves carbon from its reservoirs in rocks and seawater, through its gaseous form (CO_2) in the atmosphere, and through living organisms.

Carbon dioxide and nitrogen oxides are **greenhouse gases**. Through the **greenhouse effect**, they trap heat in Earth's atmosphere and make life possible. An increase in greenhouse gases is causing **global climate change**.

Self-Quiz

Answers in Appendix I

1. A community consists of species _____ .
   a. that share a habitat
   b. that do not compete
   c. that share a gene pool
   d. with the same niche

2. Which of the following can be a symbiosis?
   a. interspecific competition
   b. predation
   c. commensalism
   d. all of the above

3. Match the species interaction with a suitable description.

_____ mutualism   a. A snake kills and eats a mouse.

_____ competition   b. A bee pollinates a flower while

_____ predation   sipping floral nectar.

_____ parasitism   c. An owl and a wood duck both

_____ herbivory   need a tree cavity to nest.

                         d. A mosquito sucks your blood.

                         e. A goat grazes on grass.

4. With interspecific competition, selection favors individuals of both species who are most _____ the competing species.

    a. similar to   b. different from

5. Parasitoids are _____ that lay eggs in their hosts.

    a. birds   b. reptiles   c. insects   d. fish

6. The establishment of a biological community on a newly formed volcanic island is an example of _____ .

    a. primary succession   c. competitive exclusion

    b. secondary succession   d. resource partitioning

7. Match the terms with suitable descriptions.

_____ producer   a. steals parental care

_____ brood parasite   b. feeds on small bits of

_____ decomposer   organic matter

_____ detritivore   c. degrades organic wastes and

_____ exotic species   remains to inorganic forms

                         d. captures sunlight energy

                         e. new to a community

8. Match each substance with its largest environmental reservoir. One reservoir choice will be used more than once.

_____ carbon   a. seawater

_____ water   b. rocks and sediments

_____ phosphorus   c. the atmosphere

_____ nitrogen

9. Earth's largest reservoir of fresh water is _____ .

    a. lakes   c. glacial ice

    b. soil water   d. water in the bodies of living organisms

10. _____ convert nitrogen gas to a form producers can take up.

    a. Fungi   c. Mammals

    b. Bacteria   d. Mosses

11. Land plants take up _____ for photosynthesis from the air.

    a. carbon dioxide   c. ammonium ions

    b. phosphate ions   d. nitrogen gas

12. Addition of _____ to water encourages algal blooms.

    a. carbon dioxide   c. salt

    b. phosphate ions   d. bicarbonate ions

13. A biological control agent is _____ a pest species.

    a. the prey of   c. mutualistic with

    b. a descendant of   d. a natural enemy of

14. Greenhouse gases _____ .

    a. trap heat in the atmosphere

    b. are released by burning of fossil fuels

    c. may cause global climate change if they accumulate

    d. all of the above

15. A(n) _____ species is one that arrives early in succession.

    a. keystone   b. pioneer   c. commensal   d. exotic

Critical Thinking

1. With antibiotic resistance increasing (Section 12.1), researchers are looking for ways to reduce use of these drugs. Some cattle once fed antibiotic-laced food now receive food containing harmless bacteria that can live in the animal's gut. The idea is that if these bacteria are in place, then harmful bacteria with the same resource needs are less likely to thrive. Explain why this idea makes sense in terms of species interactions.

2. Figure 17.9 shows a stingless fly that mimics a stinging wasp. Researchers have found that in such mimicry systems, mimics benefit most when they are rare relative to the well-protected model. Can you explain why?

3. Ectotherms such as invertebrates, fish, and amphibians convert more of the energy in the food they eat into body tissues than do endotherms such as birds and mammals. What allows ectotherms to make more efficient use of food energy?

Visual Question

1. Consider this hypothetical energy pyramid for a forest. Which tier shows how much of the energy captured by the forest's trees ends up in herbivores?

18

THE BIOSPHERE AND HUMAN EFFECTS

Application ❯ 18.1 Going With the Flow

REMEMBER: Earthquakes result from the movement of tectonic plates (Section 11.5). Radioisotopes break down in a predictable manner (2.2).

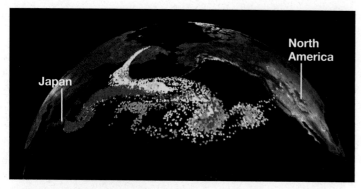

A. March 16, 2011 distribution of atmospheric ¹³⁷Cs from the Fukushima atomic power plant. Red denotes the highest concentration of radiation. Release of radioactive material into the air began on March 12, 2011.

B. A biologist takes samples from kelp growing along the coast of California in July of 2014. Analysis of the samples showed that deep sea currents had not yet delivered ¹³⁷Cs from Fukushima to California's coastal waters.

Figure 18.1 Flow of radioactive pollutants from Japan.

(A) NOAA; (B) Photo by David Nelson.

Earth's air and seawater are in constant circulation and their movements facilitate the global distribution of energy and nutrients. Movements of air and seawater can also carry pollutants released in one region far from their source. Consider what happened after a powerful earthquake that occurred off the coast of Japan triggered a huge tsunami (tidal wave). Together, the earthquake and tsunami killed more than 20,000 people and decimated some of Japan's coastal cities. In the coastal town of Fukushima, damage to a nuclear power plant caused the release of radioactive material into the air and sea.

Prevailing winds carried radioisotopes accidentally released into the air at Fukushima eastward (Figure 18.1A). Rain deposited the vast majority of this radioactive material in the Pacific Ocean, but some remained aloft and continued farther east. Radioisotopes from Fukushima were first detected along the west coast of North America about 60 hours after their release, and in Europe about a week later. Within 18 days, some radioisotopes released at Fukushima had circled the globe.

Materials move more slowly in the ocean. Think about the millions of tons of debris that was dragged into the ocean by the tsunami. Most of this material sank, but some stayed afloat and was carried along by surface currents that flow east from Japan toward the Americas. Scientists have been monitoring the movement of the floating material and recording when objects that are clearly tsunami debris turn up along the west coast of North America. Several Japanese boats lost during the tsunami came ashore on the west coast of the United States in 2013. One arrived in Washington State carrying five live fishes and a variety of invertebrates native to the tropical Pacific.

Deep sea currents move material more slowly than winds or surface currents. Such currents are carrying radioisotope-contaminated water that was released into the sea at Fukushima eastward. The water contains cesium 137 (¹³⁷Cs), a radioisotope with a half-life of 30 years. Scientists expect deep marine currents to deliver water enriched with ¹³⁷Cs to the west coast of North America from about 2015 to perhaps as late as 2020. Although this water will certainly contain more ¹³⁷Cs than normal seawater, scientists think it is unlikely to pose a threat to human health.

Scientists are closely monitoring the movement of radioisotopes from Fukushima and their effects on natural communities. For example, one group of scientists is documenting the level of radiation in kelp forests along the western coasts of North America. They collect kelp samples several times a year (Figure 18.1B) and test the samples for radioisotopes released from Fukushima. Kelp was chosen because of its important role in the coastal ecosystem and because it tends to take up and concentrate radioactive material, making it easier to detect.

Bits of kelp tested shortly after the accident at Fukushima showed the presence of iodine 131, a radioisotope produced in nuclear reactors, but rare in nature. Researchers assume the ¹³¹I traveled on the winds before falling into the sea and being taken up by the kelp. ¹³¹I has a half-life of 8 days, so it is no longer a concern and there is no evidence that exposure to it did any long-term harm to the kelp.

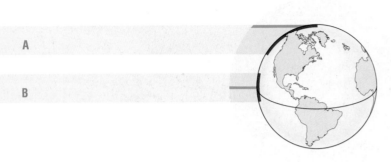

Figure 18.2 Variation in intensity of solar radiation with latitude. For simplicity, we depict two equal parcels of incoming radiation on an equinox, a day when incoming rays are perpendicular to Earth's axis.

Rays that fall on high latitudes **A** pass through more atmosphere (blue) than those that fall near the equator **B**. Compare the length of the green lines. Atmosphere is not to scale.

Also, energy in the rays that fall at the high latitude is spread over a greater area than energy that falls on the equator. Compare the length of the red lines.

18.2 Factors That Affect Climate

REMEMBER: The biosphere includes all regions of Earth where life can exist (Section 1.2).

What lives where in the biosphere depends in large part on differences in regional climate. **Climate** refers to average weather conditions, such as temperature, humidity, wind speed, cloud cover, and rainfall, over a long interval. Air circulation patterns and ocean currents influence a region's climate, which in turn affects the types of species that can live there.

Air Circulation Patterns A region's latitude (how far north or south it is) determines how much light energy it receives. On any given day, equatorial regions receive more sunlight than higher latitudes for two reasons (Figure 18.2). First, sunlight traveling to high latitudes passes through more atmosphere to reach Earth's surface than sunlight traveling to the equator. Fine particles of dust, water vapor, and greenhouse gases absorb some solar radiation or reflect it back into space, so less light energy reaches the poles. Second, the energy in any incoming parcel of sunlight is spread out over a smaller surface area at the equator than at the higher latitudes. As a result of these factors, Earth's surface warms more at the equator than at the poles.

Latitudinal differences in surface warming, and the resulting effects on air and water, give rise to global patterns of air circulation and rainfall (Figure 18.3). At the equator, intense sunlight warms the air and causes evaporation of water from the ocean. As the air heats up, it expands and rises ❶. The same effect causes a hot-air balloon to inflate and rise when the air inside it is heated. As the equatorial air mass rises, it flows north and south and begins to cool. Cool air can hold less moisture than warm air, so moisture leaves the air as rain that supports tropical rain forests.

When the air reaches about 30° north and south latitude, it has cooled and dried out. Being cool, the air sinks downward ❷. When it descends, it draws moisture from the soil. As a result, deserts form at around 30° north and south latitude.

Air that continues flowing along Earth's surface toward the poles once again picks up heat and moisture. By a latitude of about 60° it has become warm and moist, and it rises, giving up moisture as rain ❸. In polar regions, cold air with little moisture descends ❹. Precipitation is sparse, and polar deserts form.

Landforms also affect rainfall. For example, moisture-laden air masses give up water as rain as they rise over coastal mountains. The dry region on the leeward side of such mountains is referred to as a **rain shadow**.

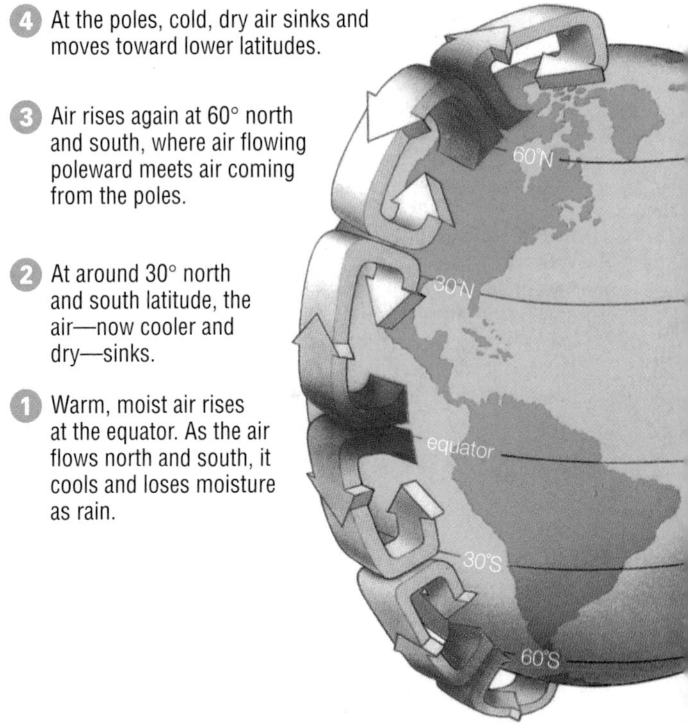

❹ At the poles, cold, dry air sinks and moves toward lower latitudes.

❸ Air rises again at 60° north and south, where air flowing poleward meets air coming from the poles.

❷ At around 30° north and south latitude, the air—now cooler and dry—sinks.

❶ Warm, moist air rises at the equator. As the air flows north and south, it cools and loses moisture as rain.

Figure 18.3 Air circulation patterns that result from latitudinal differences in the amount of solar radiation reaching Earth.

climate Average weather conditions in a region over a long time period.

rain shadow Dry region on the downwind side of a coastal mountain range.

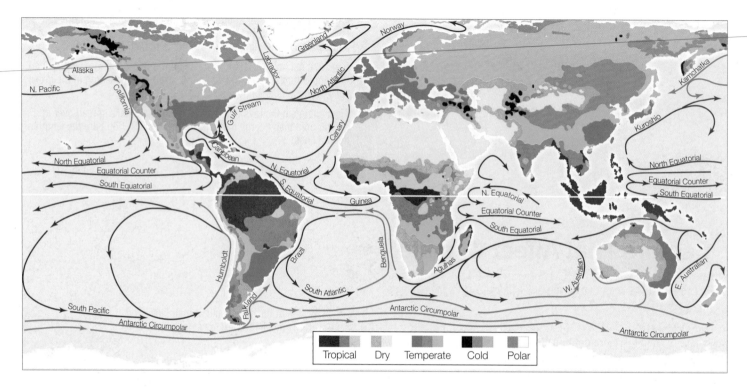

Figure 18.4 Global ocean circulation.
Warm surface currents start moving from the equator toward the poles, but prevailing winds, Earth's rotation, gravity, the shape of ocean basins, and landforms influence the direction of flow. Water temperatures, which differ with latitude and depth, contribute to the regional differences in air temperature and rainfall.

NASA.

Ocean Circulation Latitudinal and seasonal variations in sunlight cause water to heat and cool. At the equator, where vast volumes of water warm and expand, the sea level is about 8 centimeters (3 inches) higher than at either pole. Existence of this slope causes sea surface water to move in response to gravity, from the equator toward the poles. As the water moves, it warms the air above it. At midlatitudes, oceans transfer 10 million billion calories of heat energy to the air every second!

Enormous volumes of water flow as ocean currents. The force of major winds, Earth's rotation, and topography determine the directional movement of these currents. Surface currents circulate clockwise in the Northern Hemisphere and counterclockwise in the Southern Hemisphere (Figure 18.4).

Swift, deep, and narrow currents of nutrient-poor water flow away from the equator along the east coast of continents. Along the east coast of North America, warm water flows north, as the Gulf Stream. Slower, shallower, broader currents of cold water parallel the west coast of continents and flow toward the equator.

biome Any of Earth's major land ecosystems, characterized by climate and main vegetation and found in several regions.

boreal forest Biome dominated by conifers that can withstand the cold winters at high northern latitudes.

temperate deciduous forest Biome dominated by broadleaf trees that grow in warm summers, then drop their leaves and go dormant during cold winters.

tropical rain forest Multilayered forest that occurs where warm temperatures and continual rains allow plant growth year-round. Most productive and species-rich biome.

Take-Home Message 18.2

What causes winds and ocean currents that affect climate?

- Longitudinal differences in the amount of solar radiation reaching Earth produce global air circulation patterns.
- Warmed water at the equator expands and flows "downhill" toward the poles. Wind, Earth's rotation, and topography affect the movement, establishing ocean currents that redistribute heat.
- The collective effects of air movements, ocean currents, and landforms determine regional temperature and moisture levels.

18.3 The Major Biomes

REMEMBER: Different mechanisms of fixing carbon adapt plants to different environmental conditions (Section 5.4). Deciduous plants lose their leaves seasonally (14.6).

Scientists classify the world's land ecosystems into a variety of **biomes**, each characterized by its climate and main type of vegetation. Figure 18.5 shows the distribution of the major biomes. A biome typically includes multiple nonadjacent regions.

Forest Biomes Evergreen broadleaf (angiosperm) trees dominate **tropical rain forests** of equatorial Asia, Africa, and South America (Figure 18.6A). Abundant rain, warm temperatures, and consistent day length allow plants to grow year-round. As a result of continual growth, rain forests take up more carbon dioxide and release more oxygen per unit area than any other biome. Thus, they are sometimes described as Earth's "lungs." Tropical rain forest is the most structurally complex biome, with many species of trees reaching 30 meters (100 feet) in height. Vines twine up the trees, and orchids and ferns grow on tree trunks and branches. Little light reaches the forest floor, so few plants grow there. Constant warmth encourages rapid decomposition and prevents leaf litter from accumulating. Thus, soil is relatively nutrient-poor. Per unit area, tropical rain forest holds more species of plants and animals than any other biome. It is the oldest biome, and its age is key to its diversity. Some modern rain forests have existed for more than 50 million years, which has provided ample time for many evolutionary branchings to occur.

Tropical regions where rains fall seasonally support tropical dry forests. The broadleaf trees that dominate such forests are smaller than rain forest trees and most lose their leaves and become dormant in the dry season.

Trees of **temperate deciduous forests** shed leaves and become dormant in preparation for winter, when conditions do not favor growth (Figure 18.6B). In spring, deciduous trees flower and put out new leaves. At the same time, leaves that were shed and accumulated during the prior autumn decay to form a rich soil. During the growing season, a somewhat open canopy allows sunlight to reach the ground, where shorter understory plants flourish.

Boreal forest, the conifer-dominated forest that sweeps across northern Asia, Europe, and North America (Figure 18.6C), is the most extensive biome. It is also referred to as taiga, which is Russian for "swamp forest," because rains that fall during the cool summers keep the soil soggy. Winters are dry and cold. The conifers are mainly spruce, fir, and pine. The conical shape of these trees helps them shed snow, and their needlelike leaves help minimize evaporative water loss during the winter when they cannot take up water from the frozen ground.

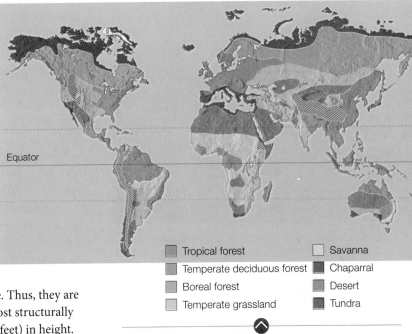

Equator

■ Tropical forest	□ Savanna
■ Temperate deciduous forest	■ Chaparral
■ Boreal forest	■ Desert
□ Temperate grassland	■ Tundra

Figure 18.5 Major land biomes.
Most biomes include areas on more than one continent.

From Russell/Wolfe/Hertz/Starr. *Biology*, 2e. © 2011 Brooks/Cole, a part of Cengage Learning®.

Figure 18.6 Examples of forest biomes.

(A) © Antonio Jorge Nunes/Shutterstock.com; (B) © James Randklev/Corbis; (C) © Serg Zastavkin, Used under license from Shutterstock.com.

A. Tropical rain forest.

B. Temperate deciduous forest in New England.

C. Boreal forest in Siberia during summer.

A. North American prairie, a temperate grassland, where bison are among the native grazers.

B. African savanna, a tropical grassland with scattered shrubs. It supports huge herds of grazing wildebeest.

C. Chaparral in California. Shrubby drought-resistant plants with small, leathery leaves predominate.

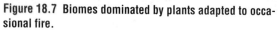

Figure 18.7 Biomes dominated by plants adapted to occasional fire.

(A) © Danny Barron; (B) Jonathan Scott/Planet Earth Pictures; (C) Jack Wilburn/Animals Animals.

Grasslands and Chaparral Grasslands are dominated by perennial grasses and other nonwoody plants that tolerate grazing, intervals of drought, and periodic fires—conditions that prevent trees and shrubs from taking over. Grasslands have highly fertile soil. They typically occur in the middle of continents and border shrublands or desert.

North America's temperate grasslands, called **prairies**, once covered much of the continent's interior, where summers are hot and winters are cold and snowy. The prairies supported herds of elk, pronghorn antelope, and bison (Figure 18.7A) that were prey to wolves. Today, these predators and prey are largely absent from most of their former range. Nearly all prairies have been plowed, and their rich soil now sustains production of wheat and other crops.

Savannas are tropical grasslands with a few scattered shrubs and trees. They lie between the tropical forests and hot deserts of Africa, India, and Australia. Temperatures are warm year-round, and there is a distinct rainy season. Africa's savannas are famous for their abundant wildlife. Herbivores include giraffes, zebras, elephants, a variety of antelopes, and immense herds of wildebeests (Figure 18.7b). Lions and hyenas eat the grazers.

Chaparral is a biome dominated by drought-resistant, fire-adapted shrubs that have small, leathery leaves. It occurs along the western coast of continents, between 30 and 40 degrees north or south latitude. Mild winters bring a moderate amount of rain, and the summer is hot and dry. Chaparral is California's most extensive ecosystem (Figure 18.7C). It also occurs in regions bordering the Mediterranean, as well as in Chile, Australia, and South Africa.

Deserts Low annual precipitation defines **desert**, a biome that covers about one-fifth of Earth's land surface. Many deserts are at about 30° north and south latitude, where global air circulation cause dry air to sink. Rain shadows also reduce rainfall. For example, the Himalayas prevent rain from falling in China's Gobi desert.

Lack of rainfall keeps the humidity in a desert low. With little water vapor to block rays, intense sunlight reaches and heats the ground. At night, the lack of insulating water vapor in the air allows the temperature to fall fast. As a result, deserts tend to have larger daily temperature shifts than other biomes.

Despite their harsh conditions, most deserts support some plant life (Figure 18.8A). Cacti are a family of plants native to Western Hemisphere deserts. All cacti are CAM plants (Section 5.4), which carry out gas exchange at night, when potential for evaporative water loss is lowest. When it does rain, cacti take up water and store it in their spongy tissues for use in drier times. Leaves modified as spines help cacti fend off animals that would like to tap this source of water.

Many deserts also contain woody shrubs such as mesquite and creosote. These plants have extensive root systems that take up the little water that is available. Mesquite roots can extend up to 60 meters (197 feet) beneath the soil surface. Desert shrubs often put out leaves only after a rain. Annual desert plants sprout and reproduce fast while soil remains moist after a rain.

Tundra **Tundra**, the biome with the coldest temperature, lies between the polar ice cap and the belts of boreal forests in the Northern Hemisphere. Most tundra is in northern Russia and Canada. Tundra is Earth's youngest biome, having appeared about 10,000 years ago when glaciers retreated at the end of the last ice age. Emergence of land from beneath these glaciers opened the way to a process of primary succession that has produced the currently existing communities. Snow blankets the arctic tundra for as many as nine months of the year. During the brief summer,

A. Mojave Desert after the winter rains. Annual plants sprout, flower, produce seeds, and die within weeks beneath slow-growing, drought-adapted, perennial cacti.

B. Arctic tundra during the summer, when shrubby plants grow under nearly continuous sunlight. Permafrost underlies the upper layer of defrosted soil.

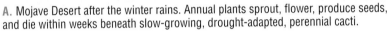

Figure 18.8 Desert and tundra, biomes with extreme climates.

(A) © George H. Huey/Corbis; (B) © Darrell Gulin/Corbis.

plants grow fast under nearly continuous sunlight (Figure 18.8B). Lichens and shallow-rooted, low-growing plants are the producers for food webs that include voles, arctic hares, caribou, arctic foxes, wolves, and brown bears. Many migratory birds nest in the tundra during the summer, when the air is thick with insects.

Even in midsummer, only the surface layer of tundra soil thaws. Below that, a layer of permanently frozen soil called **permafrost** can be as thick as 500 meters (1,600 feet). Permafrost acts as a barrier that prevents drainage, so the soil above it remains perpetually waterlogged. Cool, anaerobic conditions slow decay, so organic remains accumulate, making the permafrost one of Earth's greatest stores of carbon. As global temperatures rise, the amount of frozen soil that melts each summer is increasing. With warmer temperatures, carbon stored in the soil is entering the atmosphere and encouraging further warming.

Tropical rain forest is Earth's oldest biome, and tundra is its youngest.

Take-Home Message 18.3

How do environmental factors affect life in the major biomes?

- Near the equator, year-round rains and warmth support highly productive, species-rich tropical rain forests. At higher latitudes, the trees of temperate deciduous forests drop their leaves and become dormant during winter. At still-higher latitudes, conifers dominate Northern Hemisphere boreal forests.
- Plants of African savannas and temperate grasslands are adapted to grazing and periodic fire. Shrubby chaparral plants are also adapted to fire.
- Deserts receive little rain, so perennial desert plants have traits that reduce their water loss or allow them to store water or tap into water deep in the soil. Annual desert plants complete their life cycle in a brief period after a rain.
- The most northerly and youngest biome is arctic tundra, where low plants are adapted to a short growing season and soil with a layer of permafrost.

chaparral Biome with cool, wet winters and hot dry summers; dominant plants are shrubs with small, leathery leaves.

desert Biome with little precipitation; its perennial plants are adapted to withstand drought.

permafrost Layer of permanently frozen soil in the Arctic.

prairie Temperate grassland biome of North America. Its grasses and other plants are adapted to recover after grazing and the occasional fire.

savanna Tropical biome dominated by grasses and other plants adapted to grazing, as well as a scattering of shrubs.

tundra Northernmost biome, dominated by low plants that grow over a layer of permafrost.

Figure 18.9 Low-nutrient lake.
Crater Lake in Oregon formed when a collapsed volcano began to fill with snowmelt about 7,700 years ago. From a geologic standpoint, it is a young lake, and its clear water is a sign of its low primary productivity.

© Lindsay Douglas/Shutterstock.

Figure 18.10 Examples of coastal ecosystems.

(A) © H. Mark Weldman Photography/Alamy; (B) © Douglas Peebles/Corbis.

A. South Carolina estuary dominated by cordgrass.

B. Florida wetland dominated by red mangroves.

18.4 Aquatic Ecosystems

REMEMBER: Photosynthetic protists are important producers in many aquatic ecosystems (Section 13.6). Succession changes community structure (17.4).

We can distinguish between types of aquatic ecosystems, just as we do biomes on land. Temperature, salinity, rate of water movement, and depth influence the composition of aquatic communities.

Freshwater Ecosystems A lake is a body of standing fresh water. All but the shallowest lakes have zones that differ in their physical characteristics and species composition. Near shore, where sunlight penetrates all the way to the lake bottom, rooted aquatic plants and algae that attach to the bottom are primary producers. A lake's open waters include an upper well-lit zone and, if a lake is deep or cloudy, a zone where light does not penetrate. Producers in the well-lit water include photosynthetic protists and bacteria. In the deeper dark zone, consumers feed on organic debris that drifts down from above.

A lake undergoes succession, meaning the community of lake organisms changes over time. A newly formed lake is deep, clear, and has few nutrients and a low primary productivity (Figure 18.9). As sediments accumulate, the lake becomes shallower. Nutrients accumulate and encourage growth of photosynthetic bacteria, diatoms, and other producers that cloud the water. These producers serve as food for tiny crustaceans, which are then eaten by fish.

Streams are flowing-water ecosystems. They typically originate from runoff or melting snow or ice. As they flow downslope, they grow and merge to form rivers. Properties of a stream or river vary along its length. The type of rocks a stream flows over can affect its solute concentration, as when limestone rocks dissolve and add calcium to the water. Shallow water that flows rapidly over rocks mixes with air and holds more oxygen than slower-moving, deeper water. Also, cold water holds more oxygen than warm water. As a result, different parts of a stream or river support species with different oxygen needs. For example, only cool, well-oxygenated water can support rainbow trout.

Marine Ecosystems An **estuary** is a mostly enclosed coastal region where seawater mixes with nutrient-rich fresh water from rivers and streams. Water inflow continually replenishes nutrients, so estuaries have a high primary productivity. Photosynthetic bacteria and protists that live on mudflats often account for a large portion of an estuary's primary production. Plants adapted to withstand changes in water level and salinity also serve as producers. For example, cordgrass (*Spartina*) is the dominant plant in the salt marshes of many estuaries along the Atlantic coast (Figure 18.10A). Cordgrass can withstand immersion during high tides, and specialized glands on its leaves excrete the salt that its roots take up with seawater.

Estuaries and tidal flats of tropical and subtropical latitudes often support nutrient-rich mangrove wetlands (Figure 18.10B). "Mangrove" is the common term for certain salt-tolerant woody plants. The plants have prop roots that extend out from their trunk and help the plant stay upright in the soft sediments. Specialized cells at the surface of some exposed roots allow gas exchange with air.

Along ocean shores, organisms are adapted to the mechanical force of the waves and to tidal changes. Many species are underwater during high tide, then exposed to the air when the tide is low. Along rocky shores, where waves prevent detritus from piling up, algae that cling to rocks are the producers in grazing food chains. In contrast, waves continually rearrange loose sediments along sandy shores, and make it difficult for algae to take hold. Here, detrital food chains start with organic debris from land or offshore.

Warm, shallow, well-lit tropical seas hold **coral reefs**, formations made primarily of calcium carbonate secreted by generations of corals, which are invertebrate animals (Section 15.3). Like tropical rain forests, tropical coral reefs are home to an extraordinary assortment of species (Figure 18.11A). The main producers in a coral reef community are photosynthetic dinoflagellates that live inside the reef-building corals. These protists provide their coral hosts with sugars. If an environmental change such as a shift in temperature stresses the coral, the coral expels its protist symbionts in a response called coral bleaching. If conditions improve fast, the symbiont population in the coral can recover. But when adverse conditions persist, the symbionts will not be restored, and the coral will die.

In the brightly lit waters of the open ocean, photosynthetic protists and bacteria are the primary producers, and grazing food chains predominate. Depending on the region, some light may penetrate as far as 1,000 meters (more than a half mile) beneath the sea surface. Below that, organisms live in darkness, and organic material that drifts down from above is the basis of detrital food chains.

On the seafloor, the greatest species richness occurs along the edges of continents. There are also some largely unexplored concentrations of biodiversity on seamounts and at hydrothermal vents. **Seamounts** are underwater mountains that stand 1,000 meters or more tall, but still lie below sea surface (Figure 18.11B). Seamounts attract large numbers of fishes and are home to many marine invertebrates. Like islands, they harbor many species that evolved there and live nowhere else. There are more than 30,000 seamounts, and scientists have just begun to document species that live on them.

Hot, mineral-rich water spews out from the ocean floor at **hydrothermal vents**. When mineral-rich hot water mixes with cold seawater, the minerals settle out as extensive deposits. Bacteria and archaea that can extract energy from these deposits serve as primary producers for food webs that include invertebrates such as tube worms and crabs (Figure 18.11C). As explained in Section 13.2, one hypothesis holds that life originated near hydrothermal vents.

A. Coral reef near Fiji.

B. Computer model of seamounts on the seafloor near the coast of Alaska. Patton Seamount, at the rear, stands 3.6 kilometers (about 2 miles) tall.

C. Hydrothermal vent community. Bacteria and archaea that extract energy from minerals are the producers here. Consumers include crabs and giant tube worms (close-up in inset photo) that grow meters long without ever eating. The worms rely on bacteria that live in their tissues to produce their food.

Figure 18.11 Regions of high biodiversity in the sea.

(A) © John Easley, www.johneasley.com. (B) NOAA; (C) NOAA/Photo courtesy of Cindy Van Dover, Duke University Marine Lab, inset © Peter Batson/imagequestmarine.com.

Take-Home Message 18.4

What factors shape aquatic ecosystems?

- Fast-flowing, cooler water holds more oxygen than warmer, slower-moving water, and the amount of available light decreases with the water's depth.
- In well-lit upper waters, producers such as aquatic plants, algae, and photosynthetic microbes are the base for grazing food chains. On coral reefs, the main producers are photosynthetic protists that live inside the coral's tissues.
- Detritus drifting down from above sustains most deep-water communities in lakes and oceans. However, hydrothermal vent communities on the ocean floor are sustained by energy that bacteria and archaea harvest from minerals.

coral reef In tropical sunlit seas, a formation composed of secretions of coral polyps that serves as home to many other species.

estuary A highly productive ecosystem where nutrient-rich water from a river mixes with seawater.

hydrothermal vent Place where hot, mineral-rich water streams out from an underwater opening in Earth's crust.

seamount An undersea mountain.

18.5 Human Impact on the Biosphere

Figure 18.12 Driven to extinction by humans.
Artist's depiction of a dodo, a flightless bird, about a meter (3.3 feet) tall. It became extinct in the 1600s, after European sailors discovered its island home.

Figure 18.13 Species threatened by overharvesting.

(A) John Butler, NOAA; (B) © George Sanker/naturepl.com.

A. White abalone. There are now too few individuals left to reproduce in wild. The species was overharvested for use as human food.

B. Black rhino. Since the 1960s, their number has been reduced from about 100,000 to about 5,000. They are killed for their horn, which is used in traditional Asian medicine.

REMEMBER: Acidity is measured as pH (Section 2.5). An asteroid impact 66 million years ago caused extinction of the dinosaurs (11.1). The size of the human population has soared in the past century (16.5). Ultraviolet light is a mutagen (6.4). Increases in greenhouse gases are causing global climate change (17.6).

The rising size of the human population and its increasing industrialization have far-reaching effects on the biosphere. We begin by discussing how human activities affect individual species, then turn to their wider impacts.

Increased Species Extinctions Extinction, like speciation, is a natural process. Species arise and become extinct on an ongoing basis. The rate of extinction picks up dramatically during a **mass extinction**, an event in which many different kinds of organisms become extinct in a relatively short period. We are currently in the midst of such an event. Unlike previous extinction events, this one is not the inevitable result of a physical catastrophe such as an asteroid impact. Humans are the driving force behind the current rise in extinction rate.

There is indirect evidence that humans had a role in the prehistoric decline of many large animals, collectively referred to as megafauna. In North America, large herbivores such as camels, giant ground sloths, and mammoths and mastodons (relatives of elephants) became extinct after the arrival of humans about 15,000 years ago. Carnivores such as lions and saber-toothed cats also disappeared. By one hypothesis, humans directly caused declines in herbivorous species such as mammoths by hunting them. These declines then led to the extinction of their predators. Evidence that humans hunted some megafauna supports this hypothesis. For example, one 13,800-year-old mastodon rib bone found in Washington State had a spear point embedded in it.

More recent extinctions were certainly caused by humans. Consider what happened to the dodo (Figure 18.12) a large, flightless bird that lived on the island of Mauritius in the Indian Ocean. Dodos were plentiful in 1600, when Dutch sailors first arrived on the island, but 80 or so years later the birds were extinct. Sailors ate some, but destruction of nests and habitat by rats, cats, and pigs introduced by the sailors probably had a greater effect.

For the purposes of conservation, a species is considered extinct if repeated, extensive surveys of its known range repeatedly fail to turn up signs of any individuals. It is "extinct in the wild" if the only known members of the species are in captivity. Some species are difficult to find, so occasionally a population of a species previously thought to be extinct or extinct in the wild turns up. However, this is a rare occurrence.

An **endangered species** is one currently at a high risk of extinction in the wild. A **threatened species** is one that is likely to become endangered in the near future. Keep in mind that not all rare species are threatened or endangered. Some species have always been uncommon.

Overharvesting is one cause of current species declines. Consider the case of the white abalone, a gastropod mollusk native to kelp forests along the coast of California (Figure 18.13A). During the 1970s, harvest of this species for human food reduced the population to about 1 percent of its original size. In 2001, the abalone became the first invertebrate to be listed as endangered by the U.S. Fish and Wildlife Service. The white abalone's current population density is too low for effective reproduction in the wild. Abalones release their eggs and sperm into the water, a strategy that is effective only when many individuals live close to one another. In an attempt to save the species, white abalones are now being bred in captivity.

Species are overharvested not only as food, but also for the pet trade, and for use as ornaments or in traditional medicines. For example, black rhinos (Figure 18.13B) are threatened by poachers who kill them for their horn, which sells on the black market for nearly $30,000 per pound. The main market for rhino horn is Asia, where there is a mistaken belief that rhino horn has medicinal properties.

Each species requires a specific type of habitat, and any loss, degradation, or fragmentation of that habitat reduces population numbers. An **endemic species**, one that remains confined to the area in which it evolved, is more likely to go extinct than a species with a more widespread distribution. For example, giant pandas are endemic to China's bamboo forests and feed mainly on bamboo (Figure 18.14A). As these forests disappeared, so did pandas. Their population, which may once have been as high as 100,000 animals, is now reduced to about 1,600 animals in the wild.

Humans also degrade habitats in less direct ways. Consider what is happening to Edwards Aquifer in Texas. The aquifer consists of water-filled, underground limestone formations that supply drinking water to the fast-growing city of San Antonio. Excessive withdrawals of water from this aquifer, along with pollution of the water that recharges it, threaten species that live in the aquifer's depths. The Texas blind salamander (Figure 18.14B), which is endemic to this aquifer, is one such species.

Deliberate or accidental species introductions damage habitats too. Section 17.1 detailed the ways that red imported fire ants accidentally introduced to the United States from South America now threaten native species. Similarly, rats and domestic cats are decimating many of Hawaii's endemic bird species.

Decline or loss of one species sometimes endangers others. Consider buffalo clover, a plant that was abundant when many buffalo grazed in the American Midwest (Figure 18.14C). The clover thrived at the edges of woodlands, where buffalo enriched the soil with their droppings and helped to disperse the clover's seeds. Like most endangered species, buffalo clover faces a number of threats. In addition to the loss of buffalo, the clover is threatened by competition from introduced plants, attacks by introduced insects, and development of its habitat for housing.

The extent of the species declines is staggering. The International Union for Conservation of Nature (IUCN) reports that of the 48,677 named species they assessed, 36 percent were threatened or endangered. We do not know the level of threat for the vast majority of the approximately 1.8 million named species, or for the millions of species yet to be discovered.

endangered species A species that faces extinction in all or part of its range.

endemic species A species that evolved in one place and is found nowhere else.

mass extinction Event in which many species in many habitats become extinct in the same interval.

threatened species A species likely to become endangered in the near future.

Figure 18.14 Species currently threatened by habitat destruction and degradation.

(A) Hung Chung Chih/Shutterstock.com; (B) Joe Fries, U.S. Fish & Wildlife Service; (C) USDA Forest Service/Photo by Sarena Selbo.

A. Panda.

B. Texas blind salamander.

C. Buffalo clover.

Figure 18.15 Deforestation.
In Brazil, a tractor plows a field that once was forest. The soybeans grown in the field will be used to feed cattle and poultry.

Frontpage/Shutterstock.com.

Figure 18.16 Results of desertification.
Wind blows dust from the expanding Sahara Desert into the Atlantic Ocean.

© Geoeye Satellite Image.

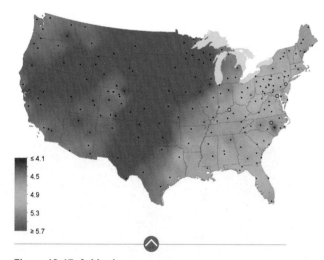

Figure 18.17 Acid rain.
Average precipitation acidities in the United States in 2012.

National Atmospheric Deposition Program/National Trends Network.

Deforestation and Desertification **Deforestation**, the removal of trees from an area, is an ongoing threat in many regions. Although the amount of forested land is currently stable or increasing in North America, Europe, and China, tropical forests continue to suffer heavy losses (Figure 18.15). Destruction of tropical rain forests is not merely a regional concern. As noted earlier, these forests have the highest primary production of any land biome, so their decline significantly affects carbon storage and oxygen production. In addition, these forests harbor the greatest number of species of any biome. Thus, destruction of an area of tropical rain forest endangers far more species than destruction of a comparable area of temperate zone or boreal forest.

Poor agricultural practices can encourage soil erosion and lead to a rapid shift from grassland or woodland to desert, a process called **desertification**. During the mid-1930s, large portions of prairie on North America's southern Great Plains were plowed, exposing the soil to the force of the region's constant winds. Coupled with a drought, the result was an economic and ecological disaster. Winds carried more than a billion tons of topsoil aloft, turning the region into what came to be known as the Dust Bowl. A similar situation now exists in Africa, where expansion of the Sahara desert causes clouds of dust to take flight over the Atlantic (Figure 18.16).

Drought encourages desertification, which results in more drought in a positive feedback cycle. Plants cannot thrive in a region where the topsoil has blown away. With less transpiration (evaporation from plant parts), less water enters the atmosphere, so local rainfall decreases.

The best way to prevent desertification is to avoid farming in areas subject to high winds and periodic drought. If these areas must be used, methods that do not repeatedly disturb the soil can minimize the risk of desertification.

Acid Rain A **pollutant** is a natural or man-made substance released into soil, air, or water in greater than natural amounts. It disrupts the physiological processes of organisms that evolved in its absence, or that are adapted to lower levels of it. **Acid rain** forms when air pollutants released by burning coal and other fossil fuels combine with water vapor and fall to Earth. The resulting rainfall can be ten times more acidic than normal (Figure 18.17). Acid rain that falls onto or drains into waterways, ponds, and lakes harms aquatic organisms. For example, heightened acidity impairs

≤ 4.1
4.5
4.9
5.3
≥ 5.7

Figure It Out: Is rain more acidic on the East Coast or the West Coast of the United States?

Answer: The East Coast

Digging Into Data

Accumulation and Transport of Radioisotopes by Tuna

Bluefin tuna are among the many fish species that were exposed to radioactive material released into the sea by the damaged Fukushima nuclear power plant. These tuna breed in the western Pacific, and some migrate to the Eastern Pacific when they are 1 to 2 years old. Daniel Madigan suspected that migrating bluefin tuna could transport radioactive material to the waters off the coast of California. To test this hypothesis, he and his collaborators analyzed the concentration of cesium radioisotopes (^{137}Cs and ^{134}Cs) in bluefin tuna and yellowfin tuna caught near San Diego in August 2011. Yellowfin tuna do not cross the Pacific; they spend their lives in California's coastal waters. The scientists compared these data with the radioisotope content of bluefin tuna that had been caught near San Diego in 2008. Figure 18.18 shows their results.

1. How did the radioisotope content of the bluefin tuna caught in August of 2011 compare to that of bluefins caught in 2008 (before the Fukushima accident)?
2. How did it compare to that of yellowfins caught in August of 2011?
3. Do these data support the hypothesis that bluefin tuna transported radioactive cesium from Fukushima in their body?
4. Bluefin tuna and yellowfin tuna are similarly sized top predators. Why was it important to compare species that were both the same size and at the same trophic level?

Tuna type, date sampled	Mean body mass (kg dry)	Mean age (years)	Mean radioisotope concentration (Bq per kilogram)	
			^{137}Cs	^{134}Cs
Bluefin, 2011 ($n = 15$)	1.5	1.5	6.3	4
Bluefin, 2008 ($n = 5$)	1.5	1.4	1.4	0
Yellowfin, 2011 ($n = 5$)	1.9	1.2	1.1	0

Figure 18.18 Radioactive cesium (Cs) concentration in two types of tuna caught along the coast of California.
^{137}Cs has a half-life of 30 years and some persists in the Pacific from weapons testing. ^{134}Cs has a half-life of 2 years and was undetectable in the Pacific prior to the accident at Fukushima. Becquerels (Bq) is a measure of radioactivity.

Adapted from the Proceedings of the National Academy of Sciences of the United States of America, Daniel J. Madigan, Zofia Baumann, and Nicholas S. Fisher, "Pacific bluefin tuna transport Fukushima-derived radionuclides from Japan to California."

---

the growth of diatoms, which are silica-shelled protists that serve as producers in many lakes. When acid rain falls on forests, it burns tree leaves and encourages loss of nutrient ions from the soil. As a result, trees become malnourished and more susceptible to disease.

In the United States, acid deposition peaked in the 1970s. Since then, federal regulations limiting sulfur dioxide emissions have helped reduce the acidity of precipitation. The world's greatest sulfur dioxide emitters are now China and India, where industrialization and coal use continue to increase.

Biological Accumulation and Magnification Two processes can cause the concentration of a chemical pollutant in an organism's body to rise far above its concentration in the environment. First, some chemical pollutants taken up by organisms accumulate in their tissues. As a result, older individuals contain more of the pollutant than younger ones. Second, by the process of **biological magnification**, a chemical pollutant passes through food chains, becoming increasingly concentrated in higher and higher trophic levels. For example, methylmercury released by coal-burning power plants is a toxic pollutant that can reach a very high concentration in large predatory fishes such as swordfish, bluefin tuna, and albacore (white) tuna. Methylmercury interferes with development of the nervous system, so pregnant women, nursing women, and children should avoid eating these fish.

The Trouble With Trash Historically, humans buried unwanted material in the ground or dumped it out at sea. Trash was out of sight, and also out of mind. We now understand that chemicals seeping from buried trash can contaminate

acid rain Rain containing sulfuric and/or nitric acid; forms when pollutants mix with water vapor in the atmosphere.

biological magnification A chemical pollutant becomes increasingly concentrated as it moves through a food chain.

deforestation Removal of all trees from a forested area.

desertification Conversion of grassland or woodlands to desertlike conditions.

pollutant A natural or man-made substance that is released into the environment in greater than natural amounts and that damages the health of organisms.

Figure 18.19 Death by plastic.
When scientists dissected this recently deceased Laysan albatross chick they found more than 300 pieces of plastic. The chick died after one piece punctured its gut wall.

Clair Fackler/ NOAA.

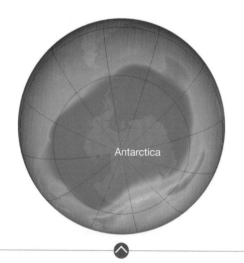

Figure 18.20 Ozone hole.
This graphic shows the September 2006 ozone hole, which was the largest ever recorded. Purple indicates the least ozone, with blue, green, and yellow indicating increasingly higher levels.

Check the current status of the ozone hole at NASA's website (http://ozonewatch.gsfc.nasa.gov/).

NASA Ozone Watch.

groundwater. The United States no longer dumps its trash at sea, but plastic and other garbage still enter our coastal waters. Foam cups and containers from fast-food outlets, foam peanuts, plastic shopping bags, plastic water bottles, and other material discarded as litter enter storm drains in coastal regions. From there it is carried to streams and rivers that can convey it to the sea. Waste plastic that enters the ocean poses a threat to marine life. For example, seabirds often eat floating bits of plastic or feed the plastic to chicks, with deadly results (Figure 18.19).

Ocean currents can carry bits of plastic for thousands of miles. These plastic bits can end up accumulating in some areas of the ocean. Consider the Great Pacific Garbage Patch, a region of the north central Pacific that the media often describes as an "island of trash." In fact, the plastic is not easily visible. Rather, the garbage patch is a region where a high concentration of confetti-like plastic particles swirl slowly around an area as large as the state of Texas.

Destruction of the Ozone Layer Between 17 and 27 kilometers (10.5 and 17 miles) above sea level, the concentration of ozone gas (O_3) is so great that scientists refer to this region as the **ozone layer**. The ozone layer benefits land organisms by absorbing most of the ultraviolet (UV) radiation in incoming sunlight. UV radiation, remember, can damage DNA and thus cause mutations.

In the mid-1970s, scientists noticed that the ozone layer was thinning. Its thickness had always varied a bit with the season, but now there was steady decline from year to year. By the mid-1980s, the spring ozone thinning over Antarctica was so pronounced that people were referring to the low-ozone region as an "ozone hole." Declining ozone quickly became an international concern. The ozone hole over the South Pole was a sign that the ozone layer was thinning.

Chlorofluorocarbons, or CFCs, were determined to be the main ozone destroyers. At the time, these odorless gases were widely used as propellants in aerosol cans, as coolants, and in solvents and plastic foam. In response to the potential threat posed by ozone thinning, countries worldwide agreed in 1987 to phase out the production of certain CFCs and other ozone-destroying chemicals. As a result of that agreement (the Montreal Protocol), the atmospheric concentrations of these pollutants is no longer rising dramatically.

Unfortunately, the ozone hole persisted (Figure 18.20). CFCs break down quite slowly, so scientists expect the existing pollutants to remain at a level that significantly impairs the ozone layer for many decades. In addition, our emission of other ozone-destroying chemicals continues to increase. The most notable of these chemicals is nitrous oxide (N_2O), which forms mainly as a result of the use of synthetic nitrogen fertilizers. Soil bacteria convert nitrogen from the fertilizer to nitrous oxide. Burning organic material, including fossil fuels, also contributes nitrous oxide to the atmosphere.

Global Climate Change The climate change resulting from our pouring greenhouse gases into the atmosphere has effects on ecosystems worldwide. For example, sea level is rising. An increase in average annual temperature elevates sea level by two mechanisms. First, water expands as it is heated. Second, heating melts glaciers and sea ice (Figure 18.21). Together, the processes have caused the sea level to rise about 20 centimeters (8 inches) in the past century. As a result, some coastal wetlands have disappeared underwater.

Until recently, ice sheets covered the Arctic Ocean year-round, making it difficult for ships to move to and from the Arctic landmass. Those ice sheets are now shrinking (Figure 18.21B). At the same time, ice on the Arctic landmass is melting.

Muir Glacier 1941

Muir Glacier 2004

A. Retreating glaciers in Alaska.

These changes will accelerate global warming because exposed soil absorbs more heat than ice, which is highly reflective. The changes will also make it easier for people to remove minerals and fossil fuels from the Arctic. With the world supply of fuel and minerals dwindling, pressure to exploit Arctic resources is rising. However, conservationists warn that extracting these resources will harm Arctic species, such as the polar bear, that are already threatened by other factors.

The temperature of the land and seas affects evaporation, winds, and currents. As a result, many weather patterns are expected to change as temperature rises. For example, warmer temperatures are correlated with extremes in rainfall patterns, with periods of drought interrupted by unusually heavy rains. Warmer seas are also expected to increase the intensity of hurricanes.

Climate change is already having widespread effects on biological systems. Arctic animals, such as polar bears and walruses, that would normally spend most of their time on ice are being forced onto land. In addition, temperature changes are cues for many temperate zone species, so warmer-than-normal springs cause deciduous trees to leaf out earlier, and spring-blooming flowers to blossom earlier. Animal migration times and breeding seasons are also shifting. Some species benefit from the warming, expanding their range to higher latitudes or elevations that were previously too cool to sustain them. Other species are harmed by the rising temperature. For example, warming of tropical waters stresses corals and has increased the frequency of coral bleaching.

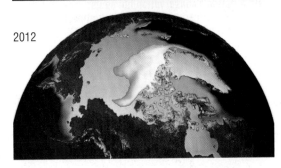
1980

2012

B. Reduced permanent ice in the arctic. The bright white is the ice that persists all year. The larger light blue-gray area shows the maximum ice coverage in the winter. Graphics are based on data from satellites.

Figure 18.21 Evidence of a warming world.

(A left) National Snow and Ice Data center, W. O. Field; (A right) National Snow and Ice Data Center, B. F. Molnia; (B) NASA.

Take-Home Message 18.5

How do human activities affect the biosphere?

- We are increasing the rate of species extinctions by degrading, destroying, and fragmenting habitats, overharvesting species, and introducing exotic species.
- Human activities turn grasslands into deserts, strip woodlands of trees, and generate pollutants and trash that kill organisms and damage ecosystems.
- Some pollutants have global effects, as when CFCs cause thinning of the ozone layer, and increasing levels of greenhouse gases bring about climate change.

ozone layer Region of upper atmosphere with a high ozone concentration; acts as a sunscreen against UV radiation.

biodiversity Of a region, the genetic diversity within its species, variety of species, and variety of ecosystems.

conservation biology Field of applied biology that surveys biodiversity and seeks ways to maintain and use it.

ecological restoration Actively altering an area in an effort to restore or create a functional ecosystem.

indicator species A species that is particularly sensitive to environmental changes and can be monitored to assess whether an ecosystem is threatened.

The extinction of a species removes its unique assortment of traits from the world of life forever.

Figure 18.22 The Klamath–Siskiyou forest, one of North America's conservation hot spots. Endangered northern spotted owls (inset) nest in old-growth trees of this forest.

David Patte, USFWS; inset, USFWS.

18.6 Maintaining Biodiversity

REMEMBER: As a result of the lack of gene flow between species, each species has a unique assortment of traits (Section 12.5).

The Value of Biodiversity Every nation enjoys several forms of wealth: material wealth, cultural wealth, and biological wealth, or biodiversity. We measure a region's **biodiversity** at three levels: the genetic diversity within its species, its species diversity, and its ecosystem diversity.

Why should we protect biodiversity? From a purely selfish standpoint, doing so is an investment in our future. Healthy ecosystems are essential to the survival of our own species. Other organisms produce the oxygen we breathe and the food we eat. They also remove waste carbon dioxide from the air and decompose and detoxify other wastes. Plants take up rain and hold soil in place, preventing erosion and reducing the risk of flooding.

Compounds discovered in wild species often serve as medicines. Two widely used chemotherapy drugs, vincristine and vinblastine, were extracted from the rosy periwinkle, a low-growing plant native to Madagascar's rain forests. Cone snails that live in tropical seas are the source of a highly potent pain reliever.

Wild relatives of crop plants serve as reservoirs of genetic diversity that plant breeders can draw upon to protect and improve crops. Wild plants often have genes that make them more resistant to disease or adverse conditions than their domesticated relatives. Plant breeders can use traditional cross-breeding methods or biotechnology to introduce genes from wild species into domesticated ones, thus creating improved varieties.

A decline in some species can warn us that the natural support system we depend on is in trouble. **Indicator species** are particularly sensitive to environmental change, and can be monitored as indicators of environmental health. For example, a decline in lichens tells us that air quality is deteriorating. The loss of mayflies from a stream tells us that the quality of water in a stream is declining.

In addition, there are ethical reasons to preserve biodiversity. As we have emphasized many times, all living species are the result of an ongoing evolutionary process that stretches back billions of years. Each species has a unique combination of traits. The extinction of a species removes its unique collection of traits from the world of life forever.

Conservation Biology Biodiversity is currently in decline at all three levels and in regions worldwide. **Conservation biology** addresses these declines. The goals of this relatively new field of biology are to survey the range of biodiversity, and to find ways to maintain and use biodiversity in a manner that benefits human populations. The aim is to conserve as much biodiversity as possible by encouraging people to value it and use it in ways that do not destroy it.

With so many species and ecosystems at risk, conservation biologists must often make difficult choices about what areas to target for protection first. To prioritize their efforts, the biologists identify conservation hot spots, places that are richest in endemic species and under the greatest threat. The goal of prioritizing is to save representative examples of all of Earth's existing biomes. By focusing on hot spots, rather than individual species, scientists hope to maintain ecosystem processes that sustain biological diversity.

Conservation scientists of the World Wildlife Fund have defined 867 distinctive land ecoregions that they consider the top priority for conservation efforts. Each region has a large number of endemic species and is under threat. The Klamath–Siskiyou forest in southwestern Oregon and northwestern California is one hot spot (Figure 18.22). It is home to many rare conifers, and two endangered birds, the northern spotted owl and the marbled murrelet, nest in the forest's old growth. Endangered coho salmon breed in streams that run through it. Logging is the main threat to this region, but a newly introduced pathogen that infects conifers also poses a concern.

Protecting biological diversity can be a tricky proposition. Even in developed countries, people often oppose environmental protections because they fear such measures will have adverse economic consequences. However, taking care of the environment can make good economic sense. With a bit of planning, people can both preserve and profit from their biological wealth. For example, Costa Rica's Monteverde Cloud Forest Reserve protects more than 100 mammal species, 400 bird species, and 120 species of amphibians and reptiles. It is one of the few habitats left for the ocelot, puma, and jaguar (Figure 18.23). Each year, about 75,000 tourists visit the reserve, and the feeding, lodging, and guiding of these tourists provides much-needed employment to local people. These people also benefit from the ecological services that their forest continues to provide.

Ecological Restoration Sometimes, an ecosystem is so damaged, or there is so little of it left, that conservation alone is not enough to sustain biodiversity. **Ecological restoration** is work designed to bring about the renewal of a natural ecosystem that has been degraded or destroyed, fully or in part. Restoration work in Louisiana's coastal wetlands is an example. Louisiana contains more than 40 percent of the coastal wetlands in the United States. These marshes are an ecological and economic treasure, but they are in trouble. Dams and levees built upstream of the marshes hold back sediments that would normally replenish sediments lost to the sea. Channels cut through the marshes for oil exploration and production have encouraged erosion, and the rising sea level threatens to flood existing plants. Since the 1940s, Louisiana has lost an area of marshland the size of Rhode Island. Restoration efforts now under way aim to reverse some of those losses and protect remaining marsh species (Figure 18.24).

Figure 18.23 Jaguars, one of the many endangered species that live in Costa Rica's Monteverde Cloud Forest Reserve.

© Adolf Schmidecker/FPG/Getty Images.

Figure 18.24 Ecological restoration in Louisiana's Sabine National Wildlife Refuge.

(A) Diane Borden-Bilot, U.S. Fish and Wildlife Service; (B) U.S. Fish and Wildlife Service.

A. Where marsh has become open water, sediments are barged in and marsh grasses are planted on them.

B. Restoration protects native species such as the roseate spoonbills that nest in the marsh during the summer.

A. Right, Bingham copper mine near Salt Lake City, Utah. The mine is 4 kilometers (2.5 miles) wide and 1,200 meters (0.75 miles) deep.

B. Below, pelican covered with oil that accidentally leaked from a deepwater oil well in the Gulf of Mexico.

Figure 18.25 Environmental effects of resource extraction.

(A) © Lee Prince/Shutterstock.com; (B) Joel Sartore/National Geographic Creative.

Reducing Human Impacts Ultimately, the health of our planet depends on our ability to recognize that the principles of energy flow and of resource limitation that govern the survival of all systems of life do not change. We must take note of these principles and find a way to live within our limits. The goal is living sustainably, which means meeting the needs of the present generation without reducing the ability of future generations to meet their own needs.

Promoting sustainable living begins with acknowledging the environmental consequences of our own lifestyle. People in industrial nations use enormous quantities of resources, and the extraction and delivery of these resources have negative effects on biodiversity. In the United States, the size of the average family has declined since the 1950s, while the size of the average home has doubled. All the materials used to build and furnish those larger homes come from the environment. For example, an average new home contains about 500 pounds of copper in its wiring and plumbing. Where does copper come from? Like most other mineral elements used in manufacturing, it is generally mined from the ground (Figure 18.25A). Mining often generates pollution and produces ecological dead zones.

Nonrenewable mineral resources are also used in electronic devices such as phones, computers, televisions, and MP3 players. Constantly trading up to the newest device may be good for the ego and the economy, but it is bad for the environment. Reducing consumption by fixing existing products promotes sustainability, as does recycling. Recycling nonrenewable materials reduces the need for extraction of those resources from the environment, so it helps maintain healthy ecosystems.

Reducing energy use is another way to promote sustainability. Fossil fuels such as petroleum, natural gas, and coal supply most of the energy used by developed countries. You already know that use of these nonrenewable fuels contributes to global warming and acid rain. In addition, extraction and transportation of these fuels have negative impacts. Oil harms many species when it leaks from pipelines, ships, or wells (Figure 18.25B). Renewable energy sources do not produce

Figure 18.26 Volunteers restoring the Little Salmon River in Idaho so that salmon can migrate upstream to their breeding grounds.

Mountain Visions/NOAA.

greenhouse gases, but they have drawbacks too. For example, dams that generate renewable hydroelectric power may discourage endangered salmon from returning to streams above the dam to breed. Similarly, wind turbines can harm birds and bats. Panels used to collect solar energy are made using nonrenewable mineral resources, and manufacturing the panels generates pollutants.

In short, all commercially produced energy has some kind of negative environmental impact, so the best way to minimize that impact is to use less energy. Shop for energy-efficient appliances, avoid incandescent lightbulbs, and do not leave lights on in empty rooms. Walking, bicycling, and using public transportation are energy-efficient alternatives to driving. Shopping locally and purchasing locally produced goods also saves energy.

If you want to make a difference, learn about the threats to ecosystems in your own area. Are species threatened? If so, what are the threats? How can you help reduce those threats? Support efforts to preserve and restore local biodiversity. Many ecological restoration projects are supervised by trained biologists but carried out primarily through the efforts of volunteers (Figure 18.26).

Keep in mind that unthinking actions of billions of individuals are the greatest threat to biodiversity. Each of us may have little impact on our own, but our collective behavior, for good or for bad, will determine the future of the planet.

Take-Home Message 18.6

What is biodiversity and how can we sustain it?

- Biodiversity is the genetic diversity of individuals of a species, the variety of species, and the variety of ecosystems. Worldwide, biodiversity is declining at all of these levels.
- Conservation biologists are working to identify threatened regions with high biodiversity and prioritize which receive protection.
- Ecological restoration is the process of re-creating or renewing a diverse natural ecosystem that has been destroyed or degraded.
- Individuals can help maintain biodiversity by using resources in a sustainable fashion.

Summary

Section 18.1 Circulation of Earth's seas and atmosphere can distribute pollutants that were released in one part of the world across the globe.

Section 18.2 **Climate** refers to average weather conditions over time. Variations in climate depend largely upon differences in the amount of solar radiation reaching different parts of Earth. The closer a region is to the equator, the more solar energy it receives. Warming of air and water at the equator sets in motion global patterns of air circulation and ocean currents. Circulating air and water distribute heat and moisture. Landforms also influence climate, as when coastal mountains cause a **rain shadow**.

Section 18.3 **Biomes** are categories of major ecosystems on land. They are maintained largely by regional variations in climate and described mainly in terms of their dominant plant life.

Near the equator, high rainfall and mild temperatures support **tropical rain forests**, which are dominated by trees that remain leafy and active all year. This is Earth's most productive and oldest biome. In **temperate deciduous forests**, trees grow during warm summers, then lose their leaves and become dormant in winter. **Boreal forest**, dominated by conifers, is the most extensive biome. It is found only in the Northern Hemisphere where winters are cold and dry, and summers are cool and rainy.

Grasslands form at midlatitudes in the interior of continents and are dominated by plants adapted to grazing. **Prairie** is a North American grassland biome; **savanna** is an African one that also includes scattered shrubs. **Chaparral**, a biome dominated by shrubby plants with tough leaves, occurs in regions with cool, wet winters and hot, dry summers. Both grasses and chaparral plants are adapted to withstand periodic fires.

Deserts occur at latitudes about 30° north and south, where dry air descends and annual rainfall is sparse. Desert plants are adapted to withstand drought or complete their life cycle fast after a rain.

Tundra forms at high latitudes in the Northern Hemisphere. It is the youngest biome and has an underlying layer of **permafrost**.

Section 18.4 Aquatic ecosystems have gradients of sunlight penetration, water temperature, salinity, and dissolved gases. Lakes are standing bodies of water. Different communities of organisms live at different depths and distances from the shore.

Streams and rivers are flowing bodies of water. Physical characteristics of a stream or river that vary along its length influence the types of organisms that live in it. A semi-enclosed area where nutrient-rich water from a river mixes with seawater is an **estuary**, a highly productive habitat. Seashores may be rocky or sandy. Grazing food chains based on algae form on rocky shores. Detrital food chains dominate sandy shores.

Coral reefs are species-rich ecosystems found in well-lit tropical waters. Photosynthetic protists that live in the coral's tissues are the main producers in this ecosystem.

The open ocean's upper waters hold photosynthetic organisms that form the basis for grazing food chains. Deeper water communities usually subsist on material that drifts down from above. However, bacteria and archaea that can obtain energy from minerals serve as the producers at **hydrothermal vent** ecosystems. **Seamounts** are undersea mountains that have a high species richness.

Section 18.5 Humans are causing a **mass extinction** by overharvesting, degrading and destroying habitats, and introducing exotic species. **Endangered species** are currently at risk of extinction; **threatened species** are likely to become endangered. **Endemic species** are more vulnerable to extinction than widely dispersed ones. Human activities can also threaten entire ecosystems, as when poor agricultural practices cause **desertification** or **deforestation** destroys a forest.

Pollutants such as those that cause **acid rain** threaten forest ecosystems. **Biological magnification** occurs when a pollutant is passed along a food chain. Trash that gets into fresh water can make its way into the oceans, where it degrades marine ecosystems. Ozone is a pollutant near the ground, but depletion of the **ozone layer** is a global threat caused by use of chemicals called CFCs.

Global warming caused by rising concentrations of greenhouse gases is melting polar ice and raising the sea level.

Section 18.6 **Biodiversity** includes the diversity of genes, species, and ecosystems. All three levels of biodiversity are declining. A decrease in biodiversity can harm humans. We rely on ecosystems to produce oxygen and decompose waste. We also benefit from

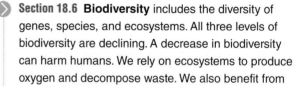

many compounds produced by wild species, and by tapping their genetic diversity to enhance our crops. Loss of **indicator species** warns us that a habitat is being degraded.

Conservation biologists document the extent of biodiversity and look for ways to preserve it, while benefiting humans. They give priority to areas with high biodiversity that are most threatened. **Ecological restoration** is the work of actively renewing an ecosystem that has been damaged or destroyed. By living sustainably, we can maintain biodiversity and ensure that resources remain for future generations.

Self-Quiz

Answers in Appendix I

1. The most solar radiation reaches the ground at _____ .
   a. the equator
   b. the North Pole
   c. midlatitudes
   d. the South Pole

2. When air is heated, it _____ and can hold _____ water.
   a. sinks; less
   b. sinks; more
   c. rises; less
   d. rises; more

3. Most North American _____ has been converted to cropland.
   a. tundra
   b. prairie
   c. desert
   d. boreal forest

4. Plants in _____ are adapted to periodic fires.
   a. deserts
   b. boreal forests
   c. tropical rain forests
   d. chaparral

5. Permafrost underlies _____ .
   a. savanna
   b. tundra
   c. desert
   d. prairie

6. The oldest and most productive biome is _____ .
   a. boreal forest
   b. tundra
   c. tropical rain forest
   d. desert

7. Bacteria and archaea that can obtain energy from minerals are the main producers at _____ .
   a. hydrothermal vents
   b. estuaries
   c. coral reefs
   d. seamounts

8. Which would hold more oxygen?
   a. a fast-moving, cool stream
   b. a warm pond

9. Match the biome with the most suitable description.
   _____ tundra
   _____ chaparral
   _____ desert
   _____ prairie
   _____ estuary
   _____ boreal forest
   _____ coral reef
   _____ tropical rain forest

   a. fresh water and seawater mix
   b. low humidity, and little rainfall
   c. North American grassland
   d. fire-adapted shrubs with tough leaves
   e. low-growing plants over permafrost
   f. most extensive biome
   g. main producers are protists
   h. broadleaf trees are active all year

10. An _____ species has population levels so low it is at great risk of extinction in the near future.
    a. endemic
    b. endangered
    c. indicator
    d. exotic

11. An _____ species can be monitored to gauge the health of its environment.
    a. endemic
    b. endangered
    c. indicator
    d. exotic

12. The 1930s environmental disaster known as the Dust Bowl is an example of _____ .
    a. deforestation
    b. desertification
    c. ecological restoration
    d. species extinction

13. The ozone layer _____ .
    a. is getting thicker
    b. helps keep Earth warm
    c. is a layer in the deep ocean
    d. screens out UV radiation

14. Acid rain _____ .
    a. harms aquatic organisms
    b. kills trees
    c. is a type of pollution
    d. all of the above

15. As a result of biological magnification, _____ tend to have the highest concentration of pollutants in their bodies.
    a. plants
    b. algae
    c. predators
    d. herbivores

Critical Thinking

1. Like the salamander shown in Figure 18.14B, many animals that live in caves are blind but had sighted ancestors. By what process might a sighted animal species that colonized a cave lose its sight?

2. In one seaside community in New Jersey, the U.S. Fish and Wildlife Service suggested trapping and removing feral cats (domestic cats that live in the wild). The goal was to protect some endangered wild birds (plovers) that nested on the town's beaches. Many residents were angered by the proposal, arguing that the cats have as much right to be there as the birds. Do you agree? Why or why not?

3. Two arctic marine mammals that live in the same waters differ in the level of pollutants in their bodies. Bowhead whales have a lower pollutant load than ringed seals. What are some factors that might explain this difference?

4. The extent of damage done by acid rain depends to some extent on the type of rocks in the area where it falls. Acid rain that falls in a region with limestone or other calcium carbonate–rich rocks does less harm than acid rain that falls in a region without such rocks. How might the presence of limestone lessen the effects of an input of acid into an ecosystem? Hint: Review Section 2.5.

Answers to Self-Quizzes

Chapter 1

1.	b	1.2
2.	c	1.2
3.	d	1.3
4.	a	1.3
5.	c	1.3
6.	Animals	1.4
7.	c	1.3
8.	d	1.3
9.	a, d, e	1.2–1.4
10.	a, b	1.2, 1.4
11.	domains	1.4
12.	b	1.5
13.	b	1.5, 1.6
14.	c	1.2
	e	1.6
	b	1.4
	g	1.6
	d	1.5
	a	1.5
	f	1.3
15.	b	1.6

Chapter 2

1.	d	2.2
2.	Hydrogen	2.2
3.	a	2.3
4.	c	2.4
5.	c	2.4
6.	a	2.5
7.	e	2.7, 2.10
8.	c	2.8
9.	e	2.8
10.	b	2.9, 2.10
11.	d	2.9
12.	c	2.4
	b	2.2
	d	2.4
	a	2.2
	f	2.4
	e	2.2
	g	2.4
13.	a	2.10
14.	c	2.8
	e	2.7
	f	2.8
	d	2.10
	a	2.9
	b	2.10
15.	f	2.9
	a	2.8
	b	2.8
	c	2.10
	d	2.7
	e	2.10
	h	2.8
	g	2.9

Chapter 3

1.	c	3.2
2.	c	3.2, 3.4
3.	c	3.2
4.	c	3.2
5.	False	3.2
6.	b	3.3
7.	c	3.5
8.	c	3.3
9.	a	3.5
10.	a	3.5
11.	c, b, d, a	all 3.5
12.	d	3.5
13.	a	3.5
14.	b	3.5
15.	c	3.5
	f	3.5
	e	3.4, 3.5
	d	3.5
	a	3.5
	g	3.4, 3.5
	b	3.5

Chapter 4

1.	c	4.2
2.	b	4.2
3.	a	4.3
4.	c	4.3
5.	c	4.3, 4.4
6.	a	4.4
7.	d	4.4
8.	b	4.5, 4.6
9.	more, less	4.5
10.	c	4.5, 4.6
11.	a	4.5
12.	b	4.5
13.	b	4.6
14.	d	4.6
15.	c	4.3
	e	4.6
	f	4.2
	b	4.3
	a	4.4
	g	4.5
	h	4.6
	j	4.4
	i	4.4

Chapter 5

1.	b	5.1
2.	a	5.1, 5.2
3.	c	5.2
4.	b	5.3
5.	c	5.3
6.	c	5.3
7.	b	5.4
8.	a	5.4
9.	False	5.5
10.	c	5.5
11.	b	5.5
12.	d	5.6
13.	d	5.7
14.	a and d	5.3 and 5.5
15.	c	5.5
	a	5.5
	d	5.5
	f	5.2
	g	5.1
	e	5.4
	b	5.2, 5.3

Chapter 6

1.	d	6.1
2.	c	6.2
3.	c	6.2
4.	b	6.2
5.	b	6.2
6.	a	6.3
7.	b	6.3
8.	b	6.3
9.	a	6.4
10.	d	6.4
11.	c	6.4
12.	d	8.3, 8.4
13.	f	6.4
14.	d	6.4
15.	d	6.2
	c	6.1
	b	6.3
	a	6.4
	f	6.4
	e	6.2
	g	6.4

Chapter 7

1.	c	7.2
2.	c	7.2
3.	a	7.2
4.	b	7.2, 7.4
5.	a	7.3, 7.5
6.	a	7.4
7.	c	7.4
8.	d	7.5
9.	d	7.6
10.	b	7.7
11.	d	7.7
12.	d	7.7
13.	c	7.7
14.	True	7.7
15.	c	7.2, 7.3
	a	7.3
	d	7.5
	b	7.2, 7.5
16.	c	7.7
	g	7.6
	h	7.3
	e	7.4
	a	7.7
	f	7.7
	b	7.7
	d	7.5

Chapter 8

1.	d	8.2
2.	b	8.2
3.	a	8.2
4.	a	8.2
5.	c	8.2
6.	a	8.2
7.	b	8.4
8.	c	8.4
9.	b	8.5
10.	b	8.5
11.	d	8.5
12.	b	8.5
13.	b	8.5
14.	b, e, a, c, d	all 8.2
15.	c	8.2
	f	8.2
	a	8.3
	h	8.2
	g	8.5
	e	8.3
	b	8.5
	d	8.5

Chapter 9

1.	b	9.2
2.	a	9.2
3.	c	9.3
4.	c	9.3
5.	b	9.7
6.	False	9.4, 9.5
7.	c	9.4
8.	d	9.5
9.	b	9.6
10.	X from mom, Y from dad	9.7
11.	b	9.8
12.	True	9.8
13.	c	9.8, 9.9
14.	b	9.3
	d	9.3
	a	9.2
	c	9.2
15.	b	9.8
	a	9.6
	d	9.8
	e	9.2
	f	9.2
	c	9.7

Chapter 10

1.	c	10.2
2.	a	10.2
3.	b	10.2
4.	b	10.2
5.	e	10.2, 10.3
6.	b	10.3
7.	a	10.2
8.	b	10.3
9.	b	10.5
	d	10.2, 10.4
	e	10.5
	f	10.4
10.	e, a, d, b, c	10.2, 10.3
11.	c	10.4
12.	True	10.4
13.	b	10.5
14.	True	10.5
15.	c	10.3
	f	10.4
	d	10.5
	b	10.1
	a	10.4
	e	10.4

Chapter 11

1. b — 11.2
2. c — 11.2, 11.6
3. d — 11.3
4. b — 11.3
5. b — 11.4
6. True — 11.4
7. c — 11.4
8. a — 11.2, 11.4, 11.5
   c — 11.5
   d — 11.4, 11.5
   e — 11.2, 11.4, 11.5
   f — 11.1
9. Gondwana — 11.5
10. c — 11.1
11. d — 11.6
12. d — 11.6
13. b — 11.7
14. j — 11.3
   i — 11.4
   e — 11.3
   k — 11.7
   f — 11.4
   c — 11.3
   b — 11.3
   g — 11.6
   h — 11.6
   d — 11.4
   a — 11.7
15. e — 11.6

Chapter 12

1. a — 12.2
2. d — 12.5
3. a, b — 12.3
4. d — 12.4
5. b — 12.4
6. b — 12.5
7. f — 12.2, 12.3, 12.5
8. a — 12.6
9. c — 12.5, 12.6
10. a — 12.6
11. d — 12.4, 12.6
12. d — 12.8
13. d — 12.7, 12.8
14. a — 12.8
15. c — 12.5
   e — 12.4
   i — 12.8
   g — 12.7
   b — 12.5
   j — 12.2, 12.3
   h — 12.8
   f — 12.7
   d — 12.8
   a — 12.7

Chapter 13

1. c — 13.3
2. b — 13.2
3. c — 13.2
4. b — 13.3
5. c — 13.4
6. c — 13.5
7. d — 13.4
8. d — 13.6
9. c — 13.6
10. b — 13.6
11. a — 13.6
12. d — 13.6
13. c — 13.5
14. d — 13.4
15. g — 13.6
   d — 13.4
   c — 13.5
   e — 13.6
   i — 13.6
   f — 13.6
   a — 13.6
   j — 13.6
   b — 13.6
   h — 13.3

Chapter 14

1. d — 14.6, 14.7
2. c — 14.5
3. b — 14.3
4. c — 14.4
5. a — 14.4
6. d — 14.5, 14.6
7. b — 14.5–14.7
8. c — 14.6
   d — 14.2
   f — 14.4
   e — 14.2
   a — 14.2
   b — 14.2
   g — 14.7
9. c — 14.8
10. a — 14.9
11. c — 14.8
12. d — 14.9
13. b — 14.9
14. d — 14.8
15. g — 14.9
   f — 14.8
   b — 14.8
   c — 14.8
   a — 14.8
   d — 14.8
   e — 14.9

Chapter 15

1. True — 15.2
2. b — 15.2
3. c — 15.2
4. b — 15.2
5. c — 15.3
6. c — 15.3
7. Notochord, nerve cord, pharynx with gill slits, tail that extends beyond anus. Pharynx with gill slits. — 15.3
8. b — 15.5
9. b — 15.4, 15.6
10. f — 15.6
11. a — 15.6
12. False — 15.7
13. d — 15.7
14. b — 15.3
   g — 15.3
   m — 15.3
   j — 15.3
   e — 15.3
   c — 15.3
   d — 15.3
   f — 15.3
   k — 15.5
   a — 15.5
   h — 15.6
   i — 15.6
   l — 15.7
15. 1-b — 15.2
   2-a — 15.2
   3-f — 15.4
   4-c — 15.5
   5-d — 15.6
   6-e — 15.7

Chapter 16

1. a — 16.2
2. c — 16.1
3. b — 16.3
4. b — 16.3
5. a — 16.4
6. a — 16.3
7. b — 16.4
8. a — 16.3
9. a — 16.5
10. d — 16.5
11. c — 16.3
12. b — 16.5
13. c — 16.3
14. d — 16.5
15. d — 16.3
   c — 16.3
   e — 16.3
   a — 16.5
   f — 16.3
   b — 16.4

Chapter 17

1. a — 17.1
2. c — 17.3
3. b, c, a, d, e — 17.3
4. b — 17.3
5. c — 17.3
6. a — 17.4
7. d — 17.5
   a — 17.3
   c — 17.5
   b — 17.5
   e — 17.4
8. b, a, b, c — 17.6
9. c — 17.6
10. b — 17.6
11. a — 17.6
12. b — 17.6
13. d — 17.3
14. d — 17.6
15. b — 17.4

Chapter 18

1. a — 18.2
2. d — 18.2
3. b — 18.3
4. d — 18.3
5. b — 18.3
6. c — 18.3
7. a — 18.4
8. a — 18.4
9. e — 18.3
   d — 18.3
   b — 18.3
   c — 18.3
   a — 18.4
   f — 18.3
   g — 18.4
   h — 18.3
10. b — 18.5
11. c — 18.6
12. b — 18.5
13. d — 18.5
14. d — 18.5
15. c — 18.5

Periodic Table of the Elements

The symbol for each element is an abbreviation of its name. Some symbols for elements are abbreviations for their Latin names. For instance, Pb (lead) is short for *plumbum*; the word "plumbing" is related—ancient Romans made their water pipes with lead.

Elements in each vertical column of the table behave in similar ways. For instance, all of the elements in the far right column of the table are inert gases; they do not interact with other atoms. In nature, such elements occur only as solitary atoms.

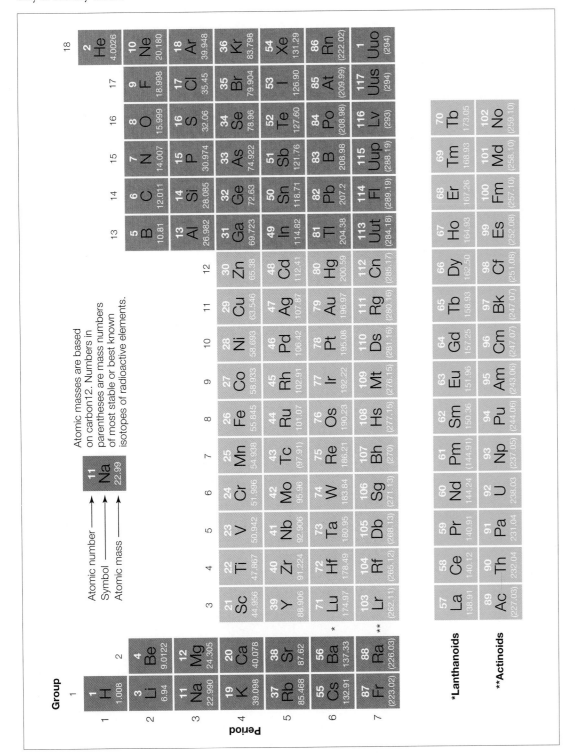

A Plain English Map of the Human Chromosomes

1
- sweet taste receptors
- Rh blood type
- marijuana receptor
- (anorexia nervosa susceptibility)
- leptin receptor
- TSH β chain
- lamin A (progeria)
- Duffy blood group antigen

2
- LH/choriogonadotropin receptor (micropenis)
- CD8; cytotoxic T cell antigen
- antibody light chain
- lactase
- (cleft palate)
- glucagon

3
- oxytocin receptor
- HIV receptor
- rhodopsin
- (alkaptonuria)
- (sucrose intolerance)
- somatostatin

4
- (achondroplasia)
- (Huntington disease)
- (Ellis-van Creveld syndrome)
- alcohol dehydrogenase (susceptibility to alcoholism)
- red hair color

5
- Cri-du-chat syndrome
- bitter taste receptor
- growth hormone receptor (pituitary dwarfism)
- interleukin-4

6
- (gluten intolerance)
- HLA/MHC
- tumor necrosis factor
- α chains of HCG, FSH, LH, and TSH
- estrogen receptor

7
- cytochrome c
- elastin
- DLX 5/6 homeotic genes
- CFTR (cystic fibrosis)
- leptin (obesity)
- (blue-deficient colorblind)
- TCR β subunit

8
- gonadotropin releasing hormone
- helicase (Werner's syndrome)
- corticotropin releasing hormone

9
- (galactosemia)
- (cerebral palsy)
- (Friedreich ataxia)
- (fructose intolerance)
- ABO blood group

10
- vitamin B-12 receptor
- mannose binding protein
- perforin
- (gluten intolerance)

11
- hemoglobin β chain (sickle cell anemia)
- insulin
- parathyroid hormone
- catalase
- PAX6 (aniridia)
- FSH, β chain
- tyrosinase (albinism)

12
- CD4
- helper T cell antigen
- oncogene KRAS2 (lung cancer, bladder cancer, breast cancer)
- keratins
- lysozyme
- (phenylketonuria)
- aldehyde dehydrogenase (alcohol intolerance)

13
- ribosomal RNA
- BRCA 2 (breast cancer)
- (gastroesophageal reflux)

14
- ribosomal RNA
- presinilin (Alzheimer's)
- TSH receptor
- immunoglobulin heavy chains

15
- ribosomal RNA
- fibrillin 1 (Marfan syndrome)
- (Tay-Sachs disease)

16
- hemoglobin α chain
- DNAse I (lupus)

17
- (Canavan disease)
- p53 tumor antigen
- NF1 (neurofibromatosis)
- serotonin transporter
- BRCA 1 (breast, ovarian cancer)
- Growth hormone

18
- B cell apoptosis regulator (B cell lymphoma)
- myelin basic protein

19
- LDL receptor (coronary artery disease)
- insulin receptor
- brown hair color
- green/blue eye color
- (Warfarin resistance)
- HCG, β chain
- LH, β chain

20
- prion protein (Creutzfeld-Jacob disease)
- oxytocin
- GHRH (acromegaly)

21
- ribosomal RNA
- interferon receptors
- (bipolar disorder, early onset)

22
- ribosomal RNA
- immunoglobulin light chains
- myoglobin

X
- dystrophin (muscular dystrophy)
- (anhidrotic ectodermal dysplasia)
- IL2RG (SCID-X1)
- XIST X chromosome inactivation control
- (hemophilia B)
- (hemophilia A)
- (red-deficient colorblind)
- (green-deficient colorblind)

Y
- sex determining region Y (SRY)
- (no sperm)
- male stature

Haploid set of human chromosomes. The banding patterns characteristic of each type of chromosome appear after staining with a reagent called Giemsa. The locations of some known genes are indicated. Also shown are locations that, when mutated, cause some of the genetic diseases discussed in the text.

Units of Measure

Length

1 kilometer (km) = 0.62 miles (mi)
1 meter (m) = 39.37 inches (in)
1 centimeter (cm) = 0.39 inches

To convert	multiply by	to obtain
inches	2.25	centimeters
feet	30.48	centimeters
centimeters	0.39	inches
millimeters	0.039	inches

Area

1 square kilometer = 0.386 square miles
1 square meter = 1.196 square yards
1 square centimeter = 0.155 square inches

Volume

1 cubic meter = 35.31 cubic feet
1 liter = 1.06 quarts
1 milliliter = 0.034 fluid ounces = 1/5 teaspoon

To convert	multiply by	to obtain
quarts	0.95	liters
fluid ounces	28.41	milliliters
liters	1.06	quarts
milliliters	0.03	fluid ounces

Weight

1 metric ton (mt) = 2,205 pounds (lb) = 1.1 tons (t)
1 kilogram (kg) = 2.205 pounds (lb)
1 gram (g) = 0.035 ounces (oz)

To convert	multiply by	to obtain
pounds	0.454	kilograms
pounds	454	grams
ounces	28.35	grams
kilograms	2.205	pounds
grams	0.035	ounces

Temperature

Celsius (°C) to Fahrenheit (°F): $°F = 1.8\,(°C) + 32$

Fahrenheit (°F) to Celsius: $°C = \dfrac{(°F - 32)}{1.8}$

	°C	°F
Water boils	100	212
Human body temperature	37	98.6
Water freezes	0	32

G

acid Substance that releases hydrogen ions in water. **32**

acid rain Rain containing sulfuric and/or nitric acid; forms when pollutants mix with water vapor in the atmosphere. **362**

activation energy Minimum amount of energy required to start a chemical reaction. **67**

active site Pocket in an enzyme where substrates bind and a chemical reaction occurs. **68**

active transport Energy-requiring mechanism in which a transport protein pumps a solute across a cell membrane against the solute's concentration gradient. **76**

adaptation *See* adaptive trait.

adaptive radiation Macroevolutionary pattern in which a burst of genetic divergences from a lineage gives rise to many new species. **227**

adaptive trait (**adaptation**) A form of a heritable trait that enhances an individual's fitness. **195**

adhesion protein Plasma membrane protein that helps cells stick together in animal tissues. Some types form adhering junctions and tight junctions. **51**

aerobic Involving or occurring in the presence of oxygen. **89**

aerobic respiration Oxygen-requiring metabolic pathway that breaks down sugars to produce ATP. Includes glycolysis, the Krebs cycle, and electron transfer phosphorylation. **89**

age structure Of a population, the distribution of individuals among various age groups. **316**

alcoholic fermentation Anaerobic pathway that breaks down sugars and produces ATP, CO_2, and ethanol. **93**

algal bloom Population explosion of single-celled aquatic producers such as dinoflagellates. **252**

allele frequency Abundance of a particular allele among members of a population; expressed as a fraction of the total number of alleles. **214**

alleles Forms of a gene with slightly different DNA sequences; may encode slightly different versions of the gene's product. **140**

allopatric speciation Speciation pattern in which a physical barrier arises and ends gene flow between populations. **224**

alternation of generations As in plants, a life cycle that alternates between a diploid spore-producing body and a haploid, gamete-producing one. **263**

amino acid Small organic compound that is a subunit of proteins. Consists of a carboxyl group, an amine group, and a characteristic side group (R), all typically bonded to the same carbon atom. **38**

amniote Vertebrate that produces amniote eggs; a reptile, bird, or mammal. **302**

amniote egg Egg with four extraembryonic membranes; allows an embryo to develop away from water. **303**

amoeba Solitary heterotrophic protist that feeds and moves by extending pseudopods. **256**

amphibian Tetrapod with a three-chambered heart and scaleless skin. Larva are aquatic and have gills, but most are air-breathing as adults. **301**

anaerobic Occurring in the absence of oxygen. **89**

analogous structures Similar body structures that evolved separately in different lineages (by morphological convergence). **205**

anaphase Stage of mitosis during which sister chromatids separate and move toward opposite spindle poles. **134**

angiosperm Seed plant that produces flowers and fruits. **272**

animal A multicelled, eukaryotic consumer that is made up of unwalled cells and develops through a series of stages. Most ingest food, reproduce sexually, and can move from place to place. **8, 285**

annelid Segmented worm with a coelom, complete digestive system, and closed circulatory system. For example, an earthworm. **290**

antenna Of some arthropods, sensory structure on the head that detects touch and odors. **293**

anther Of a flower, the part of the stamen that produces pollen grains in pollen sacs. **272**

anthropoids Primate subgroup that includes monkeys, apes, and humans. **306**

anticodon In a tRNA, set of three nucleotides that base-pairs with an mRNA codon. **119**

apicomplexan Parasitic protist that enters and lives inside the cells of its host. For example, the parasite that causes malaria. **252**

aquifer Porous rock layer that holds some groundwater. **342**

arachnids Land-dwelling arthropods with four pairs of walking legs and no antennae; for example, a spider, scorpion, or tick. **294**

archaea Group of single-celled organisms that lack a nucleus but are more closely related to eukaryotes than to bacteria. **8, 246**

arthropod Invertebrate with jointed legs and a hardened exoskeleton that is periodically molted. **292**

asexual reproduction Reproductive mode of eukaryotes by which offspring arise from a single parent. **133**

atom Fundamental building block of all matter. Consists of varying numbers of protons, neutrons, and electrons. **5**

atomic number Number of protons in the atomic nucleus; determines the element. **25**

ATP Adenosine triphosphate. Nucleotide that consists of an adenine base, a ribose sugar, and three phosphate groups. Functions as a subunit of RNA and as a coenzyme in many reactions. Important energy carrier in cells. **41**

australopiths Informal name for chimpanzee-sized hominins that lived in Africa between 4 million and 1.2 million years ago. Some are considered likely ancestors of modern humans. **306**

autosome A chromosome that is the same in males and females. **107**

autotroph Organism that uses carbon dioxide as its carbon source, and obtains energy from light or breakdown of minerals. **247**

bacteria Singular, bacterium. The most diverse and well-known group of prokaryotes (organisms that lack a nucleus). **8, 246**

bacteriophage Virus that infects bacteria. **242**

Barr body Condensed and inactivated X chromosome in a body cell of a female mammal (the other X chromosome is active). **125**

base Substance that accepts hydrogen ions in water. **32**

base-pair substitution Mutation in which a single base pair changes. **122**

bell curve Bell-shaped curve; typically results from graphing frequency versus distribution for a trait that varies continuously. **160**

bilateral symmetry Having right and left halves with similar parts, and a front and back that differ. **286**

binary fission Method of asexual reproduction in which a prokaryote divides into two identical descendant cells. **246**

biodiversity Scope of variation among living organisms. Of a region, the genetic diversity within its species, variety of species, and variety of ecosystems. **8, 366**

biofilm Community of microorganisms living within a shared mass of slime. **53**

biogeochemical cycle Cycle in which a nutrient moves among environmental reservoirs and into and out of food webs. **342**

biogeography Study of patterns in the geographic distribution of species and communities. **191**

biological magnification A chemical pollutant becomes increasingly concentrated as it moves through a food chain. **363**

biology The scientific study of life. **4**

bioluminescence Light produced by a living organism. **252**

biome Any of Earth's major land ecosystems, characterized by climate and main vegetation and found in several regions. **355**

biosphere All regions of Earth where organisms live. **5**

biotic potential Maximum possible population growth under optimal conditions. **320**

bipedalism Habitually walking upright. **306**

birds Common name for the lineage of feathered amniotes that descended from dinosaurs. **303**

bivalve Mollusk with a hinged two-part shell. **290**

bony fish Common name for a jawed fish with a skeleton composed mainly of bone. **299**

boreal forest At high northern latitudes, a biome dominated by conifers that withstand cold winters. **355**

bottleneck Reduction in population size so severe that it reduces genetic diversity. **220**

brood parasite An animal that manipulates another species into raising its young, for example a cowbird. **336**

brown alga Multicelled, photosynthetic protist with brown accessory pigments; for example a kelp. **254**

bryophyte Member of a plant lineage that does not have vascular tissue; a moss, liverwort, or hornwort. **264**

buffer Set of chemicals that can keep the pH of a solution stable by alternately donating and accepting ions that contribute to pH. **32**

C3 plant Type of plant that uses only the Calvin–Benson cycle to fix carbon. **88**

C4 plant Type of plant that fixes carbon twice, in two cell types. **88**

Calvin–Benson cycle Cyclic carbon-fixing pathway that forms sugars from CO_2; light-independent reactions of photosynthesis. **87**

camouflage Evolved body shape, color pattern, or behavior that helps an organism blend into its surroundings. **335**

CAM plant Type of plant that fixes carbon twice, at different times of day. **88**

cancer Disease that occurs when a malignant neoplasm physically and functionally disrupts body tissues. **138**

carbohydrate Molecule that consists primarily of carbon, hydrogen, and oxygen atoms in a 1:2:1 ratio. Complex types (polysaccharides such as cellulose, starch, and glycogen) are polymers of monosaccharides. **34**

carbon cycle Movement of carbon among rocks, water, the atmosphere, and living organisms. **345**

carbon fixation Process by which carbon from an inorganic source such as carbon dioxide becomes incorporated (fixed) into an organic molecule. **87**

carpel Floral reproductive organ that produces the female gametophyte; consists of an ovary, stigma, and often a style. **272**

carrying capacity Maximum number of individuals of a species that a specific environment can sustain. **318**

cartilaginous fish Fish that has jaws, paired fins, and a skeleton made of cartilage; for example, a shark. **299**

cell Smallest unit of life; at minimum, consists of plasma membrane, cytoplasm, and DNA. **5**

cell cycle The collective series of intervals and events of a cell's life, from the time it forms until its cytoplasm divides. **133**

cell junction Structure that connects a cell to another cell or to extracellular matrix; e.g., tight junction, adhering junction, or gap junction (of animals); plasmodesmata (of plants). **58**

cell theory Theory that all organisms consist of one or more cells, which are the basic unit of life; all cells come from division of preexisting cells; and all cells pass DNA to offspring. **47**

cell wall Rigid but permeable layer of extracellular matrix that surrounds the plasma membrane of some cells. **52**

cellular slime mold Heterotrophic protist that usually lives as a single-celled, amoeba-like predator. With unfavorable conditions, cells aggregate into a cohesive group that can form a fruiting body. **257**

cellulose Tough, insoluble carbohydrate that is the major structural material in plants. **34**

centromere Of a duplicated eukaryotic chromosome, constricted region where sister chromatids attach to each other. **106**

cephalopod Predatory mollusk with a closed circulatory system; moves by jet propulsion; for example, an octopus or squid. **291**

chaparral Biome with cool, wet winters and hot dry summers; dominant plants are shrubs with small, leathery leaves. **356**

charge Electrical property; opposite charges attract, and like charges repel. **25**

chemical bond An attractive force that arises between two atoms when their electrons interact; links atoms in molecules. *See* covalent bond, ionic bond. **28**

chlorophyll *a* Main photosynthetic pigment in plants. **84**

chloroplast Organelle of photosynthesis in the cells of plants and photosynthetic protists. Has two outer membranes enclosing semifluid stroma. Light-dependent reactions occur at its inner thylakoid membrane; light-independent reactions, in the stroma. **56**

choanoflagellates Heterotrophic protists with a collared flagellum; protist group most closely related to animals. **257**

chordates Animal phylum in which embryos have a notochord, dorsal nerve cord, pharyngeal gill slits, and a tail that extends beyond the anus. Includes invertebrate and vertebrate groups. **297**

chromosome Structure that consists of DNA together with associated proteins; carries part or all of a cell's genetic information. **106**

chromosome number The total number of chromosomes in a cell of a given species. **106**

chytrid Fungus that produces flagellated spores. **275**

cilia Singular, cilium. Short, movable structures that project from the plasma membrane of some eukaryotic cells. **57**

ciliate Unwalled, single-celled protist with many cilia. **251**

clade A group whose members share one or more defining derived traits. **229**

cladistics Making hypotheses about evolutionary relationships among clades. **229**

cladogram Evolutionary tree diagram that summarizes hypothesized relationships among a group of clades. **229**

cleavage furrow In a dividing animal cell, the indentation where cytoplasmic division will occur. **136**

climate Average weather conditions in a region over a long time period. **353**

cloaca Of some vertebrates, a body opening that releases urinary and digestive waste, and also functions in reproduction. **301**

clone Genetically identical copy of an organism. **100**

cloning vector A DNA molecule that can accept foreign DNA and be replicated inside a host cell. **176**

closed circulatory system System in which blood flows through a continuous network of vessels and exchanges with cells take place across vessel walls. **287**

club fungus Fungus that produces spores by meiosis in club-shaped cells. **275**

cnidarian Radially symmetrical invertebrate with two tissue layers; uses tentacles with stinging cells to capture food. **288**

coal Fossil fuel consisting primarily of the carbon-rich remains of seedless nonvascular plants. **268**

codominance Inheritance pattern in which the full and separate phenotypic effects of two alleles are apparent in heterozygous individuals. **155**

codon In an mRNA, a nucleotide base triplet that codes for an amino acid or stop signal during translation. **118**

coelom A body cavity completely lined by tissue derived from mesoderm. **286**

coenzyme An organic cofactor; e.g., NAD. **70**

coevolution The joint evolution of two closely interacting species; macroevolutionary pattern in which each species is a selective agent for traits of the other. **228**

cofactor A molecule or metal ion that associates with a protein and is necessary for its function. **70**

cohesion Property of a substance that arises from the tendency of its molecules to resist separating from one another. **31**

cohort Group of individuals born during the same interval. **320**

colonial organism Organism composed of many integrated cells, each capable of surviving and reproducing on its own. **250**

colonial theory of animal origins Well-accepted hypothesis that animals evolved from a colonial protist. **285**

commensalism Species interaction that benefits one species and has no effect on the other. **332**

community All populations of all species in a given area. **5, 330**

comparative morphology The scientific study of similarities and differences in body plans. **192**

competitive exclusion When two species compete for the same resource, the better competitor drives the weaker one to extinction in that habitat. **334**

complete digestive tract Tubular gut with two openings. **286**

compound Molecule that has atoms of more than one element. **28**

compound eye Eye that consists of many individual units, each with its own lens. **293**

concentration Amount of solute per unit volume of solution. **30**

conifer Woody gymnosperm with needlelike leaves. **270**

conjugation Mechanism of gene transfer in which one prokaryotic cell directly transfers a plasmid to another. **246**

conservation biology Field of applied biology that surveys biodiversity and seeks ways to maintain and use it. **366**

consumer Organism that obtains energy and nutrients by feeding on the tissues, wastes, or remains of other organisms; a heterotroph. **6, 339**

continuous variation A range of small differences in a shared trait. **159**

contractile vacuole In freshwater protists, an organelle that collects and expels excess water. **250**

control group Group of individuals identical to an experimental group except for the independent variable under investigation. **13**

coral reef In tropical sunlit seas, a formation composed of the secretions of coral polyps; serves as home to many other species. **359**

covalent bond Type of chemical bond in which two atoms share a pair of electrons. **28**

critical thinking The act of evaluating information before accepting it. **12**

crossing over Process in which homologous chromosomes exchange corresponding segments during meiosis. **144**

crustaceans Lineage of mostly marine arthropods with two pairs of antennae; for example, a shrimp, crab, lobster, or barnacle. **294**

cuticle Secreted covering at a body surface. In plants it is waxy and helps conserve water. **58, 264**

cytoplasm Jellylike mixture of water and solutes enclosed by a cell's plasma membrane. **47**

cytoskeleton Network of protein filaments that support, organize, and move eukaryotic cells and their internal structures. *See* microtubule, microfilament, intermediate filament. **56**

data Results of an experiment or survey. **13**

deciduous plant Plant that sheds all its leaves in preparation for a seasonal dormancy. **271**

decomposer Organism that breaks down organic material into its inorganic subunits. **247, 339**

deforestation Removal of all trees from a forested area. **362**

deletion Mutation in which one or more nucleotides are lost from DNA. **123**

demographics Statistics that describe a population. **315**

demographic transition model Model describing how human birth and death rates change as a region becomes industrialized. **325**

denature To unravel the shape of a protein or other large biological molecule. **39**

density-dependent limiting factor Factor whose negative effect on growth is felt most in dense populations; for example, infectious disease or competition for food. **318**

density-independent limiting factor Factor that limits growth in populations regardless of their density; for example a natural disaster or harsh weather. **319**

desert Biome with little precipitation; its perennial plants are adapted to withstand drought. **356**

desertification Conversion of grassland or woodlands to desertlike conditions. **362**

detritivore Consumer that feeds on small bits of organic material (detritus). **339**

deuterostomes Animal lineage with a three-layer embryo in which the mouth is the second opening to form; includes echinoderms and chordates. **286**

development Multistep process by which the first cell of a new multicelled organism gives rise to an adult. **6**

diatom Single-celled photosynthetic protist with brown accessory pigments and a two-part silica shell. **254**

differentiation Process by which cells become specialized during development; occurs as different cell lineages begin to use different subsets of their DNA. **101**

diffusion The spontaneous spreading of molecules or atoms. **73**

dihybrid cross Cross between two individuals identically heterozygous for alleles of two genes; for example $AaBb \times AaBb$. **154**

dinoflagellates Single-celled, aquatic protist typically with cellulose plates and two flagella; may be heterotrophic or photosynthetic. **252**

diploid Having two of each type of chromosome characteristic of the species ($2n$). **107**

directional selection Mode of natural selection in which a phenotype at one end of a range of variation is favored. **215**

disease vector Organism that carries a pathogen from one host to the next. **243**

disruptive selection Mode of natural selection in which traits at the extremes of a range of variation are adaptive, and intermediate forms are not. **215**

DNA Deoxyribonucleic acid. Nucleic acid that carries hereditary information; consists of two chains of nucleotides twisted into a double helix. **6, 41**

DNA cloning Set of methods that uses living cells to mass-produce targeted DNA fragments. **176**

DNA fingerprinting *See* DNA profiling.

DNA library Collection of cells that host different fragments of foreign DNA, often representing an organism's entire genome. **176**

DNA polymerase DNA replication enzyme. Uses a DNA template to assemble a complementary strand of DNA. **108**

DNA profiling Identifying an individual by analyzing the unique parts of his or her DNA. **179**

DNA replication Process by which a cell duplicates its DNA before it divides. **108**

DNA sequence Order of nucleotides composing a strand of DNA. **106**

DNA sequencing *See* sequencing.

dominant Refers to an allele that masks the effect of a recessive allele paired with it on the homologous chromosome. **152**

double fertilization In flowering plants, one sperm fertilizes the egg (to form the zygote), and a second sperm cell fuses with the endosperm mother cell (giving rise to endosperm). **273**

echinoderms Invertebrates with a water–vascular system and an endoskeleton made of hardened plates and spines. **296**

ECM *See* extracellular matrix.

ecological footprint Area of Earth's surface required to sustainably support a particular level of development and consumption. **325**

ecological restoration Actively altering an area in an effort to restore or create a functional ecosystem. **367**

ecological succession A gradual change in a community in which one array of species replaces another. **337**

ecology The study of interactions among organisms, and between organisms and their environment. **314**

ecosystem A community interacting with its environment through a one-way flow of energy and cycling of materials. **5, 339**

ectotherm Animal that gains heat from the environment; commonly called "cold-blooded." **302**

electron Negatively charged subatomic particle. **25**

electron transfer chain Array of membrane-bound enzymes and other molecules that accept and give up electrons in sequence, thus releasing the energy of the electrons in small, usable steps. **72**

electron transfer phosphorylation Process in which electron flow through electron transfer chains sets up a hydrogen ion gradient that drives ATP formation. Occurs in photosynthesis and in aerobic respiration. **87**

electrophoresis Laboratory technique that separates DNA fragments by size. **178**

element A pure substance that consists only of atoms with the same number of protons. **25**

endangered species A species that faces extinction in all or part of its range. **360**

endemic species A species that evolved in one place and is found nowhere else. **361**

endocytosis Process by which a cell takes in a small amount of extracellular fluid (and its contents) by the ballooning inward of the plasma membrane. **76**

endoplasmic reticulum (ER) Membrane-enclosed organelle that is a continuous system of sacs and tubes extending from the nuclear envelope. Smooth ER makes lipids and breaks down carbohydrates and fatty acids; ribosomes on the surface of rough ER make polypeptides that thread into its interior. **55**

endoskeleton Internal skeleton; hard internal parts that muscles attach to and move. **296**

endosperm Nutritive tissue in the seeds of angiosperms (flowering plants). **273**

endosymbiont hypothesis Mitochondria and chloroplasts evolved from free-living bacteria that entered and lived inside another cell. **241**

endotherm Animal that produces its own heat; commonly called "warm-blooded." **302**

energy The capacity to do work. **65**

energy pyramid Diagram that illustrates the energy flow in an ecosystem. **341**

enzyme Organic molecule (protein or RNA) that speeds up a reaction without being changed by it. **34**

epigenetic Refers to heritable changes in gene expression that are not the result of changes in DNA sequence. **126**

epiphyte Plant that grows on the trunk or branches of another plant but does not harm it. **267**

epistasis Polygenic inheritance, in which a trait is influenced by multiple genes. **156**

equilibrial life history Life history favored in stable environments; individuals grow large, then invest heavily in each of their few offspring. **321**

estuary Highly productive, aquatic ecosystem where nutrient-rich water from a river mixes with seawater. **358**

eudicots Most diverse angiosperm lineage; characterized by two seed leaves; includes herbaceous plants, woody trees, and cacti. **273**

eugenics Idea of deliberately improving the genetic qualities of the human race. **185**

eukaryote Organism whose cells characteristically have a nucleus; a protist, fungus, plant, or animal. **8**

evaporation Transition of a liquid to a vapor. **31**

evolution Change in a line of descent. **193**

exocytosis Process by which a cell expels a vesicle's contents to extracellular fluid. **76**

exon Nucleotide sequence that remains in an RNA after post-transcriptional modification. **117**

exoskeleton External skeleton; hard external parts that muscles attach to and move. **292**

exotic species A species that has been introduced to a new habitat and become established there. **338**

experiment A test designed to support or falsify a prediction. **13**

experimental group In an experiment, a group of individuals who have a certain characteristic or receive a certain treatment as compared with a control group. **13**

exponential model of population growth Model for population growth when resources are unlimited. The per capita growth rate remains constant as population size increases. **317**

extinct Refers to a species that no longer has any living members. **226**

extracellular matrix (ECM) Complex mixture of cell secretions, the composition and function of which vary by cell type. **58**

extreme halophile Organism that lives in a highly salty habitat. **248**

extreme thermophile Organism that lives in a high temperature habitat. **248**

facilitated diffusion Passive transport mechanism in which a solute follows its concentration gradient across a membrane by moving through a transport protein. **75**

fat Substance that consists mainly of triglycerides. Saturated types are composed mainly of triglycerides with three saturated fatty acid tails. **36**

fatty acid Organic compound with an acidic carboxyl group "head" and a long carbon chain "tail." *See* saturated fatty acid, unsaturated fatty acid. **36**

feedback inhibition Regulatory mechanism by which a change that results from some activity decreases or stops the activity. **72**

fermentation An anaerobic pathway that breaks down sugars to produce ATP. **92**

ferns Most diverse lineage of seedless vascular plants. **266**

first law of thermodynamics Energy cannot be created or destroyed. **65**

fitness Degree of adaptation to an environment, as measured by an individual's relative genetic contribution to future generations. **195**

fixed Refers to an allele for which all members of a population are homozygous. **220**

flagellated protozoan Unwalled, single-celled protist that has one or more flagella. **250**

flagellum Long, slender cellular structure used for movement. **53**

flatworm Soft-bodied, bilaterally symmetrical invertebrate with organs but no body cavity; for example, a planarian or tapeworm. **289**

flower Specialized reproductive structure of a flowering plant. **272**

fluid mosaic Model of a cell membrane as a two-dimensional fluid of mixed composition. **50**

food chain Sequence of steps by which energy moves from one trophic level to the next. **339**

food web System of cross-connecting food chains. **340**

foraminiferan Heterotrophic single-celled protist that secretes a calcium carbonate shell. **251**

fossil Physical evidence of an organism that lived in the ancient past. **192**

founder effect After a small group of individuals found a new population, allele frequencies in the new population differ from those in the original population. **220**

free radical Atom with an unpaired electron; most are highly reactive and can damage biological molecules. **27**

fruit Mature ovary of a flowering plant, often with expanded accessory parts; encloses a seed or seeds. **272**

fungus Single-celled or multicelled eukaryotic consumer that breaks down material outside its body, then absorbs nutrients released from the breakdown. **8, 274**

gamete Mature, haploid reproductive cell; e.g., an egg or a sperm. **144**

gametophyte Haploid gamete-forming body that forms in a plant life cycle. **263**

gastropod Mollusk that moves about on an enlarged "foot" at its lower surface; for example a snail. **290**

gastrovascular cavity Saclike cavity with one opening; functions in digestion and respiration. **286**

gene A part of a chromosome that encodes an RNA or protein product in its DNA sequence. **115**

gene expression Process by which the information in a gene guides assembly of an RNA or protein product. Includes transcription and translation. **115**

gene flow The movement of alleles between populations. **221**

gene pool All the alleles of all the genes in a population; a pool of genetic resources. **214**

gene therapy Treating a genetic defect or disorder by transferring a normal or modified gene into the affected individual. **184**

genetic code Complete set of sixty-four mRNA codons. **118**

genetic drift Change in allele frequency due to chance alone. **220**

genetic engineering Laboratory process by which deliberate changes are introduced into an individual's genome. **181**

genetically modified organism (GMO) Organism whose genome has been modified by genetic engineering. **181**

genome An organism's complete set of genetic material. **176**

genomics The study of genomes. **179**

genotype The particular set of alleles that is carried in an individual's chromosomes. **152**

genus Plural, genera. A group of species that share a unique set of traits; first part of a species name. **10**

geologic time scale Chronology of Earth's history; correlates geologic and evolutionary events. **201**

global climate change Wide-ranging changes in rainfall patterns, average temperature, and other climate factors that result from rising concentrations of greenhouse gases. **347**

glomeromycete fungus Soil fungus whose hyphae extend inside the cell wall of a plant root cell. **278**

glycolysis Set of reactions in which glucose is broken down to two pyruvate for a net yield of two ATP. Part of fermentation and aerobic respiration. **90**

GMO *See* genetically modified organism.

Golgi body Organelle that modifies polypeptides and lipids, then packages the finished products into vesicles. **55**

Gondwana Supercontinent that existed before Pangea, more than 500 million years ago. **201**

green algae Single-celled, colonial, or multicelled photosynthetic protists; algal lineage most closely related to land plants. **255**

greenhouse effect Warming of Earth's lower atmosphere and surface as a result of heat trapped by greenhouse gases. **346**

greenhouse gas Atmospheric gas that helps keep heat from escaping into space and thus warms the Earth. **346**

groundwater Water between soil particles and in aquifers. **342**

growth In multicelled species, an increase in the number, size, and volume of cells. **7**

gymnosperm Seed plant that produces "naked" seeds (seeds that are not encased within a fruit). **270**

habitat The type of place in which a species lives. **331**

half-life Characteristic time it takes for half of a quantity of a radioisotope to decay. **198**

haploid Having one of each type of chromosome characteristic of the species. **143**

herbivory An animal feeds on a plant, which may or may not die as a result. **335**

hermaphrodite Individual animal that makes both eggs and sperm. **288**

heterotroph Organism that obtains both carbon and energy by breaking down organic compounds. **247**

heterozygous Having two different alleles of a gene. **152**

histone Type of protein that associates with eukaryotic DNA and structurally organizes chromosomes. **106**

HIV (human immunodeficiency virus) Enveloped RNA virus that causes AIDS. **244**

homeostasis Process in which an organism keeps its internal conditions within tolerable ranges by sensing and responding to change. **6**

hominins Lineage of bipedal primates; includes humans and extinct humanlike species. **306**

Homo erectus Extinct human species with body proportions similar to modern humans; arose in and dispersed out of Africa. **307**

Homo habilis Earliest named human species; had australopith-like proportions and is known only from Africa. **307**

Homo neanderthalensis Neanderthals. Closest extinct relatives of modern humans; had large brain, stocky body. **308**

Homo sapiens Modern humans; evolved in Africa by about 200,000 years ago, then expanded their range worldwide. **308**

homologous chromosomes Chromosomes that have the same length, shape, and genes. In sexual reproducers, one member of a homologous pair is paternal and the other is maternal. **134**

homologous structures Body structures that are similar in different lineages because they evolved in a common ancestor. **204**

homozygous Having identical alleles of a gene. **152**

humans Members of the genus *Homo*. **307**

hydrogen bond Attraction between a covalently bonded hydrogen atom and another atom taking part in a separate covalent bond. **30**

hydrophilic Describes a substance that dissolves easily in water. **30**

hydrophobic Describes a substance that resists dissolving in water. **30**

hydrothermal vent Underwater opening from which mineral-rich water heated by geothermal energy streams out. **237, 359**

hypertonic Describes a fluid that has a high overall solute concentration relative to another fluid. **73**

hypha A single filament in a fungal mycelium; consists of a chain of cells. **275**

hypothesis Testable explanation of a natural phenomenon. **12**

hypotonic Describes a fluid that has a low solute concentration relative to another fluid. **73**

inbreeding Mating among close relatives. **221**

incomplete dominance Inheritance pattern in which one allele is not fully dominant over another, so the heterozygous phenotype is an intermediate blend between the two homozygous phenotypes. **155**

indicator species A species that is particularly sensitive to environmental changes and can be monitored to assess the state of an ecosystem. **366**

inheritance Transmission of DNA to offspring. **7**

insects Land-dwelling arthropods with a pair of antennae, three pairs of legs, and—in the most diverse groups—wings. **295**

insertion Mutation in which one or more nucleotides become inserted into DNA. **123**

intermediate filament Stable cytoskeletal element that structurally supports cell membranes and tissues; also forms external structures such as hair. **57**

interphase In a eukaryotic cell cycle, the interval during which the cell grows, roughly doubles the number of its cytoplasmic components, and replicates its DNA in preparation for division. **133**

interspecific competition Two species compete for a resource in an interaction harmful to both. **333**

intron Nucleotide sequence that intervenes between exons and is removed during post-transcriptional modification. **117**

invertebrate Animal without a backbone. **284**

ion Atom or molecule that carries a net charge. **27**

ionic bond Type of chemical bond in which a strong mutual attraction links ions of opposite charge. **28**

iron–sulfur world hypothesis Hypothesis that life began in rocks rich in iron sulfide near deep-sea hydrothermal vents. **238**

isotonic Describes two fluids that have the same overall solute concentration. **73**

isotopes Forms of an element that differ in the number of neutrons their atoms carry. **25**

jawless fish Fish that has a skeleton of cartilage, but no jaws or paired fins; for example, a lamprey. **299**

karyotype Image of an individual's complement of chromosomes arranged by size, length, shape, and centromere location. **107**

key innovation An evolutionary adaptation that gives its bearer the opportunity to exploit a particular environment more efficiently or in a new way. **228**

keystone species A species that has a disproportionately large effect on community structure. **338**

kidney Organ of the vertebrate urinary system that filters blood, adjusts its composition, and forms urine. **298**

knockout An experiment in which a gene is deliberately inactivated in a living organism; also, an organism that has a knocked-out gene. **124**

Krebs cycle Cyclic pathway that, along with acetyl–CoA formation, breaks down pyruvate to carbon dioxide in aerobic respiration's second stage. **91**

lactate fermentation Anaerobic pathway that breaks down sugars and produces ATP and lactate. **93**

lancelets Invertebrate chordates that have a fishlike shape and retain their defining chordate traits into adulthood. **297**

larva Preadult stage in some animal life cycles. **288**

law of nature Generalization that describes a consistent natural phenomenon that has an incomplete scientific explanation. **18**

lichen Composite organism consisting of a fungus and a green alga or cyanobacterium. **279**

life history traits Characteristics related to growth, survival, and reproduction such as life span, age-specific mortality, age at first reproduction, and number of breeding events. **320**

lignin Compound that stiffens walls of some cells (including xylem) in vascular plants. **264**

lineage Line of descent. **193**

lipid Fatty, oily, or waxy organic compound; e.g., a triglyceride, steroid, or wax. **36**

lipid bilayer Double layer of lipids arranged tail-to-tail; structural foundation of all cell membranes. **37**

lobe-finned fish Bony fish that has bony supports inside its fins. **300**

logistic model of population growth Model for growth of a population limited by density-dependent factors; numbers increase exponentially at first, then the growth rate slows and population size levels off at carrying capacity. **318**

lungs Internal saclike organs inside which blood exchanges gases the the air; the respiratory organs in most land vertebrates and some fish. **298**

lysosome Enzyme-filled vesicle that breaks down cellular wastes and debris. **55**

macroevolution Evolutionary patterns and trends on a larger scale than microevolution. **226**

mammal Vertebrate that nourishes its young with milk from mammary glands. **303**

mantle Skirtlike extension of tissue in mollusks; covers the mantle cavity and secretes the shell if one is present. **290**

marsupial Mammal in which offspring complete development in a pouch on the mother's body. **304**

mass extinction Event in which many species in many habitats become extinct in the same interval. **360**

mass number Of an isotope, the total number of protons and neutrons in the atomic nucleus. **25**

master gene Gene encoding a product that affects the expression of many other genes. **124**

medusa Bell-shaped, free-swimming cnidarian body form. **288**

megaspore In seed plants, a haploid cell that gives rise to a female gametophyte. **269**

meiosis Nuclear division process that halves the chromosome number; the basis of sexual reproduction. **142**

membrane potential *See* resting potential, action potential.

messenger RNA (mRNA) Type of RNA that carries a protein-building message. **115**

metabolic pathway Series of enzyme-mediated reactions by which cells build, remodel, or break down an organic molecule. **71**

metabolism All the enzyme-mediated chemical reactions by which cells acquire and use energy as they build and break down organic molecules. **34**

metamorphosis Dramatic remodeling of body form during the transition from larva to adult. **293**

metaphase Stage of mitosis at which all chromosomes are aligned in the middle of the cell. **134**

metastasis The process in which cells of a malignant neoplasm spread from one part of the body to another. **138**

methanogen Organism that produces methane gas as a metabolic by-product. **248**

microbiome Collection of microorganisms that inhabits a specific habitat, such as a human body. **236**

microevolution Change in allele frequency. **214**

microfilament Cytoskeletal element composed of actin subunits. Reinforces cell membranes; functions in movement and muscle contraction. **57**

microspore In seed plants, a haploid cell that gives rise to a male gametophyte (pollen grain). **269**

microtubule Hollow cytoskeletal element composed of tubulin subunits. Involved in movement of a cell or its parts. **56**

mimicry Evolutionary process whereby two or more species come to resemble one another. **334**

mitochondrion Double-membraned organelle that produces ATP by aerobic respiration in eukaryotes. **55**

mitosis Nuclear division mechanism that maintains the chromosome number. Basis of body growth and tissue repair in multicelled eukaryotes; also asexual reproduction in some eukaryotes. Has four stages: prophase, metaphase, anaphase, and telophase. **133**

model Analogous system used for testing a hypothesis. **12**

molecule Two or more atoms joined by chemical bonds. **5**

mollusk Invertebrate with a reduced coelom and a mantle. **290**

monocots Lineage of angiosperms that includes grasses, orchids, and palms. **273**

monohybrid cross Cross between two individuals identically heterozygous for alleles of one gene; for example $Aa \times Aa$. **153**

monomer Molecule that is a subunit of polymers. **33**

monotreme Egg-laying mammal. **304**

morphological convergence Evolutionary pattern in which similar body parts evolve separately in different lineages. **205**

morphological divergence Evolutionary pattern in which a body part of an ancestor changes in its descendants. **204**

mosses Most diverse group of nonvascular plants; Low-growing plants that have flagellated sperm and disperse by producing spores. **265**

motor protein Type of energy-using protein that interacts with cytoskeletal elements to move the cell's parts or the whole cell. **57**

mRNA *See* messenger RNA.

multicellular organism Organism composed of a variety of specialized cells, each unable to survive and reproduce on its own. **250**

mutation Permanent change in the DNA sequence of a chromosome. *See* base-pair substitution, deletion, insertion. **109**

mutualism Species interaction that benefits both species. **278, 332**

mycelium Mass of threadlike filaments (hyphae) that compose the body of a multicelled fungus. **275**

mycorrhiza Fungus–plant root partnership. **278**

natural selection Differential survival and reproduction of individuals of a population based on differences in shared, heritable traits. Driven by environmental pressures. **195**

neoplasm An accumulation of abnormally dividing cells. **137**

neutron Uncharged subatomic particle in the atomic nucleus. **25**

niche The role of a species in its community. **331**

nitrogen cycle Movement of nitrogen among the atmosphere, soil, and water, and into and out of food webs. **344**

nitrogen fixation Process of combining nitrogen gas with hydrogen to form ammonia. **248,344**

nondisjunction Failure of chromosomes to separate properly during nuclear division. **166**

notochord Stiff rod of connective tissue that runs the length of the body in chordate larvae or embryos. **297**

nuclear envelope A double membrane that constitutes the outer boundary of the nucleus. Nuclear pores in the membrane control the entry and exit of large molecules. **54**

nucleic acid Chain of nucleotides; DNA or RNA. **41**

nucleotide Small molecule with a deoxyribose or ribose sugar, a nitrogen-containing base, and one, two, or three phosphate groups; e.g., adenine, guanine, cytosine, thymine, uracil. Monomer of DNA or RNA; some have additional roles. **41**

nucleus Of an atom; core area occupied by protons and neutrons. **25** Of a eukaryotic cell; organelle with a double membrane that holds, protects, and controls access to the cell's DNA. **47**

nutrient Substance that an organism needs for growth and survival but cannot make for itself. **6**

oncogene Gene that helps transform a normal cell into a tumor cell. **137**

open circulatory system Circulatory system in which the circulatory fluid (hemolymph) leaves open-ended vessels and flows among tissues before returning to the heart. **287**

opportunistic life history Life history favored in unpredictable environments; individuals reproduce while young and invest a small amount in each of many offspring. **321**

organelle Structure that carries out a special metabolic function inside a cell. **47**

organic Describes a compound that consists mainly of carbon and hydrogen atoms. **33**

organism Individual that consists of one or more cells. **5**

osmosis Diffusion of water across a selectively permeable membrane; occurs when there is a difference in solute concentration between the fluids on either side of the membrane. **73**

ovary In flowering plants, the enlarged base of a carpel, inside which one or more ovules form. In animals, the egg-producing female gonad. **272**

ovule Of seed plants, chamber inside which megaspores form and develop into female gametophytes; after fertilization, becomes a seed. **269**

ozone layer Atmospheric layer with a high concentration of ozone that prevents much UV radiation from reaching Earth's surface. **240, 364**

Pangea Supercontinent that began to form about 300 million years ago and broke up 100 million years later. **200**

parasite A species that withdraws nutrients from another species (its host), usually without killing it. **335**

parasitoid An insect that lays eggs in another insect, and whose young devour their host from the inside. **336**

passive transport Membrane-crossing mechanism that requires no energy input. **75**

pathogen Disease-causing agent. **236**

PCR Polymerase chain reaction. Laboratory method that mass-produces copies of a specific section of DNA. **177**

pedigree Chart of family connections that shows the appearance of a trait through generations. **160**

peptide bond A bond between the amine group of one amino acid and the carboxyl group of another; joins amino acids in proteins. **38**

per capita growth rate The number of individuals added during some interval divided by the initial population size. **317**

permafrost Layer of permanently frozen soil in the Arctic. **357**

peroxisome Enzyme-filled vesicle that breaks down amino acids, fatty acids, and toxic substances. **55**

pH Measure of the amount of hydrogen ions in a fluid. **32**

phagocytosis "Cell eating"; an endocytic pathway by which a cell engulfs large particles such as microbes or cellular debris. **77**

phenotype An individual's observable traits. **152**

phloem Vascular tissue of plants; distributes sugars through its sieve tubes. Consists of sieve-tube members and companion cells. **264**

phospholipid Lipid with a phosphate group in its hydrophilic head, and two nonpolar fatty acid tails; main constituent of eukaryotic cell membranes. **36**

phosphorus cycle Movement of phosphorus among rocks, water, soil, and living organisms. **343**

phosphorylation A chemical reaction in which a phosphate group is transferred from one molecule to another. **71**

photosynthesis Metabolic pathway by which most autotrophs use light energy to make sugars from carbon dioxide and water. **6**

phylogeny Evolutionary history of a species or group of species. **229**

pigment An organic molecule that can absorb light of certain wavelengths. Reflected light imparts a characteristic color. E.g., chlorophyll. **83**

pilus A protein filament that projects from the surface of some prokaryotic cells. **53**

pioneer species Species that can colonize a new habitat. **337**

placental mammal Mammal in which developing offspring are nourished within the mother's body by way of a placenta. **304**

plankton Community of mostly microscopic drifting or swimming organisms. **251**

plant Multicelled, typically photosynthetic organism; develops from an embryo that forms on the parent and is nourished by it. **8, 263**

plasma membrane Membrane that encloses a cell and separates it from the external environment. **47**

plasmid Of many prokaryotes, a small ring of nonchromosomal DNA. **246**

plasmodial slime mold Heterotrophic protist that moves and feeds as a multinucleated mass; forms a fruiting body when conditions are unfavorable. **256**

plate tectonics theory Theory that Earth's outermost layer of rock is cracked into plates, the slow movement of which conveys continents to new locations over geologic time. **201**

pleiotropy Inheritance pattern in which a single gene affects multiple traits. **156**

polarity Separation of charge into positive and negative regions. **28**

pollen grain Immature male gametophyte of a seed plant. **264**

pollen sac Of seed plants, chamber in which microspores form and develop into male gametophytes (pollen grains). **269**

pollination Delivery of a pollen grain to the egg-bearing part of a seed plant. **269**

pollinator Animal that moves pollen from one plant to another, thus facilitating pollination. **273**

pollutant A natural or man-made substance that is released into the environment in greater than natural amounts and that damages the health of organisms. **362**

polymer Molecule that consists of multiple monomers. **33**

polymerase chain reaction *See* PCR.

polyp In cnidarians, a tubular, typically sessile, body form; for example a sea anemone. **288**

polyploid Having three or more ocomplete sets of chromosomes. **165**

population A group of organisms of the same species who live in a specific location and breed with one another more often than they breed with members of other populations. **5, 314**

population density Number of members of a population in a given area. **315**

population distribution The way in which members of a population are dispersed in their environment. **315**

population size Total number of individuals in a population. **315**

prairie Temperate grassland biome of North America. Its grasses and other plants are adapted to recover after grazing and the occasional fire. **356**

predation One species (the predator) captures, kills, and feeds on another (its prey). **334**

prediction Statement, based on a hypothesis, about a condition that should exist if the hypothesis is correct. **12**

primary production The energy captured by an ecosystem's producers. **341**

primary succession Ecological succession occurs in an area where there was previously no soil. **337**

primate Mammalian group with grasping hands; includes lemurs, tarsiers, monkeys, apes, and humans. **305**

primer Short, single strand of DNA or RNA that base-pairs with a specific DNA sequence in DNA synthesis. **108**

prion Infectious protein. **40**

probability Out of all possible outcomes of an event, the chance that a particular outcome will occur. **17**

probe Fragment of DNA or RNA labeled with a tracer; can hybridize with a nucleotide sequence of interest. **176**

producer Organism that makes its own food using energy and nonbiological raw materials from the environment. **6, 339**

product A molecule that is produced by a reaction. **66**

prokaryote Informal name for a single-celled organism without a nucleus; a bacterium or archaeon. **8**

promoter In DNA, a nucleotide sequence to which RNA polymerase binds; site where transcription begins. **116**

prophase Stage of mitosis during which chromosomes condense and become attached to a newly forming spindle. **134**

protein Polymer of amino acids; an organic molecule that consists of one or more polypeptides. **38**

protist General term for eukaryote that is not a fungus, plant, or animal. **8, 250**

protocell Membranous sac that contains interacting organic molecules; hypothesized to have formed prior to the earliest cells. **239**

proton Positively charged subatomic particle that occurs in the nucleus of all atoms. **25**

protostomes Animal lineage with a three-layer embryo in which the first opening to form is the mouth; includes most bilateral invertebrates. **286**

pseudopod A temporary protrusion that helps some eukaryotic cells move and engulf prey. **57**

Punnett square Diagram used to predict the genetic and phenotypic outcomes of a cross. **153**

radial symmetry Having parts arranged around a central axis, like spokes around a wheel. **286**

radioactive decay Process by which atoms of a radioisotope emit energy and subatomic particles when their nucleus spontaneously breaks up. **25**

radioisotope Isotope with an unstable nucleus. **25**

radiometric dating Method of estimating the age of a rock or fossil by measuring the content and

proportions of a radioisotope and its daughter elements. **199**

radula Tonguelike organ of many mollusks. **290**

rain shadow Dry region on the downwind side of a coastal mountain range. **353**

ray-finned fish Bony fish with fins supported by thin rays derived from skin. **300**

reactant A molecule that enters a reaction and is changed by participating in it. **66**

reaction Process of molecular change. **34**

receptor protein Membrane protein that triggers a change in cell activity in response to a stimulus such as binding a certain substance. **51**

recessive Refers to an allele with an effect that is masked by a dominant allele on the homologous chromosome. **152**

recombinant DNA A DNA molecule that contains genetic material from more than one organism. **176**

red alga Single-celled or multicelled photosynthetic protist with a red accessory pigment. **255**

reproduction Process by which parents produce offspring. *See* sexual reproduction, asexual reproduction. **7**

reproductive cloning A laboratory procedure that produces genetically identical animals; e.g., somatic cell nuclear transfer (SCNT). **101**

reproductive isolation The end of gene flow between populations. **222**

reptile Amniote subgroup that includes lizards, snakes, turtles, crocodilians, and birds. **302**

resource partitioning Evolutionary process whereby traits of competing species come to differ as a result of the selective pressure imposed by the competition. **334**

restriction enzyme Type of enzyme that cuts DNA at a specific nucleotide sequence. **175**

rhizoid Threadlike structure that anchors some plants. **265**

rhizome Stem that grows horizontally along or just below the ground. **266**

ribosomal RNA (rRNA) RNA that becomes part of ribosomes. **119**

ribosome Organelle of protein synthesis. An intact ribosome has two subunits, each composed of rRNA and proteins. **52**

RNA Ribonucleic acid. Nucleic acid with roles in gene expression; consists of a single-stranded chain of nucleotides. *See* messenger RNA, transfer RNA, ribosomal RNA. **41**

RNA polymerase Enzyme that carries out transcription (RNA synthesis). **116**

RNA world hypothesis Hypothesis that RNA served as the first material of inheritance. **238**

roundworm Unsegmented worm with a pseudocoelom and a cuticle that is molted as the animal grows. **291**

rRNA *See* ribosomal RNA.

sac fungus Fungus that produces spores by meiosis in saclike cells. **276**

salt Ionic compound that releases ions other than H^+ and OH^- when it dissolves in water. **30**

sampling error Difference between results derived from testing an entire group of events or individuals, and results derived from testing a subset of the group. **17**

saturated fatty acid Fatty acid with only single bonds linking the carbons in its tail. **36**

savanna Tropical biome dominated by grasses and other plants adapted to grazing, as well as a scattering of shrubs. **356**

scales Hard, flattened elements that cover the skin of reptiles and some fishes. **299**

science Systematic study of the observable world. **12**

scientific method Making, testing, and evaluating hypotheses about the natural world. **13**

scientific theory Hypothesis that has not been disproven after many years of rigorous testing. **18**

seamount An undersea mountain. **359**

second law of thermodynamics Energy tends to disperse spontaneously. **65**

secondary succession Ecological succession occurs in an area where a community previously existed and soil remains. **337**

seed Embryo sporophyte of a seed plant packaged with nutritive tissue inside a protective coat. **264**

sequencing Laboratory method of determining the order of nucleotides in DNA. **178**

sex chromosome Member of a pair of chromosomes that differs between males and females. **107**

sexual reproduction Reproductive mode by which offspring arise from two parents and inherit genes from both. **141**

sexual selection Mode of natural selection in which some individuals outreproduce others of a population because they are better at securing mates. **218**

shell model Model of electron distribution in an atom. **27**

short tandem repeat In chromosomal DNA, a sequence of a few nucleotides repeated multiple times in a row. Used in DNA profiling. **159**

single-nucleotide polymorphism (SNP) A one-nucleotide DNA sequence variation carried by a measurable percentage of a population. **174**

sister chromatids The two DNA molecules of a duplicated eukaryotic chromosome, attached at the centromere. **106**

SNP *See* single-nucleotide polymorphism.

solute A dissolved substance. **30**

solution Uniform mixture of solute completely dissolved in a solvent. **30**

solvent Liquid in which other substances dissolve. **30**

somatic cell nuclear transfer (SCNT) Reproductive cloning method in which the DNA of an adult donor's body cell is transferred into an unfertilized egg. **100**

sorus Cluster of spore-forming chambers on a fern frond. **266**

speciation Evolutionary process in which new species arise; e.g., allopatric speciation, sympatric speciation. **222**

species Unique type of organism designated by genus name and specific epithet. Of sexual reproducers, often defined as one or more groups of individuals that can potentially interbreed, produce fertile offspring, and do not interbreed with other groups. **8**

species diversity The number of species and their relative abundance within a community. **331**

spindle Temporary structure that moves chromosomes during nuclear division; consists of microtubules. **134**

sponge Aquatic invertebrate that has no tissues or organs and filters food from the water. **288**

sporophyte Diploid spore-forming body that forms in a plant life cycle. **263**

stabilizing selection Mode of natural selection in which an intermediate form of a trait is adaptive, and extreme forms are not. **215**

stamen Floral reproductive organ that consists of a pollen-producing anther and, in most species, a filament. **272**

statistically significant Refers to a result that is statistically unlikely to have occurred by chance alone. **17**

steroid A type of lipid with four carbon rings and no fatty acid tails. **37**

stigma Pollen-receiving part of a carpel. **272**

stomata Singular, stoma. Closable gaps formed by pairs of guard cells on aboveground plant surfaces. When open, they allow the plant to exchange gases with air. When closed, they limit water loss. **88, 264**

stroma The cytoplasm-like fluid between the thylakoid membrane and the two outer membranes of a chloroplast. Site of light-independent reactions of photosynthesis. **84**

stromatolites Dome-shaped structures composed of layers of prokaryotic cells and sediments; form in shallow seas. **240**

style Elongated portion of a carpel that elevates the stigma above the ovary. **272**

substrate Of an enzyme, a reactant that is specifically acted upon by the enzyme. **68**

surface-to-volume ratio A relationship in which the volume of an object increases with the cube of the diameter, and the surface area increases with the square. Limits cell size. **47**

survivorship curve Graph showing the decline in numbers of a cohort over time. **320**

symbiosis One species lives on or inside another in a commensal, mutualistic, or parasitic relationship. **332**

sympatric speciation Divergence within a population leads to speciation; occurs in the absence of a physical barrier to gene flow. **224**

taiga *See* boreal forest.

taxon Plural, taxa. A rank in the classification of life; consists of a group of organisms that share a unique set of traits. **10**

taxonomy Practice of naming and classifying species. **8**

telophase Stage of mitosis during which chromosomes arrive at opposite ends of the cell. Two new nuclei form as the chromosomes loosen. **134**

temperate deciduous forest Biome dominated by broadleaf trees that grow in warm summers, then drop their leaves and become dormant during cold winters. **355**

temperature Measure of molecular motion. **31**

tetrapod Vertebrate with four limbs. **298**

theory *See* scientific theory.

threatened species A species likely to become endangered in the near future. **360**

thylakoid membrane A chloroplast's highly folded inner membrane system; forms a continuous compartment. Site of light-dependent reactions of photosynthesis. **84**

tissue A collection of one or more specific cell types that are organized in a way that adapts them to a task. **286**

total fertility rate Average number of children born to females of a population over the course of their lifetimes. **324**

tracer A substance that can be traced via its detectable component. **26**

transcription RNA synthesis; process by which enzymes assemble an RNA using a strand of DNA as a template. Part of gene expression. **115**

transcription factor Regulatory protein that influences transcription by binding directly to DNA. **124**

transduction Mechanism of gene transfer. A virus moves genes from one host cell to another. **246**

transfer RNA (tRNA) RNA that delivers amino acids to a ribosome during translation. **119**

transformation Mechanism of gene transfer. A prokaryotic cell takes up and uses DNA from its environment. **246**

transgenic Refers to a genetically modified organism that carries a gene from a different species. **181**

translation Protein synthesis; process by which a polypeptide chain is assembled from amino acids in the order specified by an mRNA. **115**

transpiration Evaporation of water from a plant's aboveground parts. **342**

transport protein Membrane protein that passively or actively helps a specific ion or molecule move across the membrane. **51**

triglyceride A molecule with three fatty acid tails that is entirely hydrophobic; main component of fats. **36**

tRNA *See* transfer RNA.

trophic level Position in a food chain. **339**

tropical rain forest Multilayered forest biome that occurs where warm temperatures and continual rains allow plant growth year-round. Most productive and species-rich biome. **355**

tumor A neoplasm that forms a lump. **137**

tundra Northernmost biome, dominated by low plants that grow over a layer of permafrost. **356**

tunicates Invertebrate chordates that lose their defining chordate traits during the transition to adulthood; adults have a secreted "tunic." **297**

turgor Pressure that a fluid exerts against a cell wall, membrane, or other structure that contains it. **74**

unsaturated fatty acid Fatty acid with one or more carbon–carbon double bonds in its tail. **36**

variable In an experiment, a characteristic or event that differs among individuals or over time. **13**

vascular plant A plant that has xylem and phloem. **264**

vertebral column Backbone. **298**

vertebrate Animal with a backbone. **284, 298**

vesicle Small, membrane-enclosed organelle; different kinds store, transport, or break down their contents. **54**

viral envelope A layer of cell membrane derived from the host cell in which an enveloped virus was produced. **242**

viral reassortment Two viruses of the same type infect an individual at the same time and swap genes. **245**

virus A noncellular infectious particle with a protein coat and a genome of RNA or DNA; replicates only in living cells. **242**

warning coloration Distinctive color or pattern that makes a well-defended prey species easy to recognize. **334**

water cycle Water moves from the ocean into the air, falls as rain and snow, and flows back to the ocean. **342**

water mold Heterotrophic protist that forms a mesh of nutrient-absorbing filaments; some are important plant pathogens. **254**

water–vascular system Of echinoderms, a system of fluid-filled tubes and tube feet that function in locomotion. **296**

wavelength Distance between the crests of two successive waves. **83**

wax Water-repellent substance that is a complex, varying mixture of lipids. **37**

wood Lignin-stiffened secondary xylem of some seed plants. **269**

xylem Complex vascular tissue of plants; its dead tracheids and vessel elements form tubes that distribute water and mineral ions. **264**

zygote Diploid cell that forms when two gametes fuse; the first cell of a new individual. **144**

zygote fungus Fungus that usually grows as a mold; sexual reproduction yields a thick-walled zygospore. **275**

The letter *f* designates figure; *t* designates table; bold designates key terms; ■ indicates human health applications; ■ indicates environmental topics.

INDEX

A